ECOSYSTEM MODELING
IN THEORY AND PRACTICE

Ecosystem Modeling
in Theory and Practice:

An Introduction with Case Histories

Edited by
CHARLES A. S. HALL
Cornell University
Ithaca, New York
and
Marine Biological Laboratory
Woods Hole, Massachusetts

JOHN W. DAY, Jr.
Louisiana State University
Baton Rouge, Louisiana

A WILEY-INTERSCIENCE PUBLICATION

JOHN WILEY & SONS
New York · London · Sydney · Toronto

Library of Congress Cataloging in Publication Data:

Main entry under title:

Ecosystem modeling in theory and practice.

   "A Wiley-Interscience publication."
   Includes index.
   1. Ecology—Data processing.   2. Ecology—
Mathematical models.   3. Biological models.
I. Hall, Charles A. S.   II. Day, John W.

QH541.15.E45E26   574.5'01'84   76-57204
ISBN 0-471-34165-7

Printed in the United States of America

10 9 8 7 6 5 4 3 2 1

To the intellectual excitement that we experienced at Chapel Hill in the late 1960s, and to the people who generated that special excitement.

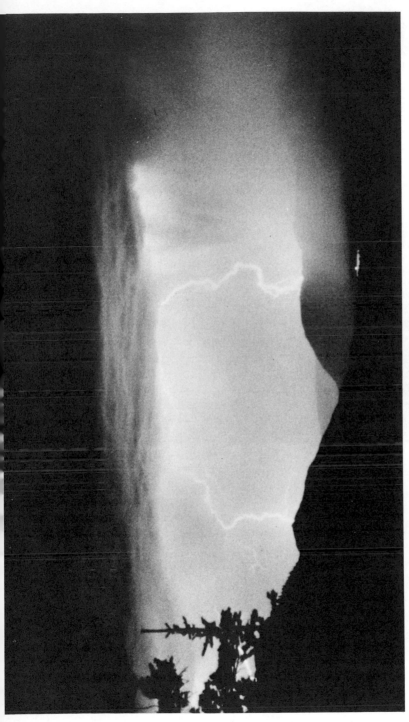

An intense electrical storm that swept through northwestern Montana in August 1973 ignited more than 200 fires in less than 2 hr. The bright spot in the right foreground is a fire caused by an earlier strike. Should this fire be suppressed because it may interfere with man's interests, or is fire a natural ecological process that should be allowed? Chapter 23 describes how computer models are used in answering this question.

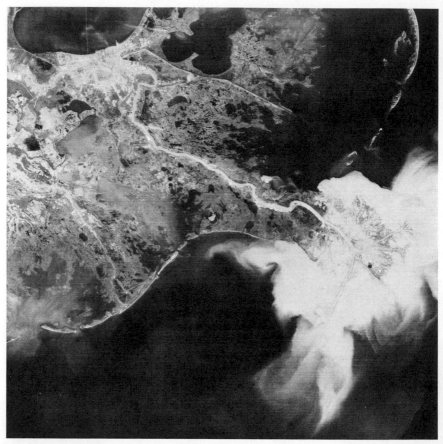

Turbid water spreads out from the delta of the Mississippi River. The River has shaped natural, economic, and social systems in Southern Louisiana. Models have been used in studies of the natural systems (Chapter 10) and in assessing the impact of man's activities on this rich and productive region (Chapter 16).

Simple controlled ecosystems such as this hot-springs algae community in Yellowstone Park can be measured and modeled easily and accurately (Chapter 12). Information gained from these simple ecosystems is useful in understanding general ecosystem processes and can be used in developing models of more complex ecosystems.

The oyster industry is the most important economic resource of Franklin County, Florida. Will new industries add to the economic well-being of Franklin County or will they disturb the sensitive oyster industry, hence threaten that county's economy? A modeling approach has been used to examine this question (Chapter 20).

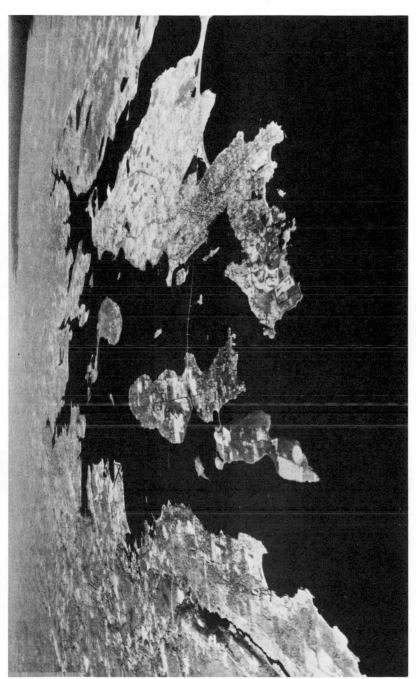

Man can be a major force in regional ecosystems such as Narragansett Bay. The interrelationships of man and nature in Narragansett Bay are considered in a hydrologic-ecologic-economic model in Chapter 25.

Contributors

Daniel B. Botkin, The Ecosystems Center, Marine Biological Laboratory, Woods Hole, Massachusetts

Walter Boynton, Hallowing Point Research Station, University of Maryland, Prince Fredrick

Mark Brown, Center for Wetlands, University of Florida, Gainesville

John W. Day, Department of Marine Sciences, Louisiana State University, Baton Rouge

Michael R. Dohan, Department of Economics, Queens College of CUNY, Flushing, New York

Edward I. Friedland, National Research Council, Washington, D.C.

Charles Gray, University of Miami Medical School, Coral Gables, Florida

Charles A. S. Hall, Ecology and Systematics, Cornell University, Ithaca, New York, and Marine Biological Laboratory, Woods Hole, Massachusetts

David E. Hawkins, Environmental Engineering Sciences, University of Florida, Gainesville

Mark Homer, Hallowing Point, Research Station, University of Maryland, Prince Frederick

Charles S. Hopkinson, Department of Marine Sciences, Louisiana State University, Baton Rouge

Bengt-Owe Jansson, Department of Zoology, University of Stockholm, Stockholm, Sweden

Robert A. Kelly, Marine Pollution Studies, Fisheries and Wildlife Division, Melbourne, Australia

W. Michael Kemp, Hallowing Point Research Station, University of Maryland, Prince Fredrick

Stephen R. Kessell, Gradient Analysis Inc., Riverside, California

James N. Kremer, Department of Zoology, University of Southern California, Los Angeles

Melvin E. Lehman, Center for Wetlands, University of Florida, Gainesville

Charles Littlejohn, Florida Department of Environmental Regulation, Tallahassee

Harold C. Loesch, Department of Marine Sciences, Louisiana State University, Baton Rouge

Angus MacBeth, United States Attorney's Office, New York

Henry N. McKellar, Belle Baruch Institute of Marine Science, University of South Carolina, Columbia

James T. Morris, The Ecosystems Center, Marine Biological Laboratory, Woods Hole, Massachusetts, and School of Forestry and Environmental Studies, Yale University

Scott W. Nixon, Graduate School of Oceanography, University of Rhode Island, Kingston

Howard T. Odum, Environmental Engineering Sciences and Center for Wetlands, University of Florida, Gainesville

W. Scott Overton, Department of Statistics, Oregon State University, Corvallis

Jeffrey E. Richey, Fisheries Research Institute, University of Washington, Seattle

Christine A. Shoemaker, Department of Environmental Engineering, Cornell University, Ithaca, New York

S. Fred Singer, Department of Environmental Sciences, University of Virginia, Charlottesville

Wade H. B. Smith, Mitre Corporation, McLean, Virginia

Walter O. Spofford, Resources for the Future, Washington, D.C.

Daniel E. Wartenberg, Department of Oceanography, University of Washington, Seattle

Richard G. Wiegert, Department of Zoology, University of Georgia, Athens

Fredrik Wulff, Department of Zoology, University of Stockholm, Stockholm, Sweden

Donald L. Young, Florida Department of Environmental Regulation, Tallahassee

Preface

This book is about the development and application of a series of tools designed to assist in the better understanding and management of systems of man and nature. It is aimed at the level of seniors in college or beginning graduate students and could be used as a text for an introductory modeling course, an introduction to the subject for someone who does not have access to such a course, or as a reference for modelers. We also hope that it will prove useful to persons involved in resource management and social decision making.

The book is divided into four sections. The first section contains introductory chapters on modeling theory and procedures. The next three sections deal with case histories where models have been used to analyze natural systems, assess environmental impact, and design optimal, or at least better, interactions between man and nature. Any one of the chapters may be read alone, and they are placed within each section more or less in order of increasing difficulty. The beginning student may prefer to read the first two chapters of each of the four sections to gain an overview.

In many respects this book is as much about ecosystems as about models and represents a series of attempts aimed at studying nature "in ecos," that is, in its entirety. Many of the chapters are based on the results of tens to hundreds of man-years of field work. In most cases the backup data presented in this book is but a very small part of the total data base that was used to construct the models because we had to eliminate most of the detailed data and justifications originally submitted with the chapters for the sake of brevity. In most cases published sources of the data used are given in the bibliographies.

Why is the development of ecosystems science and modeling important? There are academic reasons, of course. However, there are social reasons of greater importance. The problems that will be facing mankind during the next decades will be immense and complicated, and require formalized whole system procedures to deal with them. For example, it is likely that there will be major and complicated changes forced upon society shortly as readily available petroleum reserves are depleted. A large proportion of man's welfare is provided via the management of nature, and managing nature is presently an extremely energy-intensive process. Therefore it will be important to learn how to do such management using a minimum amount of industrial energy. Such management schemes would also mean less industrial pollution. Many of the chapters in this book give examples of how low-energy, computer-based information systems can be substituted for traditional energy-intensive management schemes. Thus we do not see good economic decisions and good environmental decisions as necessarily antithetical—in fact, we think that they are often synonymous.

Some of the models presented here, like science itself, will change considerably with time as more is learned about them and the systems they represent. If precise results are sufficiently important, models can be improved to the degree of accuracy required, generally to the limit of available data, or even to the limit of our ability to get data. In the meantime we like very much the statement of Daniel Botkin: "It [the computer model] forces us to see the implications, true or false, wise or foolish, of the assumptions we have made." Very often science is done and proclamations for management are issued without exploring the full consequences of that decision within the framework of what the manager knows of the system. At a minimum, models can assist in exploring such consequences.

Many of the chapters included in this book are attempts to include computer model building into societal decision making. Does the inclusion of computer models, which are opaque to most citizens, represent an intrusion of "experts" and "mystique" into public decision making? Is this an undermining of the democratic process? These are real and important questions, and the answer is a qualified yes. Average citizens do not have access to modeling techniques and indeed may be hoodwinked by unscrupulous decision makers armed with reams of computer readout. However, the potential for such mischief is no different than what characterizes other decision-making processes, which often also are inaccessible to the average citizen. If a model is to be incorporated into decision making, at least the assumptions behind the decision-making process are available to *anyone* with a little knowledge of programming and the problem at hand. We think that this may more than compensate for the addition of "mystique" to decision making occasioned by the use of "computer experts." This book should dispel a little of this "mystique," for we think that the procedures of modeling, once stripped of jargon and clearly explained, are not particularly difficult. Certainly, they are no less difficult to understand than the complicated processes of man and nature represented by these models.

We hope that you will enjoy and learn from our book.

CHARLES A. S. HALL
JOHN W. DAY, JR.

Ithaca, New York
Baton Rouge, Louisiana
November 1976

Acknowledgments

Ecosystem studies are of necessity an interactive process. We acknowledge the assistance of many teachers, colleagues, students, and friends, unfortunately too numerous to mention by name. They include especially the students in systems ecology at Cornell and in estuarine ecology at LSU. Kenneth Watt made many important suggestions about the final form of the book. Hollis Fishelson, Janice Gengenbach, Beth Hedlund, and Joette Serio typed much of the manuscript and set us straight on many matters. And finally we thank Joanna and Carolyn for very important feedback.

The following provided support in the form of secretarial work, computer time, and stimulating academic environments: The Department of Marine Sciences, Louisiana State University; Section of Ecology and Systematics, Cornell University; Biology Department, Brookhaven National Laboratory; and the Ecosystems Center, Marine Biological Laboratory, Woods Hole, Massachusetts. Funding was provided in part by the National Sea Grant Program, the U.S. Atomic Energy Commission, and the National Science Foundation.

Contents

I BASIC PRINCIPLES

1 Systems and Models: Terms and Basic Principles,
CHARLES A. S. HALL AND JOHN W. DAY, JR.   5

2 A Circuit Language for Energy and Matter,
CHARLES A. S. HALL, JOHN W. DAY, JR., AND HOWARD   37
T. ODUM

3 A Strategy of Model Construction, W. SCOTT OVERTON   49

4 Mathematical Construction of Ecological Models,
CHRISTINE A. SHOEMAKER   75

5 Values and Environmental Modeling,
EDWARD I. FRIEDLAND   115

6 Economic Values and Natural Ecosystems,
MICHAEL R. DOHAN   133

7 Energy, Value, and Money, HOWARD T. ODUM   173

8 Modeling in the Context of the Law, ANGUS MACBETH   197

II CASE HISTORIES: UNDERSTANDING NATURAL SYSTEMS

9 Life and Death in a Forest: The Computer as an Aid
to Understanding, DANIEL B. BOTKIN   213

10 A Model of the Barataria Bay Salt Marsh Ecosystem,
CHARLES S. HOPKINSON, JR. AND JOHN W. DAY, JR.   235

11 An Empirical and Mathematical Approach Toward the
    Development of a Phosphorus Model of Castle Lake,
    California, JEFFREY E. RICHEY   267
12 A Model of a Thermal Spring Food Chain,
    RICHARD G. WIEGERT   289

III CASE HISTORIES: ASSESSING ENVIRONMENTAL IMPACT

13 Baltic Ecosystem Modeling,
    BENGT-OWE JANSSON AND FREDRIK WULFF   323
14 Models and the Decision Making Process: The Hudson
    River Power Plant Case, CHARLES A. S. HALL   345
15 A Simulation that Failed: The Biospheric Productivity
    Model, DANIEL E. WARTENBERG AND CHARLES A. S. HALL   365
16 Modeling Man and Nature in Southern Louisiana,
    JOHN W. DAY, JR., CHARLES S. HOPKINSON, JR., AND
    HAROLD C. LOESCH   381
17 War, Peace, and the Computer: Simulation of Disordering
    and Ordering Energies in South Vietnam, MARK BROWN   393
18 Application of an Ecosystem Model to Water Quality
    Management: The Delaware Estuary, ROBERT A. KELLY
    AND WALTER O. SPOFFORD, JR.   419

IV CASE HISTORIES: DESIGNING OPTIMAL INTERACTIONS OF
    MAN AND NATURE

19 An Analysis of the Role of Natural Wetlands in
    Regional Water Management, CHARLES LITTLEJOHN   451
20 A Modeling Approach to Regional Planning in Franklin
    County and Apalachicola Bay, Florida, WALTER
    BOYNTON, DAVID E. HAWKINS, AND CHARLES GRAY   477
21 Energy Cost-Benefit Analysis Applied to Power Plants
    Near Crystal River, Florida, W. MICHAEL KEMP,
    WADE H. B. SMITH, HENRY N. MCKELLAR,
    MELVIN E. LEHMAN, MARK HOMER, DONALD L. YOUNG,
    AND HOWARD T. ODUM   507
22 Pest Management Models of Crop Ecosystems,
    CHRISTINE A. SHOEMAKER   545
23 Gradient Modeling: A New Approach to Fire Modeling
    and Resource Management, STEPHEN R. KESSELL   575

24 The Problem of Population Optima, S. FRED SINGER WITH
    THE ASSISTANCE OF JAMES T. MORRIS   607

25 Narragansett Bay—The Development of a Composite
    Simulation Model for a New England Estuary,
    SCOTT W. NIXON AND JAMES N. KREMER   621

AUTHOR INDEX   675

SUBJECT INDEX   677

ECOSYSTEM MODELING
IN THEORY AND PRACTICE

PROLOGUE TO SECTION I

This section includes a series of chapters designed to acquaint the reader with the basic concepts and philosophies of modeling necessary to read and appreciate the case histories presented in the rest of the book. However, it is not a prescription for modeling, for the knowledge required for good modeling is obviously more complicated than what we present here. There is no substitute for a good teacher and lots of practical experience.

Nevertheless, we think the information in this section can get you started on what we have found to be a most interesting journey. Chapter 1 is a capsule summary of what modeling is all about and includes definitions of the most commonly encountered "jargon." Chapter 2 introduces the "energy-flow language" that we have found useful in organizing and communicating the complicated interactions of an ecosystem. Chapter 3, written by Scott Overton, a forest ecologist and statistician, discusses some of the problems encountered in deciding what components to include, or not to include, in a model. He concludes with a prescription for a model-building *strategy*. Chapter 4 is an introduction to the mathematics most commonly used in modeling, written by Christine Shoemaker, a mathematician and population biologist. This chapter may be viewed as an introduction to the mathematical *tactics* of modeling. The next three chapters consider the very difficult area of the *values* that enter into models designed for use for decision making in society. Ed Friedand is a political scientist suspicious of all models, who wants us to make certain that we are being honest about the values we include in our models. Michael Dohan, an economist, summarizes the pertinent aspects of the most commonly used system of value—that of money—and adds some new concepts as to how economics can more adequately measure the values of nature. H. T. Odum, an ecologist with very wide interests and experience, presents us with the extremely interesting theory that value is determined not by man's subjective assessment, but by the process of natural selection. The final chapter in this section, written by Angus MacBeth, a lawyer with a long and impressive environmental record, tells us about the differences between scientific and legal "truth," and the ways in which societal decision-making processes can be influenced, and hopefully improved, by the use of models.

I
BASIC PRINCIPLES

1
Systems and Models: Terms and Basic Principles

CHARLES A. S. HALL
JOHN W. DAY, JR.

This chapter is designed to introduce the basic vocabulary and concepts of systems ecology and modeling. It develops the essentials of simulation modeling in preparation for later chapters in the book that use models. We begin with definitions of a system and systems analysis and then go through some basic steps generally used to develop conceptual, diagrammatic, mathematical, and computer models. A more detailed development of many of the concepts introduced here can be found in the other chapters in Section I of this book. Although we have tried to include all essentials of modeling, our treatment is rather brief and superficial for such an extensive subject. The interested reader is advised to consult the valuable introductions in Watt (1966, 1968), Patten (1971), H. T. Odum (1970), Walters in E. Odum (1971), and Maynard-Smith (1974).

WHAT IS A SYSTEM?

Any phenomenon, either structural or functional, having at least two separable components and some interaction between these components may be considered a *system*. Another more general definition is "any object whose behavior is of interest." Depending upon your philosophical point of view, systems may be either the actual building blocks of nature or, instead, an effort by man (like taxonomy) to impose order upon the seeming chaos of nature. We are not concerned with the distinction in this chapter—either or both will do for our purposes. *Systems analysis* is the formalized study of any system, or of the general properties of systems. *Holism* is the philosophy of studying the total behavior (or other total attributes) of some complicated system.

Various systems are part of our everyday life, and we are familiar with the heating system of our house or the ignition system of our automobile. Yet each of these has components that themselves could be considered systems (a spark plug is an organization of conductors, insulators, connectors, etc.), and each in turn is part of a larger system, for example, a house or an automobile. Thus any particular system that we may wish to study is part of a hierarchy of other systems. It is up to us to choose the level that we work with, and our first order of business is to define the spatial, temporal, and conceptual limits that we wish to address. Eugene Odum has suggested that we include the next-larger sized system as well as the level in which we are most interested, so that we do not miss important environmental relationships.

In this book we are concerned with the larger systems of nature, including the ways that man interacts with nature. Such systems are normally called ecological, sociological, or economic, and they display the same types of interactions and generalities of scale as physical systems display. For example, a mouse in a field is composed of a series of living systems; the nervous system interacts with the circulatory system, the endocrine system, and so on to give the whole, that is, the mouse. Yet each of these systems has subcomponents, which in turn have subcomponents, and so on down to the limits of our perception. Even the cell itself has descending hierarchies of systems. In addition, the mouse is a component of larger environmental systems. Often the importance of each of these levels is jealously advocated by the scientists involved in their study.

The systems of nature that are considered in this book begin with the individual organism and ascend toward greater complexity. Many mice of one species constitute a *population* of mice; the mouse population, together with the other populations of animals and plants with which they interact, and with which they share a spatial environment, are considered *communities*. For example, the communities of man include other species: trees, dogs, and even mice. Communities, together with their nonliving associates, form *ecosystems*, short for ecological systems. The definition of any of these levels is in practice somewhat arbitrary, but ecosystems generally comprise all components of a unit of landscape with geographical and geological continuity. Some examples are ponds, lakes, woodlots, estuaries, fields, farms, and cities. In practice, the boundaries are set by the investigator, and all structures and interactions within comprise the ecosystem. Very large ecosystems of subcontinental dimensions and strong biotic continuity are called *biomes*. All the living material of the earth collectively is called the *biosphere*.

We emphasize the ecosystem level in this book because we wish to study all the interactions, both biotic and abiotic, that may be important to an environmental problem, and because ecosystems are a convenient level to approach many environmental problems. Hence there are both logical and operational reasons for studying phenomena at the level of the ecosystem, although, of course, it is necessary to consider levels of complexity above and below. It has been said that

each level [of complexity] finds its explanations of mechanism in the levels below, and its significance in the levels above (Bartholomew, 1964).

WHAT IS A MODEL?

A model is any abstraction or simplification of a system, and, to paraphrase Von Clauswitz, modeling is an extension of scientific analysis by other means. Models of ecosystems are simpler than real ecosystems, just as model airplanes are simpler than real airplanes. A model should have the important functional attributes of the real system. Obviously it cannot have all attributes, or it would not be a model—it would be the real system. Modeling is done to aid the conceptualization and measurement of complex systems and, sometimes, to predict the consequences of an action that would be expensive, difficult, or destructive to do with the real system. Sometimes, as with model airplanes or some computer models of nature, modeling can be just plain fun.

Another definition might be that models are devices for predicting the behavior of a complicated, poorly understood entity from the behavior of parts that are well understood (Goodman, 1975). A third definition is that models may be considered the formalization of our knowledge about a system. For the moment, our first definition is the most useful—models are simplifications.

Modeling is needed for the understanding of nature because the complexity of nature is often overwhelming. However, models must be checked frequently against the real world to assure that their representation of the real world is accurate, or at least is inaccurate in ways of which we are aware. A most powerful tool is the interplay of model and empiricism, ideally in an alternating series. Most of the chapters in other sections of this book show how the use of models together with empirical science has contributed to a greater understanding of the system than either process alone.

The analysis of complex systems as systems, and the modeling of these systems, is contrary to reductionist trends in science. Isolating and controlling very small components of nature recently have been the most powerful investigative tools aiding man in understanding nature (Platt, 1968). The very success of these techniques, however, has made holistic science less acceptable to much of the scientific community, but also more necessary. Many chapters give examples where failure to consider the total system in an economic or environmental decision has resulted in undesirable consequences. To a certain extent, the "holes" in large-scale systems research (that is, subcomponents of possible importance that are excluded by accident or design) are matched by "doughnuts" in traditional science-by-isolation. In other words, the larger environment may influence the components of interest to such an extent that an isolated laboratory experiment may exclude some critical components, or the behavior of a system may not be simply the sum of the behaviors of its isolated parts. Ideally reductionist and holistic methods both can be applied to a problem. However, holistic views predominate in this book.

TWO MODELING APPROACHES

Two major types of ecological models (and modelers) exist, which can be classified for convenience as *analytic* models and *simulation* models, although the terms are not strictly antithetical. Although both approaches are, in theory, directed at increasing our understanding and prediction of ecological systems and their components, in practice the two methods are generally used for completely different questions, and they use completely different mathematical approaches. Analytic modeling is normally characterized by the use of pencil, paper, and relatively complicated mathematics. Simulation modeling tends to be characterized by the use of simpler mathematics together with computers. Until recently, there was little communication between the two groups of ecological modelers, and neither group paid much attention to the publications of the other. However, this has been changing.

The analytic approach refers to a mathematical set of procedures for finding exact solutions to differential and other equations (see Chapter 4). Since the behavior of some kinds of mathematical equations are well known, the analytic approach has great power if some aspect of nature can be described well by one or several of these equations. The analytic approach has been used successfully in physics, for example, since there are many interactions of particles that can be described well (on the average) by relatively simple mathematical expressions. As long as the particles do, in fact, respond in the same fashion as the analytic equations, the scientist using analytic procedures has tremendous predictive power because the behavior of the particles can be explored over a wide range of conditions simply by exploring the response of the mathematical equations.

There is a great deal of appeal in this approach, stemming in part from the predictive power and aesthetics of mathematics. The analytic approach has been used considerably in some aspects of ecology, especially population biology and population genetics. Analytic methods also have been useful in the development of theoretical models that are the basis for many developments in ecological theory. In some cases analytic models have been useful in the development of population management programs. Unfortunately they have been of less use in the study of whole ecosystems. The reason is that analytic equations are useful only under certain rather restricted conditions, that is, only when the equations describing biotic processes are linear and/or when there are relatively few equations to be solved at the same time. In addition, they are accurate only for describing processes in nature that are continuous (animal reproduction, for example, is not). Table 1 shows that the number of possible classes of equations that can be solved by analytic means are few. A model of an ecosystem normally includes from

Table 1   Classification of Mathematical Problems[a] and Their Ease of Solution by Analytical Methods[b]

	Linear Equations			Nonlinear Equations		
Equation	One Equation	Several Equations	Many Equations	One Equation	Several Equations	Many Equations
Algebraic	Trivial	Easy	Essentially impossible	Very difficult	Very difficult	Impossible
Ordinary differential	Easy	Difficult	Essentially impossible	Very difficult	Impossible	Impossible
Partial differential	Difficult	Essentially impossible	Impossible	Impossible	Impossible	Impossible

[a] Courtesy of Electronic Associates, Inc.
[b] After Franks (1967).

a dozen to several hundred or even more simultaneous equations, and they are as likely to be nonlinear as linear.

There is a further restriction in the effectiveness of analytical equations for use in ecosystem studies. An analytic equation is frequently a purposeful simplification of the process being modeled. For example, many population models assume that the intrinsic rate of increase ("r") will be constant over the range of interest for a given analysis. This greatly simplifies the process of model building, is correct under certain conditions or definitions, and allows for the use of some rather powerful mathematics during the investigation of certain questions. In fact, however, the actual rate of increase, or even the potential rate of increase, may vary in nature according to many factors that are a function of the population and the environment. For example, if food is seasonally limiting, or if the members of the population become on the average less healthy over the span of the investigations, the potential rate of increase is likely to change downward. However, to go from equations assuming a constant potential rate of increase to ones that would incorporate changes in the rate over time could create extremely messy mathematics, and this has not, as a rule, been attempted.

Thus the analytic approach normally is not especially useful in the construction of ecosystems models because of the need to solve simultaneously dozens to hundreds of equations, many of which may be nonlinear. Fortunately the modeler may turn to procedures frequently considered under the aegis, "simulation models." [We point out here that "simulation" actually refers to the process of putting any equation set on a computer and exploring consequences. Technically "numeric" might be the preferred antonym

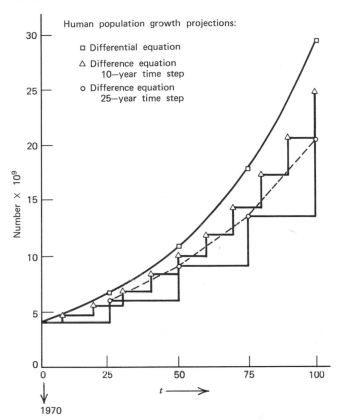

Figure 1 A comparison of predictions for the human population of the earth if present trends continue. N_0 — population in 1970, N_t = population at time t, r = the net rate of population increase. Three solutions are offered: $\square$ = analytic equation ($N_{t+1} = N_t(\exp^{rt})$); $\triangle$ = numeric solution [$N_{t+1} = N_t + (r \times 10 \text{ yr})$] with a 10-year time step; $\bigcirc$ = numeric solution [$N_{t+1} = N_t + (r \times 25 \text{ yr})$] with a 25-year time step. In this case the analytic solution gives the most nearly correct solution since human births occur at about 2-sec intervals.

to analytic (see Chapter 4), but "simulation" is more commonly used, and we use it also.] A simulation model does not give an exact solution to an equation over time as does an analytic model, and hence it may be subject to a series of errors because of the inexact nature of the solution techniques used (Figure 1).* In addition, you lose the ability to explore model behavior simply and cheaply (at least if you are good at mathematics) by simply

---

* Although we normally think of analytic solutions as more precise, this is not the case when discontinuous functions are modeled. For example, if an animal population breeds once a year in the spring, its birth rate can be better represented by a numeric than an analytic equation.

exploring known mathematics. However, you gain two very important properties. The first is that you are able to solve very many equations nearly simultaneously, and, secondly, it is possible to include all manner of non-linearity within these equations. With simulation techniques it is not necessary to leave out important model components in order to approximate a biological activity with a well-understood (but relatively inflexible) mathematical function and biotic activities can be approximated no matter how complicated. But the real benefit from simulation or numeric techniques is that they are ideally suited for digital computers.

Chapter 4 gives examples of how both approaches may be used in modeling and some consideration of the errors inherent in each. Unfortunately, there is no magic formula that allows either the beginner or the experienced modeler to model without error, or to choose whether analytic or simulation techniques are preferable for a given application. At every step of the way it is necessary to use your best judgment according to the conditions, and there are always tradeoffs. Most of the situations in this book require the use of simulation procedures, and the rest of this chapter is addressed to simulation techniques.

MECHANISTIC VERSUS DESCRIPTIVE EQUATIONS

In either analytic or simulation modeling the investigator may wish to incorporate various degrees of mechanism in the model. In other words, although it is possible simply to include an equation that *describes* the relation of two variables without any indication of causality, it is preferable to include the mechanisms relating one variable to another. For example, Chapter 10 explores the relation of marsh grass productivity in Louisiana to nutrients, sunlight intensity, and temperature. At the time of the development of this model there was not enough known about the physiology of the grasses to predict which independent variable would be the most important in determining the pattern of photosynthesis over an annual cycle. However, the authors found that temperature alone would give the best model results when compared to the actual field data, and so they used the temperature–plant growth relation as a first approximation in their model even though they did not understand the relation between the two. Subsequent studies have shown that evapotranspiration, which is closely related to temperature, is probably the most important factor limiting the primary productivity of the marsh plants over much of their range. The descriptive field studies, in this case, allowed the development of an adequate simulation model even when the mechanisms were not known. Future models can incorporate the physiological mechanisms controlling evapotranspiration and productivity to give a better model—one that will probably be better at predicting plant response to new conditions. Computer simulations that are not based on a knowledge of mechanism can be an important tool in suggesting procedures for investigating what the mechanisms in fact are. As a general rule, though, it is preferable to include mechanistic interactions.

The sociology of analytical versus simulation modelers is an interesting subject to consider for a moment. In the past the two groups have not communicated a great deal, although this has been changing recently. Analytical modelers often have entered the field of environmental modeling from mathematics or physics and are intrigued by the aesthetics and power of pure mathematics. The simulation modeler, on the other hand, generally has entered modeling from the more practical vocations of field biology or engineering. His interest tends to be not so much with aesthetic precision or mathematical power but with the inclusion of all parameters that he considers important, even where the underlying mechanisms are known only imperfectly. We look forward to a future in which the two approaches are more closely allied.

WHAT CAN WE DO WITH MODELS?

Models are useful to science in a number of ways, and one of the most important is that they assist a scientist in conceptualizing, organizing, and communicating complicated phenomena. But models can do considerably more. We may consider these: *understanding*, *assessing*, and *optimizing*. If the behavior of a number of parts of a system is fairly well understood, as well as the relation between the parts, they may be combined into a more complex model. This may give us new, or *emergent*, properties, that is, information about the behavior of the system that was not obvious from the behavior of the parts, and this may help us in the generation of new, testable hypotheses about the system. Thus as Botkin points out in Chapter 9, one of the most important uses of models is to *generate hypotheses*. In addition, once a reasonably accurate computer model is made of a complicated natural system, it is possible to check the data or assumptions that went into making the model by comparing the model behavior to that of the natural system under similar conditions. If the model and the real world disagree, then one or the other, or both, are imperfectly known, and tracing the error will almost certainly increase our understanding of the real or the model system. So another principal use of models is *to test the validity of field measurements and our assumptions derived from this data*. Models are no panacea: they are but one tool of many available to a scientist. The important goal is not necessarily the construction of the model or even the model's output, but to further our understanding of complicated systems. And they force the scientist to state his assumptions explicitly.

Prediction is the projection of future or hypothetical states that have not, at least while we have been measuring, occurred in the real world. It is related to simulation of an existing system as extrapolation is related to interpolation; that is, it goes outside the bounds of known circumstances. Thus models not only aid in the understanding of complex systems, they allow us to study this complex system under conditions that we are not yet able to observe or create, or may never be able to or want to observe or create, in the real world. There are a number of circumstances under which this is useful. For example, it may be extremely expensive or time consuming to undertake a large-scale experiment with a natural system, but very inexpensive and quick to perform the experiment on a computer, as Botkin has demonstrated with his forest growth models (Chapter 9). Another slightly different use of predictive modeling is in the prediction of some system property that you actually do measure—to see whether computer projections and field data agree.

Predictive models are used frequently in resource management situations to *assess* environmental impact or change. Assessment models estimate the effects of some activity not yet performed or the results of which are difficult to observe. For example, Kelly and Spofford (Chapter 18) attempt to assess the effect of waste-dumping into the Delaware River, and Day, Loesch, and Hopkinson (Chapter 16) attempt to assess the long-term effect of man's activities on the Louisiana coastal zone. Another example would be the assessment of the effects of building power plants on Hudson River fish populations (Chapter 14). In this case the real "experiment" could be done, and in fact may be done, but it is possible that the "experiment" will be destructive to a valuable fish population. Section III deals with many assessment models.

One important use of models is to *optimize environmental decision making*, and this inherently implies some value judgment and, in addition, some scheme for the management of the system. Optimizing monetary return, for example, does not always mean that environmental quality is simultaneously being optimized. Models may aid in the optimization, or choosing of the best pathway, for complicated conditions far into the future, although there is no guarantee that either the model is correct or that decision makers will pay attention to it. Chapters 5, 6, and 7 include three very different approaches to defining value systems, and Section IV of this book gives a number of examples of models designed to optimize environmental decision making. For example, the model presented by Kemp et al. attempts to assess many factors relating to the net worth of a power plant in Florida, and Kessell's chapter on fire modeling is an exciting account of how models are being used today to help manage forest fires. In practice and in this book, the difference between assessment and optimization is not necessarily distinct.

This brings us to the ultimate function of modeling, or indeed ecological investigations. What is it that we want to know from our models? The answer is no different from the same question directed toward science in general: We wish to know more about the structure and behavior of nature, both now and in the future. Models are one tool of many that aid in this process.

CONSTRUCTING A MODEL

Once a need for understanding, assessing, or optimizing a system has been formalized, the operational pathway that modelers often follow to create a model from a given situation may be summarized as in Figure 2. The modeler generally starts with a conception of what the important parts in his system are and some idea of how these parts interconnect. At this stage the development of the model is heavily dependent upon the modeler's experience and intuition, that is, what the modeler assumes to be important in his system and what he hopes to find out by modeling. Chapter 3 is an attempt to formalize to some extent the processes of developing model structure. In general, a knowledge of what the important components and interactions are, and the values that they have, may be obtained directly from field experience, or, with less precision, earlier studies on other systems. Then a model is constructed and "run," and the answers are checked by various means. If the answers are unrealistic, more information is needed, and the process is repeated. As Overton points out in Chapter 3, there are no magic formulae for producing a good model. However, the many case histories in this book should give the reader many suggestions.

Note the emphasis in the following diagram on *feedback* (that is, the mechanism by which the output of a process influences the input to that process at a later time) between preliminary results and later inputs to the models. This feedback is characteristic of both natural systems and the modeling process, and the right kind of feedback tends to stabilize both natural and model systems. The process by which modeling is checked with reality is called validation and is covered briefly at the end of this chapter and in more detail in Chapter 3.

The definitions of what are the most important questions, components, and relations historically have been the aspect of greatest separation between the analytical and simulation modelers. The former have tended to create models by elaborating mathematical statements that describe, or

16

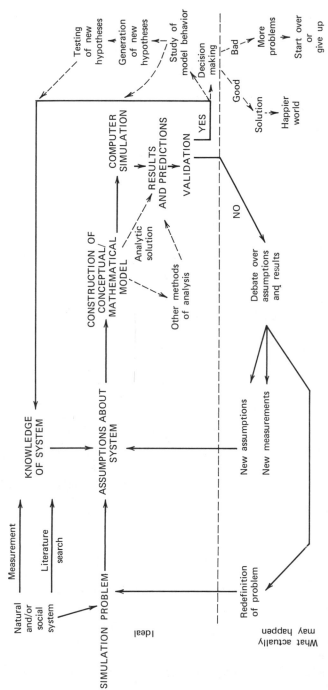

Figure 2 A diagrammatic model of the model-building process.

might describe, species or population interactions, either from an intuitive basis or from elaboration of empirical mathematical relations. The latter have tended to start from large quantities of data on standing crops, productivity, exchange rates, and so forth. The models that simulation modelers have constructed often use rather simple mathematics but have tended to agree well with field data, although perhaps simply because the models are forced to recreate what is known. Ideally ecological models may be constructed by field biologists with their data working in conjunction with more mathematically sophisticated modelers. This has been happening more frequently recently, and an increasing number of modelers today are both good field biologists and good mathematicians.

Once we have some knowledge of our system, and we have decided to construct a model of it, what steps are to be followed from this knowledge to the actual modeling? Too often modeling is thought to be necessarily connected with computers, but this may or may not be the case. There are actually a series of steps that generally are used in the construction of a model. In general, all models start at the beginning and go through some or all of these steps:

$$\text{CONCEPTUAL} \rightarrow \text{DIAGRAMMATIC} \rightarrow$$
$$\text{MATHEMATICAL} \rightarrow \text{COMPUTER}$$

CONCEPTUAL MODELS

The formation of a conceptual model is simply an extension of our other scientific processes, or indeed it may be these scientific processes themselves. One takes the components, interactions, and mechanisms that he believes to be operable in the system and considers them within the framework of the whole system and the questions he is interested in. The details of this process are beyond the scope of this book; they are the essence of science itself. However, we think that many of the case histories provide a feeling for the ways that many scientists have decided them.

To construct a conceptual model one says, "This is how I think my system is." This may be a priori (based on pure logic) and/or empirical (based on experience, on data collected for that or some other purpose, or on a previous model constructed for some other system). If, for example, we wish to model animal population interactions in a small pond, we can assume that it, like many others we know, is populated with bluegills (a fish species) and water fleas, or *Daphnia*. We then can assume that the bluegills eat the *Daphnia*, and this will be one of the interactions of the model. This seems like a logical thing for the bluegills to do, as *Daphnia* are about the right size to be bluegill food. Or we can tow some nets, and upon finding both bluegills and *Daphnia*, and *Daphnia* in the stomachs of bluegills, we can state with some certainty that the bluegills are eating the *Daphnia*. We then must *quantify* that relation if we are to understand the feeding process more fully. This can be done in several ways: (1) by finding how many *Daphnia* are in the bluegills' stomachs and how fast they pass through, or (2) by finding out how fast the *Daphnia* are declining in the pond after correcting for birth rate and other sources of mortality. If the information is not available, or would be expensive or redundant to get, we can use literature values from similar situations.

DIAGRAMMATIC MODELS

Very often a conceptual model is followed by a diagrammatic model drawn on paper or constructed from steel rods or balsa wood. This type of model is used extensively, for example, in explaining and teaching the structure of microscopic and submicroscopic entities, such as a model of an atom. An atomic model, of course, is a simplification and an idealization; yet we readily accept these models because they have become a part of our culture and because they explain adequately the most important aspects of atomic structure. The diagrammatic models presented in this book also represent abstractions of nature and are useful for conveying the essence of a system in the same way atomic models are.

   **Figure 3**   A simple box-and-arrow diagram.

One simple type of diagrammatic model, and one that has great usefulness, is the compartment or "box-and-arrow" diagram. Figure 3 is a simple *compartment model*. It is common to divide a rather complicated system into a number of "lumped" components (i.e., compartments) and consider the *inputs* to, and *outputs* from, each compartment.* The arrow represents the movement or flow of energy or materials from the *Daphnia* to the bluegills during feeding. Inputs to a compartment from outside the system of interest are called *forcing functions, driving variables,* or *exogenous variables.* Inputs to a compartment from inside the system are called *endogenous.* The con-

---

* In this case we are looking at the movement of energy, or materials; other similar-looking diagrams may be diagramming causative pathways by which given input condition cause certain behavioral outputs of the entity represented by the box. These two uses of box-and-arrow diagrams may overlap, but should be carefully defined.

tents of the boxes and the entities they represent are called the *state variables* of the system; that is, they are quantitative representations of the entities (states) that change (vary) with time. For example, we might put all of the bluegills into one box labeled "bluegill," even though we know perfectly well that there are big bluegills and small ones, hungry ones and full ones, each of which generally will behave somewhat differently. Likewise, we may conveniently place all the *Daphnia* into one box. The degree to which different entities (big bluegills or small, or perhaps bluegills and catfish) are included as part of the same state variable is an expression of the degree of *aggregation*.

Since ecological systems comprise many components that are highly interactive, the reduction of the number of components is necessary and by definition part of modeling. The complexity of real world systems is usually simplified by the aggregation of processes and components that are similar into functional groups such as trophic (i.e., food) levels, particle size, functional "guilds," and so forth. This process is often a major factor involved in setting up diagrammatic models.

For any particular model we must decide how simple or complex it should be. If too simple, we may not accurately describe the system; if too complex, we may become lost in details or ultimately develop a model no less complicated than the system being modeled. In addition, we may be no better off if we make a very disaggregated model but do not, or are not able to, have measurements of all the subcomponents. How much we aggregate and even the sorts of things we choose to use as state variables depends not only on what we know but what we want to know and the questions we ask of the model. For example, the photosynthetic process has many distinct biochemical steps, but it can be adequately described for most ecological purposes as total energy fixed, either net or gross. The biochemical details are not normally important here. However, for many ecological processes we do not know all of the details, whether important or not, and we must make assumptions with the hope that gross empirical observations will include the effects of the smaller scale interactions that are not well known.

We might consider at one extreme a model of some ecosystem in which there is but one entity of interest, and in which the only behavior of interest is the overall behavior of that entity. An example might be the following: We are interested in a model of the total amount of photosynthesis in a small pond. We have found that some algebraic expression adequately describes the rate of photosynthesis as a function of light, water clarity, and chlorophyll content of the pond:

$$P = f(L, C, Chl)$$

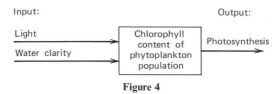

Figure 4

We may draw a diagrammatic model as in Figure 4. This model is extremely aggregated; that is, many different species, water depths, and microzonations are all lumped into one state variable. However, suppose that we had information indicating that photosynthesis was uniform in the upper meter of the lake, but that the lower section had a very different photosynthetic rate. We could build a somewhat more disaggregated model that, instead of lumping all plants into one state variable, puts them into two. If our direct or theoretical knowledge of the system allows, we may increasingly disaggregate the model into different parts until there is one state variable (or variable in a computer) for every single individual organism. This is in general quite unnecessary and defeats the concept of modeling, which is to simplify. In this case, the individual organisms do not behave sufficiently differently to make any difference to the outcome of the model, or for that matter, to be able to be measured. Obviously, in this case it makes sense to aggregate at least the individuals of the same species found in the same general location and perhaps to aggregate all individuals of all the different phytoplankton species and use the average productivity of these. In one recent model, ten taxonomic groups of phytoplankton were found sufficient to give an accurate representation of photosynthesis in several lakes (Lehman et al., 1975). However, the degree to which we aggregate is not always a modeling decision but frequently is influenced by whether or not the appropriate information is available. In addition, the degree to which one aggregates (or purposely does not) may be of secondary importance. It is a much more important question to decide what functional components and interactions to include in a model than to decide the degree to which the components should be aggregated. This concept is considered in more detail in Chapter 3.

A commonly used level of aggregation in ecological systems is trophic (feeding) level: plants are lumped together, herbivores are put together, carnivores, top carnivores, bacteria, and so forth. Another scheme is to lump organisms and other organics by size: dissolved organic matter, particulate organic matter, and so forth. In a number of examples used later, a combination of these two approaches is used. If one knows several species are particularly important, they may be treated separately and the rest of the species at that trophic level grouped together. There is no magic formula expressing the degree of aggregation acceptable under different circumstances. There is always a trade-off between portraying the system more accurately with increased disaggregation and wasting time, becoming unrealistic, or becoming sidetracked with unnecessary detail. One way to determine the best level of aggregation is for the modeler to use his intuition and see if his colleagues or critics can suggest something better. Another check is to build different levels of aggregation into different models of the same system and see if the outputs of the models are different with respect to the questions being asked. The purpose of a model may lead to lumping of variables even when there is enough information to separate them. For example, a model of world weather patterns would not need information on microclimates in a forest, and in fact such detail would probably be counterproductive. In general, the level of aggregation depends on the questions that you want to ask and the resources available to answer them.

CONSTRUCTION OF MATHEMATICAL MODELS: RELATIONS BETWEEN STATE VARIABLES

Models, whether analytic or simulation, are based on relations between forcing functions and state variables, relations among state variables, or relations between either and the behavior of state variables. For example, the level of incident sunlight (a forcing function) and the photosynthesis (a "behavior" of a state variable) in a small pond are related. When one system output always varies in some regular manner as a system input varies, we say that the output is a *function* of the input. In Figure 4, we would say that photosynthesis is a function of light (and chlorophyll quantity and water clarity). In this case, we say that sun intensity is the *independent variable* and photosynthesis is the *dependent variable*. Such a relation generally, but not necessarily, implies causality; that is, we assume that the increase in sun intensity is the cause of an increase in photosynthesis.

Once we have decided which functions and state variables are important to include in our model, we must determine (1) the *initial values* for the various state variables, (2) the forcing functions that will change the state variables, and (3) the functional relations between the state variables themselves. Ideas for such determinations are given in the various case histories; here we give a generalized overview of the procedures used to mathematically define relations between state variables.

Once we have created a conceptual model and drawn our diagrammatic model, our next step is to create a mathematical (normally algebraic) model.

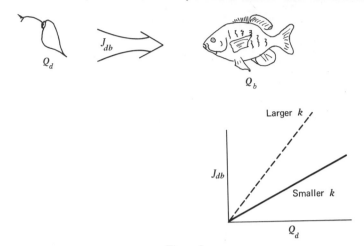

Figure 5

For example, we may wish to construct a simple mathematical model (Figure 5) of the transfer of food energy from *Daphnia* to bluegills:

$$J_{db} = kQ_d \qquad (1)$$

units: $Cal\ m^{-2}\ hr^{-1} = hr^{-1}\ Cal\ m^{-2}$

Equation 1 says that J_{db}, the quantity of energy (or carbon or whatever we are interested in) eaten by the bluegills during time interval t, is equal to k, the feeding coefficient, times the concentration of *Daphnia* in the pond.* If we increased the number of *Daphnia* in the pond by a factor of three, then the bluegills would see and eat three times as many of them. We call this a *linear* relation, since increasing the biomass of the *Daphnia* increases the transfer of energy by a proportional amount (Figure 5 and glossary of Chapter 3).

A term expressing the rate of some process as a linear function of the concentration gradient between sources (in this case, the *Daphnia*) and sink (in this case, the bluegills) is called a *linear transfer coefficient* or *proportionality constant*. A lowercase "k" normally is used to represent this constant, and it normally has units "per time." The quantity flowing in an equation using k is the quantity per the same amount of time as is used to define k. Sometimes k is normalized so that it will vary between zero and one to represent some proportion of the maximum possible rate of the process (see Chapter 9 for an example of this use of k).

---

* As a convention in this book we normally use Xs to represent forces, Qs for state variables representing quantities (i.e., biomass or concentration), Ns for state variables representing numbers (i.e., animal population numbers), Js for flows, and ks for coefficients.

If the number of *Daphnia* eaten depends not only upon the biomass of the *Daphnia* but also upon the biomass of the bluegills, we must rewrite Eq. 1. A simple way is to assume that feeding will be a direct function of the product of the biomass of *Daphnia* and bluegills:

$$J_{db} = kQ_dQ_b \qquad (2)$$

Now we are faced with an uncertain situation—which of the above formulas is a better representation of what actually occurs in nature? If the *Daphnia* are so sparse and the bluegills so dense that every *Daphnia* is eaten by a bluegill as soon as it becomes large enough to be seen by a bluegill, no matter how many bluegills are present, the first condition holds. Normally, however, the second condition, that of the feeding relation being a function of both predator *and* prey biomass, is more nearly correct. We may also wish to include additional terms in our mathematical representation of the feeding relation between the two species to represent the effects of, for example, the relative "fullness" of a predator.

If the flow of energy from the prey to the predator was a function of only the concentration of prey, it is called a *donor-controlled* interaction, in that the rate of the flow depends on only the donor, or source, of the material transferred. Similarly, where both donor and recipient are involved, as in equation 2, we call it a *donor-recipient controlled* interaction. Other, more complicated relations are considered below and in chapters such as 11 and 12.

Donor-controlled equations are very popular because they "behave" well, that is, they are mathematically tractable, flexible, and adaptable to many uses. In many physical and some biological situations they are clear representations of the processes modeled. However, they tend not to be good representations of interactions between competing or feeding organisms, and that is why work similar to Botkin's (Chapter 9) and Wiegert's (Chapter 12) is very important. In general, it is much more difficult to model accurately interorganism interactions than relations between physical forcing functions (e.g., sunlight intensity, water flows and stratification, temperature, and so on) and organisms. As a consequence, models that have emphasized seasonal variations in forcing functions (e.g. Chapters 10, 13, 14, and 25) frequently have been more successful in simulating ecosystem behavior than have models that are principally concerned with organism interactions. Both approaches are needed, however, especially as we attempt to predict ecosystem response to stress.

Our next question is how do we measure coefficients for equations of interaction? Assuming we have a relation that can be defined by a donor-controlled relation (e.g., see Chapter 19; donor-controlled equations are also used in part in most other chapters) how can we measure the transfer coefficient? First rearrange equation 1 so that $k = J/Q$. Thus if we can measure the quantity of the "donor" and the rate of flow out of the compartment, then we can solve for the transfer coefficient. This technique works where it has been demonstrated that the relation between flow and source is linear, and it has been applied successfully in, for example, many hydraulic models. But if such a simple donor-controlled model is inadequate for describing component interactions, other methods must be used. These are developed to some extent in the next section. Different types of forcing functions are introduced in Chapter 2.

This section considers several somewhat more complicated formulas representing additional types of interactions that are likely to be useful in describing relations between forcing functions and state variables, or relations between several state variables. These relations may be used for some analytical modeling as well as for simulation modeling. For the moment, we assume that we do not know the underlying mechanisms of the process that we are attempting to model and that what we want to do is find a curve shape that adequately represents the process that we are interested in. Figure 6 gives a number of equations that are particularly useful in such modeling processes.

Once we have an equation representing the relation that we are interested in modeling, it becomes necessary to determine the coefficients that will make the curve correspond as closely as possible with the real-world data. This is done best with empirical methods, that is, by observation and/or experimentation. Otherwise literature values may be used. There are two principal empirical methods that may be used: *isolation* and *field correlation*. In the first procedure attempts are made to isolate the pathway or interaction of interest, generally in the laboratory, under carefully controlled conditions. For example, Nixon and Kremer (Chapter 25) found that they needed an equation to relate the growth rate of juvenile copepods (a small shrimp-like crustacean) to water temperature. The time and money that they had available did not allow them to measure this process themselves with copepods from Narragansett Bay, where they were working. However, laboratory studies had been undertaken elsewhere on related species. These studies were done by isolating a species of copepod in a laboratory chamber where everything (e.g., food, age of copepods, competition, etc.) was kept constant except for water temperature. Then the growth rate of the copepods was measured at the different temperatures. The results are plotted as data points in Figure 7.

Nixon and Kremer then chose a mathematical (in this case, exponential) relation that gave the best representation of copepod growth as a function of water temperature (solid line, Figure 7).

But how did they get the coefficients (i.e., 0.055 and 0.097) that made their function (plotted as the smooth line in Figure 7) fit the data so well?

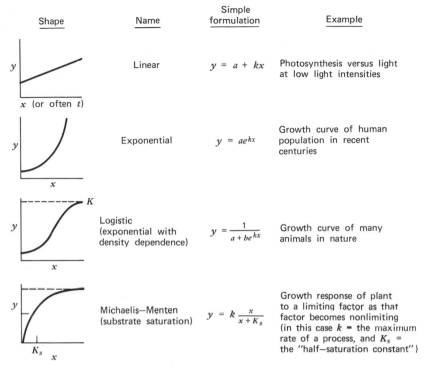

Shape	Name	Simple formulation	Example
	Linear	$y = a + kx$	Photosynthesis versus light at low light intensities
	Exponential	$y = ae^{kx}$	Growth curve of human population in recent centuries
	Logistic (exponential with density dependence)	$y = \dfrac{1}{a + be^{kx}}$	Growth curve of many animals in nature
	Michaelis–Menten (substrate saturation)	$y = k\dfrac{x}{x + K_s}$	Growth response of plant to a limiting factor as that factor becomes nonlimiting (in this case k = the maximum rate of a process, and K_s = the "half–saturation constant")

Figure 6 Mathematical relations commonly encountered in modeling. a, b, k, and K_s are constants.

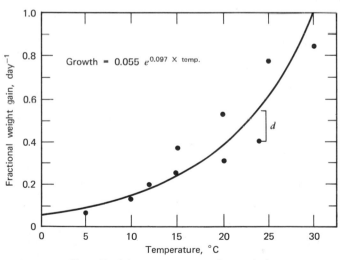

Growth = $0.055\ e^{0.097 \times \text{temp.}}$

Figure 7 A least-square regression analysis.

25

They used the technique of *linear regression analysis*, which is a statistical procedure that helps one to find the best coefficients to fit an equation to data points. In this case they did a least squares regression after transforming the data. A least squares regression is a technique whereby the values of coefficients in a linear equation are found that give the "best fit"; that is, the sums of the squared vertical distances between the predicted line and the data points, called the sum of squares, are minimized for a unique value of the coefficients. The least squares regression technique is most easily done for linear equations; this is the reason why Nixon and Kremer transformed their data from curvilinear to linear form. They expressed their original equation ($G = a'e^{b'T}$) in a linear form ($Y = bX + a$) by taking the natural log of their equation, $\ln G = \ln a' + b'T$, and letting $\ln G = Y$, $X = T$, and $\ln a' = a$. They used the natural log of their growth data to fit the linear form of the least squares regression; this is an example of what is meant by transforming the data. When temperature is plotted against the natural log of growth increment, the relation is linear.*

An extension of simple linear regression analysis is multiple regression analysis in which more than one independent variable is possible. Multiple regression analysis is similar in practice and application to the simple case we have been discussing. The subject of regression analysis is covered in detail in any good statistics book, which should be consulted by the interested reader (e.g., Dixon and Massey, 1969).

Thus far we have discussed a method for finding the values of unknown coefficients in simple linear equations, but what about equations of such a complicated form that they cannot be linearized? In this case we have to turn to the use of a computer to find the values for us through a process of iteration rather than by analytical solution. In practice, the user supplies the computer with the basic equation form, and a series of different coefficients are used in successive computer runs. The computer is programmed to "home in" on the best set of coefficients by rejecting values that give a higher sum of squares and accepting values that decrease the sum of squares. Polynomials (Chapter 4) can be used (with caution) to fit nearly any set of data, and this is used, for example, for some forcing functions in Chapter 10. The principle is quite simple, but very sophisticated procedures do exist, which we do not go into here, for solving equations in this manner.

---

* The coefficients a and b are computed according to the following formulas:

$$b = \frac{n\sum XY - \sum X\sum Y}{n\sum X^2 - (\sum X)^2} \quad \text{and} \quad a = \frac{\sum Y}{n} - \frac{b\sum X}{n}$$

where n is the number of data points. After these coefficients are computed, we notice that $b = b'$, but $a = \ln a'$; so to find the value of a' in the original equation a' is set equal to base e to the power a, or $a' = e^a$.

There is always a problem of trade-offs between the amount of control exercised in a laboratory study and the loss of realism imparted by factoring out other variables that do exist in nature. For example, the copepods in our previous example may not have grown as well in the laboratory chamber as they would have in their natural habitat. One way of avoiding the arti-ficialities of the laboratory is to use the second method of *field correlation* to examine the extent that two variables measured in nature (or their transfor-mations) vary together linearly in time or space. For example, we may measure photosynthesis during one day and find that the rate of photosyn-thesis is closely related to the temperature. We may run a correlation analysis on the data and find an r^2 of, say, 0.91. This means that 91% of the variation in productivity is associated with similar changes in temperature. An r^2 (or r) of 0.00 means that neither variable can be used to predict the other, and an r of less than zero means that the two processes tend to occur in opposite directions. For example, we would expect that there would be a positive cor-relation coefficient between the amount of sun intensity and the temperature at the earth's surface, but a negative r between sun intensity and the activity of a night-loving raccoon.

The "r" is called the *correlation coefficient*. Its value is always between -1 and 1. If its value is positive, the independent and dependent variables vary in the same direction; if negative, the variables vary in opposite direc-tions. The value of r^2, called the *coefficient of determination*, tells us how much variation is accounted for by the regression. In the example being discussed, $r^2 = (0.91)^2 = 0.83$ means that 83% of the variation in the data points can be explained by the mathematical function, or to phrase it another way, 83% of the variation in the productivity data can be explained by temperature.* The remaining 17% is unaccounted for by this analysis. Values of r^2 are always between 0 and 1. When all the points fall on the regression line, $r^2 = 1$. When the points are scattered, such that the regres-sion line becomes horizontal, then $r^2 = 0$. One seeks to find the expression and/or combination of dependent and independent variables that gives the highest value of r^2, because its value is a measure of the "goodness of fit."

Correlation techniques such as the one developed above help us to decide what processes are occurring together, and hence might be related directly or indirectly. However, it is important to note that co-occurrence does not necessarily mean any causal relation, since both may be responding to a third variable that has not been included in the analysis or of which the investigator

---

* The value of r^2 is computed from the following formula:

$$r^2 = \left(\frac{n\sum XY - (\sum X)(\sum Y)}{\{[n\sum X^2 - (\sum X)^2][n\sum Y^2 - (\sum Y)^2]\}^{1/2}} \right)^2$$

where n, X, and Y are the same as before.

is not even aware. In addition, correlation coefficients work only for two processes that have a linear (or linearized) relation. In the above example it would be tempting to say that productivity is higher because the temperature is higher. However, both temperature and photosynthesis 'may in fact be responding to sunlight intensity. Correlation techniques are quite useful in simulation modeling, but they must be used with extreme caution. A powerful method is to use relations suggested by correlation techniques backed up by experimental procedures. If two variables show a high correlation, they may be related quantitatively by regressions for modeling purposes.

ARRAYS

Since ecosystems often are composed of logically similar state variables (e.g., different species of plants or different subregions of the ecosystem), it becomes useful to consider these groups in some organized fashion. This often is done easily with a system of *arrays*, or algebraic checkerboards, which are a convenient and powerful tool for modeling when combined with linear or more complicated algebra in a computer. For example, if we have many fish species eating many food items, we may wish to model all the energy-flow pathways from the various different food items to the various different fish. We may conveniently do this by calling each species of fish a number: bluegills equal 1, catfish equal 2, and so on. For convenience, when M, an arbitrarily named variable, equals one we are talking about the biomass of bluegills. FISH(1), read as "fish sub one," means that we are considering the biomass of bluegills; FISH(2) means the biomass of catfish, and so on. In this case FISH(M) is a *one-dimensional array* that is *subscripted*; that is, it has a number of more specific definitions depending upon what "M" is at the moment. Note that we could have used $Q(M)$ or $Q1(M)$ for our fish array, since we are talking about quantities or state variables. Normally the name chosen for a state variable is arbitrary and often mnemonic.

We can create an array for the prey of the fish as well: PREY(1) might be the biomass of *Daphnia*, PREY(2) the biomass of midges, and so forth. We could call the subscripts in this case Ns; that is, PREY(N), $N = 1,10$ means that there are ten prey items that we are considering in our pond. Perhaps we have found nine important ones, and PREY(10) means "other." In practice we store a number in the *address* called "PREY(1)," and so forth. When we want to use that number we go (in the computer) to that address and read the number stored there. At some later time we may store another number in PREY(1) if, for example, the *Daphnia* have reproduced. Thus we are using arrays as state variables.

Table 2 Matrix of Energy Flow

		Daphnia	Midges	$\cdots$	Other	Name of prey
Name of Predator	One-dimensional Array	PREY(1)	PREY(2)	$\cdots$	PREY(10)	One-dimensional ← Array
Bluegill	FISH(1)	FLOW(1,1)	FLOW(1,2)	$\cdots$ FLOW(1,10)		Two-dimensional ← Array
Catfish	FISH(2)	FLOW(2,1)	FLOW(2,2)	$\cdots$ FLOW(2,10)		
Shiner	FISH(3)	FLOW(3,1)	FLOW(3,2)	$\cdots$ FLOW(3,10)		
$\vdots$	$\vdots$	$\vdots$	$\vdots$	$\cdots$	$\vdots$	
Other	FISH(10)	FLOW(10,1)	FLOW(10,2)	$\cdots$ FLOW(10,10)		

We also may use arrays to represent the exchange rates between compartments. Since we have a series of Ms that refer to different fish species and a series of Ns that refer to things that fish eat, we conveniently may consider the pathways of food from prey to predator. If we are interested in the rate of feeding (i.e., the flow of material or energy) of each of these pathways, we may say FLOW(M,N). This is an example of a two-dimensional array. Two-dimensioned arrays usually are called *matrices* (Table 2). If M is one and N is one, then we are considering the rate that bluegills are eating *Daphnia*, and so on. Arrays are extremely useful when dealing with many entities with some similarities—for example, many different species in a trophic level—in a computer. *Iterative* (repetitive) processes, in conjunction with arrays, are a basic tool of computer modeling. For example, we could calculate the total food energy transferred between predators and prey in the example by solving for all predators and all prey.

DIFFERENCE EQUATIONS

Once we have established a conceptual model, defined the important forcing functions, state variables, and coefficients, and organized them all with a series of arrays, the most common and perhaps the most powerful mathematical tool used to tie them together is the difference equation such as Eq. 3.

$$Q_{t2} = Q_{t1} + \sum_{i=1}^{n} J_i \qquad (3)$$

The quantity at time $t2$ equals the quantity at time $t1$ plus or minus all the flows during that time interval. This expresses a change in a quantity (a difference) that occurs during a given unit of time under certain conditions; for example, the above equation could represent the change in the biomass of a bluegill population that is eating *Daphnia*. Difference equations are convenient because we can easily measure changes happening over a discrete time interval or under certain conditions. We can then sum, or integrate, these changes over a long period of time or under a more complicated set of conditions to give results that would otherwise be impossible or inconvenient to achieve. Thus one of the functions of a model is to *simplify a time-dependent process*; a process is measured during a short time interval, then applied over a longer interval.

Another function of a model is to *simplify a condition-dependent process.* In practice this normally means that we measure complicated functions one at a time, then assume that the relation found for each holds true when several are working simultaneously. For example, we might find that bluegills will eat about three times more *Daphnia* in an hour at 20° than at 10°C. Also, they might eat about twice as many in an hour if they have not eaten in 24 hr than if they have not eaten in 12 hr. We would then assume that a bluegill starved for 24 hr in a 20°C aquarium will eat about six times as many *Daphnia* as one starved for 12 hr in a 10°C aquarium. However, we must remember that it is only an assumption that the two relations are acting independently, and the model output should be checked with what actually happens in nature.

In actual models the time-dependent and condition-dependent processes are combined, giving us a very powerful arrangement for describing processes that are very complicated. The results then can be checked with the real system.

COMPUTER MODELS

After a series of difference equations has been written that adequately describes your system, a *computer* model may be constructed. Computer models are normally either *analog* or *digital*. Both will give virtually the same answers to the same inputs and same model structures, and in theory either could be used for any given problem. In practice digital computers ("the computer" that lives in "the computer center" of most universities) are more commonly used, particularly where there are large amounts of input data. Analog computers (which are often desk-top sized) are much cheaper to buy and operate, they are very useful for teaching because they give answers instantaneously, and they are best adapted for theoretical modeling where the model itself is not based on large quantities of input data. The digital computer has longer "turnaround time," is more expensive, is better suited for nonlinear relations, and is ultimately more flexible. Hybrids have been developed that are particularly useful for modeling.

The basic difference between the two is that the digital computer deals with digits: something is on (1) or off (0). Inside the machine, this means that a small iron ring is magnetized or it is not. The ring cannot be $\frac{2}{3}$ magnetized. The analog computer, on the other hand, is based on continuously varying electrical currents and charges. It is possible to charge a capacitor in an analog circuit to $\frac{2}{3}$ capacity. The difference between the two computers is like the difference between a freely moving wheel that can be turned to any position and one that is constrained by a ratchet to stop only at one of certain predetermined positions. An analog is an analogy—something that is a quantitative, although not qualitative, representation of the real world. In practice, the difference between the concepts "digital" and "analog" is not especially important to the modeler, although the difference in actual hardware is.

Although it is not the purpose of this chapter to explain programming, we put formula (2) into both FORTRAN, a digital language, and analog notation. FORTRAN and other similar languages do most of their "thinking" by way of algebraic difference equations. Here is Eq. 2 written in FORTRAN: the sentences following the "Cs" are called comment statements and are not part of the operational program. The reader may ignore the first two lines of the program.

```
        INTEGER D,B,T
        DIMENSION J(2,3),Q(3)
C   MAKE DAPHNIA TROPHIC LEVEL 2, AND BLUEGILL TROPHIC
C   LEVEL 3
        D = 2
        B = 3
C   INITIALIZE BIOMASS OF DAPHNIA AND BLUEGILLS
C   Q'S ARE THE STATE VARIABLES OF THIS PROGRAM
        Q(D) = 100.0
        Q(B) =  10.0
C   INITIALIZE TOTAL ENERGY FLOW OVER THE DAY
        TOTFLO = 0.0
C   MAKE THE TRANSFER COEFFICIENT 0.001 PER HOUR PER
C   (1 CAL PRED) * (1 CAL PREY)
        K = 0.001
C   CALCULATE THE ENERGY FLOW FOR 24 HRS. BY ONE HOUR
C   INTERVALS
        DO 10 T = 1,24
        J(D,B) = K * (Q(D) * Q(B))
        TOTFLO = TOTFLO + J(D,B)
   10   CONTINUE
```

```
C   PRINT THE RESULTS
        PRINT 15, TOTFLO
C   THE FOLLOWING STATEMENT TELLS THE COMPUTER HOW TO
C   PRINT THE RESULTS
        15 FORMAT (F10.2)
        END
```

If we put this on computer cards and run it through the computer, it will print out how many calories will be transferred from the *Daphnia* to the bluegills during a time span of 24 hr. We could subtract the energy eaten from the *Daphnia* and add it to the bluegills to calculate the new values of the state variables each hour:

$$Q(D) = Q(D) - J(D,B)$$

```
C   ASSUME BLUEGILLS ASSIMILATE 50 PERCENT OF THE FOOD
C   THEY EAT
```

$$Q(B) = Q(B) + 0.50 * J(D,B)$$

This would update the biomass of bluegills and *Daphnia* each hour to give a more accurate representation of energy flow. Note that the use of subscripts with our state variables (i.e., $Q(N)$, $N = 1, 2, 3 \ldots$) allows us to deal with many species, and so forth, as discussed in the previous section.

To do the same operation on an analog computer, we would build the following circuitry by *patching* (plugging in wires) the analog board (which looks a little like a telephone switchboard). By patching the analog board we are, in essence, connecting electronic components (resistors, capacitors, and so forth) that solve equations electronically. For example, integration is done by capacitor that adds and subtracts electrical flows. Patching for analog computing performs the same function as *programming* for a digital computer; that is, both are the physical and operational construction of an *algorithm*, which is a set of logical mathematical procedures (e.g., a series of difference equations). The creation of an algorithm comes under our heading of constructing a mathematical model.

Figure 8 is a patching diagram for the same problem on an analog computer. To run this program we would patch, or plug in, wires to the correct modules and set transfer coefficients on variable electrical resistors (called potentiometers or pots). The quantity of a state variable over time would be the charge on a capacitor. Model solutions could be read on a voltmeter, an oscilloscope, or drawn on an xy plotter.

The power and usefulness of digital computers (and in a slightly different sense, analog computers) is a result of their ability to *iterate*, that is, to do a process or a series of processes again and again. This has particular applica-

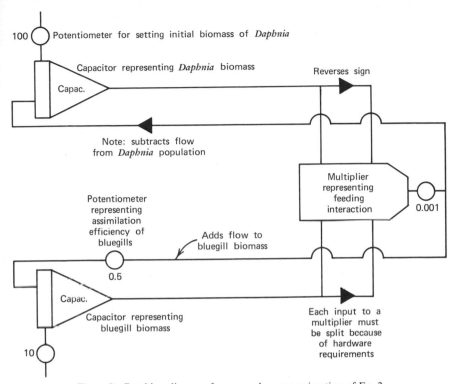

Figure 8 Patching diagram for an analog approximation of Eq. 2.

tion when one wants to do an algorithm many times, changing state variables or a coefficient each time. An ecological application might be a model of photosynthesis that is run again and again, changing the biomass of the plants each day according to the previous day's photosynthesis. Then we might add our *Daphnia* and bluegills to make a simple food-chain model. Iteration also allows the programmer to integrate many effects by solving the results of all processes as they occur over and over with short time intervals. Each new run uses the results of the previous run; for example, the little FORTRAN program just presented could be done ten times (for 10 days) by means of another iterative *do loop*.

One final step in the modeling process is to determine how well the model mimics the "real world," a process called *validation*. There are several ways to do this, and the process may be somewhat circular. Ideally, several of these methods would be used to check a given model. The first is by *simulation and comparison*. Given the same inputs, do the larger systems and the model give the same results? Is this the case for both the component parts and the whole system? In Chapter 18 the results of the model are checked against known data to see if the model is realistic. For some cases it is, for others it is not, and in certain cases there are no data to check the model results. Simulation also can be used to check the completeness of the experimental measurements of a natural system. If the model and the natural system do not agree, the model is not necessarily wrong. The difference may be because of incomplete or inaccurate measurements of the natural system, and the model may suggest which measurements are incomplete or wrong. There also have been developed a series of more formalized procedures for validating model components (e.g., Caswell, 1976).

Another way to check the accuracy of the model is by experimental *prediction.* If there is a reasonable set of conditions that could occur in the natural system that has not yet been observed, the conditions may be programmed for the model and the results recorded. If these conditions then occur, or are made to occur, in the larger system and similar results are found, some confidence in the validity of the model is warranted. If this is extended to a large set of possible cases that bracket all possible conditions that the real system is apt to be exposed to, we could perhaps have complete confidence in our model for all possible situations. In practice, this rarely or never occurs.

A third way to check on the validity of a model is to expose the model to its critics. Although this is rarely mentioned in books on modeling, it occurs often in practice both in the literature and in courtrooms. Unfortunately, due to differing philosophies of the persons involved, the issue may not have an objective solution.

Sensitivity analysis is a means of determining how important the different model components and coefficients are to the model output and aids the researcher in understanding which components should be measured most carefully. In practice, sensitivity analysis consists of making a series of runs or partial runs of a computer model while varying model structure or coefficients over expected or possible ranges and noting the importance of each on the output. Both validation and sensitivity analysis are covered more fully in Chapter 3, and a simple example is found in Chapter 15, Table 1.

SOME ADDITIONAL TERMS AND CONCEPTS

Models that give definite unique outputs to a given set of inputs are called *deterministic.* However, nearly any measurement of nature includes some statistical component; that is, there is no precise relationship between variables but rather a mean and standard error. If we include such uncertainty in our model (by, for example, using forcing functions with algorithms that produce random numbers with a statistical distribution), the models are said to be *stochastic*, and the output of one analysis may be different than that of another. For example, instead of representing the sun as an average value we could vary it from day to day in the program to represent clear and cloudy days (e.g., see Chapter 25). As a general rule, deterministic models are used as they are much simpler and cheaper to construct. However, if it is very important to explore all possible future states of a system, stochastic models are required.

A further difficulty in constructing models was pointed out by Levins (1966). He has characterized models by their relative degree of *precision*, *generality*, and *realism*. Each of these characteristics is created at some expense to one or both of the others. For example, models that are constructed to be very precise representations of a certain system tend to lose their generality; that is, they are less effective at representing a broad range of similar systems. Models that mimic certain system variables very closely are considered *precise*, and those that account for all relevant variables and relations are *realistic*. Finally, we call conclusions that are not particularly sensitive to model structure *robust*. There is no magic formula as to what are the most desirable characteristics for a given situation. However, it is important not to confuse precision (e.g., how many significant figures) with accuracy (e.g., the extent to which a measurement or simulation reflects reality).

This chapter simplifies many aspects of modeling in order to introduce the reader to the central concepts with as little pain as possible. But it also reflects our belief that the mechanics of building a model are relatively simple, particularly when compared with the complexity of determining the important attributes and relations of an ecosystem. We hope that the case histories included later in the book give you an appreciation for many of the subtleties and uncertainties of model building. In the meantime, this chapter should arm you with enough vocabulary to attack the rest of the book. The next three chapters develop in more detail some of the concepts touched upon here and should help prepare you to construct models yourself.

ACKNOWLEDGMENT

We thank Sam Bledsoe, Hal Caswell, Ray Van Houtte, Bob Howarth, Si Levin, Jim Morris, John Muratore, Scott Overton, Bob Rovinsky, Robert Shore, and Christine Shoemaker for many helpful comments. However, we assume full responsibility for the final product.

REFERENCES

Bartholomew, G. A. 1964. The roles of physiology and behavior in the maintenance of homeostasis in the desert environment. In *Homeostasis and feedback mechanisms, Symposia of the Society for Experimental Biology*. Vol. 18, pp. 7–29. Academic, New York. Pp. 7–29.

Caswell, H. 1976. The validation problem *in* B. C. Pattern, (Ed.), *Systems analysis and simulation in ecology*. Vol. 4., Academic, New York. Pp. 313–325.

Dixon, W. J. and F. J. Massey, Jr. 1969. *Introduction to statistical analysis* (3d ed.). McGraw Hill, New York.

Franks, R. G. E. 1967. *Mathematical modeling in chemical engineering*. Wiley, New York.

Goodman, D. 1975. Personal communication.

Lehman, J. T., D. B. Botkin, and G. E. Likens. 1975. The assumptions and rationales of a computer model of phytoplankton dynamics. *Limn. Oceanog.* **20**: 343–364.

Levins, R. 1966. The strategy of model building in population biology. *Amer. Sci.* **54**: 421–431.

Maynard-Smith, J. 1974. Models in Ecology. Cambridge University Press, Cambridge, England. 146 pp.

Odum, H. T. 1970. *Environment, power, and society*. Wiley-Interscience, New York.

Patten, B. 1971. A primer for ecological modeling and simulation with analog and digital computers. In B. C. Patten (Ed.), *Systems analysis and simulation in ecology*, Vol. I. Academic, New York. 607 pp.

Platt, S. R. 1968. Strong inference. *Science* **146**: 347–353.

Walters, C. 1971. Systems ecology: the systems approach and mathematical models in ecology. In E. P. Odum., *Fundamentals of ecology*. Saunders, Philadelphia. Pp. 276–292.

Watt, K. E. F., Ed. 1966. *Systems analysis in ecology*. Academic, New York. 273 pp.

Watt, K. E. F. 1968. *Ecology and resource management*. McGraw-Hill, New York. 450 pp.

2
A Circuit Language for Energy and Matter

CHARLES A. S. HALL
JOHN W. DAY, JR.
HOWARD T. ODUM

W hen the complexity of a radio circuit becomes more than trivial, the electronics engineer turns to wiring diagrams to assist in the organization and communication of information and to assist in the development of operational equipment. Similarly, in the analysis of *any* complex system it becomes necessary to create diagrams that organize and simplify the processes that are being considered. Many of the systems of man and nature included in this book are, like their electronic counterparts, too complicated to be conveyed easily by words alone, and most readers are unable or unwilling to conceptualize the system from a series of differential equations. We describe here briefly a symbolic language previously presented elsewhere in more detail (Odum, 1967a, 1967b, 1971, 1972; Odum and Odum, 1975) that we believe greatly assists both in the conveyance of complicated and voluminous ecological information and in the analysis of complex systems.

The development of these symbols follows logically from earlier summaries of whole ecosystems (e.g., Odum, 1957), from early attempts to build analog computers from radio parts before commercial models were available (Odum, 1960), and from basic thermodynamic consideration of natural phenomena (Lotka, 1922; Brody, 1945; Odum and Pinkerton, 1955). Fundamental to the use of the various symbols are (1) the requirement of conservation of matter and energy and the degradation of some energy into waste heat during useful work processes, (2) the importance of feedbacks and interactions in the functioning of complex systems, and (3) the use of some fundamental physical and biotic principles. A more complete description of the symbols with the physical basis for their operation is found in Odum (1972), and a number of principles generally, but not necessarily, associated with their use is given in Chapter 7. Suffice it to say here that the modules, and the processes that they represent, have broad generality such that we find analogies and homologies in ecology, economics, physics, hydrology, electronics, and so forth. Most of the models discussed in the book are shown using these symbols.

The energy-flow language is based on a series of modules that represent both systems processes and mathematical functions, connected by lines representing transfer pathways of energy, materials, or information. Although the symbols were originally created to diagram energy flows, they are equally useful for material-flow pathways. As a general rule it is permissible to construct flow diagrams of energy, of materials, or of the combination of the two, as long as it is explicitly stated (and remembered) what it is that is being diagrammed, and as long as the energy relations that drive the processes are kept in mind. The symbols that are the basis for this symbolic language are given in Figures 1 to 11. It is important to remember that we are simply expanding the box and arrow diagrams of Chapter 1 to include more specifically defined "boxes." One particularly useful aspect of this symbolic language is that it can be readily translated into computer languages.

Sources, or Forcing Functions

The circular module (Figure 1) represents a source of energy or materials from outside the system being considered. For example, if we were interested in modeling a forest, it would not be necessary to include the fusion processes in the sun for our model. We are simply interested that the sun does supply photons to run the system. Therefore, we draw the sun as a circle to represent a source of energy (Figure 1a), the origin of which we are not presently concerned with. Other sources might be leaves entering a stream if we are interested in studying the stream and not the forest, oil entering a city (Figure 1b), nutrients entering a lake, or water flowing over an oyster reef (Figure 1c). The output of the source with time might be constant no matter what else happens; we would call this a *constant-force* source (Figure 2a). The sun is an example of this (at least as far as *we* are concerned!). If the flow becomes less as the source becomes depleted, for example, with water draining from a tank or with oil as a well becomes depleted, the output of the source would be less with time (Figure 2b). Remember that we are using Xs to represent forces, Js flows, and Qs for quantities of materials or energy. Now if there were a *regularly occurring* but *varying* input to the system we could represent it as a sine wave (Figure 3a), a square wave (Figure 3b), or some other function. For example, the input of light (J_a) to an ecosystem from the sun on a completely clear day approximates a sine wave.

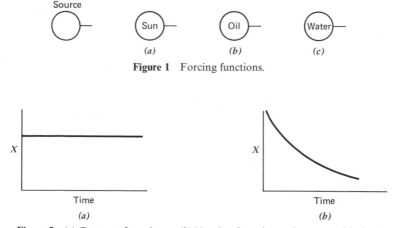

Source

(a)   (b)   (c)

Figure 1   Forcing functions.

X

Time

(a)

X

Time

(b)

Figure 2   (a) Constant force input. (b) Varying force input-decreases with time.

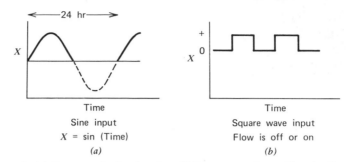

Figure 3   (*a*) Sine-wave forcing function. (*b*) Square-wave input. Flow is off or on.

We put certain sources into our diagrams when we know that they are important in controlling the functioning of the system being represented. Therefore, the source symbol is also referred to as a *forcing function*, sometimes called an *exogenous variable* or a *driving variable*. For example, a common situation in modeling is to have the input of sunlight intensity force the photosynthesis of the system. Another example would be the "pulsed" addition of organic materials to a stream when the leaves fall in the autumn.

Heat Sink

The heat-sink module (Figure 4), similar in shape and function to the ground symbol used in electronics, represents energy that must be degraded into heat (J_H) for any real process to occur, according to the second law of thermodynamics. Thus if chemical potential energy is changed into mechanical energy, as when glucose is burned in an animal to provide muscular locomotion, some of the energy originally found in the glucose is converted to heat during that process and thus is lost from the energy still available to do work. Whenever we construct a model of any process that uses energy, the heat-sink module or its equivalent must be included to represent the necessary by-product of useful work or energy storage. One of the most common uses of this symbol in ecological modeling is to represent maintenance metabolism, that is, the energy that is used to maintain an organism

Heat
sink   **Figure 4**   Heat sink.

against the random shaking apart of molecules (i.e., entropic degradation or depreciation).

Storage (A State Variable)

The next symbol we consider is the storage module (Figure 5). In the *passive-storage* module (Figure 5a) no new potential energy is generated in loading or unloading this tank, although some outside work must be done in moving material and energy in and out of storage to overcome friction. Examples are addition of gasoline to a fuel tank or flow of nitrate into a lake. The second situation (Figure 5b) represents the storage of new potential energy against some backforce, such as pumping water uphill into a reservoir. The work of storage requires the dispersal of some potential energy into heat: thus the heat sink is shown. The storage of water in a tank or of electricity in a battery are two examples of *potential-generating work*. As a general rule, the discharge of a passive storage and the loading of a storage against the increasing resistance to loading of the storage (for example, a water tower) would look like Figure 6. In each case the gravitational force (X) that either

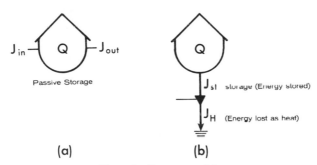

Figure 5   Storage module.

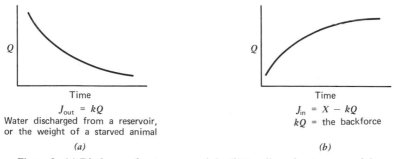

Figure 6   (*a*) Discharge of a storage module. (*b*) Loading of a storage module.

causes the flow out of the storage (Figure 6a) or resists the influx of additional water (Figure 6b) is a direct function of the quantity (Q) of water stored. A more complex situation would be loading and unloading water at the same time, or loading gasoline into a water tower, for not only would the gasoline have potential energy due to gravity, but it would also have chemical potential energy. Biological examples of "reservoirs" of energy or materials include storages as biomass, detrital materials, photosynthetically produced sugars in cells, and so forth.

Work Gate and Intersections

The *work-gate* module represents energy and/or material intersection points, that is, locations where one flow interacts with another. The interactions represented by this symbol are thermodynamic work processes and must include the loss of energy to the heat sink. There are several examples: A person standing on a dam can use his relatively small energies to turn a valve that controls the flow of a much larger flow of water through a generator. In this case the larger flow is that of the water energy, but it is the smaller human energy that controls it. Other examples include energy used by a predator to catch a prey, petroleum energy used to create fertilizer that is then applied to agriculture, or a triode in a radio. In the second case it is acceptable to look at the fertilizer either as the minerals themselves or as the energy required to make and use the fertilizer, depending on your interest. The work gate is an important symbol because it is used to show feedback loops by which systems are regulated. In Figure 7 one relatively small force (X_1) makes possible or accelerates a relatively large flow of energy (J). The result of this work gate is often a larger net flow of energy at J, since the energy required at the input X_1 may be small by comparison. Intersections may have many mathematical relations. A few examples are given in Figure 8. As shown, it is often convenient to include the nature of the interaction in the module itself.

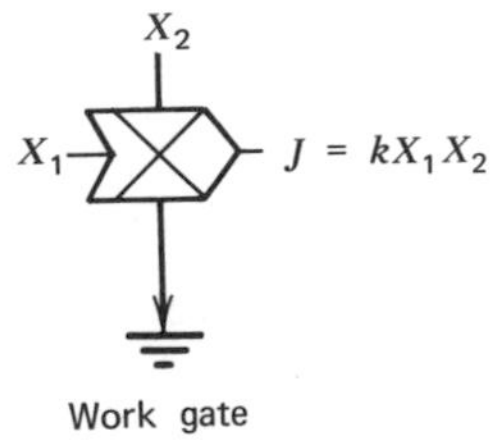

Work gate

Figure 7 Work-gate module. Flow through this module is a product of two forces acting on it.

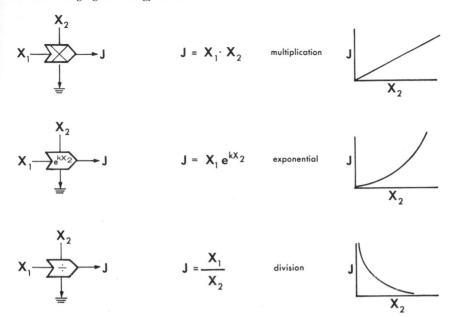

Figure 8 Work-gate relations.

Consumers (Another State Variable)

A hexagon is used to represent self-maintaining components of a system, such as animals and decomposers, or cities. This module is a combination of the storage and work-gate modules (Figure 9). Thus the module has the ability to store energy against a gradient and to use the stored energy to do self-maintenance work such as to use its own energy to obtain and use food. It represents the critical elements of life—the ability to incorporate energy and materials against a gradient and the use of energy over time to carry out

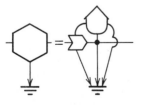

Self—maintenance

Figure 9 Module for a consumer. The three heat sinks represent (from left to right) (1) energy lost in feeding, (2) energy lost in storing energy, and (3) maintenance metabolism.

maintenance metabolism. Therefore, it must include a heat sink. The self-maintenance module can be used to represent a single living organism or a group of organisms such as a flock of birds, a school of fish, or a city.

Producer (Another State Variable)

The bullet-shaped module for a green plant (Figure 10c) is a combination of several modules. All green plants have heterotrophic functions of maintenance, as do animal populations, but include, in addition, mechanisms to capture sunlight and to use this captured energy to produce energy-rich reduced carbon compounds. Thus the primary producer module contains the self-maintenance module with a maintenance feedback to a cycling receptor system. The *cycling-receptor* system (Figure 10a) represents the receiver of pure wave energy. In plants this is light, but in other cases it might also be sound or water waves. In this module energy interacts with some cycling material (chlorophyll in green plants) producing an energy-activated state that then returns to its deactivated state passing energy on to the next step in a chain of processes. In life processes plants capture wave energy from the sun (photosynthesis) and use this energy to reduce carbon. Consumers, including the heterotrophic components of green plants, mobilize this energy by reoxidizing these carbon compounds in the process of respiration.

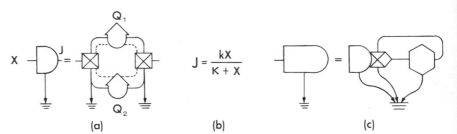

Figure 10 (*a*) "Pure-energy receptor" for example, a chloroplast. (*b*) A simple formula for photosynthesis vs. light energy input. (*c*) Green-plant module.

Flow-Control Modules

The modules are related by various flow pathways. Figure 11a shows a pathway where there is no backforce other than friction or inertia, for example, the discharge of a tank of water onto the ground. The transfer coefficient, k, is an inverse function to the resistance of the pathway. Where

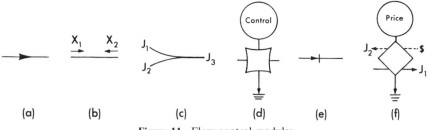

Figure 11 Flow-control modules.

there is backforce, such as with one tank filling another, Figure 11*b* is used. When two flows add together 11*c* is used. A control switch (Figure 11*d*) is used where flows have only an off or an on condition. Examples are migration, voting, digital computer memory, and seasonal reproduction. The *one-way valve* (Figure 11*e*) indicates that flow can be in one direction only, such as water in a sloping pipe without a pump. The *monetary-transaction* module (Figure 11*f*) is used where money as well as energy is flowing. Note that money flows in the opposite direction of energy or materials such as when gasoline is purchased.

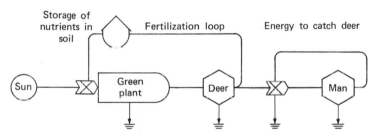

Figure 12 Simple diagrammatic model using circuit-language module.

The symbols are most useful when integrated into simple or complex circuits. Here is a simple example (Figure 12): A green plant is eaten by a deer. The deer's feces act as a fertilizer for the plant and accelerate the growth of the plant in a feedback loop to the work gate. We might also add man as a predator that feeds upon the deer, and in this case we add the energy that the predator uses to chase the deer. We might, if we want to emphasize it, also add a similar work gate for the deer, although this is already included in the consumer symbol. Figure 13 is another simple diagram showing the use of these modules to diagram some energy flows in a house,

(a)

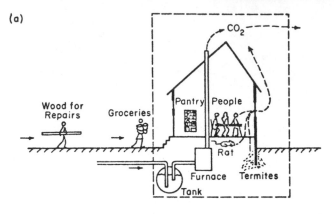

(b)

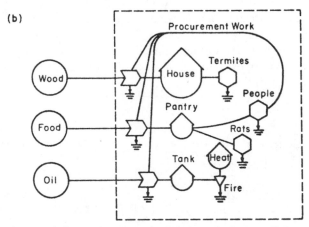

Figure 13 Diagram of a human house system. (*a*) Schematic diagram. (*b*) Energy-flow diagram (from Odum, 1972).

and Figure 14 is a diagram based on field work conducted in New Hope Creek, North Carolina. After a conceptual diagram is constructed, numbers representing flows and storage then may be conveniently inserted on the diagram as in Figure 14. Then differential (or difference) equations may be constructed to represent changes in the state variables as a function of the flows in and out. Finally a computer model may be developed from these differential equations and simulations run. These diagrams and models, of course, are no better than the conceptions, assumptions, and numbers that go into them. Many of the case histories in this book demonstrate how these

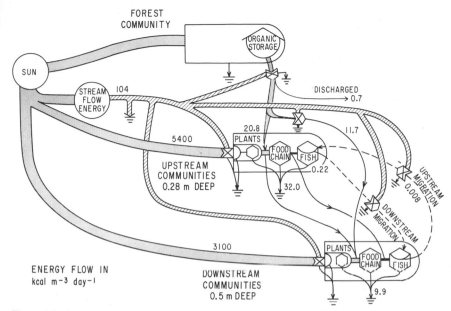

Figure 14 An example of a more complex energy-flow diagram. This one is constructed to emphasize the interactions of neighboring ecosystems and the relative magnitudes of their metabolism (from Hall, 1972). The width of lines of flow is approximately proportional to their magnitude.

symbols are used to describe very complex systems and the difficulties involved in getting an accurate representation of nature.

REFERENCES

Brody, S. 1945. *Bioenergetics and growth*. Reinhold, New York. 1023 pp.

Hall, C. A. S. 1972. Migration and metabolism in a temperate stream ecosystem. *Ecology* **53**: 585–604.

Lotka, A. J. 1922. Contribution to the energetics of evolution. *Proc. Nat. Acad. Sci., U.S.* **8**: 147–155.

Odum, H. T. 1957. Trophic structure and productivity of Silver Springs, Florida. *Ecol. Monogr.* **27**: 55–112.

Odum, H. T. 1960. Ecological potential and analogue circuits for the ecosystem. *Amer. Sci.* **48**: 1–8.

Odum, H. T. 1967a. Energetics of world food production. *In the world food problem*, Vol. III. Government Printing Office, Washington, D.C. Pp. 55–94.

Odum, H. T. 1967b. Work circuits and system stress. *In AAAS symposium on primary production and mineral cycling in natural ecosystems.* University of Maine Press, Orono. Pp. 81–138.

Odum, H. T. 1971. *Environment, power, and society.* Wiley, New York. 331 pp.

Odum, H. T. 1972. An energy circuit language for ecological and social systems: Its physical basis. In B. C. Patton (Ed.), *Systems analysis and simulations in ecology.* Vol. II. Academic, New York. 592 pp.

Odum, H. T. and E. C. Odum. 1976. Energy basis for man and nature. McGraw Hill. New York. 296 pp.

Odum, H. T. and Pinkerton R. C. 1955. Time's speed regulator: the optimum efficiency for maximum power output in physical and biological systems. *Amer. Sci.* **43:** 331–343.

3
A Strategy of Model Construction

W. SCOTT OVERTON

In the various chapters of this book you are exposed to a variety of perspectives and philosophies of models and modeling. Models are simplifications of reality, or abstractions of reality. But in this chapter I emphasize that models are also structured knowledge and that the process of modeling is the process of imposing structure on knowledge. There are many different uses of models, and all the different reasons for modeling follow the many different uses. The purpose of modeling then is to put knowledge into a more usefully structured form. If the structures are primarily mathematical, it is because mathematical structures are designed for this purpose and, therefore, constitute the proper fabric of models. The process of modeling requires a working knowledge of mathematical modeling structures, and a substantial part of this book is devoted to development of this knowledge and of basic techniques of model construction.

It is the purpose of ecosystem research to describe and understand as fully as possible the behavior and mechanisms of the various levels from organism to ecosystem. The task that faces a modeler charged with building a specific model is to draw the relevant knowledge from the scientific pool and to organize it in a manner suitable to the model objectives (Figure 2, Chapter 1). He may need to supplement the available knowledge with new data to complete the model, and he may need to invent new structures if he is trying to answer unusual questions, but by and large he will rely heavily on existing knowledge and accepted structure. The art lies in knowing what not to include and what to do with the parts that are necessary.

We emphasize that modeling is very much an art. The techniques are necessary but not sufficient. The best mathematician cannot build good models until he develops the art of drawing from the scientist the essence of the system being modeled and of representing this essence in parsimonious and tractable form. And the scientist must develop the discipline to leave out the detail that he has mastered through many years of investigation if it is not essential to the specific model being built. Intuition plays an important role, but intuition can be developed by exercise and by conscious effort to construct models according to criteria of good modeling. Adoption of a viable strategy of modeling is an essential step in this development. The purpose of this chapter is to formulate and formalize such a strategy.

A viable strategy must begin with the model objectives, expressed as a list of model specifications. These describe the properties of the finished model and the uses for which it is designed. Next, a strategy must prescribe an approach that reduces the task into tasks of manageable size. Then the submodels are constructed and validated, and the whole model is assembled and validated.

After the model is constructed and validated, one can obtain answers to the questions that the model was designed to answer, and sometimes to a number of other questions as well. Postconstruction strategy involves seeking model behaviors of interest and validation of the model with regard to those behaviors.

An outline of the strategy thus appears as follows:

1 Specify the model objectives as a list of model specifications.
2 Identify submodels and subobjectives.
3 Construct and validate submodels.
4 Assemble the submodels into the complete model, and validate.
5 Seek answers to the objective question.
6 Examine the general behavior of the model: identify behaviors of interest.
7 By sensitivity analysis, identify the structure and parameters that are causal for the behaviors of interest.
8 Validate those causal structures and parameters.

In developing these topics, I emphasize structure and the strategic aspects of the model building process with little attention to tactics. Tactical matters of solving equations and choosing the best form for an explicit relation are properly left for other chapters and authors. But availability of tactics limits the possible strategies and selection of a strategy constrains tactics, so that some tie-ins are needed and provided.

OBJECTIVES

In examining the modeling literature, we must be impressed with the fact that model objectives are seldom stressed sufficiently. It is often difficult to identify objectives from model documentation, and the great majority of criticisms of models relate to a capacity for which the model was not designed in the first place. Now if a model is built for a specific purpose, then the model specifications must adequately provide for that purpose, and they are thus subject to critique. And the model can be examined with regard to its meeting specifications; if they are not met, then presumably the purpose is not fulfilled.

In Figure 1, model specifications are shown as explicit representation of the model objectives, specifying the properties of the model. The objective question is an element of the specifications, and the *answer* addresses the *objective question*. The answer must be given consideration if, but only if, the model meets specification and if the specification fulfills the objectives. It is appropriate, then, that model documentation contain explicit identification of model specifications and that a strategy of modeling provide for this explicit identification as the model is constructed.

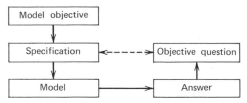

Figure 1   Overview of a modeling strategy.

But before we really get into that, a brief discussion of purpose, goals, objectives, activities, and associated concepts is in order. These words have different meanings to different people, and I want to use them in a somewhat restricted manner. Modeling is an activity, just as mountain climbing or hiking is an activity. To climb to the top of a particular peak or to build a model to a particular set of specifications is an objective. Unless the objective is specified, there is no definitive end to the activity. Even worse, without an objective there is no direction. If one is modeling or hiking for exercise, then it does not matter what direction is taken.

In my usage, objectives are achievable at least hypothetically. Achievement is recognizable. But goals may not be achievable in the same sense. The *goal*, "to build a general model structure of a forest ecosystem" has no identified termination point, but it is nevertheless a viable goal towards which to spend scientific effort. Specification of a *purpose* allows one to convert a goal statement into an objective statement. "To build a model for a particular estuary" for the *purpose* of "evaluating the impact of a particular power plant" has the structure of an objective. But before it becomes a *working objective* the word impact must be made much more explicit. This statement must be expressed in terms of model specifications.

But again, we may use the mountain climbing analogy to examine the nature of the specification. It is not necessary to specify which of the many possible routes will be taken, but it *is* necessary to restrict the routes on account of available technology, ability of the members of the party, and other conditions. In answering the objective question and in addressing the objective, specifications relate to the properties of the answer, not to the specific manner in which the answer is obtained. Only insofar as the nature of the model may have an effect on these properties is it of concern.

We cannot include in our model mechanisms that are not understood, nor exclude those which are essential to our purpose. Alternate structures that yield equivalent behavior may be selected arbitrarily, so long as nothing else in the objective specifies that one or the other is better. Within the set of specifications, there is a great deal of latitude for innovation and individual expression and art. But the specifications constrain the model to fulfill the objective.

In elaborating the elements of the specifications list, let us consider the objective question "what will be the effect of. . . ." The first step is to identify the subsystem of the ecosystem to which this question relates. Call this the *target system*. Clues are found in the identity of the perturbation, its timing and duration, its spatial extent, or whatever seems relevant. Next it is necessary to choose the appropriate "environment" for the target system. This involves higher levels of the ecosystem of which the target subsystem is an integral part and other subsystems at the same level as they relate to (interact with) the target subsystem. It also involves environment, proper, like climate and other physical factors. Then depending on the nature of the *purpose*, it may be necessary to incorporate *mechanism* by bringing in the next lower level of the target system. If the purpose requires only that a normal behavior be described, then the lower levels are not ordinarily needed. But if a perturbation is made which changes the normal behavior, then it is necessary to model the mechanisms that contain the causal explanations of how the system works and to include these mechanisms in the model (Figure 2).

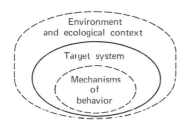

Figure 2 The target system must be considered in the appropriate context and environment, both specified by the objectives, and in many cases it is necessary to model the internal mechanisms in order to respond to the stimuli specified by the objectives

Thus far, we have identified the extent and scope of the ecological system relevant to the objective question. This is now made more explicit by the process of specifying the outputs of the model analysis and the general conditions under which the appraisal is to be made. These specifications include both resolution of the variables and criteria of realism and precision. Resolution involves matters of detail, such as organizational level, spatial and temporal representation, and the range of system variables. Realism refers to structural agreement with biological reality and theory, and precision refers to adherence to prescribed behavior. It is of particular importance that criteria of validation be included at this step, for these criteria specify the end of the activity; the specified objectives have been met when the validation criteria are met.

The elements of the specifications which make the objective explicit are thus:

1 The objective question.
2 Perturbations and stimuli to be accomodated.
3 Exact target system to which the question applies.
4 Exact higher level system of which the target system is a part.
5 Exact environment to which the question applies.
6 Region of prescribed behavior.
7 Region of extrapolation and prediction.
8 Factual and hypothetical base (data sets, assumptions, source of knowledge).
9 Criteria of empirical validation (objective functions for curve fitting and criteria for prediction testing).
10 Criteria of theoretical validation (realism, agreement with accepted knowledge and theory).

These are elaborated throughout the chapter and can be identified in many instances in other chapters of this book. We can best discuss criteria of validation and the regions of prescribed behavior in the context of the discussion of the validation process. But, in practice, it is necessary to produce the list of specifications to guide the other steps of the process of building a model. These specifications provide the properties of the model that are needed to fulfill the objective for which the model is being built.

SUBSYSTEM STRUCTURE

This step in the modeling strategy can be skipped if one is building a simple model, but is essential if the model is complex, for several reasons. The most pragmatic reason for identifying subsystems is the reduction of the modeling problem to manageable proportions. But we shall find that a considerable body of theory suggests that real systems are modular-hierarchical, so that this is also justified as a sound form for theoretical structure.

When one attempts to build a model of a large complex system, the number of variables and relations among variables can be overwhelming. In order to come to grips with problems of implementation, it is necessary to concentrate on smaller subsets of variables, or modules. But modules can be identified in several different ways, and here I would argue that the direction of general systems theory can be valuable—that modules should be identified sub-systems. One must also be aware that the structures used, if successful, may become an integral part of the body of knowledge on which they are imposed.

This is proper, as such structures properly constitute the theory of ecosystem science, but it imposes an obligation on the modeling scientist to use structures for his models that are compatable with the prevailing body of accepted knowledge. Deviations are acceptable only as a deliberate and responsible attempt to develop an alternate theory.

We may also observe that because ecosystems are special cases of systems, a theory of ecosystems must constitute a *special systems theory*. We may proceed to construct such a theory in two ways. We may build the theory from scratch, applying mathematical structures to ecological knowledge, or we may begin with a *general systems theory* and add specific content and restriction to accomodate the subject context of ecosystems. Some, including H. T. Odum, have taken the first approach, and others of us have taken the second. Hopefully we will end at the same place.

The word *structure* is used in several ways in system modeling, so it is necessary that the present use be identified. Explicitly, I use "structure" to relate to identified subsystems and identification of their couplings, as well as to the mathematical form of a relationship among variables. The distinction is made by speaking of subsystem structure and relational structure. The concern here is with subsystem structure.

The book by Pattee (1973) and particularly the contribution by Simon to that book give a good view of the current theory of modular hierarchical structure and its role in study of systems. The paper by Schultz (1967) nicely makes the case for representation of level of organization in ecosystem definition. In the tradition of hierarchical levels of organization, we may redraw Figure 2 to look like Figure 3, with each oval now representing an identified subsystem.

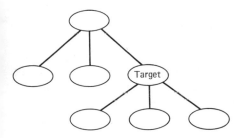

Figure 3 The target system is identified in the context of the higher-level ecological system relevant to the objectives of the model. Required mechanism of the target system is provided by explicit identification of lower level subsystems.

A point that is strongly emphasized in this chapter is that the model, and the system modeled, is defined by the objective of the model. The objective specifies the model and the activities in which it may be engaged (e.g., see Chapter 25). It follows that submodels also are defined by objectives and that subobjectives are implied by submodels. The observation that decomposition

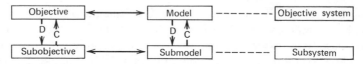

Figure 4 Subobjectives define submodels, and submodels imply subobjectives. This relationship can be used to advantage in modeling.

of a model into submodels can be accomplished by the process of decomposition of objectives into subobjectives leads to an important perspective of a modeling strategy (Figure 4).

In practice, one will usually have existing model structures that are part of the current paradigm of the system being modeled, so that there is intended no implication that the decomposition process is independent and unique in each effort. The structure should, so far as possible, comply with the existing paradigm, deviating only when necessary or for some explicit purpose.

It is necessary to emphasize the decomposition of the objectives because decomposition of the model into submodels must be constrained by the objectives. Each part has a specific function to perform in the overall model; its objective is the set of specifications that ensures that it satisfactorily performs that function. A particular part may be modeled directly or further decomposed into submodels. One part may require high resolution in time and structure, another low resolution and less detail. One may operate in continuous time, the other in discrete time. One part may require an energy-flow model, another a statistical-population model. Decomposition into parts allows each part to be modeled in the manner best suited to its fulfillment. The integrity of the *whole* is provided by the rules of composition and decomposition.

The objectives are decomposed, submodels are identified, constructed, and validated, the model is composed of the parts, and the original objectives are fulfilled by validation of the composition. This is easily stated in less abstract terms. Big models are best constructed by identifying a series of parts. Then the parts are built and individually tested before assembly. The assembled whole is also tested against the specified holistic behavior to validate the assembly. The kicker in this is the simple fact that the parts will not fit to yield the desired holistic behavior unless there was a master plan from the beginning. This amounts to identification of the parts by the process of decomposition.

CONSTRUCTION AND VALIDATION

The general process of model building can be viewed as an iterative algorithm (see Figure 2, Chapter 1). Given that the objectives are made explicit by the list of specifications, the process is:

1 Construct a model and generate output. Go to 3.
2 Revise the model and generate output. Go to 3.
3 Test against the validity criteria. Stop *or* return to 2.

That is, the activity ceases when the model achieves the validation standards set by the specifications or when one is forced to admit inability to achieve these standards.

When the model passes the validity tests, it is ready for the *purpose* for which it was constructed. This view of validation as an integral part of the modeling process is not reflected in some discussions, but actually is close to the way most modelers work and should work. In general, the objective specifies that the model will behave in such and such a manner in such and such a region of behavioral space, according to the prescribed body of data, knowledge, and assumptions, and that it will be capable of extrapolation to such and such a region. Then the modeling process is a search for sufficient refinement to yield the prescribed behavior, and sufficient realism (adherance to accepted knowledge and theory) and adequate mechanism to accomodate the extrapolation.

This latter distinction is vitally important. Validation of the capacity for extrapolation is theoretical, while validation of the behavior over the region in which the system has been observed is empirical. The model must behave according to specification over the known region and have the theoretical mechanisms to extrapolate into the prediction region (Figure 5).

The concept that a model should behave according to our experience, over those regions of behavioral and environmental space for which experience exists, is fundamental to model building. The prescribed behavior is made part of the specification of the model, and the model is deliberately constructed to yield this behavior. This specification is of the form "the output trajectory will take such and such a set of paths under such and such environmental conditions and input regimes." This can be viewed in the context of

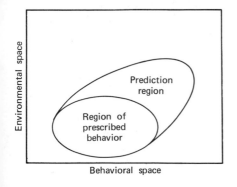

Figure 5   Validation space is partitioned into two regions. Objective validation is over the region of prescribed behavior; theoretical validation, and extended validation, over the prediction region.

curve fitting: the object is to create satisfactorily close model behavior to the prescribed behaviors and where criteria of satisfaction are also part of the model specification. This part of validation is an essential part of the model-construction activity.

Validation by prediction is at a different level. Suppose one has satisfactorily constructed a model according to specifications and then makes model predictions outside the region of prescribed behavior. The only internal validation of this prediction is the assurance that the structures involved are compatible with prevailing theory. But *extended* validation is sometimes possible, in the form of a subsequent experiment conducted to test the prediction. If the test fails, then one must examine the model to determine the structures responsible for failure. This leads either to correction of the model or of the underlying theory on which it was constructed. Many of the case histories in this book relate how additional field work and new concepts were suggested by initial model failure.

This process can easily be generalized. If one studies the behavior of a model outside the region of prescribed behavior and notes all behavior that seems surprising or unrealistic or undesirable or of interest for any number of reasons, then one can isolate the causal model structures and examine the validity of those structures. It is part of our modeling philosophy that specific behaviors *do* isolate on subsystems, at least to a large degree, and that it should be a goal of modular model construction to identify this isolation. Then a subsystem of interest can be studied experimentally to determine if the model has adequately reflected its behavior. This is the most powerful way in which models can shape the course of the experimental study of ecosystems. Validation brings us full circle to discovery of gaps in knowledge and to mechanisms for filling those gaps.

The logical nature of validation must also be appreciated. Validation is strongly related to hypothesis testing and curve fitting. Just as a hypothesis cannot (ordinarily) be proven to be true, one cannot validate a model in the absolute sense of proving that a particular form is correct, or even best. Validation is properly associated with the concept of adequacy. A model is built for a particular purpose, and that purpose is expressed in the specifications, which include validation criteria. When the validation criteria are met, the model is adequate. We can even argue that it should be *just* adequate, unless we have other purposes in mind.

Several stages of model construction can be identified. The first is associated with the selection of structure, and this is accomplished with coarse validation. At this stage, one may often attempt to achieve only qualitative agreement with the behavior of the system being modeled. Only after general qualitative validation has been achieved should one address questions of fine tuning, or achievement of closer fit by adjustment of parameters.

When we put these thoughts with the perspective of hierarchical models, we see that at the lower levels of model organization the submodels are simple. They may represent the interaction of one species with another or a simple process of one or two causal variables. At this level, choice of model (relational structure) and the process of parameter estimation follow classical curve fitting considerations. Qualitative fit may suffice; if not, an objective algorithm for minimization of an objective function can usually be used, for example, a nonlinear least squares algorithm.

It is in this context that the general parameter estimation problem can be best introduced, and the concept carries to consideration of all models and degree of complexity, even though the algorithms cannot span the hierarchy with the concept. Consider a simple elementary relation for which a set of data exists in the form of one input and one output, and for which a particular mathematical function has been postulated. The estimation problem is to choose values of the constants of the function for which the model behaves as closely as possible to the set of outputs. To implement this, it is necessary to make explicit what is meant by "as close as possible." The objective function, θ, defines close. It is then necessary to use either trial and error methods or some algorithm to converge to the minimum value of θ.

Consider a system with a single input variable, z, a single output variable, y, and a postulated relation, f,

$$y(t + 1) = f(y(t),z(t))$$

Symbolically, this fits the general system schematic.

$$z \longrightarrow \boxed{f} \longrightarrow y$$

The function, f, will ordinarily be sufficiently general that one or more constants can be chosen arbitrarily. These constants are the parameters of a system (z and y are the variables), and the process of choosing parameter values is called *estimation*. In a proper estimation circumstance, one has a set of data $\{z(t),y(t); t = 0,1,2,3, \ldots, n\}$. A particular value of the parameters will yield the predicted output sequence, $\{\hat{y}(t)\}$, and one can calculate for each occasion, t, the error, $\hat{y}(t) - y(t)$. From this, the squared-error-objective function is defined

$$\theta = \sum(\hat{y}(t) - y(t))^2$$

An example of a case in which an estimating algorithm can be used is provided by simplifying the function

$$y(t + 1) = f(z(t))$$

This form represents the well-known "regression," nonlinear or linear, and parameter estimates are provided by stock computer programs in virtually all computer centers. Returning to the general form,

$$y(t + 1) = f(y(t),z(t))$$

one often may have to rely on guessing and intuitive methods to obtain parameter values.

Sensitivity analysis is often an aid in the process of parameter estimation and fine tuning. Typically, one will determine the incremental change in θ due to an incremental change in a particular parameter, yielding a clue as to desirable parameter change. In lower-level subsystems, this is quite effective, but at intermediate and higher levels, it becomes more difficult to properly interpret, as we discuss later.

Conceptually, the problem of parameter estimation and adjustment to yield desired model behavior is identical at each level of subsystem, but the complexity and difficulty increases quite rapidly as one proceeds upwards in the hierarchy. It was in part in recognition of these problems that I began the work (Overton, 1972; 1975) that has led to the approach to modeling that I call the FLEX paradigm, briefly treated in the next section.

Let us now review the strategy. We have a model objective, decomposed into subobjectives, and a parallel structure of conceptual system model and submodels. The decomposition rules specify the couplings of the subsystems. We face, then, the problem of modeling the behavior of one of the subsystems, which can be accomplished in the isolation produced by the coupling structure. We specify subsystem inputs, outputs, and memory variables, and a *relation* between the outputs and the rest of the variables. This relation can be linear or nonlinear, discrete or continuous, deterministic or stochastic. There is little restriction to its form beyond the general needs of agreement with prevailing theory and practice and the statistical and mathematical tractability of the processes of parameter estimation and study of model behavior.

Typically, such a model will be composed of biologically identifiable processes, many of which may be well understood in isolation and for which the forms and parametric values are prescribed by the prevailing scientific paradigm. In many others, it may be possible to construct small experiments for parameter estimation and to estimate parameters either directly, as by measurement, or indirectly, as by a statistical model.

The general process of parameter estimation is usually addressed at the level of the lowest modeled relation, not at an intermediate or higher level, as we might suppose from appraisal of published treatments of sensitivity analysis. Further, parameter estimation and choice of the mathematical form of such a relation are usually integrated processes. Choice of equation may be empirical, in which case selection among several alternatives will often be made in terms of goodness-of-fit under the "best" values of parameters for the various alternatives.

But some problems *do* arise in building of system models that require attention beyond the ordinary. For example, suppose that one has hypothesized a linear differential equation model for a system of n variables. It is then desired to estimate the parameters of the matrix of the equation. If the system is in nonzero equilibrium over a specified period, then it is possible to estimate the parameters directly in terms of measured fluxes over the various paths during the period.

$$k_{ij} = \frac{J_{ij}}{tQ_i^*}, \, i \neq j$$

$$k_{ii} = -\sum_{j \neq i} k_{ij} + \frac{J_{oi} - J_{io}}{tQ_i^*}$$

where J_{ij} = total observed flux from compartment i to compartment j in period of duration t, Q_i^* = equilibrium value of the ith compartment variable, Q_i, and k_{ij} is the element of the ith column and the jth row of the transition matrix. J_{oi} and J_{io} refer to flows to compartment i from the environment and flows from compartment i to the environment.

This is nice and straightforward, but deceptively restrictive! First, equilibria are seldom found in ecological systems under study. Any seasonal fluctuation will violate this restriction. If the problem is caused by changing coefficients through the season, then it sometimes may be managed by defining a sequence of such models through the season or replacing k_{ij} by a normalized, seasonally varying coefficient. Estimation of continuously changing coefficients is more difficult, and the general problem of nonequilibrium variables persists even if the coefficients are property constant.

Second, it may be exceedingly difficult to measure total flux along any pathway; this is like the number of gallons of water passing through a pipe in a network among buckets. Most such measurements are indirect as well as difficult, and one might be much better off taking another approach. Since the solution forms of linear systems are explicit, it is possible in this case to use curve-fitting techniques with data defined as periodic measurement of the amounts of "substance" in the various compartments. Solution forms (the equations for the time trajectory of the compartment variables) are sums of exponentials, and nonlinear curve fitting has long been used (in pharmacology, for example) in estimating the parameters of the solution equation. In this, the exponential parameters are the *eigenvalues* of the transition matrix.

Even though linear systems are of limited *realism* in ecology, they are exceedingly useful in development of system "sense" and in communicating ideas. The canonical, or modal, form of linear systems is especially valuable in treatment of several related concepts. In accordance with the level of this book, we consider only the simplest case, that in which the system is "diago-nalizable." In a number of realistic circumstances, a linear system can be redefined in terms of new, "modal," or "canonical" variables.

$$X^* = BX$$

these being linear functions of the original variables and such that the differential equation for X^* is

$$\dot{X}^* = \Lambda X^*, \quad \text{where} \quad \Lambda = \begin{bmatrix} \lambda_1 & & & \\ & \lambda_2 & & \\ & & \circ & \\ & & & \circ \\ & & & & \lambda_n \end{bmatrix}$$

and where the λ_i are the eigenvalues of the equation for $\dot{X}$.

Consider one of the modal variables, x_i^*, defined by

$$x_i^* = b_{i1}x_1 + b_{i2}x_2 + \cdots + b_{in}x_n$$

With the b_{ij} the elements of the ith row of $\mathbf{B}$. Then, because of the diagonal nature of Λ, we can write

$$\dot{x}_i^* = \lambda_i x_i^*$$

and examine the behavior of this variable without regard for the behavior of the others. A new coordinate scheme has been constructed in which the system behavior is much more easily visualized, even though the variables are artificial. This is not always possible, in this form, even for linear systems, but when it is, there is a great advantage in terms of perceiving and under-standing behavior. The nature of the linear transformation, $\mathbf{B}$, and the concepts of eigenvalues, eigenvectors, and canonical forms of matrices and linear systems can be found in any text on matrix algebra or linear system theory. For the purpose of this treatment, it will suffice to say that a great many (possibly most) linear models of interest can be put into this diagonal form by the appropriate transformation.

That is, the x_i^* are *independent* or uncoupled. They can be described dynamically in isolation from each other. I have suggested (Overton, 1974) that this property of linear systems is related to the desirable properties of uncoupled systems; beneath the natural variables lies a system of uncoupled artificial variables. Out of this, we can guess that, in general, we may need to invent new, artificial variables as well as new, artificial parameters in the general process of system decomposition. This is just a way of saying that any system is better perceived and more comprehensible in one coordinate system or form of representation than in another and that one objective of any modeling activity should be to discover whatever simplifying or clarifying structures do exist. For linear systems, a general analytic method of finding the underlying uncoupled variables is available, but no such method exists for the general case.

Most models are conceptualized either in differential equations or in difference equations. In linear models choice is arbitrary, because one can translate exactly from one form to the other. But in nonlinear models exact translations are not generally possible. Since most large-scale simulation is on digital computers and most realistic ecological models are nonlinear, one must either model in discrete (difference) form, or use some numerical approximation to the continuous (differential) form (see Chapter 4).

MODEL STRUCTURE

Elsewhere in this chapter, I discuss aspects of model structure in several contexts. In this section, an attempt is made to tie together some of these considerations, while maintaining the latitude needed for various approaches taken by other authors.

Ecosystems are special cases of systems from which it follows that ecosystem theory is a special case of general systems theory. An ecosystem model must represent a general system model with the addition of particular content required by ecosystems, and a model must be a representation of the theory. All general systems theories (e.g., Klir's and Mesarovic's) have a foundation relational form that is interpretable as a *function* between two sets of variables. The simplest conceptualization is of the system (S) as a function between *inputs* and *outputs*:

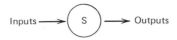

Then the model is any *rule* that generates outputs from inputs. I call inputs z variables, and outputs, y variables, but other authors use different notation. Inputs are often called driving variables or forcing functions.

In this representation, we say that the model structure is provided by the *rule* (i.e., by the function or more generally by the relation). Further, we can identify the *form* of the rule as opposed to the specific *relation*. For example, we could express a rule as

$$y = b_1 e^{b_2 z}$$

this equation expressing the relation in a parametric form, but leaving un-answered the question of parameter value. And the *exact* rule can be re-parameterized (i.e., expressed in a different *form* with different parameters), as, for example,

$$y = b_1 b_2^z$$

The model has not changed, but the representation (parameterization) has. So the distinction between parameters and *form* of the function is not sharp, but it is still useful to make the distinction because the selection of a form is selection of *coarse* behavior, and parameter adjustment is the mechanism of fine tuning.

System forms in which outputs are determined completely by inputs are of limited interest in ecosystem modeling; the systems of primary interest are called *state-determined* or *sequential* systems by various authors. That is, outputs are functions of past values of outputs and inputs as well as of current inputs. In implementing this, it is usual to identify state variables, some of which are conceptually measurable as current values of the system and some of which are past values of measurable system values. In my notation, only current system quantities are identified as state quantities, Xs (note, Qs are frequently used in an equivalent sense elsewhere in this book), and past values of those quantities are called *memory* quantities. But whatever identification is used, the output variables can now be repre-sented as a function of input, state, and memory variables, and the content of the model is the three (or four) kinds of variables and the two kinds of structure, the form of the relation and the parameters of the relation.

Then a model "run" is simply generation of model outputs according to the *rule* over a specified set of times, given a particular set of *initial states* and a particular set of inputs over the set of times.

In developing a general approach to modeling ecosystems, which I call the FLEX paradigm, I have adopted the general systems theory of George Klir (1969) as that which seems to me best suited to the needs of ecosystem modeling, although Mesarovic's 1972 theory does have some additional con-tent of great value. The above development fits Klir's definition of a system according to its *behavior*. In the FLEX modeling paradigm, it is held that the behavioral form is conceptually identified with characterization of the holistic behavior (nature) of an object system. This is the level of natural history of the object.

However, many modeling problems require the capacity to respond mechanistically to perturbation, so that mechanisms of behavior must be explicitly represented in the models built to solve these problems. This requires identification of *internal* variables of the object system, which represent the nature of the parts or subsystems. Klir's *Universe-Coupling*

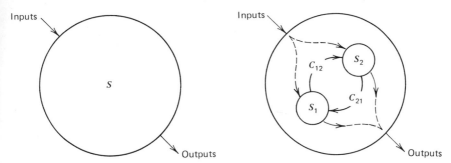

Figure 6 Hierarchial structure is generated by decomposition of a behavioral model form into its structural form and by identification of the higher-level system of which a particular system is a component.

structure provides a formal structure for this representation, being the definition of the system as a coupled collection of subsystems, each defined according to its behavior.

The FLEX paradigm then provides that each identified system will be defined (and modeled) according to its behavior and that a structural representation will also be constructed if required by the objectives of the model. The dual definition is provided by a *module form* that can operate in two modes, FLEX or REFLEX. Additional levels can be provided indefinitely by switching FLEX modules to the REFLEX mode and constructing subsystem models below. Operational details are provided in the FLEX2 Manual (White and Overton, 1975).

But one can model modular hierarchically in any modeling paradigm with recognition of the simple relationship. Couplings are provided between subsystems by common external variables; outputs of one subsystem become inputs of another. The subsystems *drive* each other in the identical sense that a system is driven by any forcing function. The illustration of Figure 6 makes the point that externally there is no difference in system definition; changes are in the identified internal mechanism.

Structure has now taken on another dimension. The external quantities are related by a rule in which we have identified the *form* and *parametric* aspects of structure. Now the system has been decomposed into subsystems by the identification of internal variables (or quantities) that are related by decompositions of the rule of the whole system. To illustrate, let the system be represented by the function, f, decomposed into two systems, f_1 and f_2,

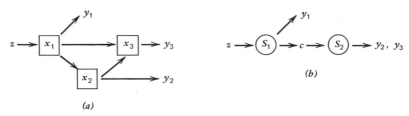

Figure 7   A system is conceived in the dual forms of variables coupled by relations, and processes coupled by variables.

where c is an *internal* coupling variable of the system. That is, the variable, c, is an output of S_1 and an input of S_2. Thus we can speak of the subsystem structure or *equivalently* of the internal-variable structure: either identifies a *decomposition* of the rule for the relation between z and y.

As a result of these distinctions, we find that the model must be conceptualized differently when addressing the behavioral (FLEX) form and the structural (REFLEX) form. The common compartment model concept is useful for FLEX; one conceptualizes a set of variables (state variables in our terminology) and then constructs a set of relations that prescribe the manner in which they change from occasion to occasion. These relations are associated with the arrows (Figure 7a). But the structural form of REFLEX emphasizes the *processes* of the system rather than the variables; subprocesses are explicitly identified and are coupled by variables. The distinction is great, and so is the value of the distinction. The concept of subdividing circles into subcircles coupled by arrows is much stronger than that of subdividing boxes into subboxes (i.e., decreasing aggregation) because all the process is outside the box. Further, it is very clear that if the coupling variables are clearly identified (in quantity and resolution), then different groups of people can work on the various pieces independently, and the parts will fit. This does not guarantee that the coupled behavior will be satisfactory, but the parts will fit and run together.

So the two modes become two distinct ways in which to conceptualize a system; they represent two of Klir's fundamental system definitions and provide the foundation for a hierarchical system theory and for modular-hierarchical model construction. And even if one does not model in this paradigm, it is still possible to recognize the three aspects of structure by recognizing parameters, form, and internal variables.

Even though each subsystem will ordinarily be modeled in the behavioral form as well as in the structural, the final version of a hierarchical model will contain a REFLEX module for any system in which subsystems are explicitly represented and a FLEX module for any system that is *terminal*, which does not have modular subsystems. The nature of the structured model is such that all REFLEX modules must be conceptualized in discrete time. But FLEX modules may be conceptualized either as discrete or continuous. It is then necessary either to provide a differential analyzer for such a module or to convert the conceptual continuous form into discrete form. The FLEX tactic of differential analysis is to reduce the temporal resolution of the module until an Euler approximation (Chapter 4) is satisfactory, but discretization of continuous conceptual structure is preferred.

Several other aspects of the differences and relations among the levels of organization are important in modeling in this manner. Again, reference to Simon and to Schultz provides depth to the following discussion. Hierarchical modeling requires the perspective that resolution in time, space, and biological detail increases as one descends through levels of organization. This implies that the holistic behavior of any identified system has a resolution and response frequency that not only is characteristic of that system, but is higher than that of higher levels of the system, and lower than that of subsystems. This forces recognition of the observation that each ecological question is related to a specific subsystem as much by its temporal and spatial aspects as by its biological orientation.

In ending this section on structure, it is well to emphasize that the *behavioral* form of system definition is related to the nature of the system and the *structural* form to the interaction of the parts—to the underlying mechanisms of how the system works. This is a perspective of "whole-system" modeling that is not yet universal. One still encounters attempts to achieve "holism" by accounting for all the details. But this aspect of the emerging system theory would hold that the model of a system must represent the rules of organization and function of that system. It is exceedingly unlikely that anyone can represent the decomposer subsystem of an ecosystem model in terms of the responses and behavior of a single hypothetical species of decomposer. Instead, this submodel should represent the process of decomposition; it must account for the rules of organization of the many species and forms of decomposition, not for the details of behavior of the many.

This point also relates to the "aggregation" problem. This is a problem only when one attempts to model the total function of a group of diverse species in the behavioral form of a single species. This is occasionally appropriate, but more often than not one should attempt to model the behavior of the assemblage. If no such behavior is discernable, then it may be appropriate to question whether it is a proper subsystem. The aggregation problem arises when one attempts to change resolution without changing levels, and this is ordinarily not appropriate. The perspective of hierarchy theory is that the model of the assemblage should represent the function of the assemblage (see Chapter 21, Figure 2). At each level behavior emerges that transcends the behaviors of the levels below, and the detail of the levels below is integrated out.

SENSITIVITY ANALYSIS AND STUDY OF MODEL BEHAVIOR

In many modeling activities, sensitivity analysis or appraisal of the effect of a perturbation, parameter change, or structural change is needed to fulfill the activity. In simple models, many such appraisals can be made analytically, that is, mathematically. In most simulation models it is appropriate to conceptualize the basic operation as a model run under some incremental change in state variable, parameter, structure, or input. One determines that a particular change in the model produces a particular change in the system behavior. The difficulty is in succinctly summarizing the necessary description of behavior in a complex model.

The first kind of study involves changes in behavior as a result of changes in initial values of system variables. This is typified by phase plots and studied by determining the direction of change from different positions. Identified features will be isoclines, equilibrium points, and other such features, which can then be used to summarize the behavior. Stability analysis takes place in this space, relating to the behavior of the system in the neighborhood of features such as equilibria and cycles.

Analyses such as this are made quite difficult to conceive in the context of ecological models by the nature of inputs or forcing functions. It is necessary to examine behavior over some realistic period of time and under a standard driving data set for such behavioral analyses to be meaningful. But study under variation in driving data is also of value and interest. This must be accomplished by meaningful changes in entire driving data sets or forcing functions over meaningful periods of time.

Sensitivity analysis, evaluating changes in behavior under changes in structure and usually with fixed initial conditions and driving data, is properly engaged more in the model construction stage than in the later stages of model characterization. The process by which changes in model structure are selected requires analysis of sensitivity at least in validation steps, and this is also the basic tool for parameter estimation and fine tuning when one operates at a level in the system above which algorithms of estimation (e.g., a Gauss-Newton algorithm for nonlinear least squares) are ineffective. But *structural stability* is another subject of concern in systems theory, and postmodeling study of system stability under structural (including parametric) change will probably emerge as an important activity when ecosystem modeling matures a bit.

Consider now the manner in which modular construction can reduce the total amount of work in sensitivity analysis. For a model with two inputs, two outputs, and four parameters, prescribe an incremental change in each parameter. Examination of this response requires four runs with the parameters varied individually, six runs with the parameters varied in pairs, four runs with the parameters varied in groups of three, and one run with all four varied together. Altogether we find $2^k - 1$ runs required for incremental perturbation of k parameters. Study of both directions ($\pm\Delta$) of the neighborhood requires $3^k - 1$ runs.

In such an analysis we seldom make the higher-order runs, simply for reasons of economy. This involves, to some degree, tacit assumption of noninteraction among the parameters or acceptance of the interactions as unimportant. If we actually make these assumptions, then we can recognize implied structure of the model. As illustrated in Figure 8, assumption that parameters 1 and 2 are not interactive with 3 and 4 implies that parameters 1 and 2 are in subsystem 1 and parameters 3 and 4 are in subsystem 2.

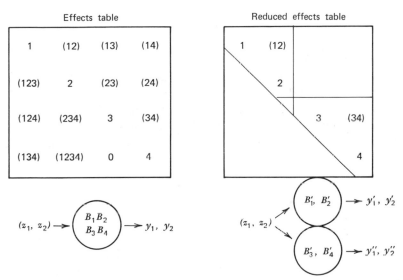

Figure 8 Recognition that certain patterns of interaction are null leads to recognition of subsystem structure. It may be necessary to identify artificial variables in order to accomplish this.

The value of such a decomposition is great. Instead of $2^k - 1$ runs, we need $2(2^{k/2} - 1)$ runs if the two subsystems are equal in number of parameters. For four parameters, we have reduced the number of runs from 15 to 6. For ten parameters, reduction is from 1023 to 62. Since each subsystem is smaller, requiring less computing, the reduction in computing is closer to 15 to 3 and 1023 to 31.

Modelers from the Grasslands Biome (a large, comprehensive ecosystem program) invented a neat way of affecting such structural decomposition

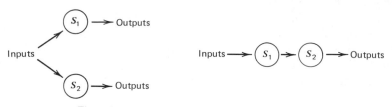

Figure 9   Parallel and series modular construction.

without actually decomposing the model: they defined macroparameters as collections of parameters to be varied simultaneously. Each macroparameter can contain as many as one parameter from each subsystem. These modelers have also used fractional factorial designs in studying the interactions among the macroparameters (R. K. Steinhorst, personal communication). One difficulty arises: if interactions exist, it is necessary to make further runs to identify them. This would be eliminated if the model were actually decomposed for study, but the approach is clearly promising as a means of discovering possible decompositions.

We may identify the decomposition of the above type as "parallel" as opposed to "series" decomposition (Figure 9). Both types afford advantages of comprehension and smallness of individual assembled model, but only the parallel forms provide a reduction in actual work of computing and processing.

VALIDATION OF CAUSAL STRUCTURES

After discovery of interesting model behavior and the identification of causal structure by sensitivity analysis, one is faced with the problem of validation of that structure. Ideally, a subsystem can be isolated and the behavior of interest identified with a behavior of the subsystem. Then one proceeds to validate the subsystem by verification of agreement with existing knowledge, theory, and data sets. It may be necessary to construct new field experiments. Too often the internal variables are not represented by data sets; indeed, many of them cannot be directly measured, so that one must relate the question to indirect measurement or estimation questions.

This entire process is facilitated if one has attempted throughout the model building process to construct forms in which the important system behaviors are identified with specific subsystems. That is, parallel modularity is a goal of model development, contributing to conceptualization of the nature of the system as well as to ease of model construction and analysis. This provides an important perspective of a modeling strategy.

SUMMARY

This chapter develops a general strategy of model building in which model objectives are decomposed hierarchically into subobjectives, perhaps at several levels. Submodels are built to achieve subobjectives and later assembled into the whole system model according to the plan specified by the decomposition. This strategy prescribes achievement of coarse model behavior through structural changes before attempts to fine tune by parameter adjustment.

Although other strategies for model building do exist, there are many advantages to the modular method presented here. Small modules greatly enhance model comprehension, locating and correcting errors, and communicating with other scientists. Additionally, parallel modeling achieves a substantial reduction in the degree of coupling among the system variables in an explicit manner that allows for a significant decrease in the dimensionality and cost of model construction and validation. Lastly, current thought regarding the nature of complex systems as nearly decomposable dictates that a goal of ecosystem modeling should be to discover the decomposition of systems being modeled.

Validation enters the study and construction of large ecosystem models in precisely the same conceptual form as is involved in simple model studies. Differences are due to complexity and dimensionality of the model and the necessary evaluation. It is emphasized here that validation is an ongoing part of the process of model construction, beginning with the first steps of the development of model structure and ending with the final steps of fine tuning. Validation and model building essentially end simultaneously, contrary to the frequent misconception that models are built and then validated. Further activities, including prediction and study of model behavior, must be constrained by the specifications of the model.

This overall strategy for model building is intended to serve as a guide for the translation of biological reality into structured knowledge *and* the actual structure of a model. It should also serve as a framework within which to use the tactics of mathematics introduced in Chapter 4. However, it must be reemphasized that modeling is not a mechanical activity, even though many mechanical activities are involved in its fulfillment. Modeling is an art, founded on general systems theory, and very much dependent on the modeler's intuition for and knowledge of the system being modeled. The strategy presented in this chapter is also intended to enhance the development of that intuition and the conceptual awareness of the role of systems science in the conduct of scientific study. But the modeler's knowledge and intuition of the system being modeled must ultimately derive from contact with the real system. At present, most modelers receive this contact second-hand, but we can anticipate a new breed of ecosystem naturalists for whom general systems theory constitutes the scientific method.

GLOSSARY

Parameter. A constant of a system. In greater generality, required in realistic use of hierarchical levels of organization, a parameter at a particular level is a slowly changing variable, possibly a system variable at a higher level.

Algorithm. A rule for obtaining a solution to a problem.

Decompose (mathematical). To express a function in terms of intermediate quantities. For example, if $y = f(z)$ is a function, then the functions, $c = f_1(z)$, $y = f_2(c)$, are a *decomposition* of this function. Then f is a *composition* of f_2 and f_1. Application to systems follows directly.

Linear equation. If an equation in several variables is such that the partial derivative with respect to each variable is a constant, then the equation is said to be linear. Otherwise, if the derivative with respect to any variable is a function of one or more variables, then the equation is said to be nonlinear. Mathematical variables may in special contexts be classified into groups. For example, in statistics and systems theory variables and parameters are identified. In such cases it is appropriate to speak of linear or nonlinear equations in the parameters or of linear or nonlinear equations in the variables.

Linear regression. A regression equation is said to be linear if it is linear with respect to the *parameters* of the equation; it is otherwise nonlinear.

Linear model. A model in which the equation of change (e.g., differential equation) is linear in the system variables; it is otherwise nonlinear.

Equation. Any mathematical statement of equality. May be used to describe a function, in which case we can identify *variables* of two kinds (argument and value, being respectively elements of the domain and range) and structure of two kinds (form and parametric value).

Paradigm. A body of definition, belief, and fact associated with some area of science or metascience. A supermodel.

Objective question. An explicit question to be answered by a model.

REFERENCES

Klir, G. J. 1969. *An approach to general systems theory*. Van Nostrand-Reinhold, New York.

Mesarovic, M. D. 1972. A mathematical theory of general systems. In G. J. Klir (Ed.), *Trends in general systems theory*. Wiley-Interscience, New York.

Overton, W. S. 1972. Toward a general model structure of a forest ecosystem. In J. F. Franklin, L. J. Dempster, and R. H. Waring (Eds.), *Proceedings: Research on Coniferous forest ecosystems, a symposium.* PNWFRES. U.S. Forest Service, USDA, Portland, Oregon.

Overton, W. S. 1974. Decomposability: a unifying concept? In S. Levin, (Ed.), *Ecosystem analysis and prediction.* Soc. Ind. and Appl. Math., Philadelphia.

Overton, W. S. 1975. Ecosystem modeling approach in the coniferous biome. In B. Patton (Ed.), *Systems analysis and simulation in ecology.* Vol. III. Academic, New York. 607 pp.

Pattee, H. H. 1973. Hierarchy theory: *The challenge of complex systems.* Brazilier, New York.

Schultz, A. 1967. The ecosystem as a conceptual tool in resource management. In S. V. Cieriacy-Wantrup (Ed.), *Resource economics.* University of California Press, Berkeley.

Simon, H. A. 1973. The organization of complex systems. In H. H. Pattee (Ed.), *Hierarchy theory: The challenge of complex systems.* Brazilier, New York.

White, C. and W. S. Overton. *Users manual for the FLEX2 and FLEX3 model processors.* Bulletin 15, Forest Research Laboratory, Oregon State University.

4
Mathematical Construction of Ecological Models

CHRISTINE A. SHOEMAKER

The purpose of building a model of an ecological system is to understand and predict the behavior of the system as a whole by quantitatively describing interactions between parts of the system. There are many steps in the development of models used for such a purpose. The goal of this chapter is to describe the fundamental mathematical steps of construction and solution of such models.

Initially, the chapter discusses mathematical forms describing relations between parts of the system. Next these relations are used to construct difference or differential equations that predict changes in the system through time. Analytical and numerical techniques for solving difference and differential equation models are explained. The last section of the chapter discusses the use of models for making decisions about the management of ecological systems. A list of symbols and their definitions is given in the Appendix.

The discussion of mathematical methods is limited to those used in the models discussed in this book, which are models designed to use data as well as ecological principles to predict the behavior of specific ecological systems. Research in theoretical mathematical ecology has been surveyed elsewhere and is not discussed here.

One of the first steps in modeling an ecological system is to define variables. Those describing the state or condition of an ecological system are called *state variables*. Examples of state variables include nutrient concentrations and population densities. The symbol, Q_i, denotes the ith state variable. In later chapters, state variables referring specifically to numbers of individuals are denoted by N_i.

A model that has more than one state variable is called *multidimensional*. If there are N state variables, they can be referred to as Q_i, $i = 1, \ldots, N$ or as $\mathbf{Q} = (Q_1, \ldots, Q_N)$. This expression is called a *vector*. By referring to $\mathbf{Q}$, we are referring to the entire collection of variables, $Q_1, \ldots, Q_N$.

A multidimensional model of an ecosystem includes state variables describing both the abiotic and biotic components of the system. Even a simple model of an ecosystem would have state variables representing plant biomass, animal biomass, and the quantity of nutrients in the ecosystem. However, in many cases such a model is too aggregated to be able to accurately predict the behavior of an ecosystem. It may be necessary to divide the animal and plant populations into orders, species, or even into age classes within species and to assign a state variable to each category. Similarly, nutrients in an ecosystem may be represented as the total weight of the material or as the weights of each of its molecular forms. For example, nitrogen can be represented as total nitrogen (one state variable) or as the amounts of ammonia, nitrate, nitrite, and organic nitrogen (four state variables).

The rates of movement of materials or biomass between factors described by state variables of a system are called *flow rates*. For example, predation is a flow of energy and material between two components of an ecological system, in this case, a predator and its prey. The movement of individuals from one age class to another within a population is also a flow rate. The symbol, $J(i,j)$, denotes the flow rate of material from state variable, Q_i, to state variable, Q_j. The symbol, $J(i,i)$, denotes the change in state variable, Q_i, that is not a result of flows from other state variables.

In a system consisting of a single variable, Q_1, the only flow rate is $J(1,1)$. This system is illustrated in Figure 1a. If Q_1 is a population, then $J(1,1)$ is the net increase in the population biomass. The net increase is the net growth rate, r (the birth rate minus the death rate), per individual times the number of individuals. If r is a constant, then

$$J(1,1) = rQ_1 \qquad (1)$$

In many cases r, the net growth rate per individual, decreases as the population size, Q_1, increases because of increased stresses from crowding, food shortages, or limited reproductive sites. In many cases, data have indicated that the decrease in the net growth rate can be modeled by replacing the constant growth rate, r, with the term

$$r' = r\left(\frac{K - Q_1}{K}\right) = r\left(1 - \frac{Q_1}{K}\right) \qquad (2)$$

The constant, K, is called the *carrying capacity*. It is the maximum population the environment can support. Note from Eq. 2 that the net growth rate, r', is zero when the population density equals the carrying capacity. However, when Q_1 is much smaller than K, r' is approximately equal to the maximum

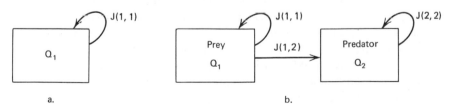

Figure 1   Flow rates in one and two dimensional ecological systems. In a) the rate J(1,1) is the net growth rate (birth minus death rate) of population Q_1. In b) the flow rate J(1,2) is the rate of predation by preador population Q_2 on prey population Q_1. The flow rates J(1,1) and J(2,2) are the net growth rates of Q_1 and Q_2 in the absence of predation.

net growth rate, r. Substituting r' for r in Eq. 1, we obtain

$$J(1,1) = r'Q_1 = rQ_1\left(\frac{K - Q_1}{K}\right) \tag{3}$$

To visualize the influence of flow rates on multidimensional systems, examine Figure 1b. Assume that Q_1 is a prey population and Q_2 is a predator population that feeds upon Q_1. Then $J(1,2)$ is the rate at which prey biomass, Q_1, is being consumed by the predator, Q_2; $J(1,1)$ is the net growth (births minus deaths) of the prey population, and $J(2,2)$ is the net growth of the predator population without prey consumption. Since the population would decrease without food, $J(2,2)$ is negative. The term, $J(2,1)$, is zero since no material moves from a predator population back to the prey population.

To construct a model of the system illustrated in Figure 1b, we must develop equations for each of the flow rates, $J(i,j)$. The rates, $J(1,1)$ and $J(2,2)$, are net growth functions and can be described by equations such as Eq. 1 and Eq. 3.

The predation rate, $J(1,2)$, has a different form because it is a function of both predator and prey densities. The simplest and most widely used description of predation assumes that the number of prey consumed is a fixed fraction of the number of prey encountered. The number of prey encountered by a single predator is the searching rate (the area or volume searched by the predator in a unit of time) multiplied by the density of prey in the searched area. The total number of consumed prey is the number of prey encountered by each predator multiplied by the predator density multiplied by the fraction of encountered prey that are consumed, that is,

$$J(1,2) = S_pQ_1Q_2\alpha \tag{4}$$
$$= aQ_1Q_2 \tag{5}$$

where α is the fraction of encountered prey that are consumed, S_p is the searching rate of the predator, Q_1 and Q_2 are the prey and predator densities, and $a = \alpha S_p$.

However, a predator's capacity to consume is finite. If the prey population density is so high that aQ_1 exceeds a predator's capacity, Eq. 4 overestimates the predation rate. Thus in such cases, we need to modify the consumption rate, α, to decrease the consumption fraction as prey become more abundant and predators less hungry. In cases of high prey density, the consumption fraction, α', often can be approximated by

$$\alpha' = \alpha \frac{K}{K + Q_1} \tag{6}$$

for an appropriate value of the constant, K. Thus if Q_1 is small relative to K, predators will not be satiated and α' is approximately equal to α, the maximum rate of consumption. However, as Q_1 increases and the limits of predator capacity are approached, the consumption fraction, α', declines. When Q_1 reaches K, the consumption fraction, α', is only one half the maximum fraction, α. For this reason, K is called the *half-saturation coefficient*. Replacing α in Eq. 4 with α' from Eq. 6, we obtain the predation rate

$$J(1,2) = \alpha' S_p Q_1 Q_2 = \alpha \frac{KS_p}{K + Q_1} Q_1 Q_2 \tag{7}$$

or

$$J(1,2) = \frac{aKQ_1}{K + Q_1} Q_2 \tag{8}$$

Note that for very large prey densities, the right side of Eq. 8 is approximately equal to aKQ_2. Thus the total consumption per predator never exceeds aK regardless of the size of the prey population. In addition to its use to describe predation, Eq. 8 has been widely used to describe the uptake of nutrients by phytoplankton. In this case, Q_1 and Q_2 refer to nutrient and phytoplankton concentrations, respectively.

The flow rates, $J(i,j)$, may be influenced by factors other than those described by state variables. For example, decisions regarding the management of an ecosystem will affect the behavior of the system. Model variables whose values are directly determined by management decisions are called *decision variables*. Since they can be controlled by human decisions, the amounts of pesticide or industrial waste released into an ecosystem are examples of decision variables. Another example of a decision variable is the choice of when and where a forest fire should be suppressed. Models using these examples are given in Chapters 18, 22, 23, and 25.

Flow rates may also be influenced by factors outside the system that affect the behavior of the system, but are not affected by the system or by decision variables. Variables describing such factors are called *exogenous variables*. Temperature, precipitation, and solar radiation are examples of factors that are described by exogenous variables and have a very important influence on many flow rates, including net productivity and nutrient uptake.

Thus it is not unusual for a flow rate, $J(i,j)$, to depend upon Q_i, Q_j, and several exogenous and decision variables. Frequently, rates that depend upon several factors are represented mathematically by multiplying together effects of factors that act independently of one another. For example, the uptake rate of a nutrient, Q_1, by phytoplankton may be represented as

$$J(1,2) = MG_1(Q_1)G_2(W_2)G_3(W_3)Q_2 \tag{9}$$

where the functions, G_1, G_2, and G_3, describe the effects of a nutrient concentration, Q_1, and the exogenous variables temperature, W_2, and solar radiation, W_3, on nutrient uptake by phytoplankton population, Q_2. The effect of the nutrient concentration often has the form, $G_1(Q_1) = \alpha' Q_1$, where α' is given by Eq. 6.

A more general form of the multiplicative relationship is

$$J(i,j) = M \prod_{m=1}^{N} G_m(X_m) \tag{10}$$

where $\prod$ means the product and X_m refers to all model variables (state, decision, and exogenous) that affect $J(i,j)$. Each of the functions, $G_m(X_m)$ is measured by holding other variables constant and calculating the fractional change in $J(i,j)$ as X_m varies. If $Y = (y_1, \ldots, y_N)$ is the base set of values at which the other variables are held, then M is the value of the flow rate when all of the variables, X_m, are equal to their base values, y_m. Usually the base values chosen are those that maximize $J(i,j)$.

Put more precisely,

$$M = J(i,j|y_1, \ldots, y_N) \tag{11}$$

$$G_m(X) = \frac{J(i,j|y_1, \ldots, y_{m-1}, X, y_{m+1}, \ldots, y_N)}{J(i,j|y_1, \ldots, y_N)} \tag{12}$$

where $Y = (y_1, \ldots, y_N) = $ the base set of values of the model variables used for comparison

$J(i,j|y_1, \ldots, y_N) = $ the value of the flow rate, $J(i,j)$, given that the model variables have the values, $(y_1, \ldots, y_N)$.

Data requirements can be significantly reduced by assuming that the effects of different variables on flow rates, $J(i,j)$, can be represented by a multiplicative relationship. For example, consider the nutrient uptake rate, $J(1,2)$, described by Eq. 9. Suppose we wish to evaluate $J(1,2)$ for four values of each of the variables, Q_1, W_2, and W_3. There are $(4) \cdot (4) \cdot (4) = 64$ possible combinations of values of Q_1, W_2, and W_3. Thus to evaluate $J(1,2)$ directly would require 64 experiments, one for each combination of values of the variables, Q_1, W_2, and W_3. Let X_1, X_2, X_3 denote Q_1, W_2, and W_3, respectively.

Only ten experiments are required to evaluate the nutrient uptake, $J(1,2)$, if it is described by the multiplicative relationship in Eq. 9. One measurement of $J(1,2)$ is required at the base set of values to estimate M. The rate, $J(1,2)$, must also be evaluated for three additional values of X_1 in order to determine $G_1(X_1)$. In addition, we need three values of $J(i,j)$ to calculate $G_2(X_2)$ and three more for $G_3(X_3)$. Thus in order to calculate $J(i,j)$ for four values of Q_1, W_1, and W_2 using the multiplicative relationship, Eq. 9, the total number of values for $J(i,j)$ that must be measured is $1 + 3 + 3 + 3 = 10$.

Therefore, by making an assumption that the variables have a multiplicative effect on a flow rate, $J(i,j)$, the number of experiments required is reduced from 64 to 10. Such a reduction is very significant, especially considering the amount of time, effort, and expense that may be involved in obtaining each value which is obtained from a number of replicates.

The multiplicative relationship becomes even more advantageous with increases in the number of variables and in the number of values of each variable. For example, if ten values of each variable are considered, the number of measurements of $J(i,j)$ is reduced from 1000 to 28. If $J(i,j)$ depends upon four variables, each of which assume ten different values, the reduction is from 10,000 to 37 measurements. Indeed, we see that because of the number of measurements required, it may be infeasible to determine flow rates, $J(i,j)$, as a function of several variables, each of which assume several values, unless the effects of each variable can be separated from the effects of the other variables by a multiplicative relationship or some other simplifying assumption.

It is important to understand under which assumptions Eq. 10 is a valid description of a flow rate. In general, the multiplicative relationship is accurate only if the effect of changes in several variables is equal to the product of the effects due to changes in the values one at a time. For example, for any two values of x_1 and x_2,

$$J(i,j|x_1,x_2,y_3,\ldots,y_n) = \frac{J(i,j|x_1,y_2,\ldots,y_m)J(i,j|y_1,x_2,y_3,\ldots,y_n)}{M} \quad (13)$$

It should also be noted that the values of $G_m(x)$ will not in general be the same if measured at different base values $(y_1,\ldots,y_n)$. For this reason measurements are usually based on the values of y that result in maximum values of $J(i,j)$. For example, if $J(i,j)$ is a growth rate, the base values of y would represent optimal values for solar radiation, temperature, nutrient concentrations, and the like.

In practice, variables usually have interdependent effects on flow rates, that is, Eq. 13 does not apply for all values of x_1 and x_2. However, the multiplicative representation is an approximation that is adequate for many purposes. An estimate of the error resulting from the assumption can be obtained by measuring the rate, $G_m(x)$, for different base sets of values, $\mathbf{Y}$. Usually the error can be decreased by reducing the range of the x_is, so that $\mathbf{X} - \mathbf{Y} = \sum_m |x_m - y_m|$ is sufficiently small.

RATES OF CHANGE

The total *rate of change* of a state variable, Q_i, is the sum of the effects due to flows into Q_i from other variables minus the flows out of Q_i. Thus

$$F_i = \sum_{j=1}^{N} \alpha_{ji} J(j,i) - \sum_{\substack{k=1 \\ k \neq i}}^{N} J(i,k) \quad (14)$$

where F_i = total rate of change of Q_i

$\quad \alpha_{ji}$ = the conversion ratio of material Q_i to material Q_j

$\displaystyle\sum_{\substack{k=1 \\ k \neq i}}^{N}$ = the sum for $k = 1, 2, \ldots, i - 1, i + 1, \ldots, N$

$J(j,k)$ = flow rate from Q_j to Q_k

The rates of change, F_i, are important because they are the basis of the difference and differential equations that describe and predict the behavior of an ecological system. We discuss the incorporation of rates of change into predictive equations in the next section.

It is necessary to include the constants, α_{ji}, in the rate of change Eq. 14, because a transfer of $J(j,i)$ units of variable Q_j does not necessarily result in the same number of units of variable Q_i. For example, if the quantity being transferred is measured in calories of Q_j biomass, only a fraction (usually around 10%) of $J(j,i)$ will be transformed into calories of Q_i biomass, because digestion is never 100% efficient.

If the flow is between two variables with like units (e.g., calories of biomass), then α_{ji} is dimensionless. If the flow is between variables with different units, then α_{ji} incorporates the change in units. For example, if Q_1 is the concentration of nitrogen in milligrams per liter and Q_2 is the phytoplankton biomass in milligrams per liter, then α_{21} is the fraction of phytoplankton biomass that is nitrogen and is in units of grams of nitrogen per gram of carbon.

For the one-dimensional case illustrated in Figure 1a, the rate of change, F_1, from Eq. 14 is simply

$$F_1 = J(1,1). \tag{15}$$

The term, $J(1,1)$, could have one of the forms listed in Eqs. 1 and 2, that is, for growth that is independent of density

$$F_1 = rQ_1 \tag{16}$$

and for growth that is limited as the density approaches a carrying capacity K,

$$F_1 = rQ_1 \left(\frac{K - Q_1}{K} \right). \tag{17}$$

Rates of change of a two-dimensional system can be similarly constructed. For example, for the plant-herbivore or predator-prey system illustrated in Figure 1b, $J(2,1) = 0$ and $N = 2$. Thus from Eq. 14

$$F_1 = J(1,1) - J(1,2) \tag{18}$$

$$F_2 = \alpha_{12}J(1,2) + J(2,2) \tag{19}$$

Since the predator population, Q_2, will decline without a food source, Q_1, $J(2,2)$ is negative.

If the growth and death rates of predator and prey populations are density independent and the consumption rate of the predator is proportional to prey density, then $J(1,2)$ has the form given in Eq. 5 and $J(1,1)$ and $J(2,2)$ have the form corresponding to Eq. 1. Thus the rates of change of Q_1 and Q_2 from Eqs. 18 and 19 are

$$F_1 = r_1 Q_1 - a_1 Q_1 Q_2 \tag{20}$$

$$F_2 = r_2 Q_2 + a_2 Q_1 Q_2 \tag{21}$$

where $a_2 = \alpha_{12} a_1$ and $r_1 > 0$. As mentioned above, $J(2,2) < 0$, so $r_2 < 0$.

If the prey population growth, $J(1,1)$, is described by the density dependent Eq. 2, then the corresponding rates of change are

$$F_1 = r_1 Q_1 \left(\frac{K - Q_1}{K} \right) - a_1 Q_1 Q_2 \tag{22}$$

$$F_2 = r_2 Q_2 + a_2 Q_1 Q_2 \tag{23}$$

PREDICTIVE EQUATIONS

Predictive equations incorporate the rates of change, F_i, into equations that predict the values of the state variables, $Q_i(t)$ through time. If time is divided into discrete units such as days, generations, or years, the value of state variable, Q_i, at time, $t + \Delta t$, is the value of the state variable, Q_i at time, t, plus the rate of change F_i multiplied by Δt, the number of time units that have elapsed. The mathematical representation of this relationship is

$$Q_i(t + \Delta t) = Q_i(t) + F_i(t) \cdot \Delta t \tag{24}$$

Equation 24 is called a *difference equation*. Usually the units of t are chosen so that Δt equals one. In this case, the difference equation is

$$Q_i(t + 1) = Q_i(t) + F_i(t) \tag{25}$$

Substituting Eq. 14, which expresses the rate of change, F_i, as a function of flow rates $J(j,i)$, into the difference Eq. 25, we obtain

$$Q_i(t + 1) = Q_i(t) + \sum_{j=1}^{N} \alpha_{ji} J(j,i) - \sum_{\substack{k=1 \\ k \neq i}}^{N} J(i,k) \tag{26}$$

Difference equations which have the form of eq. 26 predict changes at discrete points in time.

Predictive equations that describe dynamics of a system continuously through time are called *differential equations*. This type of equation is based on the *derivative* of $Q_i(t)$, which is the instantaneous rate of change of $Q_i(t)$ and is denoted by the symbol, $dQ_i(t)/dt$ (or by $\dot{Q}$). The derivative is defined to be

$$\frac{dQ_i}{dt} = \lim_{\Delta t \to 0} \frac{Q_i(t + \Delta t) - Q_i(t)}{\Delta t} = L \tag{27}$$

where the symbol, $\lim_{\Delta t \to 0}$, means "the limit as Δt goes to zero." The *limit*, which is called L, exists if $(Q(t + \Delta t) - Q(t))/\Delta t$ becomes very close to L as Δt becomes very close to zero. The precise mathematical requirement for L to be a limit is that for any positive number, ε, chosen, another positive number, δ, can be found such that

$$-\varepsilon < \left(L - \frac{Q_i(t + \Delta t) - Q_i(t)}{\Delta t} \right) < \varepsilon \quad \text{if} \quad -\delta < \Delta t < \delta \tag{28}$$

If there exists an L such that Eq. 28 is satisfied, then $(Q_i(t + \Delta t) - Q_i(t))/\Delta t$ is said to *converge* to L as Δt approaches zero. In this case the derivative exists, and the function, $Q_i(t)$, is said to be *differentiable* at the point, t.

To relate Eq. 27 to the rates of change, F_i, note that Eq. 24 can be rewritten in the form

$$\frac{Q_i(t + \Delta t) - Q_i(t)}{\Delta t} = F_i(t) \tag{29}$$

Thus from Eqs. 27 and 29

$$\frac{dQ_i}{dt} = \lim_{\Delta t \to 0} \frac{Q_i(t + \Delta t) - Q_i(t)}{\Delta t} = \lim_{\Delta t \to 0} F_i(t) = F_i(t) \tag{30}$$

The last equality follows because $F_i(t)$ does not depend upon Δt. Thus the derivative equals the rate of change F_i defined in Eq. 14, that is,

$$\frac{dQ_i}{dt} = F_i(t) = \sum_{j=1}^{N} \alpha_{ji} J(j,i) - \sum_{\substack{k=1 \\ k \neq i}}^{N} J(i,k) \tag{31}$$

The two models most widely used to describe the growth of a single species are obtained by substituting Eqs. 16 and 17 into Eq. 31 to obtain

$$\frac{dQ}{dt} = F = rQ \tag{32}$$

and

$$\frac{dQ}{dt} = F = rQ \left(\frac{K - Q}{K} \right) \tag{33}$$

Equation 32 is called the *exponential growth* equation, and Eq. 33 is the *logistic growth* equation. Exponential and logistic growth also have difference equation forms

$$Q(t + 1) = Q(t) + rQ(t) \tag{34}$$

and

$$Q(t + 1) = Q(t) + rQ(t)\left[\frac{K - Q(t)}{K}\right] \tag{35}$$

which are obtained by substituting Eqs. 16 and 17 into Eq. 25.

Models predicting the behavior of two interacting species can be constructed in a similar fashion. The derivative, dQ_i/dt, of each state variable is calculated by substituting the flow rates, $J(j,i)$, into Eq. 31. For example, the dynamics of predator and prey populations whose interactions can be described by Eqs. 20 and 21 can be predicted from Eq. 31 by the following set of differential equations:

$$\frac{dQ_1}{dt} = F_1 = r_1Q_1 - a_1Q_1Q_2 \tag{36}$$

$$\frac{dQ_2}{dt} = F_2 = r_2Q_2 + a_2Q_1Q_2 \tag{37}$$

where r_1, a_1, and $a_2 > 0$ and $r_2 < 0$. Equations 36 and 37 are called the *Lotka-Volterra equations.*

If the prey population has a growth rate that decreases with increased prey density, the set of differential equations obtained by substituting rates of change, F_1 and F_2, defined in Eqs. 22 and 23, would be a more appropriate model.

By using the same method of model construction, the predator-prey model considered above can be extended to incorporate other species and substances. An example of such an extension is given in the next section.

AQUATIC ECOSYSTEM MODELS

To illustrate the application of the modeling techniques described above, we discuss in detail a three-dimensional model developed by DiToro et al. (1971) to predict the response of phytoplankton populations to nutrient loads released into the San Joaquin River from industrial and municipal sources. We use this model as an example because it clearly illustrates the modeling methods we have discussed and because it has been a basis for a number of more detailed models of aquatic ecosystems developed later, including the models presented in Chapters 18 and 25.

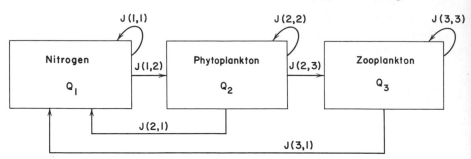

Figure 2   Flow rates in an aquatic ecosystem. See text for discussion.

The ecosystem considered by DiToro et al. consists of a zooplankton population (Q_3), a phytoplankton population (Q_2), and a nutrient (Q_1). The zooplankton feed upon phytoplankton, which in turn absorb nutrients. Through excretion and death, the zooplankton and phytoplankton populations return nutrients to the water. This system is illustrated in Figure 2.

The first step in building a model of the ecosystem described by Figure 2 is to develop equations for each of the flow rates, $J(i,j)$. Some of these flows will be zero. For example, zooplankton do not absorb nutrients directly, so $J(1,3)$ is zero. Similarly, phytoplankton do not eat zooplankton, so $J(3,2)$ is zero.

One term that is not zero is $J(2,3)$, which is the rate of grazing by zooplankton on phytoplankton. Zooplankton graze by filtering phytoplankton from water. The amount of water filtered by each zooplankton multiplied by the concentration of phytoplankton in the filtered water is the amount of phytoplankton ingested by each zooplankton. Multiplying this amount by the number of zooplankton gives the total amount of phytoplankton consumed by zooplankton, that is,

$$J(2,3) = \text{filtering rate} \times \text{phytoplankton concentration} \\ \times \text{number of zooplankton}$$

$$= C_g \times \left(\frac{Q_2}{V}\right) \times Q_3 \tag{38}$$

or

$$J(2,3) = aQ_2Q_3 \tag{39}$$

where V = volume, C_g = filtering rate, $a = C_g/V$, and Q_2 and Q_3 are the number of phytoplankton and zooplankton, respectively. Note that Eq. 39 has the same mathematical form as the general expression for predation, Eq. 5.

The rate, $J(1,2)$, is the amount of nutrient consumed by the phytoplankton population per unit of time. This rate is directly proportional to the growth rate of the phytoplankton population, which depends upon temperature, solar radiation, and the amount of nutrient available. Assuming that the effects of different factors could be represented by a multiplicative relationship, as in Eq. 9, DiToro et al. described the rate of nutrient uptake by

$$J(1,2) = G_T(X_T) \cdot G_s(X_s) \frac{a_1 K_1 Q_1 Q_2}{K_1 + Q_1} \cdot M \tag{40}$$

where $G_T(X_T)$ and $G_s(X_s)$ are functions describing the effects of temperature, X_T, and solar radiation, X_s, on phytoplankton growth. DiToro et al. estimated these functions from published data. The exact forms of the functions, G_T and G_s, are given in Table 1.

In an aquatic ecosystem, nutrients are returned to water by the excretions and death of phytoplankton and zooplankton. The flow of this material from the phytoplankton population is represented by

$$J(2,1) = Q_2 G_T^P(X_T) \tag{41}$$

where the function, $G_T^P(X_T)$, describes the effect of temperature X_T on $J(2,1)$ and is given in Table 1.

DiToro et al. assume that the rate at which nitrogen is returned to water due to zooplankton death and respiration is directly proportional to the size of zooplankton population, that is it is of the form, $K_3 Q_3$, for a given constant, K_3. However, zooplankton also return nutrients to water by excreting phytoplankton material that zooplankton cannot assimilate. Published data indicate that the amount of phytoplankton assimilated by zooplankton can be estimated by an equation of the form of Eq. 8, that is, the amount of phytoplankton assimilated equals

$$\frac{a_2 K_2 Q_2 Q_3}{K_2 + Q_2} \tag{42}$$

Thus the amount of phytoplankton biomass that is not assimilated is the total amount ingested ($a Q_2 Q_3$ from Eq. 39) minus term 42, that is,

$$a Q_2 Q_3 - \frac{a_2 K_2 Q_2 Q_3}{K_2 + Q_2} = \left[a - \frac{a_2 K_2}{K_2 + Q_2} \right] Q_2 Q_3 \tag{43}$$

The total flow of nutrients from the zooplankton population, Q_3, to the nutrient pool, Q_1, is the rate, $K_3 Q_3$, from death and respiration plus the rate of excretion of unassimilated phytoplankton (Eq. 43). Therefore,

$$J(3,1) = K_3 Q_3 + \alpha_{23} \left[a - \frac{a_2 K_2}{K_2 + Q_2} \right] Q_2 Q_3 \tag{44}$$

Table 1   Parameter Values and Functional Relationships used in an Aquatic Ecosystem Model by DiToro et al. (1971).

Parameter Description	Symbol	Value
Conversion factors		
Nitrogen to phytoplankton	α_{12}	5.88 g C/g N
Phytoplankton to zooplankton	α_{23}	0.6 g C/g C
Zooplankton to nitrogen	α_{31}	0.28 gN/gC
Phytoplankton to nitrogen	α_{21}	0.17 g N/g C
Half-saturation coefficients		
Phytoplankton uptake of nitrogen	K_1	0.025 mg N/l-day
Zooplankton uptake of phytoplankton	K_2	3 mg C/l-day
Zooplankton decay (death and excretions)	K_3	0.075/day
Zooplankton grazing rate/volume	a	$\dfrac{0.13}{V}$ liter/day-g C
Maximum saturated growth rate	M	3 day^{-1} °C^{-1}
Light saturation intensity	I_s	300 ly/day

Effects of Exogenous Variables	Functional Relationship	
Celsius temperature x_T on phytoplankton respiration	$G_T^P(X_T) = (0.005)X_T$	
Celsius temperature x_T on phytoplankton growth	$G_T(X_T) = X_T/30$	
Light intensity $I(x_s)$ on phytoplankton growth	$G_s(X_s) = \dfrac{I(X_s)}{I_s}\exp\left[1 - \dfrac{I(X_s)}{I_s}\right]$	
Solar radiation x_s and depth z on light intensity	$I(X_s) = X_s \exp(-kz)$	

(It is necessary to multiply the last term by α_{23} to convert units of phytoplankton biomass into units of zooplankton biomass.)

In an aquatic system, there are also changes in the concentrations of nitrogen, phytoplankton, and zooplankton that result from the movement of water into and out of the system. This change is incorporated in $J(i,i)$, which was defined earlier as the change in variable Q_i that is independent of other state variables. The rate $J(i,i)$ equals the amount of variable, Q_i, entering the system in incoming water minus the amount of variable Q_i

removed by outgoing water. If q is the rate at which water enters and leaves the system, then q multiplied by the concentration, Q_i/V (V equals volume), gives the rate at which substance Q_i is removed. Similarly, if c_i is the concentration of substance i in the incoming water, then qc_i is the rate at which substance i is entering the system. Thus

$$J(i,i) = qc_i - q\frac{Q_i}{V} \qquad (45)$$

In models used to evaluate the relation between nutrient loading and eutrophication, the quantity of nutrients entering the system is dependent upon decisions about the management of the surrounding drainage area, decisions such as the level of waste treatment required of industrial and municipal sewage plants and the type of land use permitted. Thus the concentration of incoming nutrients can be represented as a function of a decision variable, $c_1 = D_1(v)$. Therefore,

$$J(1,1) = qD_1(v) - q\frac{Q_1}{V}. \qquad (46)$$

The equations for all the flow rates, $J(i,k)$, are summarized in Table 2. From these equations, the rates of change F_i can be calculated as the sum

Table 2   A Summary of Flow Rates for an Aquatic Ecosystem Model.[a]

$$J(1,1) = qD_1(v) - q\frac{Q_1}{V}$$

$$J(1,2) = \frac{a_1 K_1 Q_1 Q_2}{K_1 + Q_1} G_T(X_T)G_s(X_s)$$

$$J(2,1) = Q_2 G_T^p(X_T)$$

$$J(2,2) = qc_2 - q\frac{Q_2}{V}$$

$$J(2,3) = aQ_2 Q_3$$

$$J(3,1) = K_3 Q_3 + \alpha_{23}\left[a - \frac{a_2 K_2}{K_2 + Q_2}\right]Q_2 Q_3$$

$$J(3,3) = qc_3 - q\frac{Q_3}{V}$$

[a] This model is described in Figure 2. Parameter definitions are given in Table 1

of flow into state variable compartment, i, minus those out of compartment i. By substituting equations from Table 2 into Eq. 31, we obtain

$$\frac{dQ_1}{dt} = F_1 = J(1,1) + \alpha_{21}J(2,1) + \alpha_{31}J(3,1) - J(1,2)$$

$$\frac{dQ_1}{dt} = qD_1(v) - q\frac{Q_1}{V} + \alpha_{21}Q_2 G_T^P(X_T)$$

$$+ \alpha_{31}\left[K_3Q_3 + \alpha_{23}\left(a - \frac{a_2K_2}{K_2 + Q_2}\right)Q_2Q_3\right]$$

$$- \frac{a_1K_1Q_1Q_2}{K_1 + Q_1}G_T(X_T)G_s(X_s) \tag{47}$$

$$\frac{dQ_2}{dt} = F_2 = J(2,2) + \alpha_{12}J(1,2) - J(2,1) - J(2,3)$$

$$\frac{dQ_2}{dt} = qc_2 - q\frac{Q_2}{V} + \alpha_{12}\frac{a_1K_1Q_1Q_2}{K_1 + Q_1}G_T(X_T)G_s(X_s) - Q_2G_T^P(X_T) - aQ_2Q_3 \tag{48}$$

and

$$\frac{dQ_3}{dt} = F_3 = J(3,3) + \alpha_{23}J(2,3) - J(3,1)$$

$$\frac{dQ_3}{dt} = qc_3 - q\frac{Q_3}{V} + \alpha_{23}aQ_2Q_3 - \left[K_3Q_3 + \alpha_{23}\left(a - \frac{a_2K_2}{K_2 + Q_2}\right)Q_2Q_3\right]. \tag{49}$$

The complete model developed by DiToro et al. can be obtained by substituting the parameters and functions listed in Table 2 into Eqs. 47, 48, and 49.

The development of Eqs. 47, 48, and 49 illustrates the method by which a relatively complicated model is constructed by a series of simple steps. The first step is to define state variables, exogenous variables, and decision variables. The next step is to mathematically describe the flows of material between variables. The rates of change, F_i, are determined for each variable as a function of flow rates from Eq. 14. The rates of change are then used to construct predictive equations that predict the behavior of the system.

A similar approach has been used to develop more detailed and complicated models of aquatic ecosystems. Such models have additional state variables to describe subclasses of populations and nutrients and to include additional substances and populations. For example, phytoplankton and zooplankton groups may be categorized by size or species. Higher trophic levels such as fish may be included. The nutrient component may be expanded to include not only different nutrients, for example, phosphorous and silica, as well as nitrogen, but also different forms of nutrients, for example, organic nitrogen, ammonia, and nitrates. One study (Lehman et al., 1975) also included variables describing the concentration of nutrients stored in phytoplankton biomass in order to predict the effect of luxury consumption of nutrients on phytoplankton growth. Many models, especially those used to assess environmental impact of waste loading on streams, include a state variable describing the concentration of dissolved oxygen. Some models describe spatial variation in population and nutrient concentrations by dividing a body of water into grid squares. A state variable is assigned for each population and nutrient concentration within each grid square. Chapter 25 is an example of a spatially heterogenous model of an aquatic ecosystem.

The number of state variables is larger in models that are more detailed than the three state variable model described by Eqs. 47, 48, and 49. However, in other respects, the model is constructed as before. First the flow rates between each two variables are represented by a mathematical expression. Next, the rates of change are determined from the flow rates and Eq. 14. Finally, predictive equations are formed from Eqs. 26 or 31.

SOLUTION TECHNIQUES

The purpose of predictive equations is to determine the value of Q_i at each point in time, t. An expression that gives the values of $Q_i(t)$ is called the *solution* of the predictive equations.

If a solution can be explicitly stated by a mathematical expression, it is called a *closed form solution* or an *analytical solution*. For example, the exponential equation (Eq. 32)

$$\frac{dQ}{dt} = rQ \quad Q(t_0) = c$$

has the closed form solution

$$Q(t) = ce^{r(t-t_0)} \tag{50}$$

The original Eq. 32 is not a solution because it gives only the instantaneous rate of change of Q, not the actual value of $Q(t)$. A list of closed form solutions to several of the models discussed in the sections on rates of change and predictive equations are presented in Figure 3.

However, closed-form solutions are attainable only for models that are fairly simple or which have a special form. Because of the complexity of

Name	Rate of Change	Graph of F_i	Predictive Equation	Closed Form Solution	Equilibrium Solutions	Graph of $Q_i(t)$
Exponential growth (differential equation)	$F_1 = rQ_1$	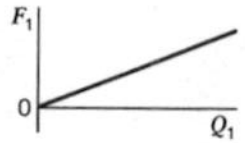	$\dfrac{dQ_1}{dt} = rQ_1$	$Q_1(t) = Q_1(0)e^{rt}$	$Q_1^* = 0$ (not stable)	
Exponential growth (difference equation)	$F_1 = rQ_1$	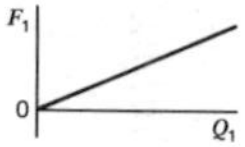	$Q_1(t+1) = (1+r)Q_1(t)$	$Q_1(t) = (1+r)^t Q(0)$	$Q_1^* = 0$ (not stable)	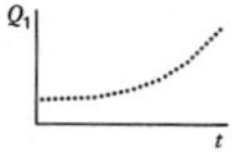
Logistic growth	$F_1 = rQ_1\left[\dfrac{K-Q_1}{K}\right]$	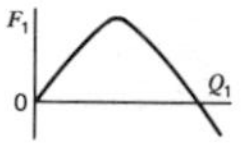	$\dfrac{dQ_1}{dt} = rQ_1\left[\dfrac{K-Q_1}{K}\right]$	$Q_1(t) = \dfrac{K}{1+ce^{-rt}}$ $c = \dfrac{K}{Q_1(0)} - 1$	$Q_1^* = 0$ (not stable) $Q_1^* = K$ (stable)	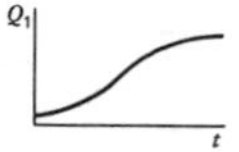
Lotka-Volterra	$F_1 = r_1Q_1 - a_1Q_1Q_2$ $F_2 = r_2Q_2 + a_2Q_1Q_2$	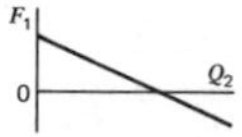 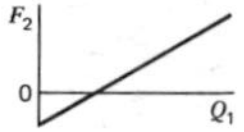	$\dfrac{dQ_1}{dt} = r_1Q_1 - a_1Q_1Q_2$ $\dfrac{dQ_2}{dt} = r_2Q_2 + a_2Q_1Q_2$	No closed-form solution	$(Q_1^*, Q_2^*) = (0,0)$ (not stable) $(Q_1^*, Q_2^*) = \left(\dfrac{r_2}{a_2}, \dfrac{r_1}{a_1}\right)$ (not stable)	$-r_2\ln Q_1 - a_2Q_1$ $+r_1\ln Q_2 - a_1Q_2$ constant 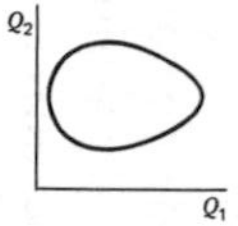
Michaelis-Menten	$F_1 = I - \dfrac{\mu Q_1Q_2}{K_1 + Q_1}$ $F_2 = \dfrac{\mu'Q_1Q_2}{K_1 + Q_1} - K_2Q_2$	 	$\dfrac{dQ_1}{dt} = I - \dfrac{\mu Q_1Q_2}{K_1 + Q_1}$ $\dfrac{dQ_2}{dt} = \dfrac{\mu'Q_1Q_2}{K_1 + Q_1} - K_2Q_2$	No closed-form solution	$Q_1^* = \dfrac{K_1K_2}{\mu' - K_2}$ $Q_2^* = I\left[\dfrac{\mu' - K_2}{\mu K_2} + \dfrac{1}{\mu}\right]$	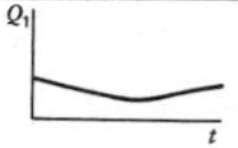 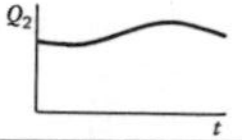

Figure 3 A summary of rates of change, closed form solutions and equilibrium solutions for several small scale ecological models.

actual ecological systems, most models that attempt to be realistic do not have closed form solutions and must be solved numerically.

To solve difference equations numerically, the values of the state variables, $Q_i(t)$, are calculated recursively; that is, the values of variables at time, t, are used to calculate values at time, $t + 1$, which in turn are used to calculate the values at time, $t + 2$. This process starts with $t = t_0$, for which an initial value, $Q_i(t_0)$, is given. For example, consider a difference equation model of a predator and a prey population whose rates of change are defined by Eqs. 20 and 21. Then from Eq. 25,

$$Q_1(t + 1) = Q_1(t) + r_1Q_1(t) - a_1Q_1(t)Q_2(t) \tag{51}$$

$$Q_2(t + 1) = Q_2(t) + r_2Q_2(t) + a_2Q_1(t)Q_2(t) \tag{52}$$

The results of recursive calculations of these equations are presented in Table 3. To numerically solve equations, initial values of the state variables and parameter values must be specified. In Table 3 we have selected the values $r = 0.5$, $a_1 = a_2 = 0.001$ and $r_2 = -1$. The initial values $Q_1(0)$ and $Q_2(0)$ are assumed to be 1000 and 400, respectively. Substituting these values into Eqs. 51 and 52, we obtain

$$\begin{aligned} Q_1(1) &= Q_1(0) + 0.5Q_1(0) - (0.001)Q_1(0)Q_2(0) \\ &= 1000 + 500 - (0.001)(1000)(400) \\ &= 1100 \end{aligned} \tag{53}$$

Table 3   Recursive Calculation of the Difference Equations.

$Q_1(t + 1) = Q_1(t) + 0.5Q_1(t) - 0.001Q_1(t)Q_2(t)$
$Q_2(t + 1) = Q_2(t) \quad Q_2(t) + 0.001Q_1(t)Q_2(t)$

Time	Prey	Predator
0	1000	400
1	1100	400
2	1210	440
3	1283	532
4	1241	683
5	1014	847
6	662	859
7	424	569
8	395	241
9	497	95
10	698	47

and

$$Q_2(1) = Q_2(0) - Q_2(0) + (0.001)Q_1(0) \cdot Q_2(0)$$
$$= 400 - 400 + (0.001)(1000)(400)$$
$$= 400 \tag{54}$$

Substituting the above values for $Q_1(1)$ and $Q_2(1)$ into Eqs. 51 and 52, we obtain

$$Q_1(2) = 1100 + (0.5)(1100) - (0.001)(1100)400 = 1210 \tag{55}$$
$$Q_2(2) = 400 - 400 + (0.001)(1100)(400) = 440 \tag{56}$$

To calculate the values of $Q_1(t)$ and $Q_2(t)$ in Table 3, this process is repeated until $t = 10$.

Numerous methods exist for numerically calculating the solution to differential equations. Many methods are based upon a *Taylor series expansion* of the function, $Q(t)$. From Taylor's Theorem, a finite real valued function, $Q(t)$, for which a finite derivative dQ/dt exists on the interval, $t \leq s \leq t + \Delta t$, can be represented by

$$Q(t + \Delta t) = Q(t) + \frac{dQ(t)}{dt} \Delta t + \cdots + \frac{d^m Q(t)}{dt^m} \frac{(\Delta t)^m}{m!} + \varepsilon(\Delta t^{m+1}) \tag{57}$$

where m is any integer greater than zero and $\varepsilon(\Delta t^{m+1})$ is an error term that is less than a constant times $(\Delta t)^{m+1}$. If $\Delta t < 1$, the error term becomes very small as m increases. If $m = 1$, then Eq. 57 becomes

$$Q(t + \Delta t) = Q(t) + \frac{dQ}{dt}(t) \Delta t + \varepsilon(\Delta t^2) \tag{58}$$

One of the simplest techniques for numerically solving differential equations, the *Euler method*, is based upon Eq. 58. With this method the value of the variable, Q, at time, $t + \Delta t$, is calculated from the equation

$$Q(t + \Delta t) = Q(t) + \frac{dQ(t)}{dt} \cdot \Delta t \tag{59}$$

$$= Q(t) + F(Q,t) \Delta t \tag{60}$$

where $F(Q,t)$ is the rate of change as in Eq. 14. We see by comparison with Eq. 58 that the error in this approximation is $\varepsilon(\Delta t^2)$, which will be quite small if Δt is small (less than one). Equation 60 is a difference equation that can be solved recursively to calculate the values of $Q(t)$ for $t_0 \leq t \leq T$ if $Q(t)$ is differentiable for $t_0 \leq t \leq T$.

In order to solve Eq. 60 for $t_0 \leq t \leq T$, we must make at least $(T - t_0)/\Delta t$ recursive calculations. If Δt is small, this number can be quite large. Thus although the error at each step of the Euler method is small, of the order of $(\Delta t)^2$, the accumulation of this error over many steps can have a significant effect on the accuracy of the final result.

Thus methods that have a smaller error than $\varepsilon(\Delta t^2)$ at each step have been developed. One of these, the *Runge–Kutta* method is based on a difference approximation of the form

$$Q(t + \Delta t) = Q(t) + \sum_{i=1}^{N} C_i F(Q(t) + a_i t + b_i) + \varepsilon(\Delta t)^{N+1} \qquad (61)$$

where C_i, a_i, and b_i are chosen so that Eq. 61 equals the Taylor series expansion, Eq. 57. There are many choices of values for C_i, a_i, and b_i that will satisfy this requirement. For example, for $N = 4$, a Runge–Kutta approximation is

$$Q(t + \Delta t) = Q(t) + \tfrac{1}{6} \cdot (m_1 + 2m_2 + 2m_3 + m_4)\,\Delta t \qquad (62)$$

where $m_1 = F(Q(t),t)$
$\quad m_2 = F(Q(t) + \tfrac{1}{2}m_1\,\Delta t, t + \tfrac{1}{2}\Delta t)$
$\quad m_3 = F(Q(t) + \tfrac{1}{2}m_2\,\Delta t, t + \tfrac{1}{2}\Delta t)$
$\quad m_4 = F(Q(t) + m_3\,t + \Delta t)$

Equation 62 is a difference equation that can now be solved recursively. The error at each step is of the order of $(\Delta t)^5$, which (for $\Delta t < 1$) is much smaller than the error of $(\Delta t)^2$ of the Euler method. Computer subroutines that solve differential equations by the Runge–Kutta method are available with most computing systems.

Regardless of the type of solution method, it is useful to know how much solutions change as parameter or initial values change. A model is said to be *sensitive* to a parameter if a small change in the value of the parameter causes a large change in a solution of the model. The evaluation of the effect of changes of parameter values on model solutions is called *sensitivity analysis*. It is an important aspect of ecosystem modeling since it identifies which parameters must be very precisely measured.

For sensitivity analysis, closed-form solutions have a considerable advantage over numerical solutions. A major disadvantage of numerical methods of solution is that numerical calculations must be repeated for each new parameter value or initial value. The effects of changes in parameter values on solutions can be predicted only by going to the effort and expense of recalculating all the solutions for each set of parameter and initial values. On the other hand, from a closed-form solution we can determine the effects of changes in parameter values from a simple examination of the mathematically explicit solution. For example, from the solution to the exponential growth, Eq. 50, we can see that doubling the initial value, c, will double the value of $Q(t)$. However, doubling the value of r to $2r$ will increase $Q(t)$ by a factor, $e^{r(t-t_0)}$. It is clear that closed-form solutions are preferable whenever they are attainable.

DYNAMIC AND STEADY-STATE MODELS

Systems in which the variables change through time are called *dynamic*. Dynamically changing systems may eventually reach a state at which the values of the variables remain steady. A system in which all the state variables stay constant through time is said to be in *steady state* or in *equilibrium*.

Long-term studies of a system may focus on its steady-state behavior rather than on its transient fluctuations. Therefore, the calculation of the values of the state variables at which the system will eventually stabilize is important. Let $\mathbf{Q}^* = (Q_1^*, \dots, Q_n^*)$ be the equilibrium values of the state variables. Since the values of the variables do not change, the rates of change equal zero or

$$F_i(\mathbf{Q}^*) = 0 \quad i = 1, \dots, n \tag{63}$$

Equation 63 is used to calculate equilibrium solutions, $\mathbf{Q}^*$. For example, to calculate the equilibrium solution to the logistic Eqs. 33 and 35, substitute Eq. 2 into Eq. 63 to obtain

$$F_1(Q_1^*) = rQ_1^* \left(\frac{K - Q_1^*}{K} \right) = 0 \tag{64}$$

By dividing both sides of Eq. 64 by rQ_1^*/K and adding Q_1^* to both sides, we obtain

$$K = Q_1^* \tag{65}$$

Thus once Q_1 reaches the level, K, it will remain at that level. This is the behavior observed in Figure 3.

To obtain the equilibrium solution to the Lotka–Volterra equations, substitute the rate of change Eqs. 20 and 21 into the equilibrium Eq. 63 to obtain

$$F_1(Q_1^*, Q_2^*) = r_1 Q_1^* - a_1 Q_1^* Q_2^* = 0 \tag{66}$$

$$F_2(Q_1^*, Q_2^*) = r_2 Q_2^* + a_2 Q_1^* Q_2^* = 0 \tag{67}$$

By dividing Eq. 66 by Q_1^*, dividing Eq. 67 by Q_2^*, and rearranging terms, we obtain the equilibrium points

$$Q_2^* = \frac{r_1}{a_1} \tag{68}$$

$$Q_1^* = \frac{r_2}{a_2} \tag{69}$$

An equilibrium point such that the system will return to it after small perturbations is called *stable*. The equilibrium point, Q_1^*, of the differential logistic growth equation (Eq. 65) is stable, whereas the equilibrium points, Q_1^* and Q_2^*, of the Lotka–Volterra equations are not stable.

A summary of equilibrium states for a number of ecological models is given in Figure 3. Note that $Q_1 = 0$ is an equilibrium solution for both the exponential and logistic growth equations, because $Q_1^* = 0$ satisfied the equation, $F_1(Q_1) = 0$. However, these are unstable equilibrium points since a slight perturbation in the value of Q_1 to a positive value results in a positive rate of change, which in turn causes Q_1 to increase and thus become even farther from 0. The equilibrium solutions to the Lotka–Volterra equations are also not stable since the system described by these equations will never reach the equilibrium values if $Q_1(0) \neq Q_1^*$ or $Q_2(0) \neq Q_2^*$. Instead, the values of Q_1 and Q_2 will oscillate, periodically returning to their initial value. This relationship is illustrated in the graph of Q_2 versus Q_1 in Figure 3. The values of Q_1 and Q_2 through time can be determined by moving counterclockwise along the egg-shaped curve in Figure 3.

STOCHASTIC PREDICTIVE EQUATIONS

All the predictive equations discussed thus far have been deterministic, that is, for a given value of $\mathbf{Q}$ and t, the rate of change, $F(\mathbf{Q},t)$, is uniquely determined. However, deterministic mathematical relations are simplifications because real ecosystems are subject to considerable variability. This can be caused by many factors, including unpredictable fluctuations in the weather, heterogenity in the abiotic environment, and differences in genetic composition among individuals of the same species.

For example, consider a population, the size of which can be predicted from the equation

$$Q(t + 1) = Q(t) + rQ(t). \tag{70}$$

The net growth rate, r, may vary because of changes in weather, in the food supply, or in the number of sites suitable for reproduction. Similarly, the relationship between a predator's rate of consumption and prey density may vary as a result of factors such as a nonuniform distribution of prey or an intraspecies variation in searching ability.

One way to model variability is to assume that the varying parameters are *random variables*. Associated with each random variable is a probability distribution that quantifies the likelihood that a variable assumes a specified value or falls within a specified range. The probability, p_i, that a random variable, Y, equals a value, k_i, is written

$$P\{Y = k_i\} = p_i \quad i = 1, \dots, N$$

The *expectation*, $E(Y)$, of a random variable, Y, is the mean value of the variable and is equal to the sum of the values, k_i, times the probability of their occurrence, that is,

$$E(Y) = \sum_{i=1}^{N} p_i k_i \tag{71}$$

It can be seen from Eq. 71 that if α is a known value (i.e., it is not a random value), then

$$E(\alpha Y) = \alpha E(Y) \tag{72}$$

It can also be shown that the expectation of the sum of two random variables equals the sum of their expectations, that is,

$$E(Y + Z) = E(Y) + E(Z) \tag{73}$$

Predictive equations, elements of which are random, are called *stochastic*. The solution to a model based on stochastic predictive equations consists of a probability distribution of $Q_i(t)$, rather than a fixed value for $Q_i(t)$ as in the deterministic case.

We can use Eqs. 70 and 71 to calculate the probability distributions and expected values of state variables that are described by stochastic difference equations. For example, consider a population, Q, the size of which can be predicted from Eq. 70. As mentioned earlier, the net growth rate, r, may vary as environmental conditions vary. Let us assume that when the environment is favorable for growth r equals $\frac{2}{3}$, and when it is unfavorable r equals 0. Assume that the probability that either of these events occurring is $\frac{1}{4}$. Let us further assume that average conditions for growth occur 50% of the time, and under these conditions r equals $\frac{1}{3}$. Then

$$P\{r = k_i\} = p_i \tag{74}$$

where $k_1 = 0 \quad p_1 = \frac{1}{4}$
$\quad\quad\quad k_2 = \frac{1}{3} \quad p_2 = \frac{1}{2}$
$\quad\quad\quad k_3 = \frac{2}{3} \quad p_3 = \frac{1}{4}$

From Eq. 71 the expected value of the net growth rate, r, is

$$E(r) = \sum_{i=1}^{3} p_i k_i = (\tfrac{1}{4})(0) + (\tfrac{1}{2})(\tfrac{1}{3}) + (\tfrac{1}{4})(\tfrac{2}{3}) = \tfrac{1}{3} \tag{75}$$

From Eq. 70 the expected value of $Q(t + 1)$ is

$$E(Q(t + 1)) = E(Q(t) + rQ(t)).$$

If the size of $Q(t)$ is known, then from Eqs. 72 and 73

$$\begin{aligned}
E(Q(t) + rQ(t)) &= E(Q(t)) + E(rQ(t)) \\
&= Q(t) + Q(t)E(r) \\
&= Q(t)(1 + E(r)).
\end{aligned} \tag{76}$$

From Eq. 75, the expected value of r is $\tfrac{1}{3}$; thus combining Eqs. 75 and 76, we obtain

$$E(Q(t + 1)) = Q(t)(1 + \tfrac{1}{3}).$$

Thus the expected size of the population at time, $t + 1$, is $\tfrac{1}{3}$ larger than the population's size at time, t.

The probability that a state variable assumes a specific value can also be determined from the probability distribution of the parameters in the difference equations. For example, if Q is described by Eqs. 70 and 74, and $Q(t) \neq 0$, then

$$\begin{aligned}
P\{Q(t + 1) = (1 + k_i)Q(t)\} &= P\{(1 + r)Q(t) = (1 + k_i)Q(t)\} \\
&= P\{r = k_i\} = p_i
\end{aligned}$$

Therefore, if $Q(0) = 1200$ and $k_i = \tfrac{2}{3}$, then

$$P\{Q(1) = (1 + \tfrac{2}{3})1200\} = P\{Q(1) = 2000\} = P\{r = \tfrac{2}{3}\} = \tfrac{1}{4}$$

Similarly, it can be shown that the probability that $Q(1) = 1200$ is also $\tfrac{1}{4}$ and that the probability that $Q(1) = 1600$ is $\tfrac{1}{2}$. Since the net growth rate, r, assumes only the values 0, $\tfrac{1}{3}$, and $\tfrac{2}{3}$, the probability is zero that $Q(1)$ assumes any other value beside 1200, 1600, and 2000.

Two random variables are called *independent* if the probability of each variable assuming specified values equals the product of the probabilities that each variable assumes a specified value, that is,

$$P\{X = k_i \quad \text{and} \quad Y = j_n\} = P\{X = k_i\}P\{Y = j_n\} \tag{77}$$

To illustrate this concept let us consider a set of difference equations describing the relationship between a predator, Q_2, and its prey, Q_1.

$$Q_1(t + 1) = (1 + r_1)Q_1(t) - aQ_1(t)Q_2(t) \tag{78a}$$

$$Q_2(t + 1) = (1 + r_2)Q_2(t) + aQ_1(t)Q_2(t) \tag{78b}$$

Assume that r_1 varies because of changes in weather and that a varies because of differences in searching ability among predators. Let us also assume that these variations are not related and that r_1 and a are independent random variables. Then if $Q_1(t)$ and $Q_2(t)$ are known values,

$$
\begin{aligned}
E(Q_1(t + 1)) &= E(Q_1(t) + r_1 Q_1(t) - a Q_1(t) Q_2(t)) \\
&= Q_1(t) + Q_1(t) E(r_1) - Q_1(t) Q_2(t) E(a) \\
&= Q_1(t)(1 + E(r_1)) - Q_1(t) Q_2(t) E(a)
\end{aligned}
\tag{79}
$$

We can also determine the probability distribution of the prey population size, $Q_1(t)$. From Eqs. 77, 78a, and 78b,

$$
\begin{aligned}
P\{r_1 = k_i\} P\{a = j_n\} &= P\{r_1 = k_j \text{ and } a = j_n\} \\
&= P\{(1 + r_1)Q_1(t) - a Q_1(t) Q_2(t) \\
&= (1 + k_j)Q_1(t) - j_n Q_1(t) Q_2(t)\} \\
&= P\{Q_1(t + 1) = (1 + k_j)Q_1(t) - j_n Q_1(t) Q_2(t)\}
\end{aligned}
\tag{80}
$$

To illustrate the use of Eqs. 79 and 80, let us assume that the net growth rate of r_1 has the probability distribution given by Eq. 74 and a assumes the values, 0.0005 and 0.0015, each with a probability of $\frac{1}{2}$. Then $E(r_1) = \frac{1}{3}$ and $E(a) = 0.001$. From Eq. 79

$$
\begin{aligned}
E(Q_1(t + 1)) &= Q_1(t)(1 + E(r_1)) - Q_1(t) Q_2(t) E(a) \\
&= Q_1(t)(1 + \tfrac{1}{3}) - Q_1(t) Q_2(t)(0.001)
\end{aligned}
$$

For example, if $Q_1(0) = 1000$ and $Q_2(0) = 400$, then $E(Q_1(1)) = 933$. If $r = \frac{1}{3}$ and $a = 0.0005$, then from Eq. 80

$$
\begin{aligned}
P\{r = \tfrac{1}{3}\} P\{a = 0.005\} &= P\{Q_1(1) = (\tfrac{4}{3})Q_1(0) - 0.0005 Q_1(0) Q_2(0)\} \\
&= P\{Q(1) = (\tfrac{4}{3})(1000) - (0.0005)(1000)(400)\} \\
&= P\{Q(1) = 1133\}
\end{aligned}
$$

Since $P\{r = \tfrac{1}{3}\} = \tfrac{1}{2}$ and $P\{a = 0.005\} = \tfrac{1}{2}$,

$$
P\{Q_1(1) = 1133\} = \tfrac{1}{2} \cdot \tfrac{1}{2} = \tfrac{1}{4}
$$

Similarly, it can be shown that $Q_1(1)$ assumes the values 400, 800, 1067, and 1467, each with a probability of $\frac{1}{8}$ and the value 733 with a probability of $\frac{1}{4}$.

It is even more difficult to obtain closed-form solutions for stochastic models than for deterministic ones. If a closed-form solution cannot be obtained, the probability distribution of $Q_i(t)$ may be estimated by *Monte Carlo simulations*. In this procedure a set of actual values, having the statistical properties of a random variable are substituted in place of the random variable in the difference equations, and the dynamics of the system are numerically calculated. Each calculation of the model will use a newly generated set of random numbers.

The values substituted are called *random numbers*, because they could have been produced by chance. Random numbers can be obtained from published tables (Kendall and Smith, 1949). In addition, algorithms exist by which a digital computer calculates a sequence of numbers that any reasonable statistical test will deem random. Such a sequence is called *pseudorandom*. The computer subroutine that calculates pseudorandom numbers is called a *random-number generator* and is available with most computing systems.

An example of Monte Carlo simulation is given in Table 4. The first two columns are the results from two different simulations. The randomly generated values of the variables, r_1 and a, are listed next to the values

Table 4   Monte Carlo Simulations of a Predator-Prey System.[a]

	Simulation 1				Simulation 2				Simulation 3			
Time	$Q_1(t)$	$Q_2(t)$	$r_1(t)$	$a(t)$	$Q_1(t)$	$Q_2(t)$	$r_1(t)$	$a(t)$	$Q_1(t)$	$Q_2(t)$	$r_1(t)$	$a(t)$
0	1000	400	$\frac{1}{3}$	0.0005	1000	400	$\frac{1}{3}$	0.0015	1000	400	$\frac{1}{3}$	0.001
1	1133	200	0	0.0005	733	600	$\frac{2}{3}$	0.0015	933	400	$\frac{1}{3}$	0.001
2	1020	113	$\frac{1}{3}$	0.0015	562	660	$\frac{2}{3}$	0.0005	871	373	$\frac{1}{3}$	0.001
3	1187	173	$\frac{2}{3}$	0.0005	751	185	0	0.0005	836	325	$\frac{1}{3}$	0.001
4	1875	103	0	0.0015	68;	70	$\frac{1}{3}$	0.0005	843	271	$\frac{1}{3}$	0.001
5	1585	289	0	0.0005	885	24	0	0.0015	894	229	$\frac{1}{3}$	0.001
6	1356	229	$\frac{1}{3}$	0.0015	854	32	$\frac{2}{3}$	0.0005	987	205	$\frac{1}{3}$	0.001
7	1341	466	$\frac{2}{3}$	0.0015	1047	14	0	0.0015	1115	203	$\frac{1}{3}$	0.001
8	1297	939	$\frac{1}{3}$	0.0005	1381	29	$\frac{1}{3}$	0.0005	1260	226	$\frac{1}{3}$	0.001
9	1121	609	$\frac{1}{3}$	0.0005	1821	20	$\frac{1}{3}$	0.0015	1396	285	$\frac{1}{3}$	0.001
10	1153	341	—	—	2374	54	—	—	1463	397	$\frac{1}{3}$	0.001

[a] The predator-prey system is described by

$$Q_1(t + 1) = (1 + r_1)Q_1(t) - aQ_1(t)Q_2(t)$$
$$Q_2(t + 1) = (1 + r_2)Q_2(t) + aQ_1(t)Q_2(t)$$

where $r_2 = -1$, r_1 is a random variable that assumes the values $0, \frac{1}{3}, \frac{2}{3}$ with the probabilities $\frac{1}{4}, \frac{1}{2}$, and $\frac{1}{4}$, respectively, and a is a random variable that assumes the values 0.0005 and 0.0015, each with a probability of $\frac{1}{2}$. The first two simulations are for different series of random values of r_1 and a. For the last set of calculations r_1 and a equal their expected values.

of Q_1 and Q_2. The last column gives the values of Q_1 and Q_2 obtained when r_1 and a are held constant at their expected values (i.e., a deterministic simulation).

As would be expected, because different values of r_1 and a have been randomly generated for each simulation, the population sizes vary between random simulations 1 and 2, and both of these results vary from those obtained in simulation 3, where the average values of r and a are used. This difference is especially noticable for simulation 2, where the final size of the predator population is considerably lower and the size of the prey population is considerably higher than in the other two cases. This difference occurs even though the means of the values of r and a generated in simulation 2 equal the expected values.

Because of the variation in results from each Monte Carlo simulation, usually a large number of simulation are made. (It is not unusual to do more than a hundred such simulations.) The statistical characteristics of the behavior of the system can be estimated by analyzing the results with such statistical indices such as the mean and variance. Monte Carlo simulation is used in Chapter 9 to describe the growth of a forest.

MANAGEMENT MODELS

Models that are used to make decisions about managing a system are called *management models*. The models discussed previously have been *predictive models*, that is, they predict the behavior of a system. Predictive models consist entirely of difference or differential equations. These equations, which predict the response of the system to decision variables, are, of course, essential to a model used to make management decisions. However, management models also include two other types of functions that are not included in models used purely for predictive purposes.

The first of these is a *criterion function*, which is a quantitative measure of the success of a management program and can be used to compare one program against another. Frequently, the criterion function is based on economic criteria such as cost or profit.

Another consideration that can affect the choice of a management program is the existence of limits either on the decision variables or on the dynamics of the system. These limits are expressed as *constraint equations*. For example, population numbers and nutrient concentrations cannot be negative. Thus

$$Q_i(t) \geq 0$$

is a necessary constraint equation.

To illustrate these concepts, let us consider a simple model for managing a single population, $Q_1(t)$, whose initial population size is K_0, that is,

$$Q_1(0) = K_0 \tag{81}$$

Assume there is a value, p, and an associated harvesting cost, c, for each individual harvested from the population. In addition, assume that the population's growth can be described by a discrete time-exponential growth equation. Thus the predictive equation has the form

$$Q_1(t + 1) = (1 + r)Q_1(t) - h(t) \tag{82}$$

where $h(t)$ is the number of harvested individuals. The number of individuals harvested cannot be greater than the total population, so the appropriate constraint equation is

$$h(t) \le Q_1(t) \tag{83}$$

One measure of success of a management strategy is the net income gained over N time periods. The total net income is the net return per harvested individual multiplied by the number of harvested individuals. Hence one possible criterion function is ϕ where

$$\Phi = \sum_{t=1}^{N} (p - c)h(t). \tag{84}$$

Thus we have developed a simple but complete management model for harvesting a population. The basic elements of a management model, the predictive equations, the constraint equations, and the criterion function, are given by Eq. 82, 83, and 84, respectively.

For a somewhat more complex example, consider the management of nutrients entering a lake that is susceptible to eutrophication. In this case, the amount of nutrients released into the lake is a decision variable. The response of the lake to this nutrient input can be predicted from Eqs. 47, 48, and 49. In such a model, management policies toward eutrophication may be expressed by constraints on the phytoplankton density. The economic criterion usually used is cost, which in this case is the cost of waste treatment or land use manipulation in order to limit nitrogen loading to the lake.

METHODS FOR SOLVING MANAGEMENT MODELS

There are two methods for using a model to choose a management decision. The first method is *trial and error*: decision variables representing several management programs are chosen, and each is substituted into the predictive equations. From the solution to the predictive equations, the criterion function for each program is evaluated and compared to the results for other management programs. The second method, known as *optimization*, directly chooses the policy that maximizes (or minimizes) the criterion function. As an illustration of the differences in solution methods, we solve the simple management model described above in Eqs. 81 to 84 by several methods.

In the trial and error approach, the predictive equations must be solved for each new value of the decision variable. This solution may be obtained either analytically or numerically. For example, Table 5 gives the results of recursive calculation of Eq. 82 for several management programs. Also given is the value of the criterion function, Eq. 84, for each policy.

The first policy examined is to harvest the maximum number of individuals at each stage. Since the only constraint on the number harvested, $h(t)$, is that it be less than the population size, then in policy 1, $h(t) = Q(t)$. Notice that because the population is so reduced at each stage, the number that can be harvested at each stage sharply declines. In policy 2 the number harvested equals the size of the population increase, $rQ = 0.1Q$. Thus the population size stays constant. In policy 3 nothing is harvested until the last stage, when the entire population is removed. Similarly, in policy 4 nothing is harvested until the last stage. However, in this case, only forty-six

Table 5   Trial and Error Simulation of a Management Model.[a]

		Policy 1		Policy 2		Policy 3		Policy 4	
		$Q(t)$	$h(t)$	$Q(t)$	$h(t)$	$Q(t)$	$h(t)$	$Q(t)$	$h(t)$
Time	0	100	100	100	10	100	0	100	0
	1	10	10	100	10	110	0	110	0
	2	1	1	100	10	121	0	121	0
	3	0	0	100	10	133	133	133	46
	4	0	–	100	–	13	–	100	–
Total harvested			111		40		133		46
Net income			$1110		$400		$1330		$460

[a] Calculations are based on Eqs. 81 to 84, which are explained in the text. Population size is $Q(t + 1) = (1 + r)Q(t) - h(t)$, where $h(t)$ is the number harvested and $r = 1$. The net return per harvested individual is $10. Thus net income equals $10 × (total harvested).

individuals are harvested, so that the final and initial populations are equal in size.

From Table 5 it is clear that policy 1, in which the maximum number is harvested each time, is not a desirable policy. Because the population is intensely harvested, the number of births is very small. The population remaining in the fourth time period is lower than for any other policy, and the net return is lower than policy 3. Policy 3 is the best policy if the criterion of success is the maximization of net income. However, if an additional constraint is included, requiring that the final population must at least equal the initial population in size, then both policies 1 and 3 are unacceptable. In this case, Policy 4 would be selected. Thus we see that constraints greatly influence both the choice of a management policy and the value of the criterion function.

The disadvantage of the trial and error approach is its expense. For a fairly realistic model that is solved numerically, each simulation usually costs at least one dollar and probably costs considerably more. The total possible number of management alternatives (each of which requires a simulation) may be very high. For example, if the initial size of a population described by Eq. 82 is ten and the number of stages is six, then the number of management alternatives (i.e., the number of harvested individuals at each stage) is greater than $10^6 =$ one million.

For a complex model for which many alternatives exist, it is obviously too expensive to test each admissible policy. For such models, one may choose a few of the policies that appear most promising and test by simulation only those. Unfortunately, the best policies may be overlooked by this procedure.

An alternate approach is to use optimization techniques that calculate directly the policies maximizing a criterion function. An optimization problem has the form

$$\underset{v_t}{\text{Max}}\ \Phi(Q,v) \tag{85}$$

where the values of v_t and $Q(t)$ are determined from the predictive and constraint equations. The symbol, $\text{Max}_{v_t}\ \Phi$, means: choose the value of v_t that will maximize the function, Φ.

Occasionally an optimization problem may have a closed-form solution. For example, the harvesting problem, Eqs. 81 to 84, has a closed-form solution. It is to harvest the entire population at the last stage.

However, most optimization models do not have closed-form solutions. For these problems, numerical methods have been developed that calculate optimal policies. There are a large number of such methods, each of which has advantages and disadvantages. The best method to use depends upon characteristics of the problem being studied, that is, whether the problem is nonlinear, stochastic, continuous, or constrained. The two methods most widely used to determine the optimal management of systems described by difference equations are linear programming and dynamic programming.

In the remaining pages of this section, linear and dynamic programming methods are discussed in detail. Readers who are not interested in the mathematical details of these optimization methods may wish to skip to the next section, which is a summary.

Linear programming can be used to solve optimization problems that have a linear criterion function of the form

$$\text{Maximize} \sum_{k=0}^{N} c_k y_k \tag{86}$$

where the variables, y_k, are subject to the following inequality and equality constraints:

$$\sum_{k=0}^{N} a_{kj} y_k \le b_j \quad j = 1, \ldots, M \tag{87}$$

and

$$\sum_{k=0}^{N} \alpha_{kj} y_k = \beta_j \quad j = 1, \ldots, M \tag{88}$$

Since the minimum of a function is the maximum of its negative, linear programming can also be used to find the minimum of a function by changing the sign of the c_ks. For most problems, many of the coefficients, c_k, a_{kj}, and α_{kj}, are zero.

We can formulate the management model, Eqs. 81 to 84 as a linear-programming problem. First, let the linear-programming variable y_i represent the population size at time, i, that is

$$y_i = Q(i) \quad i = 1, \ldots, N \tag{89}$$

The amounts harvested, $h(i)$, are also variables, so let them be denoted by the linear-programming variables, $y_{N+1}, \ldots, y_{2N}$, that is,

$$y_{i+N} = h(i) \quad i = 1, \ldots, N \tag{90}$$

Substituting Eqs. 89 and 90 into the difference equation, Eq. 82, which describes the growth of the population, Q, we obtain

$$\begin{aligned} y_{i+1} = Q(i+1) &= (1+r)Q(i) - h(i) \\ &= (1+r)y_i - y_{i+N} \end{aligned} \tag{91}$$

or

$$y_{i+1} - (1+r)y_i - y_{i+N} = 0 \tag{92}$$

Equation 92 is a linear equality constraint and can be written in the form of the equality constraint given above in Eq. 88 if we define $\beta_i = 0$ $(i > 0)$ and the constants, $\alpha_{k,i}$, as the following:

$$\alpha_{k,i} = \begin{cases} 1 & \text{if } k = i + 1 \\ -(1 + r) & \text{if } k = i \\ -1 & \text{if } k = i + N \\ 0 & \text{otherwise} \end{cases} \tag{93}$$

For $\alpha_{k,i}$ defined as in Eq. 93, Eq. 92 equals Eq. 88. Since the population's initial size, $Q(0)$, is given, we also have the constraint

$$y_0 = Q(0) = K_0 \tag{94}$$

Then if

$$\alpha_{0,0} = 1 \text{ and } \beta_0 = K_0 \tag{95}$$

Eq. 94 is identical to Eq. 88 for $i = 0$.

The inequality constraint Eq. 83 prevents the number harvested from exceeding the population size, that is, $h(i)$ must not exceed $Q(i)$. From Eqs. 89 and 90, this is equivalent to

$$y_{j+N} \geq y_j \tag{96}$$

or

$$-y_j + y_{j+N} \geq 0. \tag{97}$$

The constraint can be written in the form of the general linear programming inequality constraint

$$\sum_{k=1}^{M} a_{k,j} y_k \leq b_j \quad j = 0, M \tag{87}$$

where $b_j = 0$ and

$$a_{k,j} = \begin{cases} -1 & \text{if } k = j \\ 1 & \text{if } k = j + N \\ 0 & \text{otherwise} \end{cases} \tag{98}$$

The criterion of success of a population management program from Eq. 84 is

$$\sum_{i=1}^{N} (p - c)h(i) = \sum_{i=1}^{N} (p - c)y_{i+N} = \sum_{i=N+1}^{2N} (p - c)y_i \tag{99}$$

If c_i is defined to be

$$c_i = \begin{cases} 0 & \text{if } i = 1, \ldots, N \\ p - c & \text{if } i = N + 1, \ldots, 2N \end{cases} \tag{100}$$

then the criterion function, Eq. 84, equals the linear programming criterion function

$$\sum_{i=1}^{N} (p - c)h(i) = \sum_{i=1}^{2N} c_i y_i \tag{101}$$

Equations 86 to 88, 93, 95, 98, and 100 comprise a linear-programming formulation of the harvesting model described by Eqs. 81 to 84. Subroutines that solve linear-programming problems are available with most computing systems. When using such subroutines to obtain the optimal values of the variables, y_i, we only need to specify the values of the coefficients, $\alpha_{k,i}$, $a_{k,i}$, β_{k_i}, b_{k_i} and c_i. For the harvesting problem discussed above, the values of these constants are determined by Eqs. 95, 98, and 100. The solution obtained by using these values of the coefficients is the same as that obtained earlier by analytical means, that is, all harvesting should take place at the last stage. Thus

$$0 = h(i) = y_{i+N} \quad i = 1, \ldots, N - 1 \tag{102}$$
$$(1 + r)^N Q(0) = h(N) = y_{2N} \quad i = 1, \ldots, N \tag{103}$$

It follows that

$$y_i = (1 + r)^i Q(0) = Q(i) \quad i = 1, \ldots, N \tag{104}$$

Equations 89 to 101 were developed to illustrate the application of linear programming techniques to ecological management models. We need not have used linear programming, since a closed-form solution was available. However, linear programming is widely used for more complex problems that do not have closed form solutions. An example of such an application is given in Chapter 18.

Linear programming is a very powerful optimization tool because it can be used for systems with a very large number of variables. Its major disadvantage is that it is only suitable for models with constraints and criterion functions that are linear, that is, those of the form

$$\Phi(Q,v) = \sum_{i=1}^{N} r_i Q_i + \sum_{i=1}^{M} k_i v_i$$

and

$$\sum_{i=1}^{N} a_i Q_i + \sum_{i=1}^{M} a_i v_i \le M_i$$

Examples of functions that are not linear are those involving exponentials, trigonometric functions, or products of state or decision variables. In some cases, such nonlinear functions can be approximated by piecewise linear functions, so that a linear programming approach can be used; but in many cases, such an approximation is not possible. Another disadvantage of linear programming is that it is not applicable if the relationships among variables are stochastic.

An optimization method that can be used for determining managment policies for systems described by nonlinear or stochastic difference equations is *dynamic programming*. Dynamic programming can be used for any problem of the form

$$H_n(\mathbf{Q}^n) = \underset{v_n, \ldots, v_N}{\text{Max}} \left[\sum_{i=n}^{N} R_i(\mathbf{Q}^i, v^i) \right] \tag{105a}$$

where state variables at each time period are related by

$$\mathbf{Q}^{i+1} = \mathbf{Q}^i + F(Q^i, v_i) \tag{105b}$$

where F is the rate of change from Eq. 14. The function R_i, is the return to the criterion function over the ith time interval.

The basis of the dynamic programming algorithm is Bellman's Principle of Optimality:

An optimal policy has the property that whatever the initial state and initial decision are, the remaining decisions must constitute an optimal policy with regard to the state resulting from the first decision (Bellman and Dreyfus, 1962).

Applying this principle to Eq. (105a), we obtain

$$H_k(\mathbf{Q}^k) = \underset{v_k, \ldots, v_N}{\text{Max}} \left[\sum_{i=k}^{N} R_i(\mathbf{Q}^i, v_i) \right] \tag{106}$$

$$= \underset{v_k}{\text{Max}} \left[R_k(\mathbf{Q}^k, v_k) + \underset{v_{k+1}, \ldots, v_N}{\text{Max}} \sum_{i=k+1}^{N} R_i(\mathbf{Q}^i, v_i) \right] \tag{107}$$

But by the definition of $H_n(\mathbf{Q}^n)$ given in Eq. (105a) we see that

$$H_{k+1}(\mathbf{Q}^{k+1}) = \underset{v_{k+1}, \ldots, v_N}{\text{Max}} \left[\sum_{i=k+1}^{N} R_i(\mathbf{Q}^i, v_i) \right] \tag{108}$$

Substituting Eq. 108 for the last term in Eq. 107, we obtain

$$H_k(\mathbf{Q}^k) = \underset{v_k}{\text{Max}} \left[R_k(\mathbf{Q}^k, v_k) + H_{k+1}(\mathbf{Q}^{k+1}) \right] \tag{109}$$

where

$$\mathbf{Q}^{k+1} = \mathbf{Q}^k + F(\mathbf{Q}^k, v_k) \tag{110}$$

Equations 109 and 110 are the fundamental equations of dynamic programming. Let $v_k^*(Q^k)$ be the value of the decision variable which maximizes the right side of Eq. 109.

Notice that although $R_k(Q_k, v_k)$ is a specified function and its value therefore known, the values of the functions, H_k and H_{k+1}, are not known. Thus Eq. 109

cannot be solved directly. Instead, the equations are solved recursively, starting with $H_N(Q_N)$ and working backward to $H_1(Q_1)$. From Eq. 106

$$H_N(\mathbf{Q}^N) = \underset{v_N}{\text{Max}} \; R_N(\mathbf{Q}^N, v_N) \tag{111}$$

Since the value of $R_N(Q_N, v_N)$ is known, the value of $v_N^*(Q_N)$ that maximizes Eq. 111 can be determined analytically or numerically. From the recursive Eq. 109,

$$H_{N-1}(\mathbf{Q}^{N-1}) = \underset{v_{N-1}}{\text{Max}} \; \{R(\mathbf{Q}^{N-1}, v_{N-1}) + H_N[\mathbf{Q}^{N-1} + F(\mathbf{Q}^{N-1}, v_{N-1})]\} \tag{112}$$

Since H_N has been calculated and $F(\mathbf{Q}^{N-1}, v_{N-1})$ and $R(\mathbf{Q}^{N-1}, v_{N-1})$ are known, the value of $v_{N-1}^*(\mathbf{Q}^{N-1})$ that maximizes Eq. 112 can be determined. From $v_{N-1}^*(\mathbf{Q}^{N-1})$, $H_{N-1}(\mathbf{Q}^{N-1})$ can be calculated. We then repeat the process, using $H_{N-1}(\mathbf{Q}^{N-1})$ to find $H_{N-2}(\mathbf{Q}^{N-2})$, which in turn is used to find $H_{N-3}(\mathbf{Q}^{N-3})$. These calculations are repeated until $H_1(\mathbf{Q}^1)$ is reached.

Dynamic programming can also be used to determine an optimal management policy for systems described by stochastic transfer functions. For example, if the state variable assumes the value η_i with a probability p_i then the recursive dynamic programming equation is

$$H_n(\mathbf{Q}^n) = \underset{v_n}{\text{Max}} \; E[R(\mathbf{Q}^n, v_n) + H_{n+1}(\mathbf{Q}^{n+1})]$$

$$= \underset{v_n}{\text{Max}} \; \left\{E[R(\mathbf{Q}^n, v_n)] + \sum_{i=1}^{M} p_i H_{n+1}(\eta_i)\right\} \tag{113}$$

The probability, p_i, is usually a function of $Q(n)$ and v_n. We can solve Eq. 113 by recursively calculating the values of $H_n(\mathbf{Q}^n)$ and $v_n^*(\mathbf{Q}^n)$, starting with $n = N$ and ending with $n = 1$.

As an example, we formulate the harvesting model, Eqs. 81 to 84 as a dynamic-programming model. The predictive equation required in Eq. 105b is (from Eq. 82)

$$\mathbf{Q}^{n+1} = (1 + r)\mathbf{Q}^n - v_n = \mathbf{Q}^n + F(\mathbf{Q}^n, v_n) \tag{114}$$

where v_n is the number of individuals harvested, $h(n)$. The return to the criterion function over each state is the net return, $(p - c)$, times the number of individuals, or

$$R(\mathbf{Q}^n, v_n) = (p - c)v_n. \tag{115}$$

Substituting Eqs. 114 and 115 into the dynamic-programming Eqs. 109 and 110, we obtain

$$H_n(\mathbf{Q}^n) = \underset{v_n \leq Q^n}{\text{Max}} \; \{(p - c)v_n + H_{n+1}[(1 + r)Q^n - v_n]\} \tag{116}$$

The value of H_N is

$$H_N(\mathbf{Q}^N) = \underset{v_N \leq Q^N}{\text{Max}} \left[(p - c)v_N \right] \tag{117}$$

Since the expression inside the brackets increases as v_N increases, the maximum value is obtained when $v_N^*(Q^N) = Q^N$ and

$$H_N(Q^N) = (p - c)Q^N \tag{118}$$

The next step is to calculate $v_{N-1}^*(Q^{N-1})$ and $H_{N-1}(Q^{N-1})$ from the equation

$$H_{N-1}(Q^{N-1}) = \underset{v_{N-1} \leq Q^{N-1}}{\text{Max}} \left\{ (p - c)v_{N-1} + H_N[(1 + r)Q^{N-1} - v_{N-1}] \right\} \tag{119}$$

Substituting the value obtained for $H_N(Q^N)$ in Eq. 118 into Eq. 119, we obtain

$$H_{N-1}(Q^{N-1}) = \underset{v_{N-1} \leq Q^{N-1}}{\text{Max}} \left\{ (p - c)v_{N-1} + (p - c)[(1 + r)Q^{N-1} - v^{N-1}] \right\} \tag{120}$$

One value of v_{N-1} that maximizes Eq. 120 is $v_{N-1}^*(\mathbf{Q}^{N-1}) = 0$. Then from Eq. 120

$$H_{N-1}(Q^{N-1}) = (p - c)(1 + r)Q^{N-1} \tag{121}$$

Repeating the recursive equations for $n = N - 2, N - 3, N - 4, \ldots, 1$, we find that

$$v_n^*(Q^n) = 0 \tag{122}$$

$$H_n(Q^n) = (1 + r)^{N-n}Q^n \quad \text{for } n < N \tag{123}$$

Thus the dynamic-programming solution is not to harvest until the last stage. As expected, this result is the same as that obtained both by linear programming and by analytical means.

Because equations describing ecosystems are generally nonlinear and possibly stochastic, dynamic programming is a very useful method for management models of ecological systems. Dynamic programming has a major limitation, however. If the state vector, $\mathbf{Q}$, has a large number of components, computation of the solution by dynamic programming is impossible. Frequently this limitation can be overcome by constructing a method (such as a decomposition of the model) that reduces the dimension of the state variable. An example of this approach is given in Chapter 22.

We have described only two of the many optimization methods currently available. For example, there are also optimization methods that solve continuous-time problems. Among the most widely used are gradient methods for unconstrained optimization problems and the Maximum Principle for constrained optimization problems (Sage, 1968, and Converse, 1970).

For more detailed information about dynamic programming and linear programming, the reader is referred to an introductory text in operations research such as that by Hillier and Liberman (1975).

SUMMARY

We have discussed the procedures by which a mathematical model is formulated, solved, and used in making predictions and management decisions. The steps in constructing and solving a mathematical model are summarized as follows:

1 Define state variables, exogenous variables, and control variables.
2 Determine the flow rates, $J(j,i)$, between state variables.
3 Calculate the rate of change of each variable as the sum of flows into the variable minus the flows out of the variable or

$$F_i = \sum_{j=1}^{N} \alpha_{ji} J(j,i) - \sum_{\substack{k=1 \\ k \neq i}}^{N} J(i,k)$$

4 Describe the dynamics of the system by substituting each rate of change, F_i, into a series of difference equations, $Q_i(t + 1) = Q_i(t) + F_i$, or into a series of differential equations, $dQ_i/dt = F_i$
5 Find a closed form or numerical solution to the model's predictive equations. If only the equilibrium solution is of interest, it can be obtained by solving the equation, $F(\mathbf{Q}^*) = 0$.
6 If the model is being used to determine management policies, use trial and error simulation or optimization methods.

This chapter serves as only a brief introduction to mathematical modeling of ecological systems. Detailed examples of models of specific ecosystems are given in other chapters of this book. For more information about some of the modeling tools discussed such as numerical solution of differential equations, Monte Carlo simulation, and optimization methods, further reading is suggested in the References.

ACKNOWLEDGMENTS

The author expresses appreciation to colleagues and students who read and commented on the manuscript, including Charles Hall, Simon Levin, Ray van Houtte, Kevin Curry, and Robert Rovinsky. Special thanks are due

Keith Porter, who made detailed and very helpful comments on each of many drafts. This work was supported in part by the National Science Foundation Engineering Research Initiation Grant GK–42147.

REFERENCES

Anderson, D. R. 1975. Optimal exploitation strategies for an animal population in a Markovian environment: a theory and an example. *Ecology* **56**: 1281–1298.

Bellman, R. E., and S. E. Dreyfus. 1962. *Applied dynamic programming.* Princeton University Press, Princeton.

Beveridge, G. S., and R. S. Schecter. 1970. *Optimization: Theory and practice.* McGraw-Hill, New York.

Converse, A. O. 1970. *Optimization,* Holt Rinehart, New York.

deWit, C. T., and J. Goudriaan. 1974. *Simulation of ecological processes.* Pudoc, Wageningen, The Netherlands.

DiToro, D. M., D. J. O'Connor, and R. V. Thomann. 1971. A dynamic model of phytoplankton populations in the Sacramento-San Joaquin delta. *Advan. Chem. Ser.* **106**: 131–180.

Gerald, J. 1970. *Applied numerical analysis.* Addison-Wesley, Reading, Mass.

Hammersley, J. M., and D. C. Handscomb. 1967, *Monte Carlo methods.* Methuen, London.

Hamming, R. W. 1971. *Introduction to applied numerical analysis.* McGraw-Hill, New York.

Hillier, F. S., and G. J. Lieberman. 1974, *Operations research.* Holden-Day, San Francisco.

Lehman, J. T., D. B. Botkin, and G. E. Likens. 1975. The assumptions and rationales of a computer model of phytoplankton population dynamics. *Limn. Oceanogr.* **20**: 343–364.

Sage, A. P. 1968. *Optimum systems control.* Prentice-Hall, Englewood Cliffs, N.J.

Spiegel, M. R. 1967. *Applied differential equations,* Prentice-Hall, Englewood Cliffs, N.J.

APPENDIX   DEFINITIONS OF SYMBOLS

A list of symbols and their definitions are provided as follows. Not included in the list are all symbols such as a, k, K, r, α, β, which are used to describe constants in a number of different types of equations. The number in parenthesis after some of the definitions indicates an equation that illustrates its use.

$E(Y)$	Expected value of random variable Y. (71)
F_i	Rate of change. (14)
$G_m(X_m)$	Effect of changes in variable X_m on a flow rate. (12)
$H_n(Q^n)$	The optimal value of a criterion function for a dynamic programming model that includes the time periods n, $n+1$, ..., N. (105a)
$h(i)$	Number of individuals harvested in the ith time period. (90)

$J(i,j)$	The flow of material from Q_i to Q_j.
$J(i,i)$	Change in Q_i that does not result from flows from other state variables
p_i	The probability a random variable assumes a certain value. (71)
Q_i	The ith state variable.
$\mathbf{Q}$	The state vector ($\mathbf{Q} = (Q_1, \ldots, Q_n)$).
Q_i^*	The equilibrium value of Q_i. (63)
$R_n(Q_n, v_n)$	The contribution to the criterion function made during the nth stage of a dynamic programming model. (105a)
v	Decision variable.
$v^*(Q)$	Optimal decision variable
V	Volume.
w	Exogenous variable.
x, y	Model variables (state, decision, or exogenous).
α_{ij}	Conversion factor for units of Q_i into units of Q_j. (14)
Φ	Criterion function. (84)

5
Values and Environmental Modeling

EDWARD I. FRIEDLAND

By their very nature models are incomplete and often unrealistic. They are created that way in order to be useful. By design and of necessity they systematically ignore, disregard, or otherwise exclude from consideration features that actually are a part of the systems whose behavior is of interest. Because of the limitations of our knowledge, funding, computer capacity, and interests, we can include in the model *only* those features of the real system that we consider *significant*, whether we intend to deal primarily with inanimate objects, biological systems, or the extremely complex behavior patterns of human beings as they interact with one another and the natural environment. At the very beginning of this inquiry into the relationship between evaluation and environmental modeling, I emphasize the need to choose and, therefore, the central role of the chooser. And I establish as the first order of priority our need to appreciate the central role of the process that determines which aspects of the system to be modeled are significant.

To put the matter in a nutshell, building models requires decisions to be made: decisions as to how the complexities of reality ought to be reduced and refined into a few symbolic expressions (usually mathematical) that portray the essential features of that reality.

MODELING AND CHOICE

The best way to acquire an understanding of what the consequences of modeling decisions are is by beginning with a brief introduction to decision theory (Luce and Raiffa, 1957). Although created as a theoretical tool for guiding inquiry into choice in general, decision theory offers a set of insights into modeling decisions that are invaluable for our purposes.

Foremost among the distinctions that decision theorists assert is the separation of "facts" and "values." Both facts and values are regarded as essential ingredients of all decisions. The difference between these two notions is taken to be whether or not their validity could be "scientifically" determined. By definition facts are taken to be those statements, assumptions, beliefs, and so forth, that are amenable, actually or in principle, to scientific testing. Values are those statements, assumptions, beliefs, and so forth, that by their nature defy such testing.

Initially, we may wish to rid ourselves of the difficulty of having to deal with values, for their character renders them inaccessible to our trusted tool, science. But this frustration is a necessary one. For no amount of facts, however extensive, is sufficient to enable a model builder or any decision maker to decide which course of action to take. Somehow, the alternatives available must be ordered in terms of desirability. All the facts describing the decision alternatives and whatever other aspects of the decision situation he may care about, do not by themselves enable that collection of facts to be transformed into an ordered ranking of the decision alternatives according to their merit or worth to the decision maker. All his attitudes for or against anything, his preferences, desires, notions of right and wrong, good and evil, and so forth, enter into his decisions as values. Values alone are the instruments with which decision makers are able to reduce into a unidimensional index of merit or worth the multidimensional information describing each alternative and the possible or likely consequences of its adoption. Furthermore, values necessarily enter into all the decisions that a would-be modeler makes, whether or not he is self-conscious of or even aware that he is making decisions.

These values operate at all levels of modeling and at all stages of the modeling process. The selection of subjects to investigate, our very notions of purpose and of the significance of our modeling efforts, and our basic methodological decisions all reflect evaluative judgments. Likewise, at the other end of the spectrum, even the most seemingly minute "technical" decisions, such as the choice of variables, functional forms, statistical techniques and criteria, data sources, and so forth, also import evaluative judgments into our studies; judgements bearing a significance of which we are frequently unaware. For example, such decisions may be contrary to our beliefs, inconsistent with other evaluations incorporated into a model, or may inadvertently introduce an evaluative assumption that we might wish to exclude from our analysis. Conversely, they may obscure facets of an evaluative issue that we wish to examine explicitly.

PURPOSES AND PROCESSES

Among the many types of fact and value consideration that play a major part in the construction of environmental models, a few are preeminent in their importance to an understanding of the links between evaluation and modeling. The first of these relates directly to our prior assertion that model builders must identify the significant features of the systems they are modeling. Significance is not an absolute. Like beauty, it lies in the eyes of the beholder. It depends not only upon the object system itself, but also upon the purposes, personal philosophy, previous experiences, and capabilities of the modeler. Those features of a system that are significant to the prospective model builder are so because of their relationship to his purposes and to the resources at his disposal.

A useful distinction may be drawn between model building intended primarily to enlarge our understanding of the ways in which systems behave, or of the relationship between their structure and the functions they perform (i.e., such as are found in Section 2 of this book) and those models intended primarily to assist in policy formation (i.e., Sections 3 and 4).

Models of the former type are most familiar in science and, in accordance with the usual notions of scientific endeavor, the model builder's purpose and thus his modeling choices may be presumed to reflect a primary interest in the discovery of truth. The significance of the variables employed in such models depends first and foremost on their explanatory or predictive power. A scientific model is successful insofar as it accurately portrays the relationships among those factors that cause the phenomena of interest to occur and to the degree that it makes predictions that agree with later observations.

Models created to assist in the formation of policy must be built and judged in accordance with different standards. This does not mean that prospective builders of such models ought not to concern themselves with accuracy or patterns of causation but rather that these issues are secondary. The basic objective is not the discovery of previously unknown truths but the collection and integration of existing knowledge and its presentation in a form useful in the policy-making process. Such a model must above all be useful, useful in that it enables those who participate in policy making to estimate how their actions may affect what will happen. Precisely how useful a particular model is depends upon many things besides its accuracy. Most importantly it depends upon how complete a model is in its portrayal of the effects that are considered significant by the responsible decision makers, planners, analysts, or other participants and upon how accurately it is able to indicate the way in which these affects may be altered under a range of possible future situations and in response to the variety of conditions that may prevail.

Therefore, it is obvious that evaluative considerations must play a direct role in models built to assist in policy making, for such considerations specify, in part, the data that the model must produce and thus some of the variables that must enter into it.

Because of the nature of this type of model we can indicate another important restriction that shapes the choice of variables. The character of policy making differs in many other ways from that of science. In particular, their respective concerns with truth are quite dissimilar. Scientists emphasize the need to *discover* truth, while policy makers, and those involved in political action generally, focus primarily on the *creation* of truths. The distinction, of course, is not hard and fast but rather one of relative emphasis. Both the scientific and the policy-making communities recognize the traditional relation of knowledge and power. But in the political world concern with knowledge is clearly tied to the power it will afford its users in their efforts to shape the future. A natural and obvious consequences of this concern is that, insofar as possible, the model should incorporate variables susceptible to control. The information that an ideal model can provide is not an accurate portrait of the future but rather an accurate portrayal of how the future might be changed by using the policy-making tools which are under the policy maker's control.

The evaluative standards that go with the active as opposed to the reflective orientation underlie the choices made by the builders of models for policy analysis. The value of the model to be created resides in the power it will be able to confer upon its users: power to anticipate precisely how their actions can influence those aspects of the future deemed to be significant.

It is a methodological fact of life that no matter how sincerely a model builder may aspire to creating "value-free" models and no matter how scrupulously he endeavors to adhere to that goal, he cannot. His own notions of purpose and value will necessarily enter into all the modeling choices he must make. They do so in many ways—in terms of which phenomena are modeled, in terms of his preference for particular mathematical and/or statistical techniques, and, most obviously (when we examine the social reality of model building), in terms of the ways in which he takes into account the opinions of others involved in the modeling process. Specifically, evaluative judgments enter this process not via the choices of a hypothetical and socially abstracted "model builder" but rather through the much more complicated set of interactive processes wherein a model builder and his colleagues, assistants, clients, superiors, critics, and so forth, jointly evaluate and influence each other's views regarding the model's construction and its content. Of course, this same criticism can be leveled at all science—and most other "objective" human endeavors.

By arguing that evaluation is necessarily a part of environmental modeling we have set forth only a methodological truism to which experienced modelers and critics have long since accommodated themselves. However, a very basic problem about the relationship between evaluation and modeling remains to be solved: namely, how ought models to deal with evaluation. And that is the topic considered next. Before proceeding, however, the reader should recognize that this problem itself is an evaluative one and thus, ultimately, cannot be solved except in relation to some fundamental evaluative assumptions about the proper social role of environmental modeling.

THE STRATEGY OF MODEL BUILDING

In the best of all possible worlds, those who build environmental models would explicitly indicate the physical and social "givens" presumed to be applicable. They would indicate the functional interrelationships taken to describe how the variables behave over time, depending upon one another and upon the pertinent "given" features. Naturally, such models would be built so as to take into account whatever limitations may exist on the builders' and/or users' ability to secure the data required to drive the model. Similarly, such models should also reflect the other limits that stem from the limited resources at the modeler's disposal: for example, his limited financial resources, time, computational capability, and so forth.

The ideal model facilitates the conduct of sensitivity analyses, that is, it permits users to rapidly assess the overall significance of uncertainties in the state of their knowledge of any of the data or relationships that make the model "work." Finally, the characteristic that would confirm the Panglossian assumption that, indeed, we are discussing the best of all possible modeling situations, would be that our model states explicitly and provides us with data on our evaluative criteria. In other words, such a model would tell us not only what the results of our actions might be, but also how good or how bad these results are and according to which standards. This would allow us to identify the exact role played by the evaluative assumptions in the models that we encounter. And, where such explicit evaluative criteria operate in models built for policy analysis, a determination can be made as to how "sensitive" the selection of specific policy options are to the particular value criteria employed.

SIMULATION AND OPTIMIZATION

Depending upon their methodological predilections and the resources they command, model builders adopt one of the two principal strategies relative to societal imposition and usefulness. Both require that a model describing the way in which a system behaves be formulated and that the model portray the forces that shape that behavior. In the first strategy a model is constructed primarily to *simulate** the way in which the real system would behave under conditions of interest to the model's builder and/or users. Where such a model is utilized to search for ways to improve the system's performance, this is accomplished outside the model itself. That is, the model is used to determine what consequences may be produced in various situations and comparisons as to how satisfactory those consequences are reserved for the user.

For example, one might use such a model to estimate the results likely to be realized by the imposition of various air pollution emission standards.

The model might take into account a variety of economic considerations, geographic patterns of population and industry, relationships describing

---

* This use of simulate is equivalent to our use of assess in Chapter 1. (C. H. and J. D.)

weather patterns and the atmospheric diffusion of pollutants, descriptions of the biotic effects and health consequences for humans and other species, the extent to which anti-pollution relations are politically feasible and enforceable, and so forth. Whether such a model is simple or complex or even whether or not it is an accurate representation of reality is beside the point here. Insofar as modeling strategy is concerned, the key feature of simulation models is that their use does not require users to explicitly specify how various individual aspects of the model's results combine in their judgment to form a single measure of value. Stated simply, they need not indicate the precise way in which they appraise how good or how bad each set of consequences may be.

By contrast, a second basic modeling strategy focuses on *optimization*. Its demands upon the model builder and user are stringent, but they extend the promise of more useful results than those directly attainable by simulation alone. The basis of the optimization approach is quite simple. When the exact method of evaluation can be specified, the modeler may be able to design his model so that it indicates how those variables that can be controlled may be set in order to maximize the evaluative function. For example, we might be able to build models that show the most advantageous method of prompting an increase in the numbers of an endangered species or models that show the least expensive combination of policies capable of improving air quality to a given level.

Once a satisfactory evaluation criterion is assumed to apply, the modeler may reasonably seek to draw upon the expanding body of mathematical and/or computational techniques for optimization to make his model produce the kind of information desired (Kaufmann, 1968). Of course, even when an evaluation criterion exists, we may still be incapable of accurately describing a system's behavior in a form amenable to the use of optimization techniques.

It would be misleading to leave the impression that while optimizing models are difficult to create, those that seek merely to simulate are a simple matter. The truth is that each strategy poses different problems to the prospective modeler. To illustrate some of the uses and problems of simulation, let us briefly examine the Energy System Generator (ESYG) model being built at Brookhaven National Laboratory, a modeling effort with which I have had some experience.

The purpose of the ESYG model is to serve as a tool for rapidly assessing the impact of proposed or estimated changes in the way we acquire and use industrial energy. Essentially, the model exists in the form of a computer program and its associated data bank. Operating together, they describe in detail the processes whereby industrial energy is produced, distributed, and used in the United States. Many factors—economic, technological, environmental, and institutional—enter into this description, and, when desired, any or all can be modified to permit a complete evaluation of their overall significance to be made. Thus the model makes it possible to appraise the effect of changes (singly, or in combination) in technology, economics, and so forth, that may happen to take place or that may be made to take place by adjusting appropriate governmental policies.

In its present form the ESYG model provides data describing each of the major flows of industrial energy in the U.S. energy system. This data is indexed in several ways and presented in terms of cumulative totals and individual yearly totals for the entire system and for every major process in that system. Moreover, for each process in each year and for every year in a given "run," data are routinely generated to specify estimated emission of acids, bases, phosphates, nitrates, dissolved solids, suspended solids, BOD, COD, heat, airborne particulates, oxides of nitrogen, sulfur dioxide, hydrocarbons, carbon monoxide, carbon dioxide, aldehydes, radiation and estimates of solid waste discharge, land-use requirements, occupational deaths, occupational injuries, occupational man-days lost, annual operating costs, and the investment costs associated with each process.

In its current form the ESYG model does not attempt to optimize, to solve, or to indicate in any way how improvements may be made in existing or proposed patterns of energy utilization. To be sure, the ultimate purpose of ESYG is to serve as a tool with which to search for such improvements on a systematic basis and to enable the search to be conducted in such a way that our extensive knowledge (limited though it is) can be brought to bear at all stages of this search. Specifically, ESYG is designed to avoid the sacrifice of knowledge that commonly accompanies attempts to squeeze real systems into a form compatible with existent optimization models.

ESYG contains no built-in evaluative standard that would permit it to automatically determine whether particular patterns of energy use are "better" or "worse" than one another or than some arbitrary standard. Its purpose is "merely" to describe in as detailed a manner as possible the complete set of consequences that would result from each pattern under study. Although complicated in terms of the quantity of data that it both uses and generates, the analytic assumptions upon which ESYG is founded are extraordinarily simple. Since its purpose is descriptive rather than prescriptive, ESYG does not induce or compel analysis to make judgments that exceed the bounds of their competence. For example, no assumptions need to be made regarding:

1 Whether or how each of the various environmental consequences of energy flow ought to be combined into a single index or series of indices,
2 Whether or how to combine individual categories of other information—health effects, demand types, financial impacts, and so forth—into overall impact measures,

3 Whether or how to employ discount rate techniques in order to aggregate the financial and nonfinancial effects of energy use that occur at various points of time,
4 Whether or how to construct an "objective function," "welfare function," and so forth, by means of which to measure completely, in one number, the desirability or social worth of all the various consequences of any particular energy use pattern, or
5 Whether or how to take into account the fact that the pattern of energy use is not determined by the federal government or any other single user, but is instead a result of the combined interaction of many individuals and organizations (including governments). Each may play one or more economic roles as a buyer, seller, and/or rule-maker in the energy market and each is characterized by limited capabilities and a concern with the pursuit of interests that are self-defined, often divergent, variable over time, perhaps varying from unit to unit of governmental and industrial organizations, and, in any case, not precisely equatable with some grand public interest that the model "ought" to be optimizing. (To the extent they are discernible, these limits and differences can be realistically portrayed and their implications for policy making assessed.)

Obviously, this realistic orientation is by no means an unalloyed blessing. By striving for a faithful representation of many factors, and particularly the many distinct decision-making components which cause energy use patterns to be what they are, we abandon the comforts of more usual modeling assumptions that substantially but arbitrarily eliminate the need to consider decision-making behavior. Instead, we are forced to employ our best, although admittedly very rough, notions or observations of how the system's many decision makers act.

To put the matter in a nutshell, whichever strategy is selected the model builder must deal with our limited ability to describe how decisions are made, to prescribe how they should be made, or both.

EVALUATION IN GENERAL

To the best of my knowledge, there are no serious objections, in principle, to the creation of optimizing models where the necessary data exist. There is simply no reason to prefer inexplicit evaluation and manual, as opposed to automatic (i.e., computer-assisted), searches for the "best" solution(s) where models are to assist in policy making.

The major obstacle to the use of such optimization models is a practical one, one that hinges upon our ability to deal with evaluation in environmental modeling and modeling in general. For it is only in the most exceptional cases that, in addition to being able to adequately "capture" the pertinent features of a system's behavior within the confines of a descriptive model, we can state a suitable evaluative procedure.

Historically, of course, the problem of determining the particular evaluation metric has been the subject of the most intensive inquiry. Depending on whether our concern is with measuring "good" and "bad" as it may pertain to society, or mankind, or some particular individual or group of individuals, or even if we are concerned with the consequences of our actions upon ourselves alone, or upon our planet, or whatever entity commands our loyalty and dedication, the solution of this problem has been taken to be the province of philosophers, theologians, moralists, economists, ecologists, and all manner of thinkers. Exploration of the evaluative standards that are and/or ought to be used in the making of public policy has always been the central task of political philosophy. Unfortunately though, the answers that political philosophers supply do not solve the evaluative problems of the model builder. For the many standards offered contradict one another and also are formulated in the very imprecise language of politics. No matter how sympathetic a model builder may be to suggestions that public decisions should aim to maximize "justice" or "satisfaction" or "freedom" or the apparently more objective criteria of "equality" and "peace," he will discover that their exact meaning is elusive. They simply defy translation from the general to the specific in whatever particular circumstances his model is meant to address. Moreover, this elusiveness has not gone unnoticed by political philosophers.

Their response has assumed several forms that necessarily concern us. Many, beginning with Plato, maintain that the exact meaning of such ideas as justice is knowable in all circumstances. But he asserts it is knowable only to some and thus cannot be spelled out explicitly in a manner that is simultaneously sufficiently general and sufficiently specific as to enable any ordinary person to evaluate the various conditions that may be pertinent. In place of a precise evaluation metric, Plato and most subsequent political philosophers have adopted an indirect approach by arguing in favor of various constitutional arrangements. These arrangements, then, by establishing *who* is to rule and/or the processes by which that rule is to be exercised, are presumed to ensure that the right standards of evaluation will be used in the making of public policy.

A noteworthy exception is the work of Bentham, a key figure in the development of modern policy analysis. The essence of his prescription is that a single explicit evaluative standard should be used to appraise all public policy options (Bentham, 1961). He identifies that standard in terms of his notion of utility, arguing that public choices should be directed toward promoting the greatest happiness for the greatest number of people. Basically, the notion of utility is defined in relation to individuals and measures the extent to which the selection of particular decision alternatives may either increase an individual's satisfaction or decrease his dissatisfaction.

Bentham's efforts to establish an explicit evaluation metric, although they exert an almost overwhelming influence in the literature of the confluent intellectual disciplines currently subsumed under the rubric "policy science," have not really been successful.

There are many reasons for this failure. First of all, it is simple enough to assert theoretically that individuals act *as if* they evaluate every aspect of each decision alternative in each decision situation in accordance with a personal evaluation function. But determining the exact nature of such a function has exceeded our empirical capabilities. Only in the most artificial and contrived laboratory settings have any satisfactory descriptions of value functions been obtained, but they are at best fragments indicating only how the particular subjects reacted to a very limited range of stimuli in the particular settings studied (Edwards and Tversky, 1967).

No less troubling is the fact that even if we possess complete knowledge of the evaluation criteria, with all their possible complexity and richness of detail used by each individual, we are still in the dark as to how these individual methods of evaluation ought to be combined to form a single "social" evaluation metric. Certainly, an enormous and often contradictory literature exists on this very topic (Goodman, 1954). But no literature can resolve such issues as whether the aggregation of individual preferences ought to proceed via a straightforward arithmetic averaging of individual preferences or whether and how to take into account individual differences in intensity of feelings, social position, education, time perspective, and so forth. This aggregation problem is distinctly political, since the nature of the method used to build the social evaluation criterion will determine who is to get what and when. And, of course, the utilitarian approach has not solved that problem.

On the other hand, the emphasis that the utilitarian concept of public choice gives to using explicit evaluation criteria is of great importance to the model builder intent upon optimizing. For there are many situations where it is possible to reach agreement on the proper evaluation metric or even on some group of possible evaluation metrics appropriate to the systems and circumstances to be modeled.

For example, a model builder may be concerned with situations where evaluation is to be based on minimizing monetary costs or upon maintaining some desired numerical ratio between component subpopulations of a given species. In other cases, more complex evaluative criteria are used, criteria based upon the construction of aggregative indices alleged to portray how value is linked to a combination of individual factors. Examples are concepts like "gross national product," "merit weighted user days," or virtually every index built to inform us about some aspect of the "quality of life." Whether specific or general, these measures have been viewed as appropriate for use as explicit evaluation metrics. As used in various optimizing models these metrics are variously termed "objective functions," "value functions," and sometimes "utility functions." They are, of course, conceived of as mathematical functions that specify the quantity to be either maximized or minimized. Such functions may assume whatever form the model builder believes to be a satisfactory representation of the way in which the distinct aspects of value (the arguments of the function) ought to be combined.

The exact function selected is always of crucial importance, for it does no less than to stipulate exactly how all the "incommensurable" aspects of a decision situation, including all the pertinent personal differences in outlook on the past of those involved are actually commensurate. Obviously, this is a tall order, and even when such a function is created to explicitly portray evaluation in the most limited of instances it will still be vulnerable to criticism.

One avenue of criticism that deserves more attention than it has received by model builders is the evaluative character of assumptions that frequently occur in both simulation and policy-oriented modeling.

For instance, when attempting to treat events that occur at different points in time the notion of a "discount rate" (see Chapter 6) is often used (Zeckhauser and Schaefer, 1968). Although disagreements as to the precise meaning of discount rates exist, there is no disagreement about its basically evaluative nature. Therefore, if information is to be presented on the combined significance of events occurring at different times the modeler must recognize the evaluative character of his choice of a technique for doing so.

A major source of evaluative assumptions that frequently are not recognized as such may be found in the "constraints" presumed to apply in various types of optimization models. Such constraints delimit the range of allowable values that certain of the variables of parameters (individually and in combination) may assume. And where these constraints describe the limitations imposed by laws of nature, they are clearly nonevaluative. However, often these constraints are employed to specify a range of acceptable values. Here "values" takes on both meanings of the term "values," that is, numerical values and evaluative values.

By operating as constraints these evaluative stipulations masquerade as something very different than the evaluative assumptions incorporated into the model's explicit function. In fact, however, they constitute a superior or dominant set of values over and above the others. Consequently, it is essential that when building or using models of this type, all "constraints" be inspected in order to determine if they are legitimate as used or whether they ought to be made a part of the evaluation function.

AGGREGATION

Perhaps the most important aspect of evaluation and its actual role in ecological modeling is the way in which models builders, by neglecting to treat certain phenomena, by accident or design, either hide or obscure features of paramount concern for evaluation and decision making.

For the most part, those difficulties arise because model builders focus upon variables at too high a level of aggregation. The effect of this is to prevent model users from examining distinctions that potentially are significant to *them*. This deficiency is apparent in the Meadows' "limits to growth" model (Meadows, 1972). Such ordinary political details as the existence of sovereign states are omitted, allowing the authors to treat population as a single homogeneous entity. This forecloses from the field of study issues related to the viability of a world in which inequalities in resource distribution are maintained by the existence of those political structures. If one wished to base policy making on the Meadows' simulation it would be necessary to consider their egalitarian distributive presupposition to determine both its accuracy and/or its suitability.

Management models that are holistic inhibit users from considering policy alternatives that utilize existing diversities. Thus it frequently happens that a user's concern with environmental models does not, and should not, depend upon grand totals but rather upon patterns of distribution. For example, a model that permits the estimation of the total yearly emissions of various air pollutants cannot enable us to study improvement strategies based upon altering the location and timing of those emissions. This, of course, is particularly important where our concern stems from worry about the impact of those emissions upon humans or other life forms and not from worry about those emissions per se.

THE SOCIAL FACTOR

Because of the relatively more advanced state of our knowledge about the behavior of "natural" as opposed to "social" systems, model builders have concentrated upon the former. A frustrating consequence is that while models may provide information on the effects produced by the modeled system, they rarely undertake to identify upon whom the effects impact. Thus, for instance, one rarely encounters a model designed to inform its user as to which groups will bear the costs and/or receive the benefits portrayed by the model. Whereas it might be convenient for the model builder to think of costs and benefits as accruing to society as a whole, the plain fact is that most people care a great deal about the social pattern in which effects are distributed. In the terminology of welfare economics, people and, therefore, policy makers are concerned with issues of equity as well as issues of efficiency. When we must evaluate the relative merit of alternative government policies our attention is directed toward their influence upon the pattern of who gets what and when they get it. As we examine prospective policies aimed at environmental protection or at increasing the amount of energy available or at any goal whatever, it is reasonable and proper that we try to evaluate how fair those policies are in terms of their relative impact upon different income groups, ethnic groups, urban, suburban, and rural populations, different age groups, the disabled as opposed to the able-bodied, present as opposed to future generations, and so on.

As noted, the principal reason why most of our models do not permit these kinds of detailed assessments to be made is that social scientists cannot adequately describe the causal factors that presently determine and in the future will determine the differential impact of various policies upon each of the groups and interests considered to be important. However, this is not the only reason. Others become apparent as we examine briefly the relationship between modeling and evaluation as they actually occur in the political process.

MODELING AND THE POLITICAL PROCESS

We have already indicated that the concern of policy makers for effectively controlling events can influence a model builder's choices. One example of how the political process may operate to discourage modelers from creating models that are capable of being used to appraise sophisticated strategies, strategies that exploit existing distributional variations, is tied to beliefs about the institutions of public administration. For instance, suppose the model builder believes either that we would be unable to enforce a selective prohibition against the use of certain types of pollution-producing passenger vehicles in urban areas or he believes that to do so would not be desirable. Then he may be inclined to construct a model that allows us to study the consequences of across-the-board prohibitions or limitations at the point of manufacture of new vehicles as is currently being done. But unless he is prepared in advance to provide for it, such a model will be most unlikely to enable its users to examine strategies that impose the penalties of higher initial cost per vehicle and of reduced fuel economy only upon those vehicles that actually are used in urban areas.

A related but more difficult problem of model inadequacy, one that revolves about the issues of evaluation and administration, has recently been raised by industrial opponents of various environmental protection standards. It arises out of the fact that, administratively, regulations cannot be enforced against "the environment" but only against specific individuals and organizations responsible for their own actions that affect the environment. Now presumably what policy makers and people in general *really* care about are the ambient levels of the various polluting substances, since it is these ambient levels that constitute or cause the physical consequences against which we seek protection. Thus, in the view of some, legislation and the establishment of regulatory standards should be phrased in terms of specific measures of ambient substances. The idea is that no restrictions on behavior ought to be imposed so long as the ambient standards are not violated. This argument is one that I find persuasive in the abstract, but only in the abstract. In practice, the adoption of this position would force governments to rely upon the dubious deterrent effect of totally unenforceable protective standards or perhaps when these standards are violated, trying to stop all the activities that produce the offending substance. If administrative realities are to be considered together with the other physical and social realities that we represent in our models, then they must be built to reflect the existence of evaluation that depends upon the various individual sources of emissions as well as, or even instead of, their aggregate impact. Such models may be used to anticipate the kinds and effects of enforcement decisions that might have to be made in enforcing the range of prospective regulatory standards we wish to analyze. They obviously would have to be disaggregated in many respects. And they would have to take into explicit account a prominent shortcoming in the state-of-the-art of model building for policy analysis, namely, our frequent reliance upon surrogate standards of evaluation when we are unable to accurately measure those qualities that we "truly" value. Among the sources of this shortcoming is the time pressure that pervades political decision making (see Chapter 8). Simply put, it arises out of the overriding need to make decisions at the time they must be made regardless of whether or not the models used in predecision analysis accord a satisfactory treatment to evaluation!

Surely one of the most problematic and complex ways in which political considerations influence how evaluation is treated in our models is linked to the importance of agreements in political life. Virtually all effective political action depends upon the reaching of agreements. Indeed, it is difficult to conceive of anyone engaged in political activity who does not constantly try to secure the support of others. The significance of this fact for model builders is revealed in Lindblom's (1964) prescription for rationality in political choice: "In collective decision making do not try to clarify values if the parties concerned can agree on policies." Eloquent testimony to the functional value of imprecision is readily discernible in the ordinary vocabulary of political discourse. And it ought not to be imagined that the overriding political interest in building agreements can somehow be excluded from the analytic tasks that affect decision making. For they are inevitably subject to the same "political" pressures. As we shall presently make clear, these pressures may originate in many ways and reflect the intents and perceptions of prospective users and/or of the model builders themselves. The direction in which these pressures operate most often corresponds with Lindblom's advice. They serve to obscure rather than to clarify the evaluative assumptions at play. As noted, the most usual way in which this comes about is via their submergence within highly aggregate indices. The essential thing is that disputes and disagreements are thought to be less likely to arise among those engaged in the process of policy formation when the exact character of the values that enter that process through the use of models, or any other way, cannot be ascertained.

WHOSE VALUES?

Although some may be reluctant to admit it, the users and builders of models are engaged, both directly and indirectly, in the political processes by which a society's decisions are made. Ultimately, how evaluation is treated in our models depends far less upon the repertoire of analytic techniques at our disposal than upon the more fundamental set of views that model builders hold about the purpose of modeling, about how public policy should be made, and about their own role in policy making. For example, the extent to which a model builder strives to make his own value premises clear, to make explicit the evaluative inputs required to use his model in a policy-making context, to create a model that portrays the differential impact of outcomes for various social groups, and (if it is to be used for policy analysis) the extent to which he tries to build it to be used for estimating the impact of sophisticated strategies may vary markedly in accordance with his conception of the political process. Model builders may aim for a kind of professional detachment or choose to pursue an advocacy role. They may see themselves as the servants of those who rule or those who pay for their work or as the servants of particular groups or causes. Some perhaps see themselves as uniquely enlightened and qualified to specify the values that should govern societal choices. Still others may see their modeling activity as an instrument to enable broad and informed public participation in policy making on the most complex issues, and as an instrument to make the premises of decision making explicit.

Clearly, there is not one best way to deal with evaluation in modeling any more than there is any one right set of values to use. Even such seemingly innocuous recommendations as the injunction to be explicit in the treatment of evaluation may conflict with the aspirations of the modeler to be politically effective.

The observations offered here are far from being a how-to-do-it guide to the prospective model builder. Instead they are aimed at providing him an appreciation of complexities that must be faced. The art of modeling is an arduous enterprise and ought not to be entered upon in the mistaken belief that the treatment of evaluation is a routine matter or that we currently possess a methodological arsenal equal to the task. Unfortunately, the building of environmental models depends upon the creation of individual solutions tailor-made to specific situations; solutions that are made up of a series of compromises between our desires for a truly satisfactory way of handling evaluation and the limitations of our intellectual tools.

EDITORS' NOTE

The next two chapters, one by an economist, the other by an ecologist, develop evaluative procedures that, although imperfect, are useful and are being used in the construction of optimization models. The limitations of such evaluative procedures as used in modeling are no greater than the limitations of such procedures in *any* decision-making process, and, at a minimum, it may be easier to point out the incorrect, biased or inexplicit assumptions in an optimization model than in our general political decision-making process.

REFERENCES

Bentham, J. 1961. *An introduction to the principles of morals and legislation.* Doubleday, New York.

Edwards, W., and A. Twersky (Eds.). 1967. *Decision making.* Penguin, New York.

Goodman, L. 1954. On methods of amalgamation. In M. R. Thrall, D. Coombs, and R. Davis (Eds.), *Decision processes.* Wiley, New York. Pp. 49–59.

Kaufmann, A. 1968. *The science of decision-making.* McGraw-Hill, New York.

Lindblom, C. 1964. Some limitations on rationality. In C. Friedrich (Ed.), *Rational decision.* Atherton, New York.

Luce, R. D., and H. Raiffa. 1957. *Games and decisions.* Wiley, New York.

Meadows, D. H., P. L. Meadows, J. Randers, and W. W. Behrens III. 1972. *The limits to growth.* Universe Books, New York.

Zeckhauser, R., and E. Schaefer. 1968. Public policy and normative economic theory. In R. Bauer and K. Gergen (Eds.), *The study of policy formation.* Free Press, New York. Pp. 27–102.

6
Economic Values and Natural Ecosystems

MICHAEL R. DOHAN

M any ecosystem models are intended to show policymakers how natural ecosystems might be better used for man's well-being. These models, however, often fall short of this goal because of the "metric problem." That is, the model's variables are not or can not be interpreted in terms of criteria widely accepted in society as measures of "better or worse." This chapter introduces to the noneconomist both the basic reasoning and the methods for using economic values as one such criteria to compare the social benefits and social costs of natural ecosystems.* The first part of this chapter describes the theoretical relationship between economic values and man's well-being.[†] The second part shows how economic values are determined in a market and why the market often fails to establish values for natural ecosystems. The third part introduces the basic economic techniques by which economic values of natural ecosystems can be estimated and used in social decision making. The last part indicates the theoretical and philosophical boundaries to such an approach when used as a basis for deciding among alternative policies of ecosystem management.

The Problem

In the absence of a common measure or "unit of account," environmental scientists frequently evaluate the social significance of their findings in subjective nonquantitative terms [e.g., Woodwell's (1974) attitude that any destruction of ecosystems is harmful], or in terms of "objective criteria" such as energy units that in fact do not have a consistent quantitative relationship to man's traditional concepts of well-being and are relatively remote to most

* The editor's constraints on this chapter are four. It should (1) be easily understood by the noneconomist, (2) identify and explain the key economic concepts about the determination of economic values and their relationship to man's well-being, (3) illustrate how cost-benefit analysis might be used to evaluate social losses and gains from man's impact on ecosystems, and (4) be short. This chapter is to be a "bridge" from ecology into economics. For those preferring a more advanced treatment of cost-benefit analysis, its application to environmental management, and extensive references, see Prest and Turvey (1965), Mishan (1971), Kneese and Bower (1972), Kneese (1970), Steiner (1969), Chase (1968), and Herfindahl and Kneese (1974). Two excellent introductory texts on "environmental economics" are Seneca and Taussig (1974) and Hines (1973). Key articles are collected in Dorfman and Dorfman (1969).

[†] Economic analysis of environmental problems to date has focused on the management of *residuals* or pollutants as they directly affect man and his immediate environment (Kneese and Bower, 1972). Economists have not devoted much attention to the contributions of *natural ecosystems* to man's well-being. Focusing ecological analysis on "man's well-being" is an important assumption because the approach specifically rejects any value system concerned with maximizing variables independent of their relationship to man's welfare. This assumption is not as restrictive as it appears. If for reasons not readily comprehensible to others, some individuals value "nature for nature's sake" or "feel for the plight of the blue whale," such feelings must be entered into our decisions. For ultimately *costs are perceptions of something lost*. See Passmore (1974) for an outstanding analysis of social attitudes towards nature.

134

people's way of thought (Odum, 1971, and in this volume*).

Consider the very real social decision problem of choosing between two competing uses of wetlands—retaining the wetlands in their natural state or filling them for construction. In such a situation ecologists might suggest that ecological stability, the impact on flora and fauna, and natural energy flows be important considerations. For society in general, however, these criteria are usually "means to ends" and not "ends" themselves. As such, they can not be compared directly to conventional indicators of man's well-being (such as health, leisure, and net economic output). For these reasons, the ecologists' criteria fall outside the conventional framework within which people compare and make choices. Such criteria, especially when they can be attained only at the cost of these other social goals, remain difficult to integrate directly into social decision making. The assumption here is that in a pluralistic society social policy strives not to achieve remote goals, "scientifically-determined" by this or that group of wise men, but rather to maximize man's overall well-being in a manner understood and perceived by most individuals in that society.[†]

---

* Odum's provocative paper (Chapter 7) goes beyond the trivial "energy is necessary for man's well-being" to show the relationship between energy flows, value, and money. Odum proposes that the value of goods, service, and money is determined by the energy flow (adjusted for quality) associated with them. If we seek in the concept of value a measure for the *relative* contribution to man's well-being by different goods and service produced by both the economy and ecosystems, then this proposed energy theory of value appears to me to have deficiencies common to any theory of value, such as Ricardo's labor theory of value, based on a single "objective" criteria. Such measures of relative value fail to consider (1) man's subjective evaluation of the relative importance of additional units of these goods and services and (2) the contributory role of other factors—labor, skills, natural endowments, technological progress, economic organization—that determine the production costs for various goods and services. Thus examining the energy flows of goods and services can give only rough prediction of their social value.

In particular, the relationship between money, its value (as determined by the price level), and energy units depends on the relationship between energy units and the quantity of outputs. This relationship is not linear (Schurr, 1972, pp. 182–193), so that the suggested policy to stabilize the value of money by limiting growth of money to the growth of energy might result in either inflation or deflation.

[†] This does *not* say that scientists should refrain from pointing out the ecological consequences of man's economic activity in terms of man's well-being, from demonstrating the inconsistencies among societal values, and from suggesting new social values as well as new policies to enhance social well-being. In truth, changes in values can be a fruitful approach to increasing man's sense of well-being. But in a democratic pluralistic society a scientist's views must and should compete with other value-forming institutions and even with views of other scientists, or we run the risk of Lysenkoism. The resulting decisions may not be the "best decisions" in some grander view of man's well-being—how often has the Church complained about this problem—but these are the decisions that are going to be made in Western society. The problem at hand is not to revolutionize the world but to move values and policy making toward a more rational use of natural ecosystems (Passmore, 1974).

With these assumptions and purposes in mind, I pose the question: How and to what extent can monetary values be used to evaluate and integrate the benefits and costs of natural ecosystems (and their disruption) into a decision-making framework so that the resulting choice among activities affecting ecosystems can be judged "better" or "worse" according to reasonable criteria widely accepted by society (or at least by those with decision-making power). Indeed, the National Environmental Policy Act of 1969 requires that all federal agencies

develop procedures which will insure that presently unquantified environmental amenities and values may be given appropriate consideration in decision-making along with economic and technical considerations [and] utilize a systematic interdisciplinary approach which will insure the integrated use of the natural and social sciences . . . in decision-making which may have an impact on man's environment. National Environmental Policy Act of 1969, Title I, Sec. 102.

What contribution can be made by economics to this task?

Economic theory concerns the principles, methods, and social institutions for allocating scarce resources among alternative uses for man's well-being. *Natural ecosystems are scarce resources.* In a modern industrial society, however, social institutions such as the market system often fail to establish "economic values" for natural ecosystems and to allocate them to their socially optimal uses. Why are economic values important, and how do they relate to the well-being of man?

AN ECONOMIC THEORY OF SOCIAL VALUE

To justify use of monetary value for evaluating social benefits and losses from natural ecosystems, we must first understand how money values relate to the well-being of individuals and society.

Individual Well-being

The basic method of economics is the analysis of trade-offs or *opportunity costs.* That is, the comparison of *what must be foregone (the losses) in order to obtain a desired end (the gains)* (Boulding, 1967). But how are such gains and losses of well-being for individuals to be measured and compared?

An individual's over all well-being, called *utility,* can be analyzed in terms of "states of being," which together give him a *greater or lesser sense of well-being* (see Boulding, 1966; Becker, 1971; Lancaster, 1966). Being well fed and housed, healthy, free to speak, having opportunities, good personal relationships, pleasant social and physical environments—all these component states of being and more enter into an individual's utility function (Mack, 1974). A *utility function* describes the relationship between an individual's well-being and his use of goods (food) and services (doctor) and experiencing certain conditions (clean air). Several comments about utility functions are in order.

First, the increase in an individual's well-being obtained from each additional unit of a particular good or service becomes smaller and smaller as the individual has or experiences more and more of the good, service, or condition per period of time. This important characteristic, called the law of *diminishing marginal utility*, is valid for life-supporting goods such as water and bread and for amenities such as scenic environments as well as for run-of-the-mill commercial goods such as TV sets and motor cars. Second, individuals are assumed to be the best judge of their own interests in most cases (more on this later). And third, this particular interpretation of the "utility function" stresses that man's well-being is determined only in part by the flow of *private goods* and by "economic goods" (i.e., by the goods and services provided by economic activity). An individual's well-being depends to a great extent on goods, services, and conditions that are provided or exist independently of an individual's abilities to provide goods, services, or money in exchange for them (see Mack, 1974, and Boulding and other essays in Hook, 1967). Some, such as public education, roads, the justice system, and religious services, require "economic resources" of labor, capital, and land, but are provided to individuals by governments and other organizations at no fee or at nominal fees.* [The provision of such goods is the subject of a new field called "grants economics" (Boulding, 1973).] Others, such as oxygen to breath, the climate, and scenic amenities exist independently of, and some would say in spite of, the economic system. This occurs either for technical and natural reasons or through widespread consensus embodied in custom or law. For these reasons decisions about the provision of many economic goods and services are made by social institutions other than the market system described below (Haefele, 1974). For example, the provision of public education is determined by voting in a community. This reflects a widespread belief that the distribution of some goods, services, and maintenance of certain social conditions *should not* be subjected routinely to the individual's "ability to pay" or similar tests of "economic rationality," for they provide social benefits that are difficult,

---

* Enjoyment of such goods, services, and conditions is not completely independent of income, because their enjoyment often requires complementary inputs of leisure, tastes, and private goods such as cars, lawyers, expensive vacations, and good housing location.

undesirable, or irrelevant to evaluate in terms of money (Mishan, 1971; Boulding, 1967; Tribe, 1972; Mack, 1974). This is so even though some individuals could be induced to pay for these goods. Some benefits and costs of natural ecosystems may fall into this noneconomic category, but as we indicate in the following, natural ecosystems also affect the economic aspects of man's well-being and should be considered in an economic decision-making framework.

An individual confronted with numerous alternative actions is assumed to choose the combination of actions that he believes will maximize his own well-being.* In choosing among mutually exclusive alternatives he compares the gains and losses of each, and then chooses the alternative that he feels subjectively gives him the largest net gain in well-being. For example, in the economic theory of consumer behavior, a "rational" person spends his income on the available goods such that the net increase in his well-being (called *marginal utility of income*) from the last or marginal dollar spent on each good becomes equal.† (For example, if the marginal utility per dollar spent on the last unit of fish purchased by the individual is greater than the marginal utility per dollar spent on the last unit of books purchased, then this individual could increase his well-being by buying fewer books and more fish.) Although most comparisons of losses and gains are ultimately made in subjective terms, that is, feelings rather than in money terms, money prices are often an intermediate step in the process of choice. This is because *money prices reflect the trade-offs available* to the buyer in the market. If fish sell for $4, and books sell for $2, then other things being equal, the buyer can obtain one fish by giving up ("not purchasing") two books and, of course, vice versa. The buyer then can ask himself, "Is one more fish worth two less books?," and act accordingly. Assuming that consumers behave in this manner, then the money price of each good provides an indicator of the *relative* importance to individuals of additional quantities of this good when compared with additional quantities of other goods and services.

---

* An individual's limited knowledge about products, his limited ability to reason, and his poor foresight in fact prevent him from making choices that he himself would regard as "optimal" under perfect conditions. Furthermore, some options, such as "clean air," can be chosen only on a collective basis. This is because an individual's actions alone may have only a small adverse impact on his environment and society. Thus he can ignore it. But when summed over many individuals such effects may be very harmful to all. Thus a consumer need be not only ecologically informed but also compelled to make decisions collectively in the proper institutional framework.

† Formally a consumer will distribute his income among all alternative purchases until $MU_x/P_x = \cdots = MU_n/P_n$, where MU is the extra satisfaction from the last unit bought (marginal utility) and P is the price.

From Individual Well-being to Social Welfare

A conventional assumption of modern economics—and that of this chapter—is that a society's well-being is made up only of the well-being of those individuals comprising that society (Mishan, 1969). This implies on one hand that society is not an entity existing apart from its component individuals, and on the other that the well-being of other entities in the universe are not considered independently of their relationship to man's well-being (Boulding, 1966; Passmore, 1974). If decision-makers are instructed to make choices that increase social welfare, that is, the well-being of their society, then they need criteria with which to judge "better or worse." Their two major problems are (1) to determine how individuals "feel" about the outcomes of the alternative actions and (2) how to "add up" or compare the "social value" of these individuals' feelings (say, of back packers and of timber users). That is, the decision makers must usually make "interpersonal utility comparisons." But there is one apparent exception!

Economists generally assume that, in the absence of *envy*, if a proposed action makes at least one person better off without making anybody worse off, then such action is usually viewed as increasing social welfare and should be undertaken (Herfindahl, 1974). This criterion of better and worse is called the *Pareto criterion* (Bator, 1957). Actions that meet this criterion are usually thought to avoid the problem of interpersonal utility comparisons and hence avoid the problem of determining "social value."* Due to the problems of obtaining information about such possibilities, however, practical opportunities that directly meet the Pareto criterion are few and far between.

To find a way around this impasse to policy making, the related compensation criterion was developed. This states that if gainers from an action can compensate the losers, and still be better off, then the action is a *"potential Pareto improvement"* and should be undertaken, given that an acceptable income distribution is established (Mishan, 1969, 1971).† Acts of voluntary exchange between two individuals can be seen as routine realizations of "potential Pareto improvements." Given the initial distribution of commodities and resources among individuals, any voluntary exchange of goods is

---

* To be at Pareto optimum certain production and exchange conditions must be fulfilled (called Pareto conditions). These conditions are described in Bator (1967). Being at a Pareto optimum may coincide, however, with a bad distribution of "well-being" among individuals and hence may not coincide with a "social welfare optimum" (Mishan, 1969).
† Although the compensation criteria leads to ambiguous results under some assumptions concerning income distribution and relative prices (Mishan, 1969, pp. 37–69), Mishan concludes that it still forms a reliable basis for decision making as long as the changes in prices and income distribution are relatively small and acceptable (Mishan, 1971, pp. 316–321).

possible only if it leaves at least one person better off while leaving the other person no worse off.*

Establishing Money Values

How are money values determined for goods and services, and in what sense do they reflect man's well-being? Let us define *private goods* as goods, services, or resources whose gains accrue entirely to the buyer and whose losses (opportunity costs) are borne entirely by the seller. In Western market economies such goods are usually allocated, produced, and distributed according to the principle of voluntary exchange among individuals and producing firms. That is, each participant in the market economy evaluates the gains and losses from any exchange as indicated by his willingness to enter into voluntary exchange carried out under certain conditions. This social institution of voluntary exchange has the further virtue that the decision makers' problem of making interpersonal utility comparisons for each specific good has been converted into a more general problem of income and wealth distribution.

This process of voluntary exchange of goods and services is facilitated by another social institution, *money*, which can best be defined as a *generalized claim by its possessor on economic goods and services.* Thus, instead of direct exchange of goods for goods, the institution of money simplifies the process of exchange by enabling the sellers to accept money from the buyers of their goods or resources and then to give the money to other sellers for goods or resources they desire to buy. Because money functions as a common unit of account, each seller and each buyer of a good need only express a single "condition" of exchange, namely, the money price. This single price expresses how much of the other economic goods and services *in general* the seller or buyer is willing to accept or forego in exchange for a specific good. People need only compare money prices of various goods to find out the "barter ratios or trade-off" at which goods, labor, and other services can be exchanged

---

* This conclusion assumes satisfactory distribution of initial resources and wealth, acceptable rules of exchange, competition, and use of market power, and the absence of unsanctioned coercion to force a "voluntary exchange." In some circumstance voluntary exchange may be judged by *nonparticipants* as not enhancing society's well-being. Objections are directed at (1) the initial conditions leading to the terms of the exchange, (2) the distribution of the gains of trade, or (3) the social nature of the objects of exchange. For example, the exchange of some goods—such as the sale of slaves and babies or the purchase of justice—is usually judged by most individuals in society to be immoral. This dependence of an individual's well-being on the social conditions (including his feelings about the behavior and well-being of others) is not at all reflected in the "price system" but rather in laws and customs. It has always been a difficult area of social policymaking (Boulding and others in Hook, 1967; Mishan, 1971, pp. 307–315).

for each other in the market. Remember that it is the *relative* money prices that are important for decision making. How are money prices determined in a market, and how do they relate to social welfare?

Economic Value and Supply and Demand

In a competitive market, with many buyers and sellers, money prices for a private good are determined by the interaction of supply and demand. A *demand curve* shows the relationship between the prices of a good and the total quantity demanded by a buyer at each of those prices during a given period of time. The market-demand curve is the sum of each buyer's demand curve. In Figure 1, the market demand curve *dd* shows that buyers are willing

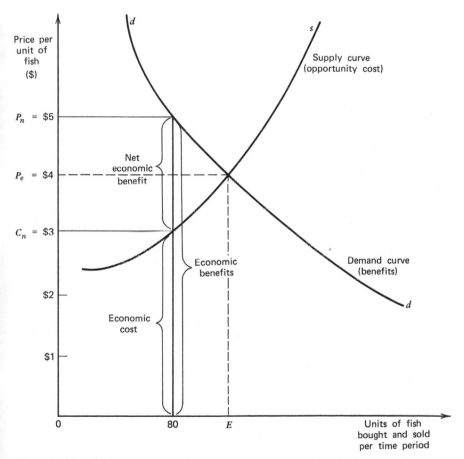

Figure 1   Hypothetical supply and demand curves for fish and economic costs and benefits.

to buy up to 80 fish at the price of $5 per fish (of a specific type, quantity, weight, say 10 lb of tuna). Why? Because the added satisfaction (well-being) obtained by buyers from each fish up to the 80th fish was greater than or equal to the added satisfaction that the buyers could get from any bundle of other goods that could be purchased for $5, the cost of each fish. The price $5 then is a monetary indicator of the *added* satisfaction, or the value that the buyers obtained from the 80th fish relative to the satisfaction from additional other goods purchasable with $5.

The downward slope of the demand curve reflects the *law of diminishing marginal utility* stated earlier. The 81st fish does not add quite as much to the buyer's well-being as the 80th fish; thus buyers are only willing to forego a smaller quantity of "other goods" to get one more fish (the 81st). Hence they would buy the 81st fish only at a lower price, say $4.80, than that of the 80th fish. The demand curve for most beneficial functions of natural ecosystems also reflects the law of diminishing utility. Thus if wetlands are abundant, the value to society of another acre of wetlands is small, but as wetlands become scarcer the value of keeping another acre as wetlands becomes larger and larger.

A supply curve for a good shows the relationship between its selling price and the total quantity that sellers (under perfect competition) are willing to supply at each of those prices during any given period of time (given the initial supply of goods and resources, preferences, technical knowledge, and the natural conditions of production). In Figure 1 sellers are willing to supply 80 fish at the price $3 because the (opportunity) cost to the sellers of supplying each unit of fish up through the 80th fish was less than or equal to $3 of other goods and resources. The price $3 is a dollar measure of the *opportunity cost* of supplying the 80th fish. This is because either the satisfaction lost by the seller giving up the 80th fish just equals the added satisfaction to him from $3 worth of other goods, or the seller must spend $3 to acquire the added resources required to produce the 80th fish. In this case, $3 is the *marginal cost*, that is, added monetary cost of producing another unit of fish. The marginal cost represents in monetary terms the opportunity cost of using the additional resources to produce the 80th unit of fish rather than using them to produce other goods for which people would have paid $3. In either case, monetary costs ultimately reflect satisfaction foregone by individuals in society. If some losses of satisfaction to individuals in society are not reflected in the seller's costs, for example, from poorer quality environment, then this private monetary cost will inadequately reflect *social* marginal costs (more on this point later).

An upward slope in the supply curve indicates an increasing opportunity cost of supplying this product to the buyers. It may rise either because the (marginal) satisfaction lost by each seller from giving up another fish increases as their supply becomes smaller, or because the marginal cost of producing more fish increased.* In some industries the marginal costs of added output are constant or may even decline as output rises.

In what way is society's well-being increased by producing the 80th fish? In Figure 1 buyers get a good for which they would willingly forego the consumption of $5 worth of other goods. But the sellers of the 80th fish only had to give up a good or use resources capable of producing other goods for which people were willing to pay only $3. Thus the *net social benefit* of producing or exchanging the 80th fish can be measured as the difference between P_n and C_n, where P_n is the value of other goods that individuals (buyers) in society were *willing to forego* to get the 80th fish, and C_n is the opportunity cost, or the value of other goods that individuals (sellers) in society in fact *had to forego* to get the 80th fish (Figure 1). In our case, P_n is $5, C_n is $3, and the net social benefit of producing the 80th fish is $2. One aim of socioeconomic policy is to undertake policies that *maximize this net social benefit*. This is called the *efficiency objective*. The above conclusions, however, rest on the assumptions that (1) buyers receive all the benefits, (2) the sellers bear all the costs, (3) that society agrees that such goods should be produced for and distributed to those who are willing to pay, and (4) benefits and costs are counted as being of equal social importance regardless on whom they may fall (a key ethical judgment).

Optimal Level of Output and Use

How much fish should be produced in a society? The quantity produced and exchanged should be expanded as long as the price buyers are willing to pay for another fish exceeds the marginal (opportunity) cost of another fish. Fish output should be expanded until the buying price has fallen to and has become just equal to the marginal cost of supplying another unit (P_e or $4 in Figure 1). This quantity, E, is a socially optimal level of output. To produce less would be to forego net social benefits obtainable by using resources for catching fish. To produce more would be to incur net social costs, because the value of the additional fish is less than the value of other goods obtainable from the same resources. Here we have a version of the general rule for optimal resource allocation. Namely, resources should be allocated to an activity up to the point where the social marginal benefits of another unit of resources in this activity is just equal to the social marginal benefits obtainable by using these same resources elsewhere.

A competitive market system (under very restrictive assumptions) will attain this optimal level of output. At this output level the price is called

---

* Two other conditions are necessary: (1) total benefits of an activity must exceed total costs and (2) social marginal benefit should be declining relative to social marginal cost. For a more rigorous statement see Herfindahl and Kneese (1974) or Seneca and Taussig (1974).

the market clearing price (P_e) because the quantity demanded equals the quantity supplied. At this equilibrium of supply and demand the market price measures both the benefits and the opportunity costs of providing this unit of good in terms of other economic goods (evaluated at their respective prices). Thus the social value of *small* changes in quantity supplied, other things being equal, is closely approximated by the market price times the small change in quantity. This property of market prices is useful for cost-benefit analysis. The common reality, however, is imperfect competition of few sellers and large corporations. Here the observed market price is higher than the marginal cost; the price, however, still reflects the relative value of the last units purchased by the buyers. Many environmentalists and some economists would argue, however, that in modern Western economies consumers are no longer sovereign, for their tastes are distorted by the corporate advertising and trendy fads. Why should such wants have claims on man's scarce natural resources? (See Boulding, 1966; Doyle, 1968; Swagler, 1975; Hansen, 1972, for consumer behavior and advertising.)

Market Value, Social Value, and Consumer Surplus

The market value of the total supply of a specific good such as clams in a market is simply the market price times the quantity. After some thought it is obvious that the market value of this *total* supply always underestimates its total importance to society [if we measure total social value by the quantity (value) of other goods individuals would give up to get the total supply of clams on an all-or-nothing basis]. For as seen above, the market equilibrium price, P_e, measures only the marginal benefits of the last unit bought. If confronted in this one case with the choice of buying all the supply or getting nothing, buyers would have been willing to pay much more. In fact, in the hypothetical clam market in Figure 2, they would have been willing to pay an amount represented by the entire area ($AOEB$ or $630) underneath the demand curve between zero and 70 bushels of clams. For they were willing to pay $12 for the first unit . . . and $9.43 for the 30th unit and . . . and $6 for the 70th unit. This difference of $210 between the total social value ($630) and market value ($420) is called *consumer surplus*, because it represents the additional amount consumers would have been willing to pay for exactly 70 units of clams, but which they did not have to pay.* For some essential types of goods (water, food, etc.) the prices that

---

* Strictly speaking, they in fact might buy a little less if confronted with paying for the entire area under the demand curve (an income effect). See Mishan (1971, pp. 325–344) for a technical discussion of consumer surplus. Similarly, we can define the area between the supply curve and the market value as "producer's surplus," because they are willing to supply the entire Qe goods for less than the market value they received.

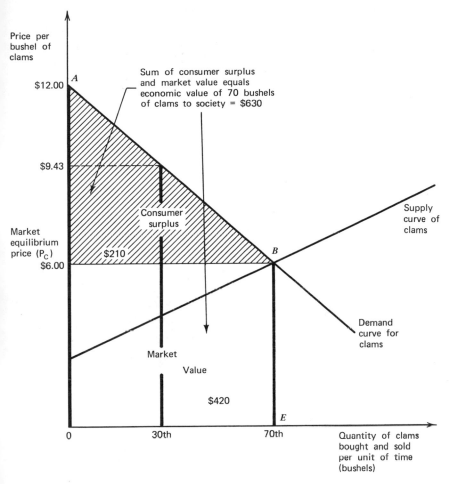

Price per bushel of clams

$12.00 — A

Sum of consumer surplus and market value equals economic value of 70 bushels of clams to society = $630

$9.43

Supply curve of clams

Consumer surplus

Market equilibrium price (P_C)
$6.00

$210

B

Demand curve for clams

Market

Value

$420

E

0

30th

70th

Quantity of clams bought and sold per unit of time (bushels)

$12 is the value of the 1st bushel of clams to buyers.

$9.43 is the value of the 30th bushel of clams to buyers.

$6.00 is the value of the 70th bushel of clams to buyers.

All clams are sold to buyers at market equilibrium price.

Figure 2 Market value, social value, and consumer surplus.

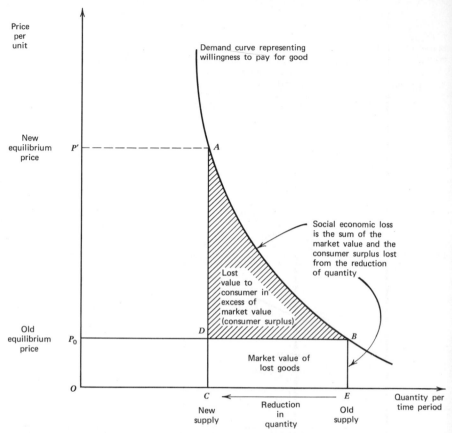

Figure 3 Measuring the social loss from large reductions in economic goods or environmental amenities.

buyers would pay for the first units necessary to sustain life are so high as to be undefined in monetary terms (as shown in Figure 1), so that total social value and total consumer surplus can not be defined even though the economic value of the marginal unit supplied in equilibrium is well defined (and may in fact be quite low, e.g., water).*

These distinctions among *total value*, *market value*, and *marginal value* are also of crucial importance in assessing the social importance of ecological change. In particular, it warns environmentalists away from the earlier mistake of parlaying ecological despoilation of small parts of an otherwise large

---

* With extreme scarcity (water in a desert) other methods of distribution such as physical force or nomadic tradition often replace the market mechanism.

area into an inevitable threat to our entire life-support system. Society is quite willing to sacrifice some benefits of ecosystems to provide themselves with economic goods of even greater benefit (Boulding, 1966).

Some of man's adverse impacts on ecosystems–DDT, acid rain, loss of wetlands, hunting of whales, deforestation, overgrazing, and so forth—have progressed far beyond the "marginal," however. For such large losses of environmental amenities, as seen in Figure 3, the social value of each additional unit of amenity lost is no longer closely approximated by the old equilibrium market price P_o, for, as the amenity becomes scarcer, society values each remaining (marginal) unit more and more. Thus, other things being equal, when society suffers large reduction in the quantity of amenities (or any good) we must estimate the loss in social benefits as the loss in market value plus the loss in consumer surplus. In Figure 3 this is the entire area under the demand curve between old quantity "E" and the new lesser quantity "C" (the area $ABCE$) instead of just the smaller market value (the area $DBCE$). This concept of consumer surplus (or willingness to pay) forms the theoretical basis for evaluating the economic value of goods that the market system and social system fail to evaluate properly or at all in monetary values.

ECOSYSTEMS AND THE FAILURE OF THE MARKET

Why the Market Fails

The market mechanism of supply and demand is at its best when evaluating and allocating the private goods mentioned above. In many circumstances, however, the market mechanism fails to accurately reveal the social benefits and costs of a good or a resource and to allocate them to their best uses (Bator, 1958; Steiner, 1969). The reasons traditionally cited for the "market failure" are (1) incomplete specification of property rights, (2) collective enjoyment of goods, (3) transaction costs, (4) option demand, and (5) imperfect knowledge or ignorance. (We define these terms in the following discussion.) All play a role in the frequent failure of the market system to establish prices that properly reflect benefits of ecosystems to man's well-being. That is, the low or zero market prices for these ecological benefits exist not because people are not willing to give up goods in order to enjoy these benefits; on the contrary, they often would pay a good sum. But due to the physical characteristics of the goods and the absence of the necessary social institutions, payment is not usually required and in some cases not even possible.

Incomplete specification of property rights often refers to the absence of ownership of a scarce or limited resource such as ground water or the fish in the ocean. This absence of ownership occurs by law, tradition, or is due to the transaction costs of determining and enforcing these rights. Such resources are called *common property* and can be exploited (captured) by anybody at zero or nominal charge (Gordon, 1954). These are usually overused, and since the benefits can be "harvested" by anybody, too little investment is made to expand or maintain common property resources and to allocate them optimally over time (Demsetz, 1967; Pejovich, 1972). Natural ecosystems or their components (such as ocean fisheries, birds, and insects) frequently have no recognized owner, and in the absence of legislation their use tends to be ignored by the market system with the above noted consequences. The overhunting of the blue whale provides a classic example of overexploitation (Clark, 1973). (For more on the optimal management of such common property resources see Herfindahl and Kneese, 1974; Institute of Animal Resource Ecology, 1970).

Often only some part of a resource's benefits have not been assigned property rights. For example, nobody owns the aesthetic benefits of an attractive building or water-retention benefits from forested land. This absence of property rights is often attributable to the "collective good" character and the high "transaction costs" for such benefits (Dales, 1972).

Collective goods (also called *public goods* or *social goods*) are the goods, services, or "conditions" for which the enjoyment by one individual does not preclude simultaneous enjoyment of these very same goods, and so forth, by other individuals (Samuelson, 1954; Steiner, 1969). They are enjoyed collectively. (Contrast this to private goods such as food that can not be consumed by more than one person.) As it is clear from the examples, there are *zero opportunity costs* of another person enjoying the good, service, or condition, given that it has already been supplied for one individual. In fact it is often costly to prevent other people from enjoying such goods once provided. Classic examples are lighthouses and defense. Modern examples are "knowledge," clean air, natural scenery, and the preservation of many species as well as favorable social and natural environmental conditions. Such a good should be provided up to the point where the added social benefit of another unit equals its added social cost. What is the *social benefit* of a collective good? In contrast to private goods, the value of one more unit of collective good to society is not just the amount that a single individual would pay for this unit but rather the amount that all individuals together would be willing to pay to have another unit made available for all to enjoy. The (private) market mechanism, however, will undervalue these collective goods and hence undersupply them, because once a collective good is provided for one individual, there is no need for others to pay in order to enjoy it and they probably will not pay voluntarily. This latter phenomenon is called the "free rider" problem (see Baumol, 1965; Steiner, 1969 for basic references). For this reason collective goods are often provided by the government or quasipublic agencies.

Even if the property rights to all benefits of a resource or good were completely specified (owned), the owner usually has the costs of identifying beneficiaries, measuring benefits, and then collecting compensation. Such costs of enforcing property rights are called *transaction costs* (Steiner, 1969). The higher the transaction costs, the lower the net benefit of exercising one's property rights. Thus for resources such as wetlands whose benefits are realizable to their owner only at high transaction costs, it is likely that the market prices of these resources understates their economic value to society.

Transaction costs may be high because of physical characteristics of the resource's benefits. For example, if the benefits are small, diffuse, difficult to contain, or enjoyed indirectly at some distance from the resource by large numbers of unknown individuals, then it may be costly to harvest, to identify, or to exclude others from directly or indirectly enjoying these benefits. The detritus, habitat, and other benefits of wetland vegetation are examples. Despite their obvious social value such benefits have neither markets nor market values, and they are treated by society as collective goods or common property (Steiner, 1969). Many benefits of ecosystems have all the above characteristics. For example, detritus from wetlands is a major contributor to the food chain of fisheries; these in turn are exploited by commercial and sport fishermen. Even if the rights to the food chain were assigned to the owner of the wetlands, the cost of enforcing such property rights would be prohibitively expensive and impossible for an individual to bear.

How and for whom the property rights are specified can also affect transaction costs. In some cases, reassignment or collectivization of property rights may reduce costs (Mishan, 1971). For example, collectivization of wetland property rights under a government or semipublic agency might permit a fee to be collected from all commercial fisheries and be paid out to owners of wetlands (Dales, 1968).

An Example of Market Failure: Estuaries and Wetlands

The extensive ecological contributions of the natural wetlands to social and economic well-being are well established and are illustrated in Figure 4 (Fiske, 1969; Fiske et al., 1967; Douglas and Stroud, 1971; Jerome et al., 1968; Gosselink et al., 1973; see also in this volume Chapters 10, 19, 21, and others). These benefits include (1) a basic link in the food chain for fisheries (*Spartina* grasses, the detritus derived from *Spartina*, algae, and phytoplankton), (2) habitat for marine life of value both directly to man and as interrelated parts of the food chain, (3) habitat both for permanent and transient wildlife (waterfowl, mammals, birds) of recreational, aesthetic, and possible scientific

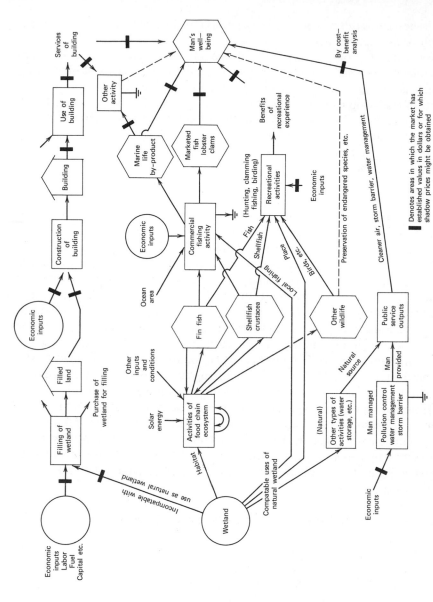

Figure 4 Two alternative uses of wetlands to man and their economic values. Symbols adapted from Chapter 2.

importance, (4) storage and release mechanism for water, (5) a source of plant nutrients for coastal waters, (6) processor and assimilator of civilization's waste, and (7) a buffer against storms. Wetlands also provide multiple recreational benefits (boating, fishing, hunting, clamming, and nature study) and aesthetic pleasures. These benefits are all provided simultaneously; a wetland in its natural state is a *multiple benefit resource* whose real benefits are diffuse, difficult to quantify, and without market prices commensurate with its social value.

A wetland filled for construction is a single benefit resource whose social benefit is readily identifiable as a place to build with a market price reflecting this benefit. If we use the wetlands for construction, however, we lose all the services of the natural wetlands; these ecosystem losses are the social opportunity costs of filling the wetlands. Figure 4 illustrates these two alternatives.

From the earlier discussion it is clear that if an area of wetland were marketed to private buyers under the constraint that it be held in its natural state, it would fetch a sum far less than would be obtained if its ecological benefits were actually being paid for by those who were benefiting from the natural wetland. Here the market mechanism of supply and demand fails in two ways. First, the market mechanism *fails to generate information* about the social value of natural wetlands. Second, even if the social benefits of the wetlands were well known, the market system usually *fails to provide incentives* to allocate the wetlands to the uses providing the greatest benefit to society, for the "private benefits" of owning natural wetlands are less than the social benefits. A typical owner of natural wetlands is rarely able to realize or to obtain its entire benefits either by direct use, by "harvesting" and sale of the benefits (as a farmer does corn) or by charging a fee to beneficiaries of the wetlands for the services rendered by his particular ecosystem. Why? First, the transaction costs of "selling" the many benefits of the wetland are prohibitive (imagine a market for mud algae?). Second, many benefits are collective goods (clean air, landscapes, water management). Third, social institutions (property rights) usually do not exist for paying the wetland's owner for such benefits [as a trout club might pay to a stream owner for fishing (Dales, 1968)]. Indeed, the tradition is often that of free access to benefits of natural ecosystems (common property of ocean fisheries). Under these circumstances many of the real benefits to society of holding a wetland in its natural state will not be taken into account in the owner's decision-making about the use of the wetland.* They remain outside of and external

* Developing an "ecological conscience" in the owner (the goal of many ecologists) does not help much. For if the owner were himself an avid naturalist, fisherman, and so forth, the estimate of the total social benefits now must include the owner's great personal enjoyment plus those benefits of the wetlands enjoyed by others. Thus even an owner's high personal evaluation fails to reflect (signal) completely the value of that wetland to society.

to the decision-making process of the wetland's owner and hence are called *externalities*. They are also called external economies, spillovers, or neighborhood effects. (The literature on externalities is voluminous—see classic article by Coase, 1960, and references and discussion in Steiner, 1969; Mishan, 1969, 1971).

Externalities and Resource Use

Externalities are real benefits and costs to society, but since they are not considered in the decision-making process, suboptimal decisions about the use of such resources are often made in the sense that some other decision would yield a potential Pareto improvement. Thus an activity with external costs usually will be undertaken on too large a scale (e.g., pesticide use by farmers), and an activity with external benefits on too small a scale (e.g., water-pollution control by private firms and municipalities). By "too small a scale" we mean that the social marginal benefit still exceeds the social marginal cost of the last resources devoted to this activity. Resources tend to be underallocated to uses whose benefits are primarily external compared to uses whose benefits are almost completely internalized (i.e., reflected in the market price). Thus there is likely to be excessive filling of natural wetlands because the private economic gains of filling the wetlands for construction usually reflect all the possible social benefits from that use. But the private economic gain of retaining wetlands in their natural state is almost always less than the social benefits of natural wetlands. This usually leads to a *suboptimal* decision. For in this case, society may be better off leaving the wetlands unfilled and unbuilt on. Such a conclusion can not be reached without measuring the economic value of benefits sacrificed to filling the marsh.

EVALUATING THE BENEFITS OF AN ECOSYSTEM

Cost-Benefit Analysis

The economist's usual method of integrating external effects into decision-making is cost-benefit analysis (Prest and Turvey, 1965; Mishan, 1971; Kneese and Bower, 1972; Steiner, 1969; Chase, 1966; Herfindahl and Kneese, 1974; Seneca and Taussig, 1973; Hines, 1973; Dorfman, 1965; Eckstein, 1958). Cost-benefit analysis is basically a method of (1) systematically evaluating (usually in monetary terms) all the private and external benefits and costs of alternative resource-use projects and (2) choosing the projects that yield the greatest net benefit (that is, benefits minus costs) after taking into account that today's society gives a lower value to benefits and costs occurring in the future than to those occurring today (for reasons discussed further in this chapter). Thus cost-benefit analysis differs from private profit-and-loss calculations primarily because it incorporates estimated economic (monetary) values for the external gains and losses of a "project" and for resources for which no values or wrong values have emerged in the market (McKean, 1958, 1966). If done correctly, the money value of all benefits—including the external benefits of a project—is the amount that individuals would be willing to pay to obtain these benefits if confronted with the choice. The money value of all the losses would be the minimum amount of compensation that individuals would willingly accept in exchange for having the losses imposed upon them (Mishan, 1971). Using this comprehensive monetization of benefits and costs and the concept of potential Pareto improvment, the criteria usually applied to cost-benefit decisions is the *efficiency objective* to maximize the difference between willingness to pay and the cost of providing it, expressed in monetary terms without regard to distribution of benefits and costs or to other objectives (Herfindahl and Kneese, 1974). To do so, the project analyst uses rather straightforward decision rules for choosing those projects that yield the largest sum of net benefits, evaluated in terms of current economic values and given the constraints placed on the decision-maker. In reality, of course, there are multiple objectives in social decision making—income distribution, national defense, social stability, and cultural development—that do not fit neatly into this calculus of value. Thus choices based strictly on benefit-cost analysis may not always coincide with broader concepts of social benefits. (We return to this problem later.)

The basic methodology is simple even if its application becomes complex. There are essentially four basic steps to cost-benefit analysis: (1) identification of alternative projects including the "null" alternative of doing nothing, (2) quantification of all of the benefits and costs from each project, (3) finding "shadow prices" and determining the economic value of benefits and costs, and (4) choice of projects (McKean, 1958). We describe briefly these steps as they might be applied to our problem of choosing between two mutually exclusive uses of wetlands.

Quantification of Gains and Losses

An ecological-economic cost-benefit analysis must be based on quantitative models (such as those described in this book) of the natural ecosystem showing the relationship between the set of man-managed activities, the outputs of the ecosystem, and the quantitative effect on man and productivity of his activities. [See Isard's (1972) pioneering effort to evaluate systematically the economic benefits of a wetland using input-output methods.] The basic guideline here is that the socially relevant output of the modeled ecosystem be completely enumerated and quantified. If they are not, the resulting cost-benefit analysis will be biased (but still quite useful in policy decisions). For our purposes assume that we have modeled successfully the relationship between loss of a given area of natural wetlands and the reduction in the productivity of commercial finfish fisheries, recreational opportunities, its public service functions (Hall, 1975), and so forth.

How does one assign money values to such gains and losses? This aspect of cost-benefit analysis remains in part an art (McKean, 1958). There is, however, one basic guideline to be observed. Given the efficiency objective, the concept of social value to be used in cost-benefit analysis is the theory of value developed in the first part of this chapter: namely, that the economic value is determined by the benefits as evaluated by the "buyers" and opportunity costs as evaluated by the "sellers."*

Finding Shadow Prices

There are several approaches to estimating the economic value of an externality (McKean, 1966; Steiner, 1969; Mishan, 1971; Herfindahl and Kneese, 1974; Krutilla and Fisher, 1975). One method is to trace through the effect of an externality until it affects activities of society that have economic value or economic (opportunity) cost associated with them. Such "interfaces" are identified in Figure 4. Negative impacts might be the economic value of output lost from reduced productivity of resources used in an activity (e.g., fisheries), from the loss of resources (e.g., by erosion or flood), or from the diversion of resources from other productive uses to offset damages (e.g., to restore "conditions" such as health or clean air). Positive impacts can often be evaluated in a similar manner. For example, the air cleansing capacity of vegetation permits smaller expenditures on air pollution controls. Natural predators reduce the need for pesticides. Maunder (1970) used this approach to estimate the value of weather patterns. Note that such shadow prices are often cost oriented and based on the monetary cost of added resources required or saved by the effects of the externality. These costs may fall short of, or exceed the amount that "individuals" might be willing to pay to enjoy the external benefits, or to avoid the external costs.

A second method to estimate the economic value of an externality is to use the market-based values of close substitutes or proxies as the "shadow prices." For example, the differences in property value of otherwise similar properties are often interpreted as reflecting the (present discounted) economic value (and hence as measuring the social importance) of external costs associated with different levels of air pollution, airport noise, and similar disamenities

---

* There are other theories of economic value such as the labor theory of value (Wolfson, 1966; Robinson, 1963).

in an area (Freeman, 1974). The value of a new recreation area often is estimated by using differential transportation costs as the shadow price for the price that people would pay to visit a recreation area (Clawson and Knetsch, 1966; Levenson, 1971; Herfindahl and Kneese, 1974). A third method is the questionaire approach of asking individuals about "how much they would pay for" a benefit, even though there are serious theoretical problems concerning the accuracy of the answers (Randal et al., 1974). All else failing, an investigator may simply use hypothetical or "contingency prices" to find what values of shadow prices would be necessary to change the decisions and then evaluate the reasonableness of such contingency prices (Mishan, 1971). In fact, good practice suggests performing a similar type of *sensitivity analysis* to find out how sensitive the benefit-cost results are to changes in the model's parameters (McKean, 1958).

Shadow Prices for Wetland Ecology: the Fisheries

Filling a given area of wetland reduces the nursery area for young fish and the food supply to offshore fisheries (see Chapter 16); this reduces the quantity and size of fishes, which in turn reduces the productivity of the "economic" resources used in commercial fishing. This lost flow of ecosystem output from the wetlands "into" the commercial fisheries has, therefore, an economic value because (individuals in) society would be willing to "pay" (i.e., forego other goods) to avoid this loss in productivity. But how much?

Applying our first method, we would conclude that a minimum estimate of the annual loss to the fisheries is the market value of the reduction in the annual fish catch that would have resulted from filling this wetland while still using the same resources in the fishing industry. Society lost goods for which it had been willing to forego at least this value in other goods, yet there was no corresponding reduction in resources used for fishing. For large losses in wetland areas, the economic value of the lost production will be even larger than its market value, estimated at the initial price, because the economic value of fishes rises as they become scarcer (as shown in Figure 2).*

---

* More technically, where production function, $F = f(L_f, B_f, R_f, W)$, relates the quantity of fish (F) produced by using resources L_f, B_f, R_f (given the initial stock of fish and the previous rate of fishing) and given the wetlands, W, then the lost productivity of the fisheries may be technically described as $\Delta F/\Delta W$. When less plentiful, the value of the (marginal) fish is larger. The initial impact of reduced productivity of fisheries is higher costs and prices per fish. Some buyers drop out of the market. If demand is very price elastic, demand for fish will drop sharply as they attempt to raise prices. In this case, the fishing industry may withdraw resources because costs cannot be shifted entirely to the consumer but must be suffered by the resources working in the fishing industry. But if demand is inelastic so that prices rise by more than enough to cover the costs of the market factors of production, additional economic resources may be attracted to fishing. In either case, it can be shown that the loss is greater than the market value (Institute of Animal Resource Ecology, 1970; Herfindahl and Kneese, 1974).

Cost and Benefits Over Time

The filling of the wetlands, however, is not just a 1-year affair. It has a permanent impact on the fisheries resulting in this annual loss year after year "forever" and has a permanent gain of "forever" using the land for building. This raises the problem of how to compare these two streams of gains and losses occurring over long periods of time. One method is to calculate and compare the current economic value of the future gains and losses for each alternative. We describe this method next.

How does a buyer decide how much to pay today in one lump sum for a "property," such as this wetland, that yields its benefits to the buyer only over a long period of time. [Most books on investment decision such as Van Horne (1974) describe the techniques outlined here. An excellent analysis of the use of natural resources over time is Herfindahl and Kneese (1974).]

For illustration assume that net economic benefits from a given use of specific plot of land are reflected in net rental payments of $1000 per year, which are expected to continue for 100 years. For simplicity we also assume that the land has zero economic value at the end of that period. (It soon becomes clear that it makes little difference what value we give to land at such a distant date.) How much money could a buyer pay today for this land and still come out as well as if he had invested the same sum of money in his best alternative investment opportunity? Assume that his best alternative investment yields 10% interest per year over a similar period of time (e.g., $10 per $100 invested per year). Then the buyer needs to find out what sum of money he would have to invest today in his best alternative investment (which, remember, yields 10%) to get exactly the same stream of payments as the rental payments of $1000 per year from the land for 100 years. This sum of money is called the *presented discounted value* (PDV) or *capitalized value* of a stream of income, and it is $9999.17 in this case. If the buyer can buy the land for less than this amount of its PDV, then the buyer should buy the land, because he will get a higher return per dollar of investment in this land than in his best alternative investment (because the $1000 per year is being earned with a smaller investment in the land than the $9999.17 needed in the next best alternative). The difference between the cost of an investment and its PDV is its *net presented discounted value*.

The usual decision criteria in cost-benefit analysis is to choose the combination of projects that have the largest net present discounted value. This yields the stream of net economic benefits judged to be most valuable from the perspective of those individuals who can influence the allocation of current resources among various capital and consumer goods through their participation in the economy (see Mishan, 1971).

The Rationale of Discounting Future Benefits

Why might it be possible to get a yield of 10% per year? One reason is *net productivity of "capital."* This simply means that if we take a bundle of resources that could have produced consumption goods valued at $100 today, we could organize them in such a manner (e.g., into a machine or building) so as to produce goods and services valued at more than $100 tomorrow (net of the cost of associated inputs and in constant prices). That is, by foregoing the consumption of goods today and using the released resources for capital investment, we can increase the possible consumption of goods in the future by an amount greater than foregone today.

A second reason is that consumers have a *positive rate of time preference.* This means that consumers, when given a choice of having a bundle of goods and services today or having these same goods and services 1 year later, will choose now rather than later, all other things being equal. In fact, they are willing to pay "interest" for this time preference, that is, accept a slightly smaller bundle now compared with the bundle of goods later.

As we see in the following, these two phenomena provide the philosophical basis for discounting (i.e., assigning a lower value to economic benefits and costs occurring in the future when making decisions today). It can be shown that under some rather strict assumptions the interest rate reflects both the net economic productivity of the last capital project undertaken (where remaining investment opportunities have lower yields) and the positive rate of time preference of consumers for economic consumer goods (Samuelson, 1974; Mishan, 1971). Hence the market rate of interest is often used as the social rate of discount (note the qualifications discussed in the following).

Calculating Present Discounted Value (PDV)

To continue our example, our prospective buyer of land needs to know what sum of money invested at 10% annual yield or interest would grow to $1000 at the end of one year. It can be calculated that $909 invested at 10% interest gives $91 interest payment, so that the principal plus interest is worth $1000 at the end of the year (approximate figures). The $909 is the *present discounted value* of $1000 payable 1 year from today if it is discounted back to the present at the current interest rate (yield on alternative investments) of 10% per year: hence the term "rate of discount." Similarly the PDV of $1000 two years from today is $826, three years from today is $751, and so forth.*

---

* The formula for figuring out the present discounted value of any sum, R, receivable at time, t, where the interest rate is i is the following: $PDV = R(t)/(1 + i)^t$. Because the PDV invested at i% per year grows to be R by the end of t years: $R(t) = PDV \times (1 + i)^t$.

As a benefit occurs further in the future, its PDV becomes smaller! With a rate of discount of 10%, this $1000 of benefits occurring 20 years in the future is worth only $143.60 today, 50 years in the future is worth $8.50, and 100 years in the future is worth only 7¢ today. One economic and social justification for discounting of future benefits is simply this: By investing resources of these lesser values today into capital projects with a net productivity of 10% per year, we would obtain the same $1000 of benefits in the future (e.g., $8.50 invested today at 10% compound interest grows to $1000 in 50 years). Thus it would be inefficient for society to consider this $1000 of future benefits any more valuable today than the cost of obtaining future benefits of equal value through capital investment.

PDV of a Stream of Payments

These rental payment are like an annuity; that is, they are to be made year after year, not just once at some future date. Estimating the PDV of an annuity such as the continuous stream of benefits of $1000 per year from land for 100 years when the rate of discount is 10% is simply the sum of the PDV's of $1000 occurring in each year.

$$PDV = \$909 + \$826 + \cdots + \$0.07$$

$$PDV = \frac{\$1000}{(1 + 0.1)} + \frac{\$1000}{(1 + 0.1)^2} + \cdots + \frac{\$1000}{(1 + 0.1)^{100}}$$

This series sums to $9999.17, which is the present discounted value of the entire stream of $1000 annual rentals for 100 years.* This amount is the most a buyer could pay for the land and not earn any less on this investment than using these same funds in his best alternative project that yields 10% each year for 100 years. Present discounted value calculations really are rather straightforward—the underlying assumptions are more difficult.

Some Problems with PDV

The estimate of the present discounted value is affected by the economic values (shadow prices) given to the stream of benefits and costs, choice of

---

* More generally, to obtain the PDV of a given stream of returns, $R(1)$, $R(2)$, ... $R(N)$, where the variable $t = 1, 2, 3, \ldots n$ is time and i is the rate of discount, we simply sum the PDV of each of the Rs. $PDV = R(1)/(1 + i) + \cdots + R(n)/(1 + i)^n = \sum_{t=1}^{n} R(t)/(1 + i)^t$, and where $R(t)$ are all the same in each year, this simplifies to $PDV = R \sum_{n=1}^{n} 1/(1 + i)$, where $\sum_{t=1}^{n} 1/(1 + i)^t$ is called the present discounted value of an annuity of $1 payable for n years where the interest rate is i%. In our case with an i of 10% the present discounted value of $1.00 for 3 years is $2.49, for 20 years is $8.51, for 50 years is $9.91, and for 100 years is $9.99. Note that extending our time horizon for $n = 50$ to $n = 100$ adds very little to the PDV of that stream of income.

socially relevant time horizons, and the choice of appropriate interest rates (called the social rate of discount). Since these questions are discussed thoroughly in the literature, their effect on cost-benefit analysis is simply indicated (Herfindahl's and Kneese's survey (1974) is excellent; see also Steiner, 1969; Mishan, 1971).

The effect of the shadow prices on PDV is obvious: higher prices for benefits increase PDV; lower prices reduce PDV. A very important problem in the analysis of ecosystems is that shadow prices based on today's prices may not reflect future scarcities. For example, if in the future the demand for recreational opportunities rises faster than other goods because of larger population, changing tastes, and higher real incomes, while the supply of recreational opportunities diminishes due to their use for other purposes, then recreational amenities have become scarcer relative to other goods, and their (market clearing) price will be higher relative to other goods. In such cases the structure of relative prices today no longer provides a guide to relative scarcities in the future. The proper correction is to adjust upward the shadow price of those goods that are expected to become relatively scarce in the future; the practical problem is "how much." Then the PDV would reflect this shift in the relative value of benefits and costs over time (Krutilla et al., 1972).

The choice of the social discount rate greatly affects the PDV. A higher social rate of discount results in a smaller PDV for any given benefit in any given year and implies that today's society gives a relatively low value to future benefits.

The above discussion suggests that extending the time horizon of a project beyond a certain point has little effect on the total PDV of its stream of benefits. This is because the discounting of far distant benefits reduces their PDV today to insignificant levels. Higher discount rates automatically shortens the time horizon relevant to decision making. For example, one future dollar is reduced to less than a nickel of PDV in about 100 years at 3%, but in only 32 years at 10%. Extending the time horizon beyond these dates, therefore, has little effect on PDV and hence the outcome of the analysis.* Many cost-benefit analyses of ecosystems are particularly sensitive to the choice of the discount rate, for ecosystem benefits characteristically occur over long periods of time.

The issue of the "correct" social rate of discount for a society has been debated for many years by economists. Some argue that the interest rate for long-term government borrowing (usually low) reflects the current society's rate of time preference. Others state that the rate of return on private

---

* For at a 10% rate of discount, the PDV of $1 paid each year for 50 years is $9.91, paid for 100 years $99.99, and paid for 1000 years about $10.00.

investment (usually much higher) reflects the current productivity of capital and hence is the appropriate rate of discount for today's decisions. The practitioner of cost-benefit analysis should acquaint himself well with these and other issues in the choice of the social rate of discount. [Mishan (1971), Herfindahl and Kneese (1974), and Fisher and Krutilla (1974) cite the relevant literature.]

Noneconomists often question the very validity of discounting future values of ecosystems (often to its detriment). Their argument is that future values from the ecosystems, which we are leaving to the coming generations, should be valued equally with the near-term economic goods that we might obtain by their sacrifice. The reply is that if "economic efficiency" is accepted as the sole decision-making criteria for allocating these resources, then discounted future economic value is the proper approach (Herfindahl and Kneese, 1974). The discounting issue actually conceals the more fundamental issue: namely, to what extent should or can economic values be the sole criteria used to decide on the use and fate of such resources. This question is considered in the following.

PDV of Lost Fishery Productivity

Estimating the present value of (avoiding) the reduction in the productivity of commercial fisheries requires a rather straightforward calculation using the annual money value of the lost productivity of the fisheries as projected into the future, the appropriate discount rate, and a time horizon of "infinity." If the annual loss were $100 per acre of filled wetlands per year and the social rate of discount 6%, then the PDV of this stream of foregone benefits is $1662. This analysis can be repeated for many benefits of natural wetlands. What might be possible shadow prices for other ecosystems benefits?

PDV of Other Wetland Benefits

Figure 4 shows some relationships between the benefit of wetlands and man's well-being for which economic value has been or could be estimated. The economic values of recreational benefits based on wetlands have been estimated in several ways. For example, the benefits of hunting, fishing, clamming, and boating have been measured by (1) user's fees and (2) equivalent commercial value of the recreational catch of fishes, clams, and so forth (Fiske et al. 1967, 1969; Gosselink, 1973). The general recreational benefits have been estimated by the willingness to incur transportation costs, questionaires, and use of reasonable hypothetical values for visitor days of different types (Texas Water Study; US Senate Document 97; Clawson and Knetsch, 1971; Mack and Myers, 1965). Differential property values be-

tween otherwise similar homes close to and increasingly distant from wetlands may also be viewed as representing the present discounted value of
the aesthetic and recreational gains of living near the wetlands (Levenson,
1971).

Shadow prices for the public service benefits of wetlands such as cleaning
air, storm moderation, waste processing, and water retention are more
difficult to determine, even if physically measured, because the market does
not directly value such benefits. The key is found in the concept of *public
opportunity costs*. In many instances society is already employing economic
resources to enhance these environmental amenities (e.g., expenditures on
air and water pollution control). Even where expenditures are not being
made now, the loss of natural ecosystems could reduce environmental
quality sufficiently to force public agencies to spend economic resources
(the "industrial equivalent") to replace in part the public-service functions
once provided by natural ecosystems (Hall, 1975).

What is the value of a public service amenity such as water-waste processing that would be lost by filling the wetlands? Figure 5 shows two hypothetical total-cost curves for the economic resources required to achieve
different levels of environmental quality (EQ) such as water quality. The
lower curve, TC_0, shows the total cost given an initial endowment of natural
ecosystems that also process wastes and improve water quality. The upper
curve, TC', reflects the increase in total cost after disrupting the natural
ecosystem and its processing of waste. The total benefit of EQ is the curve,
TB.* In Figure 5 the public agency has chosen the environmental quality
level, EQ_0. Perhaps they chose this level because they believe that it maximizes net social benefit from EQ. This implies that they believe that the
social marginal benefits (SMB) of moving to even higher EQ are outweighed
by the social marginal costs (SMC). (At EQ_0, SMB just equals SMC.) In
reality, of course, factors other than the efficiency objective effect the choice
of EQ by public agencies and legislatures.

After public service functions of ecosystems are lost, any given level of
environmental quality (EQ) now requires more economic resources. At
these higher costs public agencies may choose a lower level (EQ_e), because
the added benefit regained by restoring EQ all the way back to the original
level (EQ_0) is likely to be less than the added costs. This lower level is optimal
in the sense that under the new circumstances no other level could now yield
larger net benefits—but remember the new EQ_0 is a less satisfying level.

* The discussion of demand curves suggests that since the first units of water are vital to life
itself, these first units cannot be given a meaningful economic value. Hence we do not know
the exact value (height) of the *total* benefit curve in Figure 5. But at some larger quantity we
become able to measure the social *marginal* benefit of *added* water quality even if we cannot
measure its total benefit. Hence the interruption in the vertical scale in Figure 5.

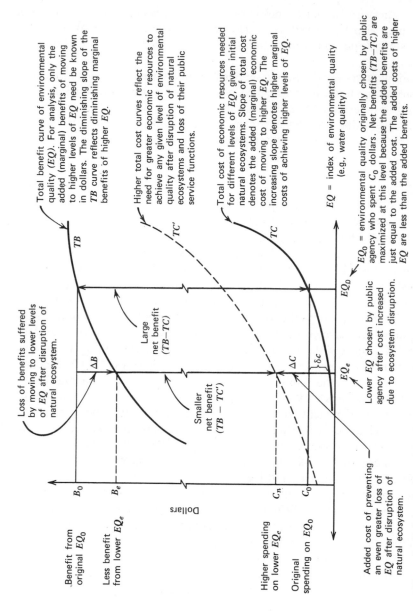

Total benefit curve of environmental quality (*EQ*). For analysis, only the added (marginal) benefits of moving to higher levels of *EQ* need be known in dollars. The diminishing slope of the *TB* curve reflects diminishing marginal benefits of higher *EQ*.

Higher total cost curves reflect the need for greater economic resources to achieve any given level of environmental quality after disruption of natural ecosystems and loss of their public service functions.

Total cost of economic resources needed for different levels of *EQ*, given initial natural ecosystems. Slope of total cost denotes the added (marginal) economic cost of moving to higher *EQ*. The increasing slope denotes higher marginal costs of achieving higher levels of *EQ*.

EQ = index of environmental quality (e.g., water quality)

EQ_0 = environmental quality originally chosen by public agency who spent C_0 dollars. Net benefits $(TB-TC)$ are maximized at this level because the added benefits are just equal to the added cost. The added costs of higher EQ are less than the added benefits.

Loss of benefits suffered by moving to lower levels of *EQ* after disruption of natural ecosystem.

Benefit from original EQ_0

Less benefit from lower EQ_e

Higher spending on EQ_e

Original spending on EQ_0

Added cost of preventing an even greater loss of *EQ* after disruption of natural ecosystem.

Lower *EQ* chosen by public agency after cost increased due to ecosystem disruption.

Dollars

TB

TC'

TC

Large net benefit $(TB-TC)$

Smaller net benefit $(TB-TC')$

B_0

B_e

C_n

C_0

ΔB

ΔC

δc

EQ_e

EQ_0

EQ_D

Figure 5 Valuing the loss of public service amenities due to the disruption of an ecosystem.

162

This loss of benefits can be estimated conservatively as the original economic cost (denoted by δc in Figure 5) of raising EQ from EQ_e to EQ_0 prior to the loss of the ecosystem. But this underestimates the loss, because the social benefit (denoted by ΔB) of moving initially from EQ_e to EQ_0 exceeded the social cost (ΔC) of doing so.

We now can see how to estimate the social value of the lost public resource function. The loss to society in this case is the sum of (1) the added economic resources (ΔC in Figure 5) used to partially restore environmental quality after the loss of the ecosystem and (2) the loss of benefits (ΔB in Figure 5) suffered in moving down to a lower less satisfying level of EQ. This evaluation is consistent with the fundamental principles of cost-benefit analysis; namely, values should be based on the beneficiaries' willingness to pay. In Figure 5 society would have been willing to pay as much as $\Delta C + \Delta B$ for this public service function so as to avoid both the loss of benefits from lower EQ and the higher economic costs necessary to keep EQ from falling further.

Common Mistakes

Several aspects of determining economic values are often overlooked by practitioners of cost-benefit analysis. First, the same benefits or costs must not be counted twice (e.g., nutrient value of marsh grasses *and* the value of the fishes). Second, only the net value attributable to an ecosystem should be counted. The error of using gross economic values is often made in evaluating the benefits of ecosystems. (See, e.g., Jerome et al., 1968; Fiske et al., 1967, 1969; Gosselink et al. 1973). For example, the economic value derived from the wetlands from clamming is not the gross market value of the clams but rather the gross value minus the cost of the boats, labor, fuel, and so forth, used for clamming. For these resources could be used elsewhere in the economy. As is often the case, however, the labor used in clamming these wetlands is not likely to be employed elsewhere owing to its lack of other skills or mobility, its recreational nature, or widespread unemployment. In this situation the social opportunity cost of using this labor for clamming is much lower, and hence the net value derived from the wetlands by clamming is much higher. This reflects still another important principle of cost-benefit analysis; if a resource is otherwise unemployed, then the social opportunity cost of using that resource is lower (Havemen and Krutilla, 1968).

Third, one should not attempt to assign monetary values to benefits that are distantly related to economic aspects of life and without close economic substitutes (Herfindahl and Kneese, 1974; Mishan, 1971; Stein, 1969; McKean, 1958). It is in this area that public policies often run the risk of being biased by the values implicit in cost-benefit analysis. One such bias is the *excessive* reduction (aggregation) of vastly different qualities into a single measure (Tribe, 1972). Instead, such nonmonetizable benefits and costs should be boldly listed along with the monetizable benefits and costs (and not dismissed in a footnote as is often the practice). Fourth, if the project causes large changes in the affected variables (e.g., losing half the local recreational area), initial market prices will no longer suffice; the loss (or gain) in consumer surplus also must be estimated (as described above). Fifth, the distribution of the benefits and costs among various groups should be described. In the absence of any compensation to the losers by the gainers (e.g., to the clammers by the builders), the decision makers may decide to reject an otherwise desirable project because of its undesirable impact on the distribution of income or noneconomic effects among the people. [Herfindahl and Kneese (1974) summarize this literature. See also Freeman (1967), Weisbrod (1969), Mishan (1971), Chapter 5.] Finally, the assumptions about the "assignment of property rights" to the benefit and costs should be made explicit. This should be done because the amount an individual is willing to pay for a benefit, being limited by income, can differ greatly from the amount he must be paid in order to voluntarily forego having the benefit [see discussion of consumer surplus and laws in Mishan (1971]. Thus birdwatchers may not be able to pay much to prevent the destruction of their favorite marsh, but they may have to be paid much more to surrender their bird-watching rights at the marsh.

Behind the apparent simplicity of many cost-benefit analyses lurk complex assumptions!

Decision on Wetland Use

The hoped-for product at the completion of the economic evaluation is a table listing the PDV of the monetized benefits and costs of the wetlands (1) in their natural state and (2) when filled for construction. A hypothetical table is presented in Table 1. This table is then used to compute the net present discounted value (benefits minus costs) of each alternative use of the wetlands. Using the decision rule described above, if the PDV of retaining the wetlands in their natural state exceeds the PDV of filling the wetlands, then the decision maker should choose to retain wetlands in their natural state. In our hypothetical example in Table 1, use of the wetlands for construction yielded benefits with a PDV of $10,000 and no nonmonetized benefits or costs, while use of wetlands for their ecological and related benefits yielded benefits with a PDV of $22,000 plus significant nonmonetized benefits. In this hypothetical case, the results of a cost-benefit analysis can justify, on the bases of economic criteria and the principle of potential Pareto improvement, that the continued preservation of the wetlands in their natural state maximizes man's well-being. Such results imply that by retaining

Table 1   Hypothetical Table of Costs and Benefits:[a] Two Alternative Uses of Wetlands.

	Present Discounted Value ($)
I. Use of Wetlands for Construction	
Market value of unfilled wetland to be filled for construction	10,000
External benefits or costs from filling and building on land	00
PDV of total net economic benefits	10,000
Nonmonetized social benefits and costs	Negligible
II. Use of Wetlands for Ecological and Related Benefits	
Productivity of fishing industries external to wetland	2,000
Finfish (flounder . . .) ⎫ Through	
Crustacean (lobster, crab) ⎬ food	
Shellfish (oyster, clam, mussel) ⎭ chain	
Productivity of fishing industries on wetland	10,000
Finfish ⎫ Harvested	
Crustacean ⎬ on	
Shellfish ⎭ wetland	
Recreational benefits on wetland	5,000
Hunting, boating, fishing, etc.	
Public amenities	5,000
Water management	
Air cleansing	
Storm barrier	
PDV of total net economic benefits	22,000
Nonmonetized social benefits and costs	
Habitat for transient wildlife (in flyway of . .) ⎫	Not
Visual amenity and "open space value" (distant from urban area) ⎪	valued
⎬	but
Contribution to preservation of endangered species ⎪	judged
Option demand (irreversible change) ⎭	significant

[a] Present discounted value per acre at 6% interest rate for 100 years.

this area of wetlands the beneficiaries of the natural wetlands would receive more than enough additional goods and services to be able to compensate those individuals who would have benefitted from filling the wetlands for construction.

More generally then, we see that in principle at least it is possible that a careful cost-benefit analysis of the economic values of ecological systems may justify the retention of many natural areas according to the same criteria we also use to justify their sacrifice to the production of economic goods. One practical barrier to performing such an analysis, however, is the absence of data on socially relevant output of ecosystems. Some innovative approaches have been suggested by ecologists including the assigning of dollar values to the total energy evaluation of ecosystems which can be used as a proxy for the many diverse benefits provided by ecosystems to man (Chapter 7). Such an approach, despite its shortcomings, may provide a useful first estimate of the social value of a particular ecosystem. Again, however, some of the social benefits of ecosystems are related neither to their energy flows nor their economic values.

THE LIMITATIONS OF COST-BENEFIT ANALYSIS OF ECOSYSTEMS

Cost-benefit analysis for the management of ecosystems has several important limitations. Assume, for example, that all benefits of the natural wetlands directly or indirectly affecting man's economic life have been evaluated. If the resulting PDV of natural wetlands is less than the PDV of the land for construction, then cost-benefit analysis indicates that filling the wetland would yield a net increase in economic benefits for society. Should we now with equally ruthless speed conclude that the natural wetlands be sacrificed to construction? Not always.

Noneconomic Values

If it is believed that the social benefit of filling wetlands for construction is almost entirely reflected by the PDV of the land, but that the economic value for several benefits of natural wetland cannot be determined with any confidence, then we should not yet accept as optimum the choice indicated by the cost-benefit analysis. In such situations the economic criteria of cost-benefit analysis is assymetrical—it may be a sufficient criteria for rejection of a project such as filling the wetlands, but it may not be sufficient criteria for acceptance of the project. Now other social criteria must be used to decide whether the nonmonetized benefits of the wetlands outweigh the net economic gains from filling them in (Herfindahl and Kneese, 1974; Mishan, 1971). Here the net economic gain of filling the natural wetland (the difference between the PDVs of the filled and natural wetlands) should be compared with those nonmonetizable benefits foregone by filling. Such social decisions are not conceptually different from any government decision to provide public goods (defense, schools, courts, parks, etc.) whose contribution has long been recognized by the public, but poorly measured by the market (Steiner, 1969; McKean, 1968). If, for example, the wetlands were one of the few nesting places of an endangered species, we may decide on political and philosophical grounds that the relatively small gain in economic goods is meager compensation for the potential loss of a species. In this case, we are not rejecting the economic criteria, but rather integrating it into a broader context of social decision making. For by using cost-benefit analysis we now know more precisely what the real trade-off is between net economic benefits and those many other elements that enter into man's well-being.

The basic reason is found in our initial discussion of man's well-being, namely, that man's well-being is affected by many factors not usually thought of in terms of economic goods. Life itself, justice before the law, unique natural areas, good health, religious tradition, honest government, and so forth are examples. Many people are reluctant and often unable to consider these aspects of life in terms of economic values and often feel that these are "rights," "freedoms," and heritage to be neither subjected to an individual's ability to pay nor routinely surrendered for ordinary economic goods. For example, some people believe that hunting the blue whale or any relatively harmless species to extinction is morally wrong. It violates man's ethical responsibility for stewardship of the earth for future generations and for God. [See Passmore's (1974) excellent analysis of man's responsibility for nature.] A cost-benefit analysis of blue whale hunting should (properly) avoid any attempt to monetize such feelings of unhappiness and dismay in people. But such costs are very relevant to policy, for what costs could be more real than violated beliefs and ethics?*

Option Demand and the Future

Conventional cost-benefit analysis is also limited by the interrelated problems of the value of choice at some future time (*option demand*) and the obligations

---

* We cannot estimate (through observation of economic behavior) the economic gain from respecting these beliefs, but we can identify the economic costs of observing these beliefs. This cost is the difference in net present discounted value between the economic benefits foregone by not undertaking this project and those of the most profitable alternative consistent with these beliefs. Thus in Figure 5, if the benefits of filling the land were completely monetized at $30,000 PDV, while those benefits of natural wetlands which could be reliably monetized were $22,000 PDV, then the opportunity cost of keeping the wetlands in their natural state to obtain the nonmonetized benefits would be $8000 PDV.

of the current generation for the welfare of future generations. Some changes in ecosystems are irreversible, so that to permit such changes now eliminates the option of making a choice in the future. The option to pick from a greater range of choices in the future is of value in itself (hence the term option demand). That is, individuals are willing to pay to preserve options because of (1) shifts in tastes and preferences of future generations, (2) uncertainty about future benefits and costs as viewed from the present generation, hence (3) the possibility of obtaining more information to make better decisions, and (4) the desire to exercise the option in the future, for which they do not have resources at the present. The concept "option demand" becomes relevant whenever the results of a project (1) are irreversible for economic or technical reasons, (2) are large changes rather than marginal changes, and (3) have no close substitutes (Fisher and Krutilla, 1974). Filling in the Grand Canyon would be such a project.

The current economic value of such future "options to choose" is difficult to establish, having few counterparts in the market, and, therefore, option demand can not be accounted for in a cost-benefit analysis in any way that irrefutably leads to better decisions. In such cases, Fisher and Krutilla (1974) council the conservative approach of avoiding such actions.

The very real possibility of underestimating today the demand for and benefits of irreplaceable goods tomorrow may foreclose opportunities for which a slightly larger capital stock is but poor recompense for future generations (Herfindahl and Kneese, 1974; Krutilla et al., 1972). Economic decisions concerning the future can not be separated from a central issue of social philosophy—namely, the responsibility of the current generation for the well-being of future generations.

ACKNOWLEDGEMENT

The author acknowledges the constructive guidance of the editors, an anonymous reviewer and my colleague, Harvey Gram, and the editorial assistance of Stephanie Tessmer and Blanche B. Dohan, who also prepared the manuscript.

REFERENCES

Arrow, K. J., and A. C. Fisher. 1974. Environmental preservation, uncertainty, and irreversibility. *Quart. J. Econ.*, **LXXXXVIII**: 312–319.

Barkely, P. W., and D. W. Seckler. 1972. *Economic growth and environmental decay: The solution becomes the problem.* Harcourt Brace, New York.

Bator, F. M. 1958. The anatomy of market failure. *Quart. J. Econ.*, **LXXII**: 351–379.

Bator, F. M. 1967. The simple analytics of welfare maximization. *Amer. Econ Rev.* 22–58.

Baumol, W. J. 1968. On the social rate of discount. *Amer. Econ. Rev.*, **LVIII**: 788–802.

Baumol, W. J. 1965. *Welfare economics and the theory of the state.* Harvard University Press, Cambridge.

Becker, G. S. 1971. *Economic theory.* Knopf, New York.

Boulding, K. 1967. The basis of value judgements in economics. In S. Hook (Ed.), *Human values and economic policy*, New York University Press, New York. Pp. 55–72.

Boulding, K. 1966. The economics of the coming spaceship earth. In H. Jarrett (Ed.), *Environmental quality in a growing economy*. Johns Hopkins Press, Baltimore.

Boulding, K. 1966. Economics and ecology. In F. Darling and J. P. Milton (Eds.) *Future environments of North America*, Natural History Press (Doubleday), Garden City. Pp. 225–234.

Boulding, K. 1973. *The economics of love and fear: A preface to grants economics.* Wadsworth, Belmont, Calif.

Chase, S. B., Jr., Ed. 1968. *Problems in public expenditure analysis.* Brookings Institution, Washington, D.C.

Cicchetti, C. J., and A. M. Freeman. 1971. Option demand and consumer surplus: further comments. *Quart. J. Econ.* **LXXXC**: 528–539.

Clarke, C. W. 1973. Profit maximization and the extinction of animal species. *J. Polit. Econ.* **LXXXI** (4): 950–961.

Clawson, M. and J. L. Knetsch. 1971. *Economics of outdoor recreation.* Johns Hopkins, Baltimore.

Coase, R. 1960. The problem of social cost. *J. Law Econ.* **3**: 1–44.

Dales, J. H. 1972. *Pollution, property, and prices.* University of Toronto Press, Toronto.

Demsetz, H. 1967. Toward a theory of property rights. *Amer. Econ. Rev.* **57**: 347–359.

Doyle, P. 1968. Economic aspects of advertising: a survey. *Econ. J.*, **LXXVIII** (311): 570–602.

Dorfman, R., Ed. 1965. *Measuring benefits of government investments.* Brookings Institution, Washington, D.C.

Dorfman, R., and A. Dorfman 1972. *Economic of the environment selective readings.* Norton, New York.

Douglas, P. A., and R. H. Stroud, Eds. 1971. *A symposium on the biological significance of estuaries.* Sport Fishing Institute, Washington, D.C.

Eckstein, O. 1958. *Water resource development.* Harvard University Press, Cambridge, Mass.

Fiske, J. D. 1969. *Report of the department of natural resources relative to the coastal wetlands in the commonwealth.* Massachusetts Division of Marine Fisheries, Department of Natural Resources, Boston.

Fiske, J. D., C. E. Watson, and P. G. Coates. 1967. *A study of the marine resources of Pleasant Bay.* Division of Marine Fisheries, Department of Natural Resources, Commonwealth of Massachusetts, Monograph Series No. 5.

Fisher, A. C., and J. V. Krutilla. 1974. Valuing long run ecological consequences and irreversibilities. *J. Environ. Econ. Man* **1**: 96–108.

Freeman, M. A. 1974. On estimating air pollution control benefits from land value studies. *J. Environ. Econ. Manag.* **1**: 74–83.

Gosselink, J., E. Odum, and R. M. Pope. 1973. *Value of the tidal marsh.* Center for Wetland Resources, Baton Rouge, La.

Gordon, S. 1954. The economic theory of common-property resource: the fishery. *J. Polit. Econ.* **62**: 124–142.

Haefele, E. T. 1973. *Representative government and environmental management.* Johns Hopkins Press, Baltimore.

Hall, C. A. S. 1975. The biosphere, the industriosphere, and their interactions. *Bull. Atom. Sci.* **31**: 11–21.

Hansen, F. 1972. *Consumer choice behavior: a cognitive theory.* Free Press, New York.

Hines, L. G. 1973. *Environmental issues, population, pollution, and economics.* Norton, New York.

Haveman, R. H., and J. V. Krutilla. 1968. *Unemployment, idle capacity, and the evaluation of public expenditures.* Johns Hopkins Press, Baltimore.

Herfindahl, O. C., and A. V. Kneese. 1974. *Economic theory of natural resources.* Merril, Columbus, Ohio.

Hook, S., Ed. 1967. *Human Values and Economic Policy.* New York University Press, New York.

Institute of Animal Resource Ecology. 1970. *Economics of fisheries management: A symposium.* University of British Columbia, Vancouver.

Isard, W. 1972. *Ecologic-economic analysis for regional development.* Free Press, New York.

Jerome, W. C., A. P. Chesmore, and C. Anderson. 1968. *A study of the marine resources of the Parker River–Plum Island Sound Estuary.* Division of Marine Fisheries, Department of Natural Resources, The Commonwealth of Massachusetts, Monograph Series No. 6.

Kneese, A. V., R. U. Ayres, and R. C. D'arge. 1970. *Economics and the environment.* Resources for the Future, Washington, D.C.

Kneese, A. V., and B. T., Bower, Eds. 1972. *Environmental quality analysis.* Resources for the Future, Washington, D.C.

Krutilla, J. V., and A. C. Fisher. 1975. *The economics of natural environments: Studies in valuation of commodity and amenity resources.* Johns Hopkins Press, Baltimore.

Krutilla, J. V., C. Cicchetti, M. Freeman, and C. Russel, 1972. Observations on the economics of irreplaceable assets. In A. V. Kneese and B. T. Bower (Eds.), *Environmental quality analysis*, Resources for the Future, Washington, D.C. Pp. 69–112.

Lancaster, K. 1966. A new approach to consumer theory. *J. Polit. Econ.* **74**: 132–157.

Levenson, A. 1971. *Evaluation of recreational and cultural benefits of estuarine use in an urban setting.* Hofstra University Center for Business and Urban Research, Hempstead, N.Y.

Mack, R. 1974. Criteria for evaluation of social impacts of flood management alternatives. In C. P. Wolf (Ed.), *Social impact assessment.* Vol. II of Daniel Carson (Ed.), *Man-environment interaction: The state of the art in environment design research.* Environmental Design Research Assoc., New York.

Mack, R., and S. Myers. 1965. Outdoor recreation. In R. Dorfman (Ed.), *Measuring benefits of government investment.* Brookings Institution, Washington, D.C. Pp. 71–116.

Maunder, W. J. 1970. *The value of the weather.* Methuen, London.

McKean, R. N. 1958. *Efficiency in government through systems analysis with emphasis on water resources development.* Wiley, New York.

McKean, R. N. 1966. The use of shadow prices. In S. B. Chase, Jr. (Ed.), *Problems in public expenditure analysis*, Brookings Institution, Washington, D.C. Pp. 33–77.

Mishan, E. J. 1969. *Welfare economics.* Random House, New York.

Mishan, E. J. 1971. *Cost-benefit analysis: An introduction.* Praeger, New York.

Odum, H. T. 1971. *Environment, power, and society.* Wiley-Interscience, New York.

Passmore, J. 1974. *Man's responsibility for nature.* Scribner, New York.

Pejovich, S. 1972. Towards an economic theory of the creation and specification of property rights., *Rev. Social Econ.* **XXX** (3): 309–325.

Prest, A. R., and R. Turvey. 1965. Cost-benefit analysis: a survey. *Econ. J.* **Dec.:** 683–735.

Randal, A., B. Ives, and C. Eastman. 1974. Bidding games for valuation of aesthetic environmental improvements. *J. Environ. Econ. Manag.* **1**: 132–149.

Rawls, J. 1971. *A theory of justice.* Harvard University Press, Cambridge, Mass.

Robinson, J. 1963. *Economic philosophy.* Aldine, Chicago.

Samuelson, P. A. 1974. *Economics.* 9th edit. McGraw Hill, New York.

Samuelson, P. A. 1954. The pure theory of public expenditure. *Rev. Econ. Stat.* **36** (4): 387–389.

Schurr, S. H., Ed. 1972. *Energy, economic growth, and the environment.* Johns Hopkins Press, Baltimore.

Seneca, J., and M. Taussig. 1974. *Environmental Economics.* Prentice-Hall, Englewood Cliffs, N.J.

Stankey, G. H. 1972. A strategy for the definition and management of wilderness quality. In J. V. Krutilla (Ed.), *Natural environments: Studies in theoretical and applied analysis,* Johns Hopkins Press, Baltimore.

Steiner, P. O. 1969. *Public expenditure budgeting.* Brookings Institution, Washington, D.C.

Swagler, R. M. 1975. *Caveat emptor.* Heath, Lexington, Mass.

Texas Water Development Board. 1968. *The Texas water plan.* Austin.

Tribe, L. H. 1972. Policy science: analysis or ideology. *Phil. Publ. Affairs* **2** (1): 66–110.

U.S. Senate Document 97. 1964. *Policies, standards, and procedures in the formulation, evaluation, and review of plans for use and development of water and related land resources.* Supplement No. 1, Evaluation Standards for Primary Outdoor Recreation Benefits. Water Resources Council, Washington, D.C.

Van Horne, J. C. 1974. *Fundamentals of financial management.* Prentice-Hall, Englewood Cliffs, N.J.

Victor, P. A. 1972. *Pollution: economy and environment.* University of Toronto Press, Toronto.

Weisbord, B. A. 1966. Income redistribution effects and benefit-cost analysis. In S. B. Chase (Ed.), *Problems in public expenditure analysis.* Brookings Institution, Washington, D.C. Pp. 177–222.

Wolfson, M. 1966. *A reappraisal of marxian economics.* Penguin, Baltimore.

Woodwell, G. M. 1974. The threshold problem in ecosystems. In S. A. Levin (Ed.), *Ecosystems analysis and prediction.* Proceedings of SIAM-SIMS Conference, Alta, Utah, July 1–5, 1974.

7
Energy, Value, and Money

HOWARD T. ODUM

The recent development of energy flow concepts and diagrams has led to a new understanding of the principles of energy and the relation of these principles to value, to money, and in addition to a more quantitative understanding of the varying qualities of different energy sources in support of useful work. This paper develops a general theory of energy in living systems beginning with the first and second energy laws and the Lotka principle of natural selection for systems with maximum power. The integration of these concepts shows how the requirement for generation of order is as much a necessary part of energy laws as the degradation of order. All systems that undergo natural selection maintain this order through characteristic closed, autocatalytic loops of high-quality energy interacting with low-quality fuel energy. Such loops are accompanied by a circulation of money where human economic affairs are concerned. This paper includes: (1) a review of basic energy principles, (2) a restatement of some relatively new and original propositions about energy, (3) a development of the concept of energy quality, (4) a further illustration of the use of energy diagrams to represent complex systems, and (5) provides an introduction to energy cost-benefit procedures. The application of these principles to a variety of real world problems is found in other papers in this book (e.g., Chapters 19, 20, and 21).

BACKGROUND

There have been many attempts to use energy to measure value, starting with Marx in the last century, at a time when neither the physical concepts of energy nor the conventions of measurement were developed. The technocrat movement of the early part of the century was based on energy ideas, but apparently without a clear measurable connection between the work of man as an individual and that of his machines, his environment, or the man-catalyzed activities by which human energy was amplified. The energy certificate was a much-publicized value system based on energy, although its means of calculation in terms of physical work was not clear (Technocrat Magazine, 1937). Orson Welles suggested an energy-based society based on "aerodollars"—an energy assessment.

Our effort to clarify the relationship of energy and money started with a realization of the counter-current systems relationship, that is, in money-based societies money flowed proportionately, but in opposite directions, to the energy required to produce the purchased goods and services (Odum,

Work supported by Rockefeller Foundation, Contract, RF73029 National Science Foundation grant, G138721, and Energy Research and Development Administration, Contract E (40-1)-4398.

1967). These concepts were developed further in a recent book (Odum, 1971), which included a formulation of the energy basis for money and the suggestion that useful work was the natural basis for value if contribution to survival of one's system is the ultimate criterion for value. Hannon (1973) recommended an energy standard of value while examining the energy uses and essential needs of consumers; however, his presentation did not consider the different qualities of various forms of energy. We attempted to overcome earlier shortcomings in a basic simulation model of agricultural production and consumption that relates energy flow to money flow value (Odum and Bayley, 1974). We summarize our earlier work in this paper and also suggest a means for adjusting money to energy flow as a national means of eliminating inflation. For a new book on these methods see Odum and Odum (1976).

ROOTS OF ETHICS IN ENERGY LAWS AND ECOLOGICAL PRINCIPLES

Ecological and other systems that survive and prosper have characteristic patterns adapted to the realities of energy laws and ecological principles. Since human systems are subject to the same energy constraints as any other system, any ethic for the survival of man must meet the same requirements. According to this analysis culture is formed by the trial, selection, and survival process that produces a structure of information that tends to retain successful patterns, at least until conditions change. Nine laws may be pertinent:

1 The *first energy principle* states that energy is neither created nor destroyed. In any diagram of a system, such as Figure 1, all energy* inflowing to the system must be accounted for either in storage increases or in outflows. Energy flows out as work, dispersed heat, or as exports of potential energy capable of driving processes elsewhere. This law is illustrated in Figure 1, where 100 kcal of energy incoming to a system (dashed-line box) per day (measured as heat equivalence) is accounted for as 100 cal heat outgoing (90 + 10). Another example is given in Figure 2, where two different kinds of energy flows interact in a process that requires them both and also generates a third and fourth kind of energy outflow. Here again the sum of the inflowing rates (300 + 10) must equal the outgoing rates (280 + 30).

2 The *second energy law* requires that some potential energy be degraded as dispersed heat in all processes. The energy-degradation principle states

---

* Unless specifically noted otherwise, energy units are Calories of heat equivalence. Later in this chapter we use Calories of Fossil Fuel Equivalence, a different measure.

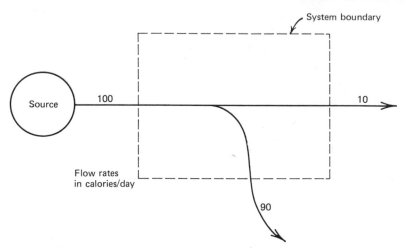

Figure 1   Energy flow illustrating energy conservation principle.

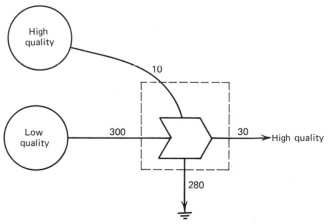

Figure 2   Energy flow where one flow controls a larger flow both of which are essential to a high quality output to the right.

that any time there is an energy transformation some energy must be degraded in quality and ability to do further work because of its dilution and dispersal into the random motions of heat in the environment. From this comes the concept that the creation or maintenance of order requires work, since without an input of ordering energies, molecules tend toward random structure. Such energy flows are irreversible, and this loss of energy available to do work is indicated by a pointed arrow into the environment—a

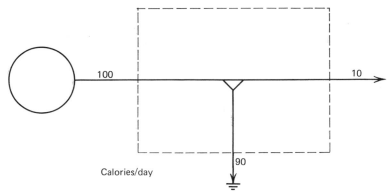

Figure 3 Energy flow illustrating energy degradation principle.

symbol we refer to as a *heat sink*. A heat-sink symbol is required on energy diagrams for each component process and pathway (Figure 3). In Figure 3, 90 kcal of heat dispersion accompany an energy transformation that generates 10 kcal of high-quality energy flow. In Figure 2, 280 kcal of energy are dispersed as heat, while 30 kcal of "high-quality" energy output are generated from the two interacting inflows. We consider the meaning of energy "quality" in subsequent sections.

Sometimes we measure the energy degradation flow by the ratio of a heat flow to the absolute temperature of the environment into which the energy is dispersed. This ratio—called *entropy*—is a measure of the loss in energy quality. Since the energy degradation principle requires a continual dispersal of heat, there is an increase in the overall entropy of the universe. In Figure 3 the entropy increase is computed as 0.3 kcal per degree.

Any storage of quality energy such as order, concentrated matter, structure, or information is potential energy that tends to disperse into the general disorder of the environment. Random motion tendencies disperse the storage into a more uniform pattern through degradation processes usually called deterioration or depreciation. A storage of order can be maintained only by supplying a flow of ordering energy. For example, in Figure 4, the storage (tank symbol) is maintained by an inflow of energy which is transformed into a higher quality of energy flowing at a rate of 10 kcal per day, replacing an equivalent amount lost in normal depreciation.
3   Alfred Lotka (1922) presented an energy codification of Darwin's principle of natural selection—the principle of *maximum power selection*. Systems develop variety as choices and the system retained by natural selection is the one that develops more power and channels this power into adaptive mechanisms.

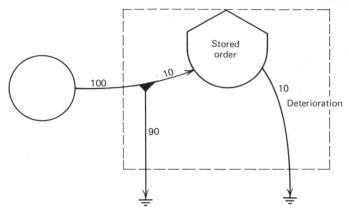

Figure 4 Energy flow that stores high-quality energy replacing that energy storage that deteriorates into randomness according to the energy degradation principle for all ordered storages.

For example, in the simple energy storage processes represented by Figure 5, more power is stored when the backforce loading is half of the input force. An example is the pumping of water into lakes at higher elevation to store energy. Since energy storage, such as accumulation of biomass, is essential to control the system's feedback operation, then adjustment to maximum power loading will contribute to competitive advantage. These principles necessitate particular energy criteria for any surviving ethic. Since

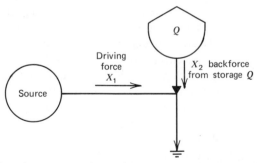

Figure 5 Generation of new storage of potential energy under influence of driving force against the backforce due to storage. Flow into Q is

$$J = \frac{1}{R}(X_1 - X_2) = \frac{1}{R}(X_1 - \frac{Q}{C})$$

where R is resistance and C is the packaging coefficient defined as force X_2 per unit quantity stored.

selection and evolution are as important to man as the first two laws of thermodynamics, we may wish to call the energy-ordering principles of Darwinian selection the "fourth law of thermodynamics."

4 *Obligation to develop order and interaction feedback.* Lotka's principle implies that the systems most likely to survive are those developing structures and functions that maximize the power flows to be used in the competitive process, because energy is a general resource that can be diverted to any competitive process. Another aspect of this process pertains to the selection for certain "high-quality" energy flows that are used to amplify larger, but lower quality, energy flows. Figure 6 shows a system that develops such a quality energy storage and then uses this to amplify and control the principal inflowing energy source. The systems that develop the kind of feedback amplifier shown in Figure 6 draw more power because they are more effective at exploiting energy resources. The feedback is called a *reward loop* because the high quality energy fed back upstream "helps" the energy unit it drains draw more energy from the principal source. One simple biotic example would be the evolution of colonial jellyfish. The feeding individuals divert a certain proportion of their food intake to sensory individuals that enable the jellyfish to more fully exploit available food.

Whereas the energy degradation principle dictates a tendency for concentrated energy storages to disperse, Lotka's principle says that because of

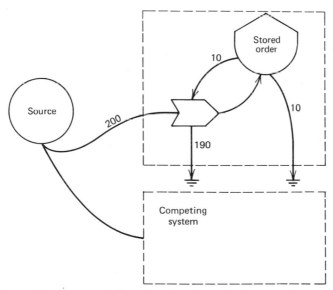

Figure 6 Energy flow that uses high quality of energy as feedback amplifier and control on energy inflows.

competition that develops in any energy flow, some order must be developed as part of the overall degradation process, since whichever systems develop this loop will be selected for. As illustrated in Figure 6 the feedback must be an interaction rather than a sum so that high quality feedback can have a multiplier effect, otherwise the loop will drain more energy from the system than it adds to it.

5 *Competitive exclusion* is a property of parallel systems that draw on a common energy source (Figure 7). In simple competitive systems that do not have population control mechanisms, one competitive unit will grow at the expense of the other(s) and cause its extinction. This is because of the ability of organisms to undertake Malthusian growth. This is an energy principle that explains cancer, the dominance of a single species of beetles in grain, and the survival of only one species of microorganism in many chemostat cultures. Competitive exclusion is an adaption useful in exploiting new energy resources. However, once new energy resources have been colonized, single-species systems tend to be replaced by diverse and controlled systems. Competitive exclusion does not have selective advantage in diverse well-developed ecosystems using many energy sources. Instead, there are roles for competition and cooperation that help maximize power, depending on the availability of energy for growth.

6 *Compensation with reward loops.* Any unit that draws potential energy from another (for example, a predator from a prey) diminishes the energy resources of the supplier because of this drain. Such a drain may put the supplier and hence recipient at a competitive disadvantage unless the supplier benefits in proportion to the drain. Since some potential energy is lost as heat along the pathway downstream, the downstream unit *must return its services* to the upstream with an *amplification factor* for the upstream flow to be helped as much as it was drained. If this amplification is not done,

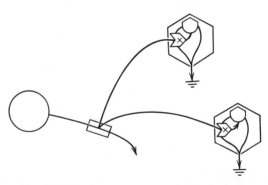

Figure 7 Competition for limited energy source which leads to competitive exclusion.

that entire energy pathway will have less competitive ability and will be selected against. For example, the downstream units may be animals which do services for the plants, such as population control or nutrient regeneration, in exchange for food. Farms and cities have this relationship as shown in Figures 8 and 9. We can also call such self-organizing processes self-design or even learning. We see mechanisms by which there is an allegiance of parts to the whole within this loop principle. In Figure 6, 10 cal of higher

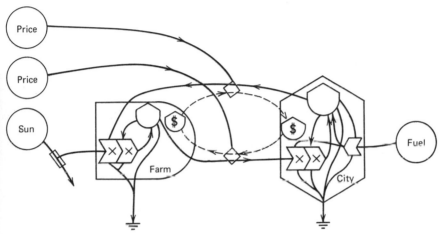

Figure 8 Farm and city economy with solar and fossil-fuel sources.

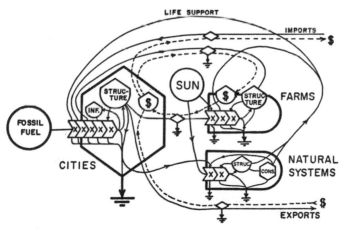

Figure 9 General system diagram for energy basis for a country aggregated into urban, agricultural, and environmental components.

quality energy acting as a 2 × amplifier compensate the upstream pumping intersection for 20 cal drained.

7 *Backward cycle of money in human relationship.* There are the same down-stream drains and feedbacks of work services from downstream within human systems, but there is the additional feature of a circulating currency that goes in the opposite direction. In Figure 8 we show farms and a city. The currency is the dashed line that has a closed and continuous pathway flowing in proportion to energy flows. There is a high ratio of money to energy on the downstream side of the loop compared to the upstream part of the loop (i.e., on the left of Figure 8). There is an *average* ratio of dollar flow to total energy flow. In the U.S. economy it has been in the vicinity of 17,000 kcal of fossil fuel energy per dollar for the economy as a whole. However, higher-quality energies cost more dollars than lower-quality energies.

Additional energy flows in support of the economy from renewable natural inputs such as sun, wind, tide, and rain. The money circle, plus the societal programming of people to think of money as valuable, provides a mechanism for adjusting the feedback work loop to match the upstream to downstream power transfer. The money is used to buy goods and services that operate as a feedback, accelerating principal energy flows. We learn from the relationship of money and work that the root of value in cir-culating money is energy. If all energy sources are cut off from an economic system, the dollar loses value until it buys nothing, since there is no energy inflowing to produce goods and services once storages in the system have been exhausted. As we have seen recently, our entire economic system is very sensitive to changes in inflowing energies.

Money circulates in closed loops, whereas energy moves in from outside, makes loops, and leaves as degraded energy. If the ratio of the flow of money to flow of energy overall were kept constant, the dollar would be-come an energy certificate. Only energy can be used to measure inflows of energy from geologic deposits or from the environment. If the energy flowing decreases relative to the number of dollars money inflates.

When we examine systems that include both man and nature, such as the general diagram for our planet in Figure 9, we find that a large part of the earth's energy budget is not included in the economy of money. It is the energy systems of the vast areas of forests, steppes, and seas that is the general one; money is involved only in certain relationships among people. From these principles we see the limitations of money and previous theories of value based on only a part of the basis for man's survival.

8 *System tracking by self-design of culture, religion, and behavior.* Human behavior programs are the means by which stored energies are developed and reinvested (Figure 6) by a society. Culture and behavior may be re-

programmed with each generation during periods of rapid change making the social system much more responsive to change than a biological system that requires genetic selection over many generations. If those patterns of ethics, religion, and culture that accompany the successful emerging energy patterns are accepted and taught, then the behavioral patterns may be regarded as an informational tracking system—always adapting with a slight lag, however, to those systems that are surviving. That behavior follows, rather than leads, the established success of a system design in no way lessens its importance. However, for making future plans in times of change it always would be late to ask people what they want or to look at existing behavior as a guide to need. The prediction of future societal needs requires the prediction of the future energy patterns. Highly flexible reprogramming of social and individual behavior can change attitudes, and if there are proper educational programs people will demand those programs the system needs for survival as planning actions are implemented. These educational programs are shown in the circuit language by the feedback system that has the self-designing property of learning, and they reinforce those pathways that work at providing continued power flow. In the system of Figure 6, for example, diversity of information and behavior can be reinforced if its feedback accelerates power as shown.

9 *Surviving systems develop the property of resisting accelerations.* In Newtonian physics, a force of acceleration is always matched by a counter-impeding force from the item being accelerated, whether it be the inertial force of a heavy weight, the turning inertia of a heavy wheel, or the magnetic backforce of an electric current being accelerated. Many systems of other types also have this property. Let us generalize the concept of inertial backforce to energy circuits containing populations of flows, including those that have the delicate flows and high amplifying potential of information. One may use the same differential equation as for the physical systems, although frequently the state variables are not as precisely defined. In Eq. 1, N, the population backforce is given as proportional to the acceleration of flow, J, with $\mathscr{L}$ a coefficient of the relationship that is a measure of the properties of stubbornness in the network.

$$N = \mathscr{L}\frac{dJ}{dt} \tag{1}$$

Some systems are self designed to have more inertia than others. For example, at the social level $\mathscr{L}$ is a measure of conservatism. During a period of application of force to a system, there are impeding back forces that represent input energies going into energy and information storages. After the application of a force is terminated the storages may affect forward flow. As with the simpler physical analogs, energy is stored and the system affected most when it resists most. It remains an open question as to when large inertial properties are desirable in adapting to the larger environment. In electrical systems inertial properties may be manipulated to gain stability or to produce oscillations or instability depending on the timing characteristics of the network being designed.

ENERGY THEORY OF VALUE

The principles enumerated above guide us in the development of a general theory of value and ethics. The first premise of the theory is that the patterns of surviving systems are right to those surviving with these systems. Moral ethic for man comes from institutionalizing surviving patterns as ultimate values, because other patterns that have been selected against are no longer defended. We hasten to warn the reader not to jump to erroneous conclusions about what properties go with surviving systems, as many people did at the time of Darwin. Selfish or destructive behavior of components leads to the extinction of both the system and the components. Survival of a unit requires survival of its support system. Any unit must contribute special work to its system or hasten the extinction of itself along with its suppliers. The requirement for survival of a part, therefore, is service to at least one other unit of its system. The value of a part to the system is its contribution to useful work where useful work is work that maximizes further energy gains and effectiveness of energy use. We can describe a state of value equilibrium when energy cost equals energy effect.

The diagram of energies in the reward loops (Figure 6 or 10) suggests the amount of such work that is necessary. A great deal of potential energy is degraded to heat as energy flows to the right, but this energy makes possible

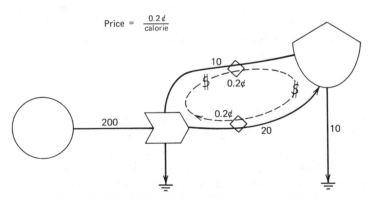

Figure 10   Constant value loop with currency circulating.

the development of the high-quality energy of the downstream unit—which may be a top carnivore in the forest or a specialized city worker in the human system. The total amount of energy necessary to produce the high-quality flow is the total energy degraded. A feedback must have at least as much amplifying effect as heat loss for survival under competition. Through amplifying the main energy flow, the feedback exerts a multiplier action upstream at the intersection of the feedback flow and the main power flow. The energy value of the downstream unit may be measured either by the calories degraded as heat in the system up to that point or by the magnitude of the amplifier action of the feedback on the main flow, because these two effects are equal when the system is competitive and in equilibrium. Therefore, the values of the services of the downstream units can be measured by the upstream energy flows responsible for generating them. If part of the downstream operation involves information storage, as with the programming of a specialized human ability, the integrated total energy involved in that development over time is a measure of that value to the system, providing the reward loop (feedback) has been established. In summary, energy value for the work of upgrading energy is the energy flux in the productive process upstream. For stored structure, the accumulated energy cost of developing the structure is the capital value in energy units.

The energy theory of value allows the evaluation of all kinds of services and accumulated works whether they be by man or not and whether they involve money or not. The assessment of values by this system allows each loop of a system, as in Figure 9, to receive appropriate consideration in planning and protection. Survival of the combined system of man and his supportive natural and industrial systems requires maximum utilization of all available energies. The importance of each subsystem to the whole is in proportion to the whole quantity and quality of its energy contribution.

QUANTITATIVE MEASURES OF ENERGY QUALITY

In Fig. 11 a high-quality flow interacts with a larger but lower quality energy flow, the high-quality one serving both to control the other and to amplify its own role. The ratio of energy flow harnessed to quality energy used in feedback is 30-fold, and the amplifier ratio of quality energy inflow to managed output is 3 to 1. The same kind of intersection and amplification is involved when stored energy is fed back to amplify inflow into the storage as shown in Figure 6.

If natural selection is to favor the system relative to competitors, the feedback pathway must operate economically. The feedback of quality energy needs to have as much amplifier action as necessary, no more no

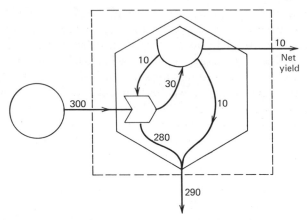

Figure 11   Energy flow that generates a net yield export of the high-quality energy of the type generated in the system for maintenance.

less, to generate the high quality energy used in the storage, in the replacement of depreciation losses, and for the amplification work. To feed back more is to waste high quality energy; to feed back less is to make a net drain on the system. Both are noncompetitive. The competitive state is a value equilibrium.

One may measure energy-flow-amplifier value either as the energy required to generate the high-quality feedback, or one may use the amplifier ratio of the intersection as a measure of the value of the feedback. In the example in Figure 10, 20 kcal go into storage in order to develop a feedback of 10 kcal. Thus 2 to 1 is the energy quality increase. A calorie in the feedback loop by this consideration is twice as valuable as a calorie in the main flow. Or, using the alternative calculation, 10 kcal generates 20 kcal of new energy inflow if it interacts with the energy source. Thus the feedback amplifier effect is 2 to 1.

If the system is in value equilibrium, the ultimate energy value of an energy flow, that is, its quantity times quality, maintains a constant value in a loop. Energy value is defined as the total energy flow developed as a given flow interacts with principal energy sources (Figures 6 and 10).

THREE ENERGY VALUES

It should be clear now to the reader that there are two energy values associated with each pathway that has amplifier intersections. The first is the *heat-equivalent value* of the flow, were it dispersed into heat without inter-active work on other energy flows. For example, in Figures 6 and 11 the feedback alone has a heat equivalence value of 10 kcal per day. The second energy value of the same flow is the *energy-release value* of the flow if and as it actually interacts with other flows. In the example of Figure 6 the feedback loop has the energy amplifier or release value of 20 kcal per day. The ratio of these two energy flows may be regarded as an *energy quality ratio*. For values greater than one we may regard the energy as of high quality, and when the value is very high we may call the flow information, following the common custom of calling items of high energy release value information. These two calorie values of the same energy flow are emphasized in Figure 12. The "energy alone" value decreases down the pathway and around the loop, but the "energy amplifier release" value is constant around the loop. For every intersection between a high-quality energy and a low-quality energy there must be one flow that has a ratio of energy flow to output greater than one and one that has a ratio less than one. The flow that has a less-than-one energy quality ratio is usually regarded as the fuel energy source. Most layman's practice regards only this flow as energy.

If a system receives enough energy it may develop an excess of the ordered high-quality product over and beyond that necessary for feedback to aug-ment principal flows. Such a system may accumulate the high quality product or export it to other processes as shown in Figure 11. In that example, there is a net yield of 10 kcal per year of net high-quality energy yield. Libraries and the export of agricultural technology represent common examples of high quality storage and export.

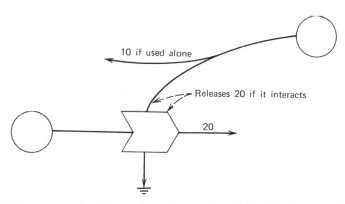

Figure 12 Two energy values for a pathway: energy alone (10 kcal) and energy-release value (20).

A third energy value is the energy expended upstream to develop the flow to that point. Figure 11 shows that 300 calories of one kind of energy develop the net yield of 10 of another type.

ENERGY QUALITY SCALE

The behavior of complex energy systems results in the degradation of some flows, while other flows are upgraded. These processes occur in chains, so that a succession of transmissions degrades initial energies quantitatively while repeatedly upgrading small amounts of the initial energy. Examples include food chains, chains of information processing, hydrological eddy chains, and so forth. It is possible to determine empirically the observed caloric equivalents for generating one type of energy from another by using data for known energy processes that have developed in real conditions of natural selection and by arranging them in chains. These chains help us develop a conversion scale from the most abundant, but dilute, low-quality energy like sunlight through energies of intermediate quality like food or petroleum, to the most concentrated energy-releasing flows that we call information (Figure 13). Each stage in the energy-quality chain allows us to calculate the calories of the input-quality energy required to generate the higher quality energy output. Ultimately, one may use one particular energy quality as a standard and refer other energy qualities to that standard. The ratios of energy flows in Figure 13 are another kind of energy quality ratio.

When all energies are expressed in units of similar quality, we can compare their general ability to do work. In some previous energy calculations the

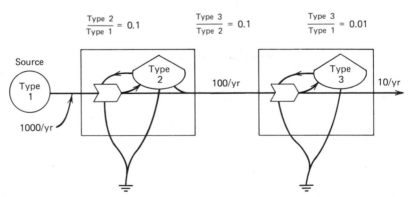

Figure 13 Energy-conversion factors are the ratios of conversions in steady-state flows where energy costs of maintaining structure are paid from the inflow so that the energy chain is in a simple line.

convention has been to express other energy qualities in units of chemical potential energy of carbohydrate food and fiber. This standard is easily measured and is widely used in general nutrition, and it is at an intermediate energy quality in the scale between sunlight and information. However, the energy scale in Figure 14 allows one to convert easily any energy quality to the source energy that would be required to generate it from any other quality of energy. We use *fossil fuel* equivalents as a common denominator when dealing with questions of public policy.

Of course, the value of a flow of high-quality energy is not realized unless the high-quality items of material or information reach their useful intersection in some other part of the system, since high-quality information depends on its intersection with a fuel for its value to be realized. Written information, for example, is only valuable if it is read by the person able to use it in other energy roles. High-quality energy is energy costly to store and degrades readily.

One corollary of the maximum power principle is the principle that high-quality energies are not used directly for low or intermediate quality roles, but instead are used as feedback amplifiers on lower quality fuel energies to generate maximum energy flow. Thus one should not use protein as fuel, but should use protein for enriching the nutrition of a carbohydrate source. One should not use electricity, a high-quality energy, to heat a house alone, but should use it to operate heat pumps or other devices that can tap a second low-quality energy source. One should not burn cookbooks to heat one's lunch. When we say "should not," we mean that systems that survive will not do these things because they will be uneconomic compared to competitors and will be eliminated by competition.

Our culture has not appreciated the universality and severity of energy laws because during times of rapid expansion of new energy resources, such

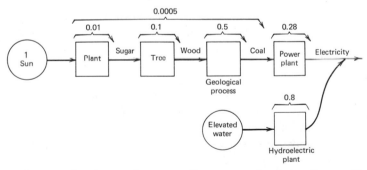

Figure 14 Some rough estimates of energy quality concentration factors. Low-quality energy is on the left.

as petroleum, there is a relatively brief period when growing systems that are in energetic lead over competitors can and must generate many energy choices—including many that are wasteful and ultimately are selected against. The energetically ineffective choices are maintained during that period when the system is energetically richer than others up until the time that all available energy flows are being used and there is no longer a competitive race for expansion of energy-using capacity. A sharp premium on competition against waste is reasserted as soon as the energy resources of competitors are similar. The generation of choice and competing alternative is a property of any energy flow and is part of the continual hand-in-glove refitting of systems to their varying energy opportunities.

TABLES OF COST EQUIVALENTS AND
ENERGY COST-BENEFIT DECISIONS

Since flows of potential energy are the source of all value, then the energy values may be used as a measure of value that includes both items of nature without man's economy and those that are accompanied by the flow of money in man's economic system. As we have learned from the scale of energy quality one must convert all such values into the same quality of energy equivalent before making a comparison. Having done this one may examine the various flows of a system of man and nature to determine the relative energy effectiveness of the various flows to survival. For example, if there are changes proposed in public environmental projects, one may draw the energy network of the present system and the network of the system with the proposed changes and calculate the total energy values for both cases to see which will generate the most useful work and thus which will be the most competitive in natural selection. The system with the most useful work will be the one that will have the most viable and competitive economy. Since human economies depend in large measure on the unmonied flows of nature as well as the monied flows, the system that maximizes the total energy value uses the most natural subsidies to the economic system to make it viable, its products cheap, and its balance of payments favorable.

Sometimes we call the tabulating of work equivalents as "*energy cost-benefit analysis*" since it is like the money cost-benefit analysis, except that this energy procedure includes all the work of the whole system and it evaluates external inflows that money does not. In making decisions about which system is most competitive one must also consider the energy costs of making the transition and energy flows that may be lost during transition periods.

EQUATIONS

The energy-circuit language shown here has mathematical equivalents for the modules, pathways, and energy conventions and constraints. Some of these are treated more fully in other chapters and in other publications. The diagrams are visual mathematics with the constraints of the real world. The translations to more traditional mathematics of differential equations are needed only for communication of definitions in traditional terms with those new to the visual mathematics or when some procedure such as patching standard analog computers requires going through regular equations. Equations for flows of money and energy are given for several price conditions in the next paragraph.

EXPRESSIONS FOR MONEY EXCHANGE AND PRICE

Depending on the situation, there is more than one relationship possible between the flow of money and the flow of energy that provides goods and services. When we consider a small economic system imbedded in a larger one, the price is not changed by the behavior of the small system and becomes an externally defined variable (Figure 8). In such a small system the flow of money generates a proportionate flow of energy, goods, and services (Figure 15). However, if we consider a large economic system such as a large nation the flows of energy and the flows of money have their own independent driving tendencies. The ratio of the two is the price generated by their relationship. As we have seen recently, a decline in driving energies relative to money in circulation is considered inflation.

Competition among a variety of alternatives provides for selection for the pathway with the largest quantity (energy goods and services) for sale per dollar since it has the lower price. Since price is inversely related to the quantity being sold, the resulting equation turns out to be a product of the purchaser's money and seller's quantity (Figure 15). Where a balance of payment is required, the price of sales must be adjusted to the price of purchases.

In situations where the tendency to spend money and the price are both influenced by the quantity of goods for sale, then the flow of money increases as the square of quantity of goods offered for sale and the money circulation becomes much more noisy (Figure 15). The last price equation in Figure 15 includes the effect of diminishing returns. Which specific equations are pertinent depends on the energy conditions controlling the system.

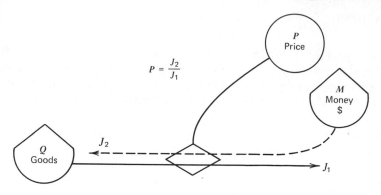

$$P = \frac{J_2}{J_1}$$

Money tends to be spent four times per year

$$J_2 = kM$$

Price controlled externally

$$J_1 = \frac{kM}{P}$$

Price inverse to quantity of goods for sale

$$P = \frac{K_1}{Q} \qquad J_1 = \frac{k}{K_1} MQ$$

Spending inverse to price also

$$k = \frac{K_2}{P} \qquad J_1 = \frac{K_2 MQ^2}{K_1{}^2}$$

Price variable with limit

$$P = \frac{k + Q}{KQ} \qquad J_1 = \frac{kKMQ}{k + Q}$$

Figure 15   Equations for money-energy interactions.

ENERGY INVESTMENT RATIO FOR ECONOMIC VITALITY

Sometimes we are asked why the maximum power principle would not cause energy-intensive fossil fueled urban economies to spread completely over the land. Many have thought that such a pattern would be wrong and would not lead to economic vitality, and they are correct. The answer has to do with protecting the natural energy contributions of that system as well as maximizing the power of the next larger system. Each system can and must send out goods and services to external areas in exchange for a balance of payments of special return services. The development of the symbiosis enables the two areas to do more collectively than either alone, since one complements the other if they are different in their production and offerings. Notice Figure 15, which includes the sale of goods and services for money that must then be used to purchase special, high-quality, external energy. The competitive position of the system is enhanced with the extra energy, but it cannot continue to develop additional external inflows beyond the point at which its sales are competitive. If the ratio of energy from outside to that from inside is greater than that of external competitors, the prices of that system's goods and services will no longer be competitive and growth will stop. If there is overshoot, there will be depressed economic conditions, and in places where urban growth has been extensive it will become dependent on subsidies from other systems.

Of course, all energies in such comparisons must be converted to one quality: for example, fossil fuel cost equivalents. The investment ratio for the United States as shown in Figure 16 is 2.5. Any development that has a ratio of purchased energies to internal (free) energies greater than 1/0.4 will be noncompetitive. In other words, it must charge more for its goods and services. As readily available world industrial energy supplies become depleted this ratio will decrease with less fossil fuel attracted per unit of natural energies.

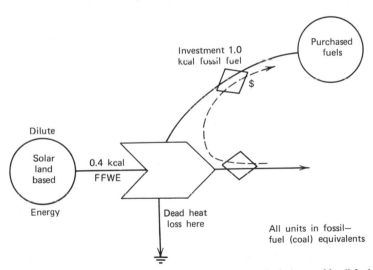

Figure 16 Diagram showing competitive ratio in United States for balance of fossil-fuel investment to output that is augmented by use of solar dilute free energy. Value of the outflow is estimated by its total energy cost as 1.4 Kcal.

AN ENERGY-MATCHING SYSTEM FOR THE WHOLE PLANET

Since man and his planet are still in the peculiar regime of trying to expand usage of fossil fuels, some special procedures presently are needed to establish a more stable pattern of man and nature and to provide for the survival of the whole system. The present pattern of competitors racing to tap the fossil fuel and other energy sources has the competitive exclusion and instability properties characteristic of cancer. In the 1970s man's industrial fuel burning is about 10% of the total global organic budget using heat units, but 72% of the total budget using fossil-fuel cost equivalents. How much further can our industrial system grow without overloading the planet's natural systems and developing energy usage patterns that are not competitive in the long run? Should we squander our rather small remaining oil supplies to continue industrial expansion?

It appears that the oil-producing nations did the right thing for the world in 1973 when they set limits on the rate of oil production. This could be the first step in an international plan for control of planetary energy that would limit the fossil fuel burning rate to that for which the present natural life support system could provide matching energy on the investment ratio principle. Such a pattern would stabilize our present system and allow energy ethics to develop a survival pattern. We need to arrange a global distribution formula for total fossil fuel (and nuclear fuel) through international treaties. In the previous section we introduced the energy investment ratio as the criterion for predicting the amount of economic development based on purchased energy that would be competitive. We extend this principle to the planet as a whole, as a guideline for international planning. Energies could be allocated for sale to each part of the globe according to the level of free unpurchased energies, all as expressed in fossil fuel equivalents. Thus energies would be sold and economic development encouraged only up to the point at which the purchased fuels ratio to free fuels was that of the world average. At present the ratio of purchased fossil fuel equivalents to free local fossil fuel equivalents in the United States is 2.5 to 1. Over the long run the ratio will become lower as readily available fossil fuels are exhausted. Such a plan automatically would induce conservation of free natural resources and inputs, as these would have to be maximized in order to maximize the fuels that could be bought. By having such a formula, no one country could gain excess power by virtue of technological development.

Could such a scheme be implemented in today's political climate? Agreement from less developed countries could be achieved as they would receive a higher prorating of the fossil fuel and nuclear power of the world, rather than the present distribution. In this way a premium is put on doing a good job with both natural and industrial sectors. Agreement from the most industrialized countries may be achieved by showing that the prorating of power will eliminate the fear and waste of war and defense spending. If no country has excess power or opportunity to gain it, it has no opportunity for large scale aggression nor the fear of it. These measures could bring all countries an equal share of the planet's industrial power subsidies while utilizing existing solar patterns too.

By this reasoning, population is best regulated by controlling the systems input power rather than by influencing individuals in a billion bedrooms. Births will adjust to economic realities, which in turn are adjusted by the ceilings on fuel available for economic expansion. In short ecological language, the world with its input power flow stabilized would shift from its recent successional pattern to a climax (steady state). An even pattern of man and nature can develop again on the planet. The costs of achieving a steady-state planet will be immense, but not nearly as great as the cost of not achieving it.

POWER EXPLOSION IF AN UNLIMITED ENERGY SOURCE DEVELOPS

Those with hopes for fusion and other new industrial energy ideas often speak of the desirability of an unlimited energy source. Unfortunately, the maximum power principle suggests that an unlimited power source would set off unlimited growth, since those with such power excess would run over those attempting to limit growth. The principle of competitive exclusion and the characteristics of early successional growth that we have observed in microbial cultures started with large food supplies or that we observe in human society during the past 200 years would be set off again. With unlimited energy, the power consumption of the planet could move from its present 10% of organic budgets to a larger, much less stable percentage. It is not likely that man's present ways, or the life support system that has been capable of maintaining biosphere stability so far, could do so with disturbance energy greater than the regular energy budget. Our life-support protection from the seas and forests would be eliminated, and as the planetary temperature and power increased so would the disorder and randomness that generates from high power flows. Something between the energy condition of the earth and the sun would result.

Fortunately, there are already limitations being placed on world power; the world's industrial energy growth may already be cresting into change towards a steady state or one of slowly declining energy as we have utilized the most accessible supplies of fossil fuels. Although a steady state based on renewable resources has different characteristics from our recent growth economies, the analogies in the natural world suggest that it can be a good state and one likely to be a good pattern for the life of man.

REFERENCES

Hannon, B. 1973. An Energy Standard of Value. *Ann. Amer. Acad. Polit. Soc. Sci.* **410**: 139–153.

Lotka, A. J. 1922. Contribution to the energetics of evolution *Proc. Natl. Acad. Sci.* **8**: 147–155.

Odum, H. T. 1967. Biological circuits and the marine systems of Texas. In Burgess and Olson. (Eds.) *Pollution and marine ecology*, Wiley, New York.

Odum, H. T. 1971, *Environment, power, and society*. Wiley, New York. 331 pp.

Odum, H. T., and S. Bayley. 1974. An energy standard in the model of economics and the ecological system. In R. O'Conner and E. Loehman, (Eds.), *Economics and decision making for environmental quality*, University of Florida, Press.

Odum, H. T. and E. C. Odum. 1976. *Energy basis of man and nature*, McGraw-Hill, New York. 292 pp.

Technocracy Magazine, 1937. The energy certificate, A–10. 22 pp.

8
Modeling in the
Context of the Law

ANGUS MACBETH

T his chapter discusses the use of models of natural systems within the context of legal proceedings, makes suggestions as to how modelers can be more effective within the legal and administrative framework of society and how good modeling can improve social decision making.

THE CONTEXT OF THE LAW

Each society tends to incorporate its mores and perceptions into the body of law by which it regulates its public and private activity. In one sense, the law may be thought of as a model of the society's conception of itself and the rights and responsibilities of the constituent parts of the society.

Immense growth in population over the last two hundred years and rapid expansion of a highly industrialized culture have greatly increased the degree and complexities of human interaction with the natural environment in the United States and the world. We have found ourselves steadily moving from the assumption that the supply of resources is essentially unlimited to the realization that we operate increasingly in a world of short supplies that are competitively sought out. At the same time, our knowledge of the effects of human activity on the environment has grown enormously, so that each act of pollution or disturbance of the existing system is known to have more ramifications, both social and biological, than was thought possible not many years ago. Today a thorough analysis of the environmental effects of automobiles does not stop with a discussion of smog-producing pollutants or of leaded gasoline, but must look at the impact of highway development on land-use patterns and the mix of physical and economic strategies needed to control or discourage the use of cars.

These historical changes have produced a growing perception of environmental problems as public rather than strictly private. Consequently, the modern pattern of environmental control consists of regulation by the administrative agencies of government. The agencies essentially manage the activities of private parties either through the development of explicit plans, as with zoning and land use planning, or through the granting or denying of permits for private activity that affects the environment, such as permits for the discharge of pollutants to waterways. The planning or review procedures and standards vary greatly, depending on which of the many statutes in the area is called into play by the planning or reviewing function.

The first major area of environmental regulation was land-use control. In the 1920s, the United States Supreme Court upheld the exercise of zoning powers by local governments that were properly authorized by the states to undertake the regulation. Since that time, local land-use regulation has proliferated, and the forms of planning have become comprehensive and complex. Recently, states such as Vermont, Oregon, and Florida have reasserted greater control over land use decisions within their borders. The tendency has increasingly been to view the larger areas as complete systems and to build into all land planning a recognition of areas of particular biotic and environmental value, such as wetlands, and to develop more of the basic planning on environmental notions such as the carrying capacity of the land. Chapter 19 is a good example of how Collier County, Florida is using such considerations in zoning.

Governmental regulation of air and water quality did not occur as a rule until long after land zoning was implemented, but today it is an important governmental activity. The latest stage in the increasing regulation of air and water has been the establishment of comprehensive federal statutes that control both the ambient quality of the medium and the discharge of pollutants from individual sources.* Regulation of air and water has been more thorough going than of land, both because it is less traditionally enmeshed in local control than are land use decisions and because the constitutional prohibition against the taking of private property without just compensation has much less application to the publicly owned resources of the air and water.

The federal government also has established a system to assure the accountability of its own agencies through the National Environmental Policy Act (NEPA), which requires an analysis of the environmental impact of each major federal action significantly affecting the quality of the human environment.† The Act applies to direct federal action such as the building of flood control dams by the Army Corps of Engineers and to federally licensed activities such as the abandonment of railway routes that can be undertaken only after approval by the Interstate Commerce Commission. This act builds environmental consideration into the full and pervasive range of federal activity.

Each of these major modern schemes of law embodies its own form of management and its own standards that the administrative agency must adhere to. For instance, the 1972 Federal Water Pollution Act Amendments require that standards of technological control be established for each industry, so that by July 1, 1983, "the best available technology economically achievable" for each type of discharge will be imposed. The assessment of this technology is to take into account

the age of equipment and facilities involved, the process employed, the engineering aspects of the application of various types of control techniques, process changes, the cost of achieving such effluent, non-water quality environmental impact (including energy requirements), and such other factors as the Administrator deems appropriate, 33 U.S.C. 1314 (b) (2) (B).

* Clean Air Act Amendments of 1970, 42 U.S.C. 1857 *et seq.*; Federal Water Pollution Control Act Amendments of 1972, 33 U.S.C. 1251 *et seq.*
† 42 U.S.C. 4321 *et seq.*

It can be seen that the standards for management are not simple and that proper attention to the terms of the statutes requires careful analysis and review.

It is the responsibility of the agency, primarily the agency staff, to see that the management schemes or reviews required by the statutes are fairly and impartially developed on the basis of a competent professional approach. Staff determinations are open to review within the agency, often in a forum whose procedures are patterned on those of the courts. Further, because the requirements of the statutory law are involved, review of agency decisions under the law may be had in the courts.

Obviously, the systematic nature of the analysis required by the modern statutes and the complexities of the subject matter that is being managed has made scientific modeling a crucial part of the legal system which governs in the environmental area. This makes the operation of modeling in the courts and in the administrative agencies a matter of importance, both for the effective and rational management of the country's resources and for the competent and persuasive presentation of the factual analysis on which such management must rest.

LITIGATION AND THE PROBLEMS OF PERSUASION

An adversary trial in a court in which models of natural systems are used poses the issue of the presentation of the model. A trial lawyer will not venture into a courtroom intending to present his evidence within the framework of a model unless he believes it will be persuasive to judge or jury.

It is important to remember that the task of a court when presented a dispute by two litigants is to reach a decision that puts the dispute to rest on the basis of interpreting the evidence presented in accordance with the requirements of the law. This structure of litigation is quite unlike the procedures of science or of scientific modeling in a number of respects and frequently makes the fruitful conjunction of the disciplines difficult. The central problem is to overcome the difficulties of translating or elucidating the model so that it will be useful in the courtroom. First, the courts with their system of juries and unspecialized judges are purposely lay institutions without expert knowledge. To a lesser extent, this is true of hearing examiners in administrative agencies as well. This means that the attorneys and the witnesses must undertake a process of education and it must often be done against a suspicion that the initially incomprehensible aspects of expertise are nothing more than self-serving jargon designed to hide the fact that the Emperor has no clothes. This is not a problem peculiar to science; anyone who has sat through a marine personal injury case and watched a jury with no seagoing experience first try to comprehend the ancient but forgotten language of seamen and then try to reach an intelligent judgment on standards of care and behavior aboard ship, will have a vivid memory of the difficulty of making the specialists' knowledge and standards apparent to a layman.

Second, the legal profession instinctively distrusts numbers appearing to give too clear a ring of certainty to issues that appear inherently uncertain and obscure, and the profession also possesses some latent feeling that mathematics and statistics are essentially shell games. These reactions are focused by the difficulty of interpreting what any set of numbers means, long experience with the self-serving qualities of witnesses, and the feeling that if some evidence is put in "hard" mathematical terms while the rest is "soft" and qualitative the apparent precision and reliability of the mathematically expressed evidence will weigh with a layman in an improper manner. A recent exchange in the *Harvard Law Review* that looked at the use of probabilities in evidence in more conventional cases spelled out at length the range of effects that lawyers believe the presentation of mathematical evidence may have on litigation.* Whatever the merits of the various positions taken in that controversy, it does make clear the difficulty of a fair presentation of a mathematically demonstrated position and the possibility of improper weighting of evidence that because of its mathematical expression appears much clearer than it truly is.

Third, there is the structure of scientific investigation that is at variance with legal fact finding. The job of the courts is to take the evidence at hand and settle the dispute that is before it once and for all; the principles of finality in litigation such as *res judicata* and the prohibition against double jeopardy prevent the constant reopening and reanalysis of disputes. In contrast, scientists, especially in basic research, move in a tentative fashion setting up hypotheses that they try to disprove, rather than prove, and constantly change and refine in a process that is not arranged to have a finite end point at which a final answer must be provided.‡ Examples of

---

* Finkelstein & Fairley, *A Bayesian Approach to Identification Evidence*, 83 Harv. L. Rev. 489 (1970); Tribe, *Trial by Mathematics: Precision and Ritual in the Legal Process*, 84 Harv. L. Rev. 1329 (1971); Finkelstein & Fairley, *A Comment on "Trial by Mathematics,"* 84 Harv. L. Rev. 1801 (1971); Tribe, *A Further Critique of Mathematical Proof*, 84 Harv. L. Rev. 1810 (1971).

‡ The problem is, of course, worse in conjunction of the law and those natural sciences that are relatively undeveloped and in which the realm of the tentative and the unsure is greatest. Douglas M. Johnston has described this well in describing what the problems of the world's fisheries are for the different disciplines:

For the natural scientist, and especially the fisheries biologist, the problem is seen as an unending inquiry into the nature of things. The sea is still a great mystery and in many regions marine observations and experiments have scarcely begun on a significant scale. . . . For the lawyer the problem of the world's fisheries lies in the present difficulty of settling fairly specific disputes or issues in such a way as to satisfy the canons of consistency, uniformity, certainty, objectivity, and justice. Johnston, D. M., *International Law of Fisheries*, 129–130 (1965).

scientists trying to develop legally or socially important "answers" with imperfect scientific knowledge can be found in Chapters 14 and 17.

Relatively few scientists are willing to set forth incomplete or only partially conclusive results to assist lawyers or courts; indeed, many consider the practice of science and the adversary system of the law to be antithetical. The entry of science into litigation through the testimony of expert witnesses occurs in markedly different fashion from the general practice of science. The critical difference is that there is no peer group review of expert testimony rendered in a legal proceeding, and while opposing experts may have an opportunity to present their findings, the deliberative process of the practice of science is missing. Thus while the scientific investigator is shocked at the lawyer making judgments with intemperate haste, the attorney is dismayed at the researcher's unwillingness to take a firm position on anything and only suggesting that what is needed is a great deal more research into this very complicated subject.

Finally, under conditions where the legal game is played for high stakes, money, coercive tactics, or covert peer group pressure can drive out accurate scientific findings or at least be used to obfuscate an issue with the sheer number of witnesses. Hence some scientists come to view the adversary proceeding as a rat's nest and, having expressed their views in a single forum, tend to shy away from future proceedings. For these and other reasons, the most highly qualified scientist may be an unwilling or even undesirable witness.

These points are not made to suggest that it is impossible to present a model persuasively in a court, but to try to illuminate some of the natural areas of friction and difficulty in the presentation. Courts have experienced lengthy and factually complicated trials that rely on a great deal of mathematical presentation. Antitrust and school desegregation litigation are two of the obvious examples. The crucial problem in the presentation of any case—or any model—is to make it intelligible and to assure that the evidence given is candid and forthright in nature.

Making a model intelligible to a court requires understanding how the model will function in terms of the court's work. The first function that a model can serve for a court is to organize and make sense of apparently disparate scraps of evidence, building up a system by which the parts become integrated into a whole. Thus in the area of nuisance law, a court may be faced with the claim that the defendant is causing a nuisance by dumping certain amounts of pollutants into a stream or marsh causing any number of perturbations in the natural system from killing fish to eventually exposing men to heavy metal poisoning. A model in such a case provides a framework by which to connect the parts of the evidence and form a coherent and rational judgment as to the importance of various parts of the evidence. The major job in the presentation of this sort of model is educational—to lay before the court an intelligible account of the workings of the system affected by the action in dispute.

But it must also be borne in mind that the court is not interested in a general education, but in specific instruction aimed at informing it what the effect of the disputed action has been or will be. To be of any use to a court, the model must show how a system will or will not be altered by actions which the court can control. A suggestion that someone upstream in another jurisdiction cease his pollution may be altogether meritorious but will not be relevant to the issues which the court must decide.

In addition, the courts are accustomed to making judgments on partial knowledge based on conflicting human memories and judgments. It is not unusual in a trial that it is impossible to do more than reconstruct the barest outline of the actions that led to a lawsuit, or that two parties engaged in the same action come away from it sincerely confident that quite different things transpired. The court may not postpone a decision until an unassailable answer is found; its task is to base its judgment on the best evidence and information that are available. In these circumstances, a court will not dismiss a model that does not rest solely on definitive investigations that leave no room for doubt or argument, and the modeler should not be hesitant to present a model that rests on assumptions and contains uncertainty so long as the assumptions are fairly spelled out and the uncertainty clearly reflected in the expression of the results. Sensitivity analysis is extremely helpful in making clear to the audience the importance of particular assumptions and identifying the crucial leverage points in the system being analyzed.

It cannot be emphasized enough that clarity and candor of presentation are crucial. No witness should attempt to give unassailable final answers when all that is honestly available are estimates of what will result from the given course of action. Clarity and candor are the keys that allow the modeler to bring the laymen in the courtroom into the world of his specialty and permit him to overcome the difficulty of jargon and the distrust of methodologies and expertise that the layman is not independently able to judge accurately. At every important point, the elements and workings of the model should be made as tactile and concrete as possible, and the extent of knowledge as to the workings of the system should be made clear. There is every reason to facilitate and encourage the trip from the laboratory to the courtroom, but the journey will always be painful unless one is honestly prepared to cope with the differences between legal and scientific procedures.

As controversies grow over the management and development of our natural systems and as the opposition of competing interests brings those controversies into the courts, the need for a continuing scientific contribution and commitment to the legal process will grow. The differences of disciplines will always make the contribution a difficult one, but if scientists and their models are to have a practical effect on the world around them, the contribution is indispensable. The achievements of science have allowed us to increase enormously the speed at which we can change the world. If those changes are to be rational, scientists must be willing to inform lawyers and courts as best they can of the consequences of the courses that lawyers and courts will be choosing for us all.

THE ADMINISTRATIVE AGENCIES AND
THE OPPORTUNITIES FOR MODELING

Unlike the courts, the administrative agencies have a direct role in formulating the policies that govern the use of the country's natural resources. Each agency acts under the mandate given by Congress, and that mandate may grant wide discretion to the agency in reaching the general goals set by Congress. The range of the agency's discretion is, of course, a matter for Congress to decide, but with great frequency Congress has given the agencies vast and comprehensive discretion. Thus the Forest Service is to administer the national forests for multiple use and sustained yield. Multiple use is given a broad and inclusive definition without sharp delineation of what actions are to be taken in any particular situation. The forest service, is to achieve

the management of all the various renewable surface resources of the national forests so that they are utilized in the combination that will best meet the needs of the American people; making the most judicious use of the land for some or all of these resources or related services over areas large enough to provide sufficient latitude for periodic adjustments in use to conform to changing needs and conditions; that some land will be used for less than all of the resources; and harmonious and coordinated management of the various resources, each with the other, without impairment of the productivity of the land, with consideration being given to the relative values of the various resources, and not necessarily the combination of uses that will give the greatest dollar return or the greatest unit output. 16 U.S.C. 531 (a).

Both the discretionary mandates and the requirement of environmental impact analysis under the National Environmental Policy Act give the agencies the duty to take an active role in searching out all the facts in any dispute or application presented to them and on the basis of their expertise chart out the course which meets the responsibilities given them by the legislature. Judge Hays summarized it in his famous instruction to the Federal Power Commission that the agencies are to be more than umpires blandly calling balls and strikes. Rather they have an active and affirmative responsibility to protect the public interest as expressed in their statutory mandate.*

* *Scenic Hudson Preservation Conference v. Federal Power Commission*, 354 F. 2d 608, 620 (2d Cir 1965).

One must recognize that the typical instruction to an agency requires the weighing of social, economic, and political factors as well as strictly scientific or biological ones. For instance, at the Law of the Sea Conference, consideration is being given to a treaty that would extend the exclusive fishing rights of coastal states to 200 miles from the shore. The State Department has undertaken an analysis under the National Environmental Policy Act of the American position. Such an analysis must, of course, consider knowledge of our coastal fisheries. But it must do much more. The extension of fishing claims outward from the coast also raises the question of how much of its national resources the country is willing to put into enforcing the claim. It demands analysis of the effects on the balance of payments. These in turn inevitably raise political and perhaps moral questions of the ordering of internal priorities. In addition, if a nation claiming exclusive fishing rights supports a fishery focused on one or a few preferred fish while not exploiting a number of other edible but unprized fish, which would otherwise be taken by other nations, clear moral and political questions of world food supply and the equity of international relations are raised.

The point here is not to suggest that modeling is a limited tool that cannot be expanded to encompass the complexity of the real world so as to tell the State Department negotiator or another administrative agency what it seeks to know, but rather to emphasize that even when the pattern of a physical system can be described adequately there remains a host of subjective value judgments that will have a profound impact on the use to which the models are put. Undoubtedly modeling will be enlarged in numerous instances in order to take account of the more roughly measured judgments by which the relations between, say, a nation's fishery policies and its energy production and water supply policies are set forth. An effort along these lines has been made in reviewing Canadian water policy.* These efforts have the very real value of heightening the consciousness of the participants to the parameters to which the system is most immediately sensitive, and by working and reworking they aid in the development of clarity in the formulation of policy. Nevertheless, the introduction of a series of weighted parameters indicating possible policy courses as an adjunct to a model of a natural system inherently alters the modeling exercise.

The broad mandates given to the agencies should make the federal administrative agencies a paradise for both modelers of strictly natural systems and for modeling that mixes natural and social science. Here should be the real opportunity to support the research that will allow one to understand a system like a vast national forest or the comprehensive workings of a waterway, and modeling should be well suited to predicting and examining

---

* Kane, Vertinsky, and Thomson, *KSIM: A Methodology for Interactive Resources Policy Simulation*, 9 Water Resources Research 65 (1973).

the effects on the natural systems of various purposeful or accidental human intrusions. Unfortunately the agencies have not yet fully exploited these opportunities. In the meantime many of the case histories presented in this book are real efforts to integrate science and social decisions.

One of the most pointed examples of the difference in approach taken by the various federal agencies and the failures of many of them to exploit the situations in which modeling could be of great use is provided by examining the actions of the agencies involved in licensing power plants on the reach of the Hudson River that covers the spawning and nursery grounds of the striped bass (Chapter 14). Each power plant using once-through cooling poses a threat to the estuary's striped bass population that can be understood only through a knowledge of the life cycle of Hudson striped bass and the relation of the fish population to the entrainment of eggs and larvae in the water withdrawn for use in the power plants.

Under the National Environmental Policy Act, the Nuclear Regulatory Commission (formerly the Atomic Energy Commission) was required to review the effect of water withdrawal by three nuclear plants, and the Army Corps of Engineers was required to review the effects of two large fossil fuel plants for which the Corps must give construction permits for the intake and discharge systems. Under the terms of the Corps' construction permits, the Department of the Interior has the power to recommend measures for the protection of the river's fish; under the Fish and Wildlife Coordination Act, both Interior and the Department of Commerce must be consulted on the effects on fish and wildlife of licensing actions by other federal agencies. Under both NEPA and the Federal Power Act, the Federal Power Commission was required to review the effects of a proposed pumped storage plant at Storm King on the Hudson. In addition, the utilities have applied to the federal Environmental Protection Agency (EPA) for permits under the 1972 Federal Water Pollution Control Act Amendments to discharge heated water to the Hudson, and under the terms of that act EPA is now reviewing the effects of the plants' operation on the Hudson fishery. Finally, the State of New York, which holds the fish of the Hudson in trust for its people, regulates the maintenance of water quality in the Hudson and through the Attorney General and the Department of Environmental Conservation has attempted to control the entrainment and impingement of fish at the plants.

These official bodies frequently have failed to put together the known information about the Hudson in order to arrive at a model of the striped bass life cycle, which is related to the hydrology of the Hudson and water withdrawal at the power plant sites. Most of the agencies do not have the biologists, experts in hydraulics and computer programmers who could take on the work. Many contend that they do not have the budget to hire competent analysts to undertake this task. Thus the first attempt that the Army Corps made to address the problem left the reader with the impression that there wasn't the slightest problem from the entrainment of nonscreenable organisms, and the Interior Department did nothing in its backup role to force the Corps to carry out the analysis that was needed. In fact, only the Atomic Energy Commission initially produced the kind of full-scale modeling effort that the situation demands, probably only because the AEC had at its disposal the expertise and investigative determination of its National Laboratories. The importance of the modeling technique is brought home by the fact that the AEC analysis of the Hudson showed the potential for major damage to the fishery, while the other agencies saw no cause for alarm. The result of the failure of the agencies has been that both the collection of information and its sophisticated analysis was largely left to the utilities; Consolidated Edison of New York, for instance, is spending over $3 million per year on Hudson research and analysis. This is too much of the cat watching the canary for the comfort of anyone concerned with the general public interest.

What is needed in both the administrative agencies and in the Congress, which must see that the bill is paid, is proselytizing work that will bring home to those with a responsibility for the management of natural resources the value that competent systems modeling and analysis can have for them. If the broad mandates of Congress are to be carried out so that the choices of alternatives are, in fact, rationally made with a consciousness of what the choices entail, every strategy for increasing the disinterested scientific capability of the agencies must be pursued: the expansion of the National Laboratories and other centers of scientific research and their use in reviewing resource management, the increase of license fees to pay for the research necessary to analyze major intrusions into natural systems, the introduction of trained researchers into the daily workings of agencies with resources management responsibilities (but not their submersion in the trivia of daily management), and the use of public pressure and litigation to require that agencies discharge their duties in keeping with the highest level of the art in analysis and review of natural systems.

This is the real challenge and opportunity for natural-systems modeling in the context of the legal structure of public policy formulation. If the great administrative agencies with their vast discretionary powers can be encouraged to use our capability for comprehensive scientific analysis wisely, we shall all be more wisely governed. If lawyers, administrators, and scientists fail in this task, our knowledge will have little effect on the management of our affairs, and the disciplines of policy formulation and analytic management will have served us poorly.

PROLOGUE TO SECTION II

Our view is not that models are some kind of magic that supplant and sit outside of traditional procedures of science. Instead, we view models as but one of many tools available to science, to be used in conjunction with traditional tools of observation, correlation, and experimentation. The following four chapters make the point rather nicely that models alone can tell you relatively little about ecosystems unless used in conjunction with the traditional methods and procedures of science. For example, Botkin's model of forest growth is based on extensive quantities of field data gathered over many decades by field botanists. One of the more important and elegant aspects of this model, missing from many other models, is an emphasis on interspecific competition. As we learn more about this process in other systems, we may increasingly include such detail in other models. The Louisiana marshes studied by Hopkinson and Day, on the other hand, are in some respects a bit easier to model than forests, since there are only a few dominant species (e.g., virtually the only important land plant is *Spartina alterniflora*). Yet the simplicity of this aspect of their ecosystem is matched by a much greater degree of complexity in the interaction of the marsh ecosystem with neighboring ecosystems and the Gulf of Mexico. Again, we see the importance of getting good field data to build a model and conversely, the way in which a model can assist in deciding which field data is the most important to get. The relation of this vast and important ecosystem to man's activities is considered in the third section of this book.

Jeffrey Richey has undertaken a most ambitious and frustrating project. Anyone who has seen the rather neat and simple diagrams of the phosphorus cycle in an introductory ecology textbook will be impressed at how much more complicated the cycle is in a real ecosystem. All the problems of building the model were complicated by the additional problems of having inadequate chemical methods for making measurement. We think that the most important aspect of his paper is not the construction of the model per se, but that the process and failures of building his first model were important in the development of his next, more ambitious study of phosphorus at Lake Washington.

Finally, Richard Wiegert shows us the relative precision that can be built into a model when you have the advantages of 8 years of field data and a relatively simple system to work with. There are two very important contributions to modeling theory contained in this chapter. The first is a general feeding equation, and the second is an investigation of how different levels of model complexity effect the results of the model. More recently, the basic concepts that were worked out for the relatively simple hot-spring communities have been applied to much more complex ecosystems, including the salt marshes of the coast of Georgia.

210

II CASE HISTORIES
Understanding Natural Systems

9
Life and Death in a Forest: The Computer as an Aid to Understanding

DANIEL B. BOTKIN

In 1748 Peter Kalm, a Swedish naturalist, visited North America, arriving at Philadelphia and traveling through much of what is now the northeast of the United States and the southeast of Canada. On his way to Montreal, he passed through virgin forests in northern Vermont. "Almost every night," he wrote in his journal, "we heard some trees crack and fall while we lay here in the woods, though the air was so calm that not a leaf stirred "(Kalm, 1966). Knowing no reason for this, he suggested that perhaps the dew loosened the roots of old trees at night or that immense flocks of passenger pigeons settled on the branches unevenly. Whatever the reason, he wrote that "they made a dreadful cracking noise." Reading his account, one imagines him confronting the deep woods with a sense of mystery and perhaps awe, for the basic life processes of birth, growth, and death, poorly understood today, were even less known for the trees in a forest in Peter Kalm's time. But we find also within this traveler a sense of utility. "All the land we passed over this afternoon was rather level," he wrote elsewhere, "without hills or stones, and entirely covered with a tall and thick forest in which we continually met trees which had fallen down, because no one made the least use of the woods."

This dual attitude holds on the one hand a feeling of awe, reverence, mystery, or beauty towards trees and forests, and on the other a sense of utility, a desire to cut the damn things down, make houses and bridges, and clear the land for farming. Such an attitude runs throughout man's history. Today, when much of the world's primeval forests have been cut down and the population and aspirations of man are rapidly increasing, this dualism has created a major conflict of interest between those who want standing trees and those who want stacks of boards. The stands of trees are valuable not only for man's soul—for his recreation and aesthetic appreciation—but forests also have a multitude of indirect benefits, such as helping control erosion and floods, aiding pollution abatement, and protecting watersheds used for water supplies. Yet the conflict between keep and cut can exist within each of us, as anyone who has wished to build a house of boards in a clearing in the forest can testify. The time seems past when we can take a haphazard approach toward these conflicts. If we want our future to have both wood and woodlands, an element of planning, management, and wise use seems necessary. Wise use of our woodlands demands knowledge and understanding. The purpose of this chapter is to point out that our search for these can profit from the assistance of some modern tools including computers.

Why huge trees fall over in the dead calm of a summer's night is still a mystery for us as it was for Peter Kalm. In fact, how a collection of many kinds of species, of trees, shrubs, and herbs, of grazers like the deer and the caterpillar, predators like the wolf and the spider, and decomposers like the puffball and soil bacteria, function together as a unit to make a forest ecosystem is poorly understood and a subject with much room left for imaginative investigation.

Forest ecosystems are difficult to study, because they are such a complex collection of species, they are large in human terms, and events are slow in terms of human generations. But as objects of ecological study, forests do have certain advantages. These include the growth rings of trees, which contain not only an account of a tree's age but also how well it was able to grow during each year of its life. Thus the history of the forest can be read in the growth rings of the trees giving us insight into the general laws that govern the dynamic operation of natural ecosystems, which hopefully one can apply to other, less stationary, natural communities.

The primeval forest that Peter Kalm saw, although little disturbed by man, was far from uniform. "I was told, in several parts of America," he noted in his journal, "that the storms or hurricanes sometimes pass over only a small part of the woods and tear down the trees in it; and I have had opportunities of confirming the truth of this observation by finding places in the forests where almost all the trees had crashed down, and lay in one direction." Where such a catastrophe occurs, the forest is reestablished by a process known as *secondary succession*.

To understand the process of succession, one must realize that trees compete at all times for essential resources—light, water, minerals, and a place to grow. In any particular spot in a forest, the tall trees shade smaller ones and suppress their growth. A tree "wins" the local competition by growing faster and shading its neighbors before they shade it. No one species wins the competition everywhere all the time as each is adapted to specific conditions. As a result of long periods of competition, different species have evolved to take advantage of different spacial and temporal conditions. Some species (called pioneer species) are adapted to the conditions immediately following a catastrophic clearing like the ones Peter Kalm observed. In northern New England, pin cherry is such a species. Its seeds are spread widely by birds and mammals who eat its fruits. The cherry seeds do not sprout in the deep forest shade, but wait until a clearing occurs. Then from some signal that is not yet understood, all the cherry seeds sprout during the first year that a large well-lit clearing occurs. If soil conditions are right, the seedlings grow and produce a dense, even-aged stand of cherry trees. The young trees grow rapidly but, as trees go, live a short time. This rapid growth and short life is typical of early successional tree species. White birch is another such early pioneer, but its seeds can sprout in shadier conditions that would not suit pin cherry (Fowells, 1965).

Since seedlings of pin cherry or white birch cannot survive under the shade of their parent trees, a stand of these species is temporary in any one location and gives way slowly to trees whose seedlings can persist in deep shade. In northern New England, sugar maple, beech, and red spruce are among these species. In contrast to the pioneer types, these tend to be slow growing and long lived. They are the dominant trees of old undisturbed forests, and one can imagine an idealized forest with constant climate and no catastrophes where these species would dominate most of the landscape, with all ages— seedlings, saplings, mature, dying, and dead trees—present indefinitely in any one local area. Although earlier in this century ecologists believed that such idealized forests had existed for long periods before European settlement of North America, the evidence available now suggests that in the last million years the only permanent attribute of forest species composition has been change itself. Along with the small catastrophies, such as sudden violent storms, slow but critical changes in climate have occurred accompanying the advances and retreats of the pleistocene glaciers.

Each tree species has a limited climatic range, and as the climate has changed, the species have marched north and south across the landscape. Thus we can view events in a forest from three different levels of detail. From a small scale point of view, we can examine the competition occurring in a single year in a single place among individual trees for light, water, and nutrients. On a larger temporal scale, we can examine the process of succession in a forest community. On a larger scale still, we can view the changes in the distribution of kinds of trees and kinds of forest communities that accompany changes in climate. These changes occur slowly over time and occur spacially at any one time with changes in topography—elevation, slope, and the compass direction of a slope.

As has been pointed out, the growth, reproduction, and death of trees are complex functions of light intensity, moisture, temperature, and nutrients. Because trees strongly affect these factors in a local area, the survival of each individual depends on its interactions with its neighbors and whether they are of the same or different species, size, age, or vigor.

Rarely are these important biological interactions well understood. Even when they are known, the interactions become very complex, and it is difficult for one to predict the consequences of the interactions among even a handful of species. In the past, an understanding of such consequences when it existed at all lay only in the minds of a few experienced naturalists. It was a matter for experience and intuition, difficult to communicate. Today we need to put this understanding on firmer ground and be able to predict behavior of a forest following our manipulation or after natural catastrophe. How can one predict the effects of perturbations on a biological community? How can one communicate these insights and predictions to those who are not experienced naturalists, but who have the power to make decisions that may determine the keep or cut strategy?

It has been apparent to many ecologists that computer models could provide useful tools for these problems. The reasoning behind this idea is as follows: One can create a computer model of a forest ecosystem, consisting of a group of assumptions and information in the form of computer language commands and numbers. By operating the model the computer faithfully and faultlessly demonstrates the implications of our assumptions and information. It forces us to see the implications, true or false, wise or foolish, of the assumptions we have made. It is not so much that we want to believe everything that the computer tells us, but that we want a tool to confront us with the implications of what we think we know.

As part of an ecological investigation called the Hubbard Brook Ecosystem Study involving the observation and manipulation of watersheds in northern New Hampshire, we have begun to examine a variety of questions with the aid of a computer model of forest growth, called Jabowa (Botkin et al., 1972a). The model is the result of a cooperative effort between scientists from Yale University and James F. Janak, a theoretical physicist, and James R. Wallis, a hydrologist, from the IBM Thomas J. Watson Research Center. How this has been done is the main point of this chapter.

The development of such a model goes through different stages. First one creates a conceptual model based on one's assumptions about how a forest works. This conceptual model then is translated into a set of mathematical equations representing quantification of these assumptions, and these equations are translated into a computer algorithm. Then one tests the computer model to see if it works—whether it can simulate events that one knows occur. Perhaps modifications are necessary at this point. Subsequently, the model can be used as a way of synthesizing information about the ecosystem, as an educational device for teaching what is known about how the ecosystem behaves, and as a tool for testing new hypotheses.

The Jabowa Model simulates the growth of individual trees on small forest plots. The size of these plots can be changed by the user of the program, but the standard size is a square 10 m on a side.

The model is operated on a digital computer and was originally designed for *interactive* use, meaning that a user of the program can interact with the computer during the execution of the program. The model thus becomes a kind of game where the user tells the computer the kind of site where he wants to grow a forest and what trees are present at the beginning. When the computer begins to grow the forest, the user can stop the program at any time and make changes: he can remove or add trees, revise the climate, or change the characteristics of the species. Thus the game played can be a scientific one where the user can ask questions such as what is the best combination of species characteristics to win out in the forest competition or how important is any particular characteristic in the competitive success of a species. The program can also be used as a management game—a user can ask a question such as what interval between cutting results in the greatest production of lumber or what is the effect of a certain kind of perturbation on the forest. Of course, the results must be taken with extreme caution, merely as suggestions of patterns and possibilities to be thought through carefully, and the user must be well aware of the limits of the assumptions and information that form the basis of the model. Through the process of working with the model the user develops new hypotheses about the forest community which he can then test against the real world. Each success or failure enables the model to be made more accurate.

Development of the model was begun in 1970, and it now exists in a variety of forms. One version operates with interactive television graphics. A user sits at a console that includes a television screen, a typewriter keyboard, and a "light pencil." The user interacts with the image on the screen by typing commands on the keyboard and by pointing the light pencil at the screen. Each tree is represented on the screen by a rectangle and two-letter code. The height and diameter of the rectangle represent the tree's height and diameter, and the code tells the tree's species. The user can plant trees of any size and species, selected from a master list. "Logging" trees is done by typing a "Log" command on the keyboard and pointing the light pencil at the trees that one wants to remove, which then disappear from the screen. The growth of the forest can then be continued with the computer taking into account the changes in conditions resulting from the logged trees.

Two other versions in two different computer languages (FORTRAN and APL) have been written for use on interactive typewriter terminals. In these, the state of the modeled forest (in this case, the species and size of each tree on simulated sampling plots), is represented by numerical tables (Figure 1), but the same interactive potential is available as with the television graphics. Other versions for long-term, large-scale simulated experiments are used in "batch" modes—without interation between user and computer during program execution.

We have tested the model enough to know that it does work at least in the sense that it reproduces the general events at each of the three levels of investigation mentioned before (competition among individual trees, succession within a community, and changes in the community accompanying changes in climate). That the computer model can mimic competition among individual trees is shown in Figure 2. Here at the beginning of the simulation a large sugar maple tree dominates the plot. This tree grows well, but shades the other, smaller trees so much that they grow little if at all. When the user of the program cuts down this large tree, the computer shows that the other trees grow much better. They are released from suppression,

```
load jabowa
q jabowa
at 100;set ix=1065786486;set log=0;set kpnt=15;set nyr=60;set ktimes=1;set till=4.0;set ielev=2000
00001
jabowa


                    THE JABOWA -- A NORTHEAST FOREST GROWTH SIMULATOR (VERSION 1)

DO YOU WANT A LIST OF SPECIES NAMES? REPLY Y OR N

y
THE 13 TREE SPECIES USED ARE THOSE FOUND ON THE 208 EXPERIMENTAL PLOTS OF
THE HUBBARD BROOK ECOSYSTEM STUDY
 1  SUGAR MAPLE      ACER SACCHARUM
 2  BEECH            FAGUS GRANDIFOLIA
 3  YELLOW BIRCH     BETULA ALLEGHANIENSIS
 4  WHITE ASH        FRAXINUS AMERICANA
 5  MOUNTAIN MAPLE   ACER SPICATUM
 6  STRIPED MAPLE    ACER PENSYLVANICUM
 7  PIN CHERRY       PRUNUS PENSYLVANICA
 8  CHOKE CHERRY     PRUNUS VIRGINIANA
 9  BALSAM FIR       ABIES BALSAMEA
10  RED SPRUCE       PICEA RUBENS
11  WHITE BIRCH      BETULA PAPYRIFERA
12  MOUNTAIN ASH     SORBUS AMERICANA
13  RED MAPLE        ACER RUBRUM
DO YOU WANT INFORMATION ON PARAMETERS? REPLY Y OR N
n
00001

                          ****************************************      IX=   1065786486

PLOT     ELEVATION    SOIL      PERCENT    GROWING        INDEX OF
NUMBER   (METERS)     DEPTH     ROCK       DEGREE DAYS    ACTUAL ET
  0        610         1.2        0          2549.4         423.1
```

Figure 1 Output of JABOWA version as it appears on the user's terminal, showing 60 years of secondary succession at 15-year intervals on the test plot (Plot 0). In this case the user specified the Elevation and Soil Depth (in meters), the minimum size tree to be cut, the number of years of the simulation, and the interval between printouts. Index of Actual ET is an index of growing season evapotranspiration in millimeters; Spec. is the species number, and refers to the list in Table 1; Num. is the number of trees of each species; Basar is the basal area contributed by each species; numbers under DBH are the diameters in cm of each tree; Leaf Area is an index of the total leaf area on the 10-m × 10-m plot and is obtained by dividing the total leaf weight of all species on the plot by 45 (dividing the total index by 100 gives a rough approximation of Leaf Area index of Leaf Area on the plot); IX is the number initiating the pseudo-random-number sequence.

219

YEAR
0

SPEC.	NUM.	BASAR.	DBH
1 /	4 /	21.991 /	2.000 2.000 2.000 4.000
2 //	4 /	62.832 /	2.000 2.000 6.000 6.000
3 /	6 /	130.376 /	2.000 2.000 8.000 9.000 3.000 2.000
	14	215.200	LEAF AREA = 6.681

DIAMETER LIMITS FOR LOGGING BY SPECIES

1	2	3	4	5	6	7	8	9	10	11	12	13
0.0	0.0	0.0	0.0	0.0	0.0	0.0	0.0	0.0	0.0	0.0	0.0	0.0

ALL TREES WERE CUT
YEAR 0 NO TREES LIVING

YEAR
15

SPEC.	NUM.	BASAR.	DBH
1 /	4 /	7.234 /	1.890 1.591 1.434 1.024
2 //	3 //	9.991 //	2.574 1.873 1.608
3 /	13 /	139.496 /	3.722 3.731 3.722 4.417 3.713 3.717 3.814 4.202 3.368 3.889 3.392 3.022 3.094
4 /	2 /	1.347 /	0.970 0.879
5 //	2 ///	0.554 /	0.590 0.598
6	2 //	1.034 //	0.843 0.779
7 /	4 /	283.334 /	9.625 9.264 9.907 9.172
	30	442.988	LEAF AREA = 22.638

YEAR
30

SPEC.	NUM.	BASAR.	DBH
1 /	5 /	118.857 /	6.675 6.175 5.894 5.088 2.833
2 //	3 //	40.943 /	5.261 4.860 0.913
3 /	13 /	795.008 /	9.737 9.746 9.717 10.689 9.673 9.799 9.179 9.891 9.206 8.674 8.771 1.611 1.609

```
 4  /       3.825 /     1.789  1.058  0.744
 5  ///     1.157 ///   1.213
10  ///     4.781 ///   1.476  1.564  0.868  0.842
12  //      0.688 //    0.936
13  /       1.487 /     1.376

        31       966.745     LEAF AREA =   19.433

YEAR  SPEC.  NUM.  BASAR.         DBH
 45
       1    5  /     363.035 /    12.373 11.789 10.492  7.643  1.298
       2    4  ///   142.066 ///   9.172  8.704  3.406  3.066
       3   11  ///  1926.324 ///  16.474 16.476 16.334 17.710 16.210 16.341 16.433 15.584 15.001 5.555 5.547
       4    1  ///     5.828 ///   2.724
       6    1  /       2.609 /     1.822
       9    3  ///    74.750 ///   6.713  6.723  2.215  2.399  1.192  0.733  0.609
      10    7  ///    29.539 ///   3.370  3.493  2.442  1.010
      13    4  /      15.453 /     3.829  1.846  0.766

        36      2559.603     LEAF AREA =   60.018

YEAR  SPEC.  NUM.  BASAR.         DBH
 60
       1    3  /     604.325 /    17.833 17.154 12.536
       2    4  ///   167.922 ///  12.846  6.395  2.382  1.486
       3    9  ///  3200.747 ///  22.939 22.829 22.485 24.378 22.038 22.166 21.186 20.438 9.194
       4    2  ///    10.893 ///   3.604  0.938
       5    1  ///     0.736 ///   0.968
       9    3  ///   325.418 ///  17.080 11.037  0.887
      10    6  ///    31.617 ///   5.329  2.622  1.690  1.134  0.576  0.710
      12    3  //     11.352 //    2.551  1.994  1.993
      13    3  /      50.946 /     6.576  4.047  2.291

        34      4403.949     LEAF AREA =  110.214

CHCRW400   TERMINATED: STOP
```

Figure 1 (*Continued*)

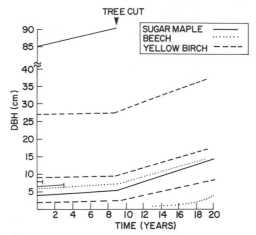

Figure 2 Simulated growth in diameter of individual trees on the same plot. Each line represents a single tree; the end of a line is that tree's death. At year 9 the large sugar maple was removed, and the remaining trees, no longer suppressed by the large maple, show increased growth rates.

and they respond according to their relative sizes and species. For example, the largest remaining tree grows very well, better than smaller remaining trees. Also, a smaller sugar maple grows slightly faster than a similar-sized beech, consistent with expectations for these species under the environmental conditions of the hypothetical plot.

That the model reproduces important aspects of the process of succession is illustrated in the next figure, (Figure 3a) which shows that numbers of the species characteristic of old stages in the forest, such as beech, increase with time. A graph of "basal area," the cross-sectional area of the tree stems (the

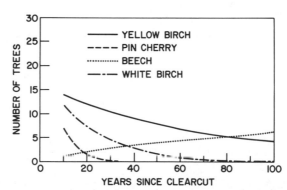

Figure 3a The average number of trees per plot for 100 plots with identical site conditions, DEGD = 2550, and deep, well-drained soil.

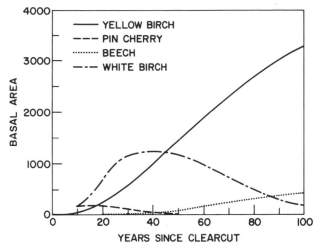

Figure 3b   The average basal area per plot for 100 plots with identical site conditions, DEGD = 2550, and deep, well-drained soil.

area of stumps if all the trees were cut down), which is an index of the amount of tree biomass on the average plot, shows that each species reaches a peak at a different stage in succession with the pin cherry first, then white birch, while yellow birch is still on its way to a peak at year 100, and beech is just beginning its increase (Figure 3b). The model was validated by comparing model predictions with data from real forests as found in a number of publications (e.g., Fowells, 1965; Siccama, 1968; Bormann et al. 1970; Marks, 1971). Details of this validation can be found in Botkin et al. (1972a).

Figure 4 shows the model's prediction of changes in two species with elevation and, therefore, under different climatic conditions. This figure compares the roles of two contrasting species, sugar maple and white birch, during the reestablishment of a forest following clearcutting at different elevations. These predictions are also consistent with studies of the actual behavior of these species in northern New England (Bormann and Buell, 1964; Bormann et al., 1970; Siccama, 1968; Marks, 1971; Braun, 1972).

How is it possible for a computer program to mimic all of these complex events? As in a real forest, the simulated forest contains tree species adapted to different successional stages. Taller trees shade smaller ones, but species with more leaves for a given individual size shade smaller competitors more than other species. Under shaded conditions, photosynthesis is higher for shade tolerant species than for intolerant ones. Species are further differentiated by species-specific survival probabilities and by differential addition of new saplings in relation to light, temperature, and moisture conditions at the forest floor.

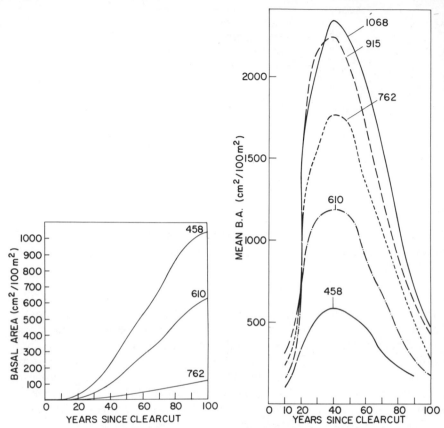

Figure 4   Average basal area as a function of time since clearcut and elevation for each of the six major species: (*a*) sugar maple, (*b*) white birch. Each line represents the average for 100 plots at a single elevation with identical site conditions including a deep, well-drained soil.

Let us now examine how these mechanisms are simulated in a little more detail, beginning with growth.

What determines how fast a tree grows? Consider a tree growing under the best possible conditions with as much light, water, and nutrients as it can use. The sugar produced by photosynthesis in the leaves is first used to maintain the living portions of existing stems, branches, and roots. New tissue can only be made from the sugar not required for maintenance of existing tissue.

Thus a tree growing in the open collects an amount of radiant energy roughly proportional to its leaf area, and its growth will be proportional to this leaf area, derated by a factor that takes into account the amount of nonphotosynthetic tissue that must be maintained. The nonphotosynthetic living tissue is concentrated in the vascular tissues of the inner bark. The thickness of these tissues remain relatively constant, and, therefore, the amount of this tissue is roughly proportional to the stem surface area. The Jabowa growth rate equation that states these assumptions for a tree growing under optimal conditions is

$$\delta(D^2 H) = RLA \left(1 - \frac{DH}{D_{max} H_{max}} \right) \tag{1}$$

where D is the diameter of the tree, H is its height, D_{max} and H_{max} are the maximum values of these quantities known for a given species, LA is the leaf area or weight, and R is a constant. The equation states that the change in volume $(D^2 H)$ of a tree over a period of 1 year is proportional to the amount of sunlight the tree receives, derated by $(1 - DH/D_{max} H_{max})$, which is proportional to the surface area (DH) of the tree stem.

In the actual operation of the computer program, the height of the tree and the leaf weight are determined from other equations as a function of the diameter, and the equation is solved for changes in diameter rather than in volume [those interested in more information about these details should consult Botkin et al. (1972a; 1972b)].

The reader should note that the factor $(1 - DH/D_{max} H_{max})$ approaches zero as DH approach $D_{max} H_{max}$ so that a tree cannot exceed the maximum height and diameter observed for its species. This equation allows species to be distinguished in terms of their maximum sizes and in terms of their optimal rates of growth at any diameter below the maximum.

The growth curve (Figure 5) resulting from this equation has a biologically satisfying shape, called the logistic, which is characteristic of many real growth curves.

The right-hand side of the growth equation is multiplied by additional factors that take shading, climate, and so forth into account. Light is necessary for photosynthesis, and the more light the greater the rate of photosynthesis—up to a point. At low light intensities, a small increase in light makes a big difference, but eventually, as the light gets brighter a green plant becomes "saturated," its machinery can work no faster, and an increase in light has no effect. Hundreds of studies have been made of the relationship between photosynthesis and light for green plants. These suggest that, generally speaking, there are two basic types of trees, called shade tolerant and shade intolerant. The shade-tolerant ones grow well under low light conditions, but are saturated at relatively moderate light levels. The shade-intolerant ones grow poorly in the shade, but much better than the shade-tolerant ones in bright light. Studies based both on theoretical considerations

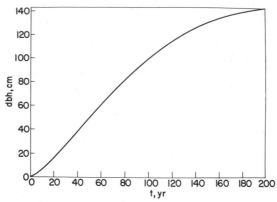

Figure 5 Time-dependent solution of the diameter-growth equation (Eq. 1) using the parameters for sugar maple (species 1).

involving the physical and chemical mechanisms of photosynthesis and on observations of plants suggest that these relationships can be represented by an equation of the form

$$r = a_1(1 - e^{-a_2(AL - a_3)})$$

where r is the relative rate of photosynthesis, AL is the light available to the tree, and a_1, a_2, a_3 are constants. The constants chosen for shade-tolerant and intolerant species (Figure 6) give reasonable fits to measured photosynthesis curves (Kramer and Kozlowski, 1960).

The amount of light available to each tree is determined by comparing the height of each tree to that of all other trees on the plot and decreasing the light intensity available above the forest by the amount of shading that occurs from the leaves of these taller trees.

Tree species have limited geographic ranges. Some grow in warm climates, others in cold. Maps of climate and of the distribution of trees suggest that the north and south boundaries of a tree species follow closely lines of uniform average temperatures, called temperature isotherms (Figure 7). Laboratory studies suggest that a species has an optimum temperature for growth and that its growth increases up to that optimum and decreases above it.

A rough index of these thermal effects is obtained from the number of growing-degree-days per year (40°F base) for the site. This quantity is defined as the sum of $(T - 40)$ over all days of the year for which the average temperature T exceeds 40°F. For each species, the relative growth $T(DEGD)$ is

$$T(DEGD) = \frac{4(DEGD - DEGD_{min})(DEGD_{max} - DEGD)}{(DEGD_{max} - DEGD_{min})^2} \quad (2)$$

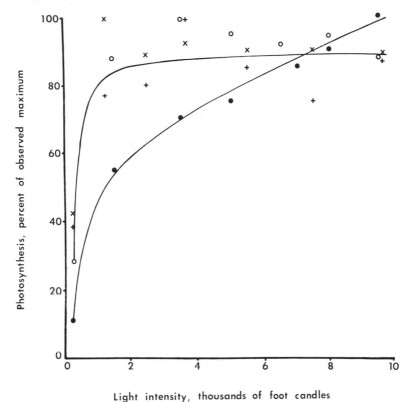

Light intensity, thousands of foot candles

Figure 6 Photosynthetic rate as a function of the available light (fraction of annual insolation) for shade-tolerant and -intolerant trees.

where $DEGD$ is the degree days for the climate of the site where the tree is growing, $DEGD_{min}$ is the degree days at the northern boundary of its species, and $DEGD_{max}$ is the degree days at the southern boundary of its species. The function is a parabola having the value zero at minimum and maximum values of $DEGD$ between the extremes.

Trees die from disease and catastrophies like fire and storms. Unlike man, a tree must continue to grow to survive, and a tree that is not growing well is much more likely to die than a vigorous one. The computer program takes this into account by keeping track of trees that do not grow well and giving them a smaller chance of surviving. Healthy trees also have a chance of dying from causes such as severe storms, lightning, and defoliation by insects. For human beings, the chance of dying is high for infants, decreases for adults, and then increases with age. People interested in selling life-insurance premiums have made studies that show that this is true. No one

Figure 7 Geographic range of yellow birch and growing-degree-day contours that closely approximate the northern and southern boundaries of this species: Degree-days are factors used by the computer program calculated from annual average temperature conditions. See text for further explanation.

seems to have made such a study for forest trees. When one has little knowledge in science, one makes the simplest possible assumption and tests it out to see if it agrees with observations. For our model we made such an assumption. We assumed that a tree that remained healthy throughout its life would have a small but definite chance of reaching the maximum known age for its species and that its chance of dying in any year would not be affected by its actual age. There is some unpublished data that supports this assumption (Harper, personal communication). From these assumptions, the chance of dying for each tree in each year can be determined. The computer determines whether each tree dies by picking a random number between zero and one. If this number is less than the chance of dying (which is also a number between zero and one), then the tree dies.

The model does not consider seeds or tiny seedlings until they survive early stages and thus are added to the population of trees on the forest plot. In each year, new saplings of each species are added on the basis of relative tolerance to shade and whether temperature and soil moisture conditions allow growth of that species. Each year the light available at the forest floor in the model is checked against thresholds for each of the shade-intolerant species. If light is above a high threshold, a large number of saplings of the very shade-intolerant cherry species are added. If the light is between this threshold and a second lower one, birches are added, and the number of individuals is increased with the light intensity. If the light is below the second threshold, only shade-tolerant species, such as beech, are added. The exact number of trees added in any year is also allowed to vary randomly within a specified range. This range is consistent with the observed variability of the real forest.

The Jabowa model allows for random variation in mortality and regeneration and is thus a stochastic model. Since the range of variation in the model is consistent with observed ranges, the projections of the model can tell us not only the average condition to be expected for a plot, but also the variability one can expect. Natural ecosystems are inherently variable, and it is important for us to be able to make projections that take this variability into account. For example, if we want to know whether a fertilizer will improve the yield of a forest, we need to know whether the inherent variability in tree growth will obscure the effects of the fertilizer.

The above discussion briefly summarizes the basic operation of the Jabowa program [those interested in more details should consult Botkin et al. (1972a; 1972b)].

Now that we have considered the basis of the model and have also established that the model reproduces the major dynamic characteristics of a forest community, we can consider its use to conduct simulated experiments about the response of a forest to perturbations and manipulation. These experiments give us insight into the importance of species interactions to the success and survival of any one species and to the persistence of the entire forest community.

For example, the model has been used to simulate somewhat idealized kinds of fertilization and stress. In one experiment, the fertilization was assumed to act in a uniform way upon all trees and to affect only the growth rates. The fertilization was implemented in the program by directing the computer to increase the diameter growth of each tree in each year by a fixed percentage of what it would have been under the normal conditions for that tree in that year, given the other normal biotic and abiotic conditions.

Such uniform fertilization is not implausible. It has been known since the 18th century that green plants grow better in air enriched with carbon dioxide. One of the results of the industrial revolution has been to increase the CO_2 content of the atmosphere through the burning of fossil fuels and the destruction of large areas of old age forests and virgin soils. Although one would expect the effects of such an enrichment to be species specific, almost no experiments have been conducted to define these effects for tree species. Thus it is reasonable to begin a consideration of this problem by assuming the simplest, that is, uniform, effect. How would the development of a forest fertilized in this manner following clearcutting compare with a "normal" one? Suppose the treatment were an analogous uniform "stress"

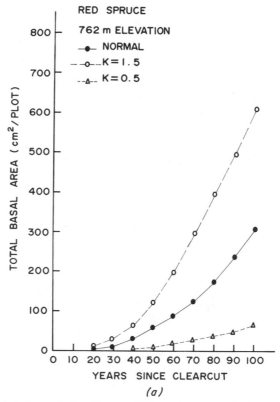

Figure 8 Simulated change in basal area per 10 × 10 m plot during the first 100 years of succession following clearcutting for spruce (a), and white birch (b). Shown are the means for 100 replicates in each treatment. Climatic conditions are those of 762-m elevation in Northern New Hampshire. K is the constant multiple of an annual diameter increment of each tree. Thus $K = 1.5$ means that the growth of each tree was increased 50% above what it would have been for that year, and represents a uniform "fertilization," while $K = 0.5$ represents a uniform "stress."

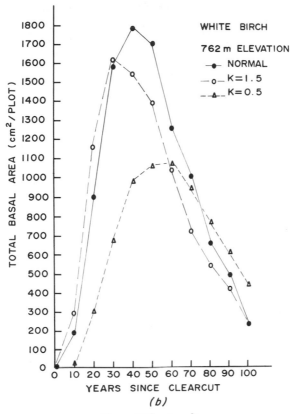

Figure 8 (*Continued*)

or decrease in growth? Our naive expectation is that, if all the growth of all trees is changed by a fixed percentage, the productivity of the entire forest would change by that percentage, and no species would benefit or lose more than any other. Our computer experiment, however, produces somewhat different results.

Species characteristic of mature forests that are shade tolerant and long lived, such as red spruce, show a very strong response to the treatments (Figure 8a). Just as one would expect, the spruce grows much more poorly under the "stress" conditions and much better under the "fertilized" conditions. However, the response to the treatments of species characteristic of early stages in forest development is considerably different. White birch, which is typical of early successional species, is shade intolerant and short lived. Under the "stress" treatment, this species does show considerable growth reduction in the first seven decades following clearcutting, but its contribution to the forest is greater than normal in the last three decades. Under the "fertilized" treatment, white birch does better during the first three decades but does poorer than normal afterwards (Figure 8b).

The result seems paradoxical, but a consideration of the assumptions of the model leads to a clear explanation. Under the "fertilized" treatment, the shade-tolerant species do so much better that they become stronger competitors against the white birch. Spruce and sugar maple grow taller and develop more leaves faster than normal, and their resultant ability to suppress white birch at least cancels the fertilization effect of the birch. Under the "stress" treatment, the canopy develops more slowly, so that light conditions near the forest floor benefit the birch and at least cancel the effect of the "stress" on this species.

As a result of the interactions among species, the productivity of the entire forest changes less than we would have guessed from our naive expectation, and in this way the species interactions seem to buffer the community against the perturbing influence.

What would happen if the fertilization affected only a single species? In another experiment, Dr. Edward G. Brittain of the Australian National University, along with myself, used the model to test the effects of changes in photosynthetic rates of a single species on the role of that species in the community. When the photosynthetic rate of sugar maple was increased greatly, the model predicted quite expectedly that the volume of sugar maple in the forest increased greatly at the expense of its most similar competitor, beech. Both of the species are long lived, shade tolerant, and characteristic of old age forests. The surprising result of this treatment is that it also resulted in an increase in the volume of pin cherry, a short-lived, shade-intolerant species characteristic of the early stages of forest development following a catastrophic clearing. This seems at first a puzzling result. Why should improving the competitive position of one species improve that of another adapted to quite a different stage in forest development? Examining the results of the simulation, it became clear that the altered sugar maple species, because its established individuals grew so well, created a forest stand of a comparatively few large trees. When one of these very large trees died, an opening was created in the forest large enough to allow the introduction of pin cherry. In the "normal" forest, sugar maple trees generally remained small enough so that the simultaneous deaths of two or more neighbors were required to allow enough light to reach the forest floor so that pin cherry could grow—a much rarer event.

Thus a simulated change which at first glance one would expect to benefit a single species had indirect benefit to a second very different one. This result was not obvious or predicted beforehand, but in retrospect it is clearly a consequence of the assumptions of the model. Here the computer model has been a guide that we think could help us in analogous situations to understand the potential vulnerability of different species to indirect effects arising from interactions among species. Thus we have accomplished our original goal: we have gained insight into the behavior of a biological community. There are no doubt many paths to the same understanding. Ours was reached with a modern tool. Peter Kalm sought understanding of natural history by traveling through primeval forests that are no longer available to us. No doubt the model that has been presented is far from definitive. It is only a beginning, helping us at least in a small way to better understand birth, growth, and death in a natural forest community, without eliminating for us who still enjoy walking through real forests a sense of awe and pleasure. Let us hope that such efforts in the future will help us to use our forest resources wisely and to heed the admonitions of Peter Kalm, who while observing the forests also observed the settlers' attitudes toward them. "People are here (and in many other places) in regard to wood, bent only upon their own present advantage, utterly regardless of posterity," he wrote in his journal. The people take "little account of Natural History, that science being here (as in other parts of the world) looked upon as mere trifle, and the pastime of fools."

REFERENCES

Bormann, F. H., and M. F. Buell. 1964. Old-age stand of hemlock-northern hardwood forest in central Vermont. *Bull. Torrey Bot. Club* **91**: 451–65.

Bormann, F. H., T. G. Siccama, G. E. Likens, and R. H. Whittaker. 1970. The Hubbard Brook ecosystem study: composition and dynamics of the tree stratum. *Ecolog. Monogr.* **40**: 373.

Botkin, D. B., J. F. Janak, and J. R. Wallis. 1972b. Some ecological consequences of a computer model of forest growth. *J. Ecol.* **60**: 849–872.

Botkin, D. B., J. F. Janak, and J. R. Wallis. 1972a. Rationale, limitations and assumptions of a northeastern forest growth simulator. *IBM J. Res. Devel.* **16**: 101–116.

Braun, E. L. 1972. *Deciduous forests of eastern North America*. Hafner, New York. 596 pp.

Fowells, H. A. 1965. *Silvics of forest trees of the United States*. U.S. Dept. of Agriculture Handbook 271, Government Printing Office, Washington, D.C.

Kalm, P. 1963. *Travels in North America*. Dover, New York.

Kalm, P. 1966. *The America of 1750: Peter Kalm's travels in North America*. A. B. Benson (Ed.). Dover, New York.

Kramer, P. J., and T. T. Kozlowski. 1960. *Physiology of trees*. McGraw-Hill, New York.

Marks, P. J. 1971. The role of *Prunus Pensylvanica* L. in the rapid revegetation of disturbed sites. Ph.D. thesis, Yale University, New Haven, Conn.

Siccama, T. G. 1968. Altitudinal distribution of forest vegetation in relation to soil and climate on the slopes of the Green Mountains. Ph.D. thesis, University of Vermont, Burlington.

10
A Model of the Barataria
Bay Salt Marsh Ecosystem

CHARLES S. HOPKINSON, JR.
JOHN W. DAY, JR.

W e are including in this book two chapters dealing with the estuarine zone of the Mississippi River, its geology, biology, and human importance (frontpiece 2). In order to develop meaningful management models of the Louisiana coastal zone, it is necessary to understand not only the structure and function of the present systems but also the historical processes that have given rise to them. This chapter presents a model of the present estuarine system, emphasizing the flow of organic carbon and major interactions in the system. Chapter 16 presents two additional models that were developed partly as a result of insights gained from the first model and partly because of important management decisions that must be made for this region. In the first of these two models, we have simulated geological processes, biotic productivity, and the potential long-term impact of man's activities on the region; the second model is concerned with alternative short-term land use policies and methods of food production. As the subject matter of our three models is closely related, a general description of the natural system is included here that serves both chapters.

DESCRIPTION OF THE NATURAL SYSTEMS

Geological Processes of Delta Formation

Coastal Louisiana represents the largest and most productive estuarine zone in the continental United States. The size and productivity of this area is closely related to the geological history of the Mississippi River, which drains 40% of the country's land area and carries tremendous amounts of water, sediments, and nutrients into and through the coastal zone. It empties into the Gulf of Mexico onto a wide, shallow continental shelf.

Since sea-level stabilization about 7000 years ago, 1.6×10^6 hectares of estuaries and 1.8×10^6 hectares of swamps and marshes have been created by sedimentation associated with delta building and overbank flooding. These areas together represent about 40% of the coastal wetlands of the continental United States.

The most important mechanism governing land building in this region is delta formation. Since sea-level stabilization there have been seven major Mississippi River distributaries with their corresponding deltas along the Louisiana coast, resulting in an extremely broad deltaic plain (Frazier, 1967; Figure 1). Active distributaries build new land while abandoned deltas undergo erosion. On a historical basis, land building has exceeded erosion, resulting in the existing large deltaic plain.

Construction of artificial levees (dikes) to protect against annual overbank flooding and to aid in navigation followed settlement by Europeans. Because of the levees, land building associated with overbank flooding and

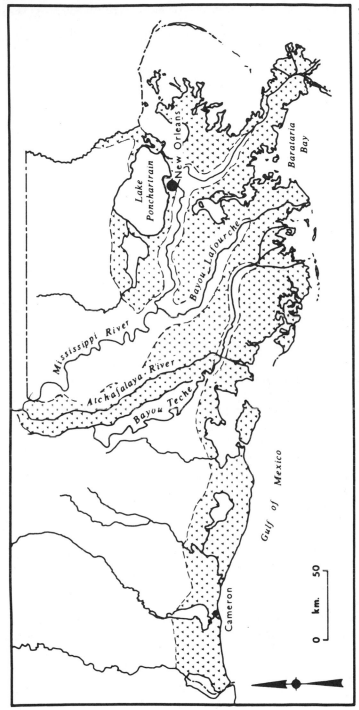

Figure 1 The Coastal Zone of Louisiana. Shaded areas are wetlands. Note high ground surrounding distributaries and the interdistributary position of estuaries such as the Barataria Bay basin.

237

distributary switching has been eliminated. The modern delta has extended to the edge of the continental shelf with most sediment now being deposited in the deep water beyond the shelf. Thus the land building phase of the historical cycle has largely been arrested while the erosion process continues. As a result, Louisiana presently is losing about 4300 hectares (16.5 miles^2) of land per year.

The Barataria Bay Saline Ecosystem

The Barataria hydrologic unit is a typical Louisiana interdistributary coastal basin that exhibits well-developed physical gradients with corresponding vegetation zones. It appears to be the most productive unit along the Louisiana coast, accounting for almost 45% of the total commercial fisheries catch (1.3 billion pounds in 1972) of the state (Lindall et al., 1972).

The Barataria Bay saline ecosystem is located at the seaward end of the hydrologic unit. The climate of this zone is nearly subtropical as indicated by the presence of black mangrove trees. Water temperature ranges from 10 to 35°C with the annual average above 20°C. Annual rainfall is about 165 cm and is fairly evenly distributed throughout the year. Tides are predominantly diurnal, with an average range of about 0.3 m, causing daily flushing of the marshes. Mean sea level also varies annually and, as we see in the following discussion, is important in marsh-estuary interactions. Low water levels occur during winter due to low-pressure storms from the northwest pushing the bay water south into the Gulf of Mexico. As a result, the coastal marshes are flooded more frequently during the warmer months.

Personnel of the Center for Wetland Resources, LSU, have studied in detail the three most important groups of primary producers in Barataria Bay: marsh grasses, phytoplankton, and benthic plants. The marsh grasses (predominantly *Spartina alterniflora*, cord grass) are the most productive. Kirby (1972) calculated net production to be as high as 2800 g organic matter m^{-2} yr^{-1}, with a mean of 2180 g m^{-2} yr^{-1}. The net production of phytoplankton and benthic plants (mostly diatoms) is about 900 g m^{-2} yr^{-1} (Day et al., 1973). Because of the warm climate primary production is high year round, and the net productivity of the marsh grass approaches that of the most productive monoculture agriculture.

Most of the marsh grass net production enters the detrital food chain (Smalley, 1958; Teal, 1962; Day et al., 1973), and there is a lag of almost a year before a significant quantity of the carbon fixed by marsh plants is available to primary consumers. Because of the low winter water levels, most of the grass that dies in the fall remains standing until the following spring. With rising temperature and water levels, much of the dead grass is broken down quickly into organic detritus and consumed by a complex microbial-meiobenthic-macrobenthic marsh community. Several workers have estimated that about 50% of the marsh grass net production is flushed into the water as organic detritus (Teal, 1962; Golley et al., 1962; Day et al., 1973; Nixon and Oviatt, 1974). Many of the lower tropic level organisms of Barataria Bay consume organic detritus as a major part of their diet.

Our studies indicate that the water and marsh areas adjacent to a shore are much more productive than inland marsh or open water areas. Standing crop and productivity of marsh grasses and abundance of macro- and meiobenthic organisms, fish, and shrimp are higher close to the marsh-water interface. The twisting tidal streams and many small ponds in the marshes result in higher production by increasing the interface length.

In addition, the rich, shallow shore areas and tidal channels are important as nursery grounds for fish, shrimp, and crabs. The young of many species spend their rapid-growth juvenile stage in these areas. A large proportion of estuarine organisms are migratory species, allowing much higher standing crops during periods of seasonally high food availability than could exist with an entirely endemic population.

All of these factors—the Mississippi River, the climate, and the biotic and physical gradients—have interacted to create the specialized and highly productive Louisiana coastal ecosystem. It is probably the most productive natural area in the United States today, with Louisiana normally ranking first of all the states in total fisheries catch; its past biotic productivity is reflected in the fact that about 25% of our domestic oil and natural gas comes from this region. The area also supports the largest fur industry in the United States. It is the southern terminus of the Mississippi flyway, the largest waterfowl migratory route in North America. Recreational activity supported by the coastal zone is estimated to be 106,174,000 user days per year (Pope, 1973). If we estimate the value of a user day at $10, then recreation in the region is worth in excess of one billion dollars annually. It is clear to us that the coastal portion of Louisiana is one of the most productive, valuable, and beautiful wild areas in this country, and we are concerned that these resources be used as wisely as possible.

Linkage of Geological and Biotic Cycles

When the river shifts into a new channel, land is built rapidly. Many minor distributaries serve to spread the water and sediment over fairly broad areas. Coastal erosion takes place continuously, but the new delta is dominated by river deposition during early stages of the cycle. The total length of land-water interface is relatively short during this stage.

As the river begins to seek a new channel (a process that extends over a period of several hundred years) coastal erosion becomes increasingly more important. As more land is lost, the land-water interface length becomes longer due to formation of small bays, ponds, and meandering tidal channels. Since total biotic productivity is a function of both land-water interface length and total marsh area, total productivity for any one bay system reaches a maximum at some time during the erosion cycle. Once this point is reached, however, the increases in productivity due to increasing interface length are more than offset by losses due to erosion, which consumes marsh area at an increasing rate.

Thus there seem to be juvenile, mature, and senescent stages of inter-distributary bay systems. Land area is greatest in the juvenile stage and then decreases in following stages. The length of the land-water interface is low during the juvenile stage, increases to a maximum during the mature stage, and then decreases in the senescent stage. Productivity is highest in the mature stage. Because the river has continually changed channels in the past, there have always been fresh juvenile estuaries waiting in the wings. It's similar to a relay race with fresh new runners waiting as the fatigued runner finishes his lap.

Man and the Marsh

Man has been a part of this coastal system for thousands of years, as evidenced by the hundreds of Indian shell mounds found throughout the region. However, man has become a significant force modifying the coastal zone only recently and is now tampering with both the long and short-term natural cycles. Presently the river is forced to remain in its existing channel and overbank flooding is eliminated by leveeing. There is little land being built and only indirect introduction of river nutrients and fresh water into the bay systems via tidal passes. This tends to decrease the productivity of this region.

Man is decreasing estuarine productivity by other means as well. Marsh land is being lost on a large scale through dredging, leveeing, and drainage. This is especially true in the vicinity of New Orleans, where large areas have been drained for urban development. As is common in "reclaimed" wetlands, the land has subsided so that large areas of the city are now below sea level. Indeed, at floodstage the entire city is below water level! Another significant loss of wetlands has occurred because of drainage for agricultural crops and, more recently, for cattle grazing. In most cases, the soils have proven unsuitable, and these efforts have been abandoned.

The major direct loss of wetlands is due to canal dredging for navigation and especially for mineral exploration and drilling. Canaling is probably the single most devastating short-term activity of man in the marsh. Not only are wetlands destroyed directly, but hydrology is altered because spoil bank levees along the canals block tidal flow of water. This has resulted in lowered productivity in the wetlands.

THE MODELING EXERCISE

It is perhaps pertinent at this point to present a short history of the factors leading up to the modeling exercise, as these factors strongly influenced the effort. The Systems Ecology Program in the Department of Marine Sciences was initiated in 1970 (at a time when neither of us was at LSU) with an aim of understanding the ecological functioning of the Louisiana coastal zone and man's impact on it. Because of the size of the coastal zone with its broad spectrum of environmental problems, it was decided at the outset that a holistic, ecosystem approach was the only practicable way to address these problems.

The initial conceptualization process was done not for mathematical modeling but for design of research. Major ecosystem components and processes were identified and studied such as primary production, community metabolism, dominant consumers, and migration. In essence, the "pre-modeling" phase of the Barataria studies fits the diagram of modeling presented in Figure 2 of Chapter 1 fairly well, with the exception that there was no formalized modeling done. When we began our model, there were already some conceptual, diagramatic, and mathematical expressions of the Barataria system. A major part of this effort, therefore, consisted of interaction with others at LSU. Much of this discussion and interaction shows up only indirectly in this paper, yet it has been one of the most important elements of this study.

Our approach in developing the Barataria Bay ecological model was first to describe conceptual models of the major flows and transformations of carbon and nitrogen and then combine them in a third model that included the major outside forcing functions. The third model was simulated to test whether our abstraction of the ecosystem described what we had observed in the field. Our early conceptual models were very detailed (e.g., Day et al., 1973). For several reasons, we decided to simplify these early schemes as much as possible. First, we had limited time and resources. Second, we wanted to know how realistic a fairly simple model could be (partially because of our limited mathematical sophistication). And third, we were interested in ecosystem level behavior. The question then arises: How much detail of component parts is necessary for a realistic description of the whole? Dick Wiegert gives a partial answer in Chapter 12, and we think that we have included, as Dick suggests, the most important factors that are ecologically

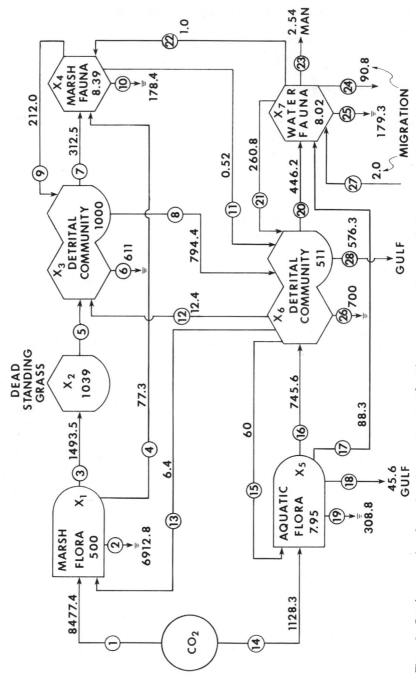

Figure 2 Steady-state values for carbon flows (g dry wt m^{-2} yr^{-1}) and storages (g dry wt m^{-2}). Numbers of pathways and storages are defined in Tables 1 and 3.

242

important. We could have constructed a less aggregated model, but we thought the errors in including components for which we did not have good measurements would be greater than the errors that would occur by aggregating these components into overall processes for which we did have good measurements. The Barataria saline system is about half marsh and half water, and these form fairly distinct ecological units. Within each of these our research suggests three basic components: primary producers, detritus decomposers, and consumers. For logical chemical subdivisions, we consulted our chemists (William Patrick and Clara Ho); our nitrogen model reflects their suggestions. We stopped at this point of aggregation because this seemed to provide enough detail for ecosystem level purposes, and for the questions we wished to ask (Chapter 3).

Biological Basis of the Carbon Model

In the carbon model we reduced the biotic system from our earlier detailed conceptual and diagrammatic models based on extensive field work (Day et al., 1973) to a simplified seven-compartment structure (Figure 2). At this level of aggregation, this is a realistic representation of the estuarine ecosystem. For the marsh unit, we added a dead standing marsh-grass compartment to our basic producer-decomposer-consumer matrix. This dead grass is the major storage of above-ground organic carbon during the winter, and it serves as the primary source material for organic carbon. We, therefore, separated the marsh subsystem into four compartments: (1) primary producers, (2) dead standing marsh grass, (3) a detritus-microbial complex, and (4) macrofauna. The detritus-microbial complex compartment includes detritus (the core material, be it of plant or animal origin) and organisms associated with it (bacteria, yeasts, fungi, protozoa, and meiofauna). The macrofauna compartment includes insects, raccoons, muskrats, birds, snails, crabs, minnows, and mussels.

We felt that we could capture the essential nature of the aquatic unit with the three basic compartments. There are two main sources of organic matter in the aquatic subsystem: that exported from the marsh and in situ production (mainly benthic diatoms and phytoplankton). The detrital complex in the water is very similar to that on the marsh, the source of most of its organic input.

Grouping all fauna into two compartments affects the realism of the model, depending on the degree of functional homogeneity of each. The marsh fauna is a fairly homogeneous functional group; practically all of the animals are herbivores or detritivores. The aquatic fauna, however, is a diverse functional group, representing all trophic levels. Simulations show, as we point out later, that this high degree of aggregation in the aquatic fauna compartment affects the realism of the model.

Nitrogen Model

Nutrient dynamics have a major controlling effect on primary production and detritus formation. Nitrogen and phosphorous fertilization of *S. alterniflora* marshes in the Barataria Bay area increased plant growth (15–20%) with NH_4^+ enrichment but not with enrichment of P or other nutrients (William Patrick, Agronomy Department, LSU, personal communication). Therefore, N was the only nutrient we modeled. The model of nitrogen incorporates the same biotic compartments as the carbon model plus six additional abiotic nitrogen compartments (Figure 3).

In contrast to our rather extensive knowledge of the carbon budget, we had only very limited field or literature data on the salt marsh nitrogen cycle. The values used in our nitrogen model were determined in several ways: from studies of other ecosystems, from ongoing LSU field and laboratory research, from calculations of nitrogen associated with known carbon flows, from educated assumptions, and from the solution of compartmental material balances.

Three atmospheric interactions are included: rain input, N fixation, and denitrification. Henderson (1974) estimated that in a temperate forest 1.0–1.3 g N m^{-2} yr^{-1} were added by rainfall, and Haines (personal communication) measured about 0.34 g N m^{-2} yr^{-1} for a Georgia estuary with less rainfall; we used 1.0 g N m^{-2} yr^{-1}. Marsh algae and bacteria fix nitrogen. Nitrogen fixation was set at 1.5 g N m^{-2} yr^{-1} based on fixation rates of 5.5 g N m^{-2} yr^{-1} in an alder stand (Witkamp, 1971) and 2.2 g N m^{-2} for rice fields during the growing season (Patrick and Mahapatra, 1968). Subsequent measurements in salt marshes in Louisiana (Patrick, personal communication) and Long Island (Whitney et al., 1975) obtained values close to our original estimates.

There are two forms of soil nitrogen, organic and inorganic. The organic fraction is a large, stable component, while the inorganic fraction is rapidly turned over. Brannon (1973) measured 1704 g m^{-2} of total N and 2.3 g m^{-2} of NH_4^+—N in the marsh soil of Barataria Bay.

Nitrogen processes operating in the soil are mineralization, immobilization, nitrification, denitrification, and fixation (Tusneem and Patrick, 1971). Mineralization is the transformation of organic N into NH_4^+—N by heterotrophic bacteria that utilize nitrogenous organic compounds as an energy source. Brannon (1973) reported a mineralization rate of 228 g NH_4^+ m^{-2} yr^{-1}. Intermittent air drying, increased temperature, pH above 9, and salt

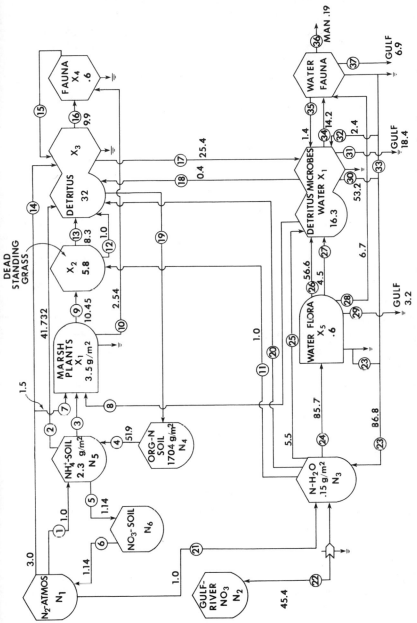

Figure 3 Steady-state flows (g N m S^{-2} yr S^{-1}) and storages (g N m S^{-2}) of nitrogen. Numbers of pathways and storages are defined in Tables I and 4.

addition are factors known to increase mineralization. Temperature is the only one of these processes for which we have data.

Immobilization is the incorporation of inorganic nitrogen into bacterial protoplasm. This happens mainly during the formation of detritus (see Odum and de la Cruz, 1967; Kirby, 1972). Our field work indicates that a large portion of this supplemental nitrogen comes from NH_4^+ in the soil. We calculated the flow from soil NH_4^+ to detritus microbes by: (1) calculating the N demand of the detrital compartment (82.7 g N m^{-2} yr^{-1}), (2) taking as much N as was available from soil NH_3 (41.7 g N m^{-2} yr^{-1}), and (3) supplying the additional detrital demand from the water N.

Nitrification and subsequent denitrification take place in the surface zone only. With fluctuating water levels there are fluctuating redox profiles. When NH_4^+ moves into an oxygenated zone it can be oxidized to NO_3^- (nitrification). If this NO_3^- then moves into a low redox zone it can be denitrified to gaseous N_2 and lost from the system. For example, 20–50% of surface supplied NH_4^+ fertilizer can be lost from rice fields in this manner (Tusneem and Patrick, 1971). Alternate wetting and drying creates ideal denitrifying conditions. We assumed that under tidal conditions 50% (maximum rate in nontidal rice fields) of the NH_4^+ present in the surface soil (1 cm deep) was denitrified.

Marsh plants take up N principally in the form of NH_4^+—N. There is about 0.7% N in *Spartina* shoots (Kirby, 1972). Multiplying the net primary production of *Spartina* by the concentration of N in their tissues gives an annual uptake by plants of about 11 g N m^{-2}.

We calculated that 85 g N m^{-2} yr^{-1} is taken up by aquatic primary producers (mainly phytoplankton) by multiplying gross production (1128 g m^{-2} yr^{-1}) by the concentration of N in diatoms (7.6%), the principle phytoplankton group. We assumed that 10% of this is released as dissolved organic matter (after Thomas, 1971). In the water, nitrogen exists primarily in the inorganic forms: NH_4^+, NO_2^-, and NO_3^-. In the Barataria Bay area Ho (1971) reported an annual average of about 0.15 g N m^{-2} of water.

The final N compartments we needed for the model were for the biota. Live *S. alterniflora* (excluding roots) contains about 0.7% nitrogen dry weight, dead standing grass contains about 0.6% nitrogen (Kirby, 1972), and the detritus microbe complex is 3.2% nitrogen (Odum and de la Cruz, 1967). Animals are about 7.6% nitrogen (Vinogradov, 1953). Nitrogen flows through the biota (unless otherwise indicated) were calculated by multiplying the carbon flow by the concentration of nitrogen associated with that carbon flow. A major pathway of N is via animal excretion and feces. Since many of the animals are detritivores we used 0.56% as the N content of their feces (after the associated detrital microbes have been stripped from the detrital core; the same as in dead standing grass). Excretion of nitrogen for the fauna was set at 7.6% of the respiratory carbon flow. In the aquatic compartments most of this N returns to the aquatic inorganic N compartment. Some of this is taken up by microbes. We calculated that 41 g m^{-2} yr^{-1} is taken up by marsh floor microbes as the tide floods over the marsh. In the marsh, N excreted by animals goes directly to microorganisms.

Simulation Model: Conceptual Base

The simulation model (Figure 4) was constructed (1) to determine whether the data base could produce simulation results that generally agreed with the patterns we observed in the field and (2) to test our understanding of the functioning of the estuarine ecosystem. The model simulates annual changes in the various carbon and nitrogen compartments. It is a composite of the carbon and nitrogen conceptual models and includes the major external forcing functions. The carbon model, incorporated in entirety, forms the core of the simulation model. Three nitrogen compartments with associated input-output flows are included and act in many places to control the carbon flow.

Forcing functions included in the model are: insolation, water level, temperature, migration of water fauna, nitrogen in the rain, nitrogen in the river, and a lag function for marsh grass decomposition (Figure 4). The input patterns of insolation, water temperature, and sea level variations are based on 10- to 20-year averages. A switching function is used to simulate the seasonal entrance of larval and juvenile fish, shrimp, crabs, and so forth into the estuary. Whenever the slope of the temperature curve exceeds a certain value the function becomes operable. Practically speaking this means the early Spring. A lag of 6 months, based on field measurements, has been used to control the timing of the flow of live grass to the dead standing compartment.

Simulation Model: Development of Algebraic Functions

The conceptual model was quantified (Figure 4) by developing the following differential equations showing the inputs and outputs for each compartment. Individual flows are designated as Jij where j represents the recipient compartment and i represents the donor compartment. Nitrogen state variables are labeled with the letter N, and a following number, for example, N_3 (nitrogen in the water) and carbon flows are labeled with a number only. As an example, $J5N_3$ means the flow of N to compartment N_3 from carbon compartment X_5. (We could have used "QN_3" and "QC_5" but the terms become too complicated.) A forcing function flow is labeled with the letter designating the particular flow incorporated, for example, JI1 means the

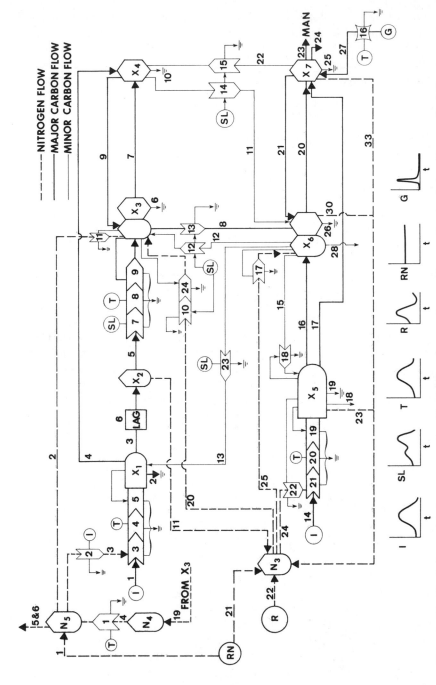

Figure 4 The simulation model. One-year patterns of forcing functions are shown at bottom. I = insolation, SL = sea level, T = temperature, R = river, RN = rain, G = migration from the Gulf of Mexico, and t = one year. Work gates are defined in Table 3 and nitrogen flows in Table 4. See text for further explanation. See also Figures 2 and 3.

flow from the sun (I) to marsh grass (X_1). An R symbolizes a respiratory flow, G symbolizes flow to or from the Gulf, and an m represents the flow of carbon to man (fisheries harvest). All compartments from Figures 2, 3, and 4 are defined and the initial steady-state values given in Table 1.

The mathematical model hinges on the following differential equations which describe the rate of accumulation of nitrogen and carbon in the various compartments.

$$\frac{dN_3}{dt} = JN_2N_3 + JN_1N_3 + J2N_3 + J5N_3 + J6N_3 + J7N_3 - (JN_33$$
$$+ \ JN_36 + JN_35)$$

$$\frac{dN_4}{dt} = J3N_4 - JN_4N_5$$

$$\frac{dN_5}{dt} = JN_4N_5 + JN_1N_5 - (JN_5N_6 + JN_53 + JN_51)$$

$$\frac{dX_1}{dt} = JI1 + J61 - (J1R + J12 + J14)$$

$$\frac{dX_2}{dt} = J12 - J23$$

$$\frac{dX_3}{dt} = J23 + J63 + J43 - (J3R + J36 + J34)$$

$$\frac{dX_4}{dt} = J34 + J14 + J74 - (J4R + J46 + J43)$$

$$\frac{dX_5}{dt} = JI5 + J65 - (J5R + J5G + J57 + J56)$$

$$\frac{dX_6}{dt} = J56 + J76 + J36 + J46 - (J61 + J63 + J67 + J65 + J6G$$
$$+ \ J6R)$$

$$\frac{dX_7}{dt} = J67 + J57 + JG7 - (J7R + J76 + J74 + J7m + J7G)$$

For example, dX_1/dt, the rate of change of the biomass of marsh primary producers, equals inputs minus outputs. The two inputs are (1) gross photosynthesis ($JI1$) and (2) the uptake and utilization of dissolved organic matter by marsh diatoms and epiphytic algae during coverage of the marsh by water ($J61$). The three outputs are (1) respiration of autotrophs ($J1R$), (2) death of

Table 1 Carbon and Nitrogen Storages from Figures 2, 3, and 4.

No.	Storage Name	Initial Condition Value	Reference
N_1	N_2 gas in atmosphere	Unlimited	
N_2	N compounds in Mississippi River	Unlimited	
N_3	(NH_4^+, NO_3^-, NO_2^-)—N in water	0.15 mg 1^{-1}	Ho, 1971
N_4	Total N in marsh soil	0.704 g m^{-2}	Brannon, 1973
N_5	Extractable NH_4^+—N in marsh soil	2.26 g m^{-2}	Brannon, 1973
N_6	NO_3^-—N in marsh soil	Negligible	
x_1	Live marsh plants, mainly S. alterniflora	500 g org m^{-2} 3.5 g N m^{-2}	Kirby, 1972
x_2	Dead standing S. alterniflora	1039 g org m^{-2} 5.8 g N m^{-2}	Kirby, 1972
x_3	Marsh microbe-detritus complex	1000 g org m^{-2} 32 g N m^{-2}	Day et al., 1973
x_4	Marsh macrofauna	8.3993 g org m^{-2} 0.638 g N m^{-2}	Day et al., 1973
x_5	Water flora	795 g org m^{-2} 0.605 g N m^{-2}	Day et al., 1973
x_6	Water microbe-detritus complex	511 g org m^{-2} 16.3 g N m^{-2}	Day et al., 1973
x_7	Water fauna	8.023 g org m^{-2} 0.6097 g N m^{-2}	Wagner, 1972
CO_2	Carbon dioxide in atmosphere	Unlimited	

Symbol	Name	Forcing Functions Value	Reference
I	Solar insolation	200–500 g cal cm^{-2} day^{-1}	Day et al., 1973
SL	Mean sea level	1.5–1.7 m	Day et al., 1973
T	Water temperature	11–28°C	Day et al., 1973
R	N supplied by Mississippi River	45.4 g m^{-2} $\pm$ 10	Assumption
RN	N added by rain	1 g m^{-2} yr^{-1}	Henderson, 1974
G	Immigration of young fish from Gulf	2 g m^{-2} yr^{-1}	Assumption

Table 2   Work Gates and Flow Functions of Figure 4.

Gate No.	Description
1	Temperature controlling mineralization rate
2	Solar insolation that directly controls marsh gross production indirectly controlling NH_4^+ uptake by marsh plants
3	NH_4^+ concentration controlling marsh gross production
4	Temperature controlling marsh gross production
5	Nonlinear self-feedback loop controlling gross production
6	0.4-year lag controlling input of grass to dead standing compartment
7	Sea level controlling breakdown of dead standing grass
8	Temperature controlling breakdown of dead standing grass
9	Microbial biomass controlling breakdown of dead standing grass
10	Sea level controlling N-rich water cover on the marsh
11	Microbial biomass controlling rate of N uptake from soil
12	Effect of sea level on washing organics onto marsh
13	Sea level controlling detritus flushing from marsh
14	Sea level affecting washout of marsh fauna into water
15	Sea level affecting the standing or washing of aquatic fauna onto marsh
16	Switch based on slope of temperature curve controlling immigration of fishes to estuary from Gulf
17	Microbial biomass controlling uptake rate of N from water
18	Algal biomass controlling uptake of DOM from water
19	Biomass of aquatic flora controlling gross production
20	Temperature controlling gross production
21	N controlling gross production
22	Floral biomass controlling uptake of N from water
23	Sea level controlling the washing of DOM onto the marsh from the bays
24	Microbial biomass controlling uptake of N from water as marsh is flooded

S. alterniflora ($J12$), and (3) consumption of live marsh plant material by marsh macrofauna ($J14$). All input and output flows of the carbon compartments are shown and referenced in the flow diagram of Figures 2 and 4. Each flow reference number is listed in Table 3, where the flow is identified and quantified. Nitrogen compartment flows are similarly shown, referenced, identified, and quantified in Figures 3 and 4 and Table 4.

Each input or output flow in the above differential equations is mathematically expressed as a constant (rate coefficient) multiplied by whatever factors control the flow (i.e., $J = kQ$). The equation is initialized by using yearly average values for all factors, including both compartment biomasses and forcing function values. The values of the constants were obtained by dividing the mean annual value of the flows by the (product of the) factors of which they are a function ($k = J/Q$). The constants assume the units necessary to

Table 3   Carbon Flows in Figures 2 and 4.

Flow	Value (g org m^{-2} yr^{-1})
1. Gross photosynthesis of marsh flora	8477
2. Respiration of marsh flora	6912
3. Net production of *S. alterniflora* going to dead standing crop	1493
4. Consumption of live marsh-plant material	77
5. Dead standing crop becoming detritus	1493
6. Respiration of marsh detrital microbes	611
7. Consumption of detritus by marsh fauna	312
8. Flushing of detritus into water	794
9. Feces and death of marsh fauna	211
10. Respiration of marsh fauna	178
11. DOM released into water from marsh fauna during high tide	0.52
12. Detrital material settled on marsh from overlying water	12
13. Uptake of DOM by marsh diatoms and epipelic algae during coverage of marsh by water	6.4
14. Gross production of water flora (phytoplankton and benthic algae)	1128
15. Utilization of DOM by phytoplankton and algae	60
16. Net production of water flora that goes into detrital food chains	745
17. Live consumption of water flora	88
18. Loss of water flora through bay passes into Gulf of Mexico	45
19. Respiration of water flora	308
20. Consumption of detritus by water fauna	446
21. Feces and death of aquatic fauna	260
22. Aquatic fauna stranded or trapped on marsh	1.0
23. Fisheries harvest by man	2.5
24. Loss of aquatic fauna to Gulf of Mexico	90
25. Respiration of aquatic fauna	179
26. Respiration of aquatic microbes on detrital complex	700
27. Immigration of aquatic fauna from Gulf into estuary	2.0
28. Loss of detritus to Gulf of Mexico via tidal exchange	576

set the flow equations equal to the flow values. Consider, for example, flow F57; the feeding by aquatic consumers on phytoplankton (Figure 4, Tables 3 and 5). Since there is a fair degree of annual variation of both of these groups we made this flow a factor of both populations: $F57 = kX_5X_7$. The value for k is determined by $k = F57/X_5X_7$. In this case, the units of k are g m^2 yr^{-1}. The equations that mathematically describe all the flows in the simulation

Table 4   Nitrogen Flows in Figures 3 and 4.

Flow	Value (g N m^{-2} yr^{-1})
1. N input via rain	1.0
2. Immobilization-microbial uptake of NH_4^+	41.7
3. NH_4^+ uptake by marsh plants	10.9
4. Mineralization by general purpose heterotrophic bacteria	51.9
5. Nitrification	0.25
6. Denitrification	0.25
7. N fixation by epiphytic algae	1.5
8. N associated with DOM taken up by epiphytic algae	0.49
9. N in grass becoming dead standing crop	10.45
10. N value of live plant material eaten by marsh fauna	2.54
11. Leaching of N into high-tide water	1.05
12. Marsh microbial uptake of leached N	1.04
13. N flow from dead standing grass to marsh detritus	8.36
14. N fixation by soil and intertidal microbes	1.5
15. N content in feces, wastes, and death of marsh fauna	12.52
16. N uptake by marsh fauna eating detritus	9.98
17. N loss associated with marsh flushing of detritus	25.4
18. N associated with detritus washed onto marsh from water	0.4
19. Immobilization in subsurface soil	51.9
20. Uptake of N in water by marsh microbes during marsh flooding	41.0
21. N input via rain	1.0
22. Riverine input to estuary	45.4
23. N excreted by water flora	23.69
24. Aquatic flora uptake of N	85.73
25. Aquatic microbial uptake from water reservoir	5.56
26. N contained in aquatic flora when becoming detritus	56.62
27. N associated with DOM which is taken up by aquatic flora	4.56
28. N taken up by aquatic fauna eating live aquatic flora	6.71
29. N lost from system when flora is flushed into Gulf	3.27
30. N excreted and leached by detritus and associated community	53.2
31. Detritus lost to Gulf N quantity	18.43
32. Assimilation by microbes of 33 when in water	2.41
33. Aquatic fauna's excreted N	12.41
34. N uptake by aquatic fauna when eating detritus	14.27
35. N associated with feces and death of aquatic fauna	1.46
36. N lost from system as man's fishery harvest	0.19
37. N lost to Gulf as fish leave estuary	6.92

Table 5 Initial Flow Averages and Flow Constants.

Flow	Initial Steady-State Values (g m^{-2} yr^{-1})	Description	Designation	Value	Comments
JN_2N_3	45.5	R = river			Sine-wave representation
JN_1N_3	1.0	RN = rain			Constant value throughout year
J_2N_3	1.05	kX_2	kN_32	0.001	
$J5N_3$	23.6	kX_5	kN_35	2.9798	
$J6N_3$	53.2	kX_6	kN_36	0.1041	
$J7N_3$	10	kX_7	kN_37	1.2464	
JN_33	41	kN_3X_3SL	$k3N_3$	0.0530	
JN_36	5.56	kN_3X_6	$k6N_3$	0.0725	
JN_35	85.73	kN_3X_5IT	$k5N_3$	71.8909	
$J3N_4$	51.9	kX_3	kN_43	0.0519	
JN_4N_5	51.9	$k.36T$	kN_5N_4	7.0883	
JN_1N_5	1.0	RN = rain		—	Constant value throughout year
JN_5N_6	0.25	kN_5	kN_6N_5	0.1106	
JN_53	41.7	kX_3N_5	$k3N_5$	0.0184	
JN_51	10.99	kIN_5X_1	$k1N_5$	0.2554	
$J11$	8477	kN_5X_1TI	$k1I$	0.0197	
$J61$	6.4	kX_1X_6SL	$k16$	0.0024	
$J1R$	6912	kX_1	k_r1	13.8139	
$J14$	77	kX_1	$k41$	0.0184	
$J12$	1493	kX_1	$k21$	2.29845	Flow lagged 0.4 yr
$J23$	1493	kX_2X_3TSL	$k32$	1.3956	
$J63$	12.4	kX_6SL	$k36$	0.0047	
$J43$	211	kX_4	$k34$	25.2361	
$J3R$	611	kX_3	k_r3	0.611	

J36	0.794	kX_3SL	k63	0.1542
J34	312	kX_3	k43	0.0372
J74	1	kX_7SL	k47	0.0242
J4R	178	kX_4	k_r4	21.2383
J46	0.52	kX_4SL	k64	0.0120
J15	1128	kIN_3X_5T	k51	24871
J65	60	kX_6X_5	k56	0.0148
J5R	308	kX_5	k_r5	41.1733
J5G	45.6	kX_5	kG5	5.7358
J56	745	kX_5	k65	93.7862
J57	88	kX_5	k75	1.3844
J76	260	kX_7	k67	32.5090
J67	446	kX_7X_6	k76	0.1088
J6G	576	kX_6	kG6	1.1279
J6R	700	kX_6	k_rR6	1.3699
JG7	2.0	G = immigration from Gulf	–	–
J7R	179	kX_7	k_rR7	22.3495
J7m	2.5			
J7G	90.7	kX_7	kG7	11.3173

Operative when slope of temp curve > +80°/yr

Constant throughout year

model are shown above; initial flow averages and flow constants are in Table 5.

Forcing-function field data for solar insolation, sea level, and water temperature were entered into the model as 24 evenly distributed points over the year. A separate nonlinear Continuous System Modeling Program (CSMP) function generator (a built-in program that comes with our computer) then interpolated the 24 bits of information and produced a continuous input to the rest of the model. A sine curve was used to represent the riverine input of nitrogen.

CSMP was used to integrate the differential equations describing the system, using the Runge Kutta fixed point method of solving integrals and an IBM model 360 digital computer. This program in essence allows a digital computer to be used as an analog computer (Patten, 1971).

RESULTS AND DISCUSSION

The results of the biological model indicate that we have a fairly realistic representation of many, but not all, of the important components of the natural system. There are some discrepancies between predicted and observed results, but generally the agreement is reasonable. Many times, however, we have no field data with which to compare our results. In several cases the results suggest new approaches for additional field or laboratory study of particular components. These points are illustrated in the following discussion. Simulations were run for 10 years and the model stabilized after 4 years; results presented here were taken from the seventh year of simulation.

Nitrogen

Results of seasonal fluctuations of total nitrogen in the marsh soil and inorganic nitrogen in the water column agree well with field data (Figure 5). The agreement is poor for NH_4^+ in the marsh soils. Soil ammonia levels are dependent on the rates of production (mainly mineralization) and utilization (mainly plant uptake and bacterial immobilization). The field results suggest that plant production and bacterial immobilization maintain a constant drain on the ammonium supply through the spring and summer, and it isn't until the fall that mineralization overrides utilization. The model results indicate that mineralization dominates and suggests that further field study of this phenomenon is needed to determine whether variable coefficients would be more appropriate in this case because of strong temperature effects.

Carbon in the Marsh

The flows and reservoirs of carbon on the marsh have previously been studied only in a piecemeal fashion, and by modeling we were able to gain

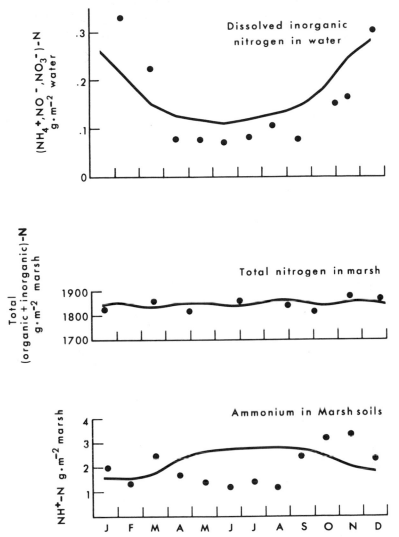

Figure 5 Predicted (solid line) versus observed (dots) values for three forms of nitrogen. See text for further explanation.

a greater insight into the system interactions that probably occur. The modeling efforts also have pinpointed areas where we need further research to understand the marsh more fully. The simulated flows and storages of carbon, from initial uptake by marsh grass to use by the marsh fauna or exportation from the marsh, are shown in Figure 6. One interesting feature of this simulation is the sequential lag that carbon exhibits as it is transferred from one compartment to the next. There is a 14-month lag in the simulations

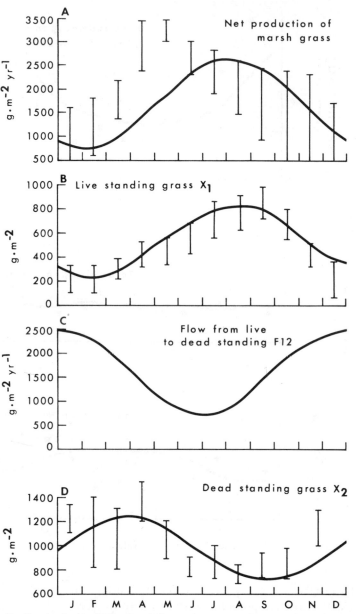

Figure 6 Predicted values (solid lines) and observed ranges (vertical bars) for various reservoirs and flows of carbon. All values are grams dry weight. See text for further explanation.

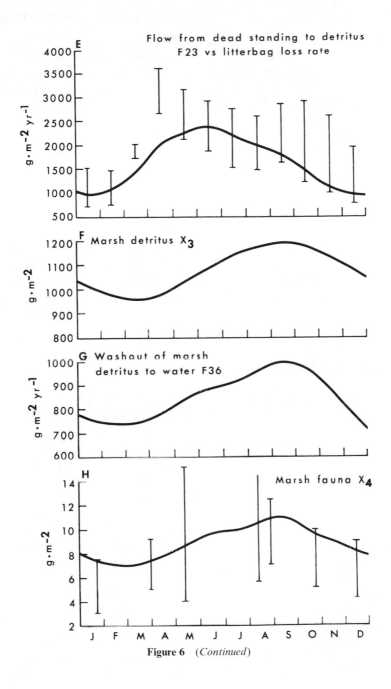

Figure 6 (*Continued*)

between peak carbon uptake by marsh plants and subsequent flushing into the adjacent bays (Figure 6G) or use by the marsh fauna (Figure 6H).

Another interesting aspect of this flow of carbon is that of detrital movement. The predicted movement of organic carbon from the dead standing plant compartment to the marsh detritus compartment ($J23$) is plotted in Figure 6E along with observed data as determined by litterbag loss measurements. A more realistic comparison should include consideration of the washout of marsh detritus to the water column ($J36$, Figure 6G). Measured litterbag loss rates are highest in spring and are mimicked, to a lesser degree, by the $J23$ simulation. The high litterbag loss rates in the fall are similar to the predicted $J36$ results. Both trends of detrital movement are linked to sea-level variations; they occur in the spring when water first begins to regularly flood the marsh and then in the fall during the highest levels of sea water. These results suggest two things: (1) the importance of including sea level in the model and (2) the fact that litterbag loss measurements reflect initial breakdown of dead standing grass as well as the flushing of detritus from the marsh. In this case, our original intention concerning model construction was supported by reasonable agreement between the model output and observed data.

A Simple Sensitivity Analysis

We investigated the effects of varying the three forcing functions, temperature, insolation, and nutrients, on the primary production and biomass of marsh grasses (Figure 7). Temperature as a single forcing function gave the best fit to field measurements of standing crop, supporting Turner's (1974) findings that temperature alone can explain 95% of the variation of peak standing crop in a variety of Atlantic and Gulf Coast marshes. This may be due to evapotranspiration being the rate-limiting process in wetlands where water and nutrients are in excess.

We tested in a similar fashion the effects of the same three forcing functions on aquatic flora productivity (Figure 8A). When production was driven by light and nutrients (I,N), higher values were obtained than when temperature effects (I,N,T) were included. Measured productivity corresponds more closely to the I, N, T simulation during the summer and more closely to the I, N simulation during the winter. This may reflect differing characteristics of two distinct populations over the year—mainly phytoplankton in summer and benthic forms in winter. The one high value for measured productivity during June and July occurred during a bloom, suggesting that nutrients or light may be more important than temperature in limiting photosynthesis during bloom periods.

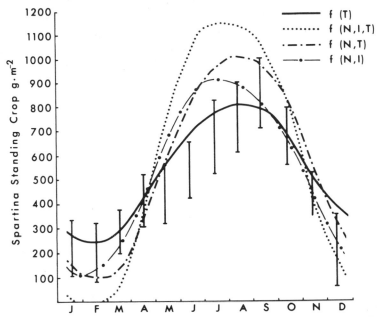

Figure 7 Effects of different forcing functions on *Spartina* standing crop. Vertical bars are range of field data. T = temperature, N = nitrogen, I = insolation. See text for further explanation.

Carbon in the Water

Simulation results of aquatic plant biomass are graphed versus field measurements of chlorophyll *a* in Figure 8B. The peak aquatic biomass occurred at the same time as the predicted peak of aquatic net production (Figure 8A) but 1 month before the highest measured level of chlorophyll *a*. This high chlorophyll level occurred at the same time as a measured phytoplankton bloom. These results suggest that our present model better reflects normal or average conditions than periodic or discontinuous phenomena such as phytoplankton blooms. In this case, a less-aggregated model that took into account individual species' response to environmental factors could perhaps give better agreement with the field observations.

The simulation results for aquatic detritus levels (Figure 8C) reflect primarily the input of detrital material from the marsh (Figure 6C). The annual variation in detrital standing crop represents a balance between inputs and the losses from consumption by aquatic fauna and flushing into the Gulf of Mexico. Although the concentration levels are about right, it is obvious that we have more to learn about seasonal patterns.

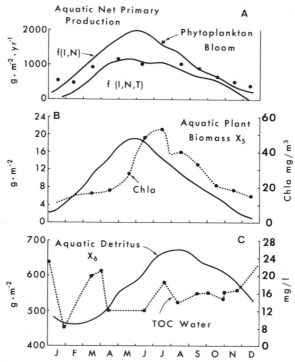

Figure 8   Predicted (solid line) versus observed data for aquatic primary production (A), aquatic plant biomass (B), and standing crop of aquatic detritus (C).

Because of the tremendous amount of lumping done to formulate the aquatic fauna compartment (Figure 9) much detail has been lost. This is evident when the predicted biomass is compared to observed biomass for zooplankton and fish. Both fish and zooplankton have measured spring peaks, whereas the simulated curve peaks in June and July. The fish biomass data presented here are trawl data from Wagner (1972), who reported that the two secondary peaks in late summer and fall represented larger fish that are probably underestimated because these fish avoid the trawl; therefore, actual fish biomass is probably higher in the late summer and fall, which would give better agreement between field data and model. In this case, it is impossible to fully validate the model because of inadequacies in the field sampling.

We ran the model alternately eliminating production from the marsh and production from the water in order to test the relative importance of these two regions as primary food sources. As expected, aquatic fauna biomass levels were reduced, but we were surprised to find that elimination of aquatic

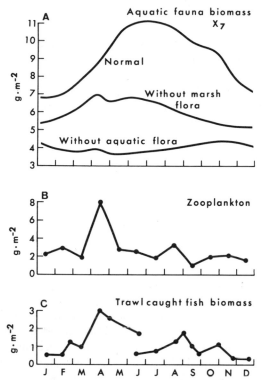

Figure 9 Predicted changes in aquatic fauna biomass under normal conditions and without productivity of marsh and aquatic flora (A). This is compared to field data for biomass of zooplankton (B) and fish (C).

primary production had the more pronounced effect (56% reduction in mean aquatic-fauna standing crop as compared to 36% reduction for the "no-marsh" case). There is also a temporal shift, with the peak biomass occurring in May for the "no-marsh" case and in November for the "no-aquatic-production" case. We can speculate as to the causes by examining the carbon and nitrogen budgets shown in Figures 2 and 3, respectively. Although the carbon flux from the marsh is a little higher than that from the aquatic flora (794 g m^{-2} yr^{-1} vs 745), the nitrogen flux through the aquatic producers is about twice as high as nitrogen in detrital material being exported from the marsh, effecting the nitrogen-dependant animal growth. Thus a primary role of unicellular algae (phytoplankton and benthic diatoms) in this estuary may be nitrogen enrichment of the organic food resources for consumers. This is a hypothesis generated by the computer that we hope to test later through empirical methods.

Some Further Thoughts

At this point, it is appropriate to ask what we have gained from this effort and whether it has been worth it.

In the formation of the carbon and nitrogen budgets and the conceptual model we were forced into careful consideration of the functioning of the estuarine ecosystem. We became aware of additional data needs and have already begun to pursue these. Thus the modeling process was not an isolated activity; rather, it encouraged us to seek from many different sources, including our colleagues in other departments, information needed in the model.

The simulation results tested our understanding of the ecosystem, raised several interesting questions, and suggested new areas for further research. Several examples are: (1) The organic soil nitrogen compartment was least accurate in predicting observed trends, and we have begun further studies on this parameter. (2) We found that the inclusion of sea level variations gave more realistic predictions, supporting the importance of hydrology in the estuarine zone. Accordingly, we have begun further field studies of the nature in which water movement affects ecological processes. (3) The model suggests temperature as the major parameter controlling marsh grass productivity. (4) Simulations of aquatic flora productivity raised questions concerning limiting factors and the role of aquatic plants in secondary consumer productivity. (5) Throughout the model we used constant coefficients and the same forcing functions. For this reason we feel the model is representative of "average or normal" conditions. The inclusion of variable coefficients and forcing functions would probably have resulted in better simulations, especially of nonperiodic phenomena such as phytoplankton blooms; however, we found that this relatively simple model gave a reasonable simulation of general ecological patterns and that it is good enough for many management purposes. As we gain more knowledge, and if specific objectives require it, we can make the model more sophisticated in the future. The modeling process and the insights gained led us to further efforts described in Part III, and we definitely feel that this has been a worthwhile task. Finally, as pointed out in Chapter 1, our modeling efforts were fun.

ACKNOWLEDGMENTS

We acknowledge many within the Center for Wetland Resources and others at Louisiana State University for help and guidance in this effort, including James Stone, James Gosselink, Hal Loesch, Gil Smith, and William Patrick. Joette Serio was especially helpful in data collection and synthesis and in typing. Johanna Saizan and Valerie Dunn prepared the figures. This research was sponsored by the Louisiana Sea Grant program, part of the National Sea Grant Program, maintained by the National Oceanic and Atmospheric Administration of the U.S. Department of Commerce. Outside LSU, we acknowledge the help and comments of Charlie Hall, Fred Wulff, Scott Nixon, and others.

REFERENCES

References for this paper are combined with the references for "Modeling Man and Nature in South Louisiana" presented in Chapter 16.

11
An Empirical and Mathematical Approach Toward the Development of a Phosphorus Model of Castle Lake, California

JEFFREY E. RICHEY

Phosphorus is an important agent in regulating primary production in lakes, but the relationship between phytoplankton growth and phosphorus utilization is perplexing. One basic problem in evaluating the relationship between nutrients and production is the lack of suitable models that can be tested in the field (Dugdale, 1967). Analyzing the way in which phosphorus limits production and determines the species composition of the plankton community requires an understanding of the forms of phosphorus available for utilization, the kinetics of uptake and regeneration of available phosphorus, and the details of the phosphorus cycle within the plankton community (Rigler, 1973). Furthermore, Lee (1973) suggested that to make quantitative predictions of the relationship between phosphate input and aquatic plant production, a reasonably detailed mathematical model describing the rates of transformation of the important forms of phosphorus present in natural waters should be developed.

This paper describes how such a model was developed for the flow of the different phosphorus compounds in Castle Lake, California, during summer steady-state conditions. This model was developed as a beginning of a quantitative tool for studying the varied and subtle phosphorus transformations and effects in lakes.

The term *model* is defined in this paper as the expression of the author's understanding of the system in question, conceptually, through schematic diagrams and through a set of mathematical equations. It indicates parameters to be measured, solves for quantities that would be difficult to measure directly, and suggests a framework for further research. Finally, the model allows hypothesis testing. For example: Is it important to know the flux as well as the concentration of phosphate in determining its importance in a lake? How important is dissolved organic phosphorus as an alternate phosphorus source? How is phosphate regenerated? With enough sophistication, such a model might be useful in predicting the degree of response of an aquatic system to a nutrient diversion or input.

The approach is threefold: (1) to identify the parameters that would be important in formulating a model of the phosphorus cycle, (2) to develop sensitive methods to measure the pertinent parameters in the field, and (3) to actually construct such a mathematical and computer model. Finally, this model is used to suggest directions for future experimental work.

CONCEPTUAL MODEL OF THE PHOSPHORUS CYCLE

The phosphorus cycle in a lake may be considered as a series of individual pools or compartments. At any particular instant in time, any compartment, i, is characterized by its volume, V_i, and quantity of solute, Q_i (hereafter, Q_i

is considered the quantity of solute per unit volume, or concentration). The solute exchanges with another compartment, Q_j, at a rate, J_{ij}. The fraction of Q_i transferred per unit time is given by the rate constant, $k_{ij} = J_{ij}/Q_i$.

The rate of change of a substance, Q_i, in some cube of water at a depth, z, at any instant in time, t, may be given by the equation

$$\frac{\partial Q_i}{\partial t} = \frac{\partial}{\partial z} K_h \frac{\partial Q_i}{\partial z} + \left(\sum J_{ji} - \sum J_{ij} \right) \tag{1}$$

where K_h is the coefficient of eddy diffusivity and $\left(\sum J_{ji} - \sum J_{ij} \right)$ is the sum of the biological transformations. The dependent variables in a lake include the rates of growth of the phytoplankton, zooplankton, and bacteria and the flux of the different nutrients. Independent forcing functions include mass transport mechanisms, temperature, and solar radiation. Equation 1 provides the basis for the development of the phosphorus model described below.

The pools and interactions important in characterizing the phosphorus cycle in the water column of a lake may be summarized in the conceptual model of Figure 1 (cf. Rigler, 1973; Richey, 1974 for more detailed reviews of the phosphorus cycle. The following discussion is based in part on references cited in Review References.) Dissolved inorganic phosphorus (Q_1) is taken up by phytoplankton (J_{13}) and bacteria (J_{14}) through active transport and becomes fixed in the metabolic pathways and cell structure as particulate phosphorus (Q_3 and Q_4, respectively). Bacteria and phytoplankton may compete for available phosphate. If there is not enough phosphate to maintain the minimum cell phosphorus necessary for photosynthesis, then algae will produce the exoenzyme alkaline phosphatase (AP) represented by S_1 in Figure 1. This enzyme catalyzes the hydrolysis of a variety of phosphate compounds, thus allowing some use (J_{23}) of dissolved organic phosphorus (Q_2) for photosynthesis. If enough phosphate builds up, S_1 "switches off"— that is, represses the formation of AP—and the dissolved phosphate alone is used. If excess phosphate is present, algae store more P than they immediately need as polyphosphate (Q_6), an orthophosphate polymer, through the process known as "luxury consumption" (S_2 in Figure 1). Then if external concentrations of orthophosphate become low, the polyphosphate may supply phosphate for photosynthetic growth (J_{61}). If none of these mechanisms can supply sufficient phosphate for photosynthesis, the system will become phosphorus limited. As part of metabolism, dissolved organic phosphorus (DOP) is excreted by phytoplankton (J_{32}). The excreted DOP may be a small molecule that polycondenses into a colloidal fraction, releasing phosphate (J_{21}) in the process. DOP may also be used by bacteria as a phosphate source (J_{24}). Zooplankton (Q_5) graze on phytoplankton and bacteria (J_{35}, J_{45}) and in turn excrete phosphate (J_{51}). The smaller the zooplankton, the

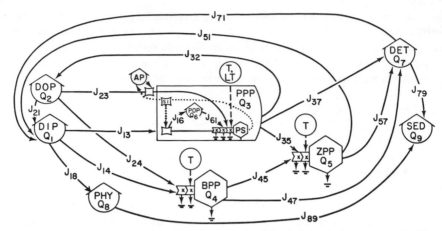

Figure 1 Conceptual model of the phosphorus cycle of Castle Lake during summer stratification (from Richey, 1974). For simplicity the forcing functions of sun, temperature, and water transport are not drawn in.

DIP (Q_1) = dissolved inorganic phosphorus
DOP (Q_2) = dissolved organic phosphorus
PPP (Q_3) = phytoplankton particulate phosphorus
BPP (Q_4) = bacteria particulate phosphorus
ZPP (Q_5) = zooplankton particulate phosphorus
POP (Q_6) = polyphosphate
DET (Q_7) = detrital phosphorus
PHY (Q_8) = ferric phosphate
SED (Q_9) = sediment phosphorus
     AP = alkaline phosphatase
      T = temperature
    LT = light
   J_{13} = phytoplankton uptake of DIP
   J_{14} = bacteria uptake of DIP
   J_{16} = polyphosphate formation
   J_{18} = precipitation of phosphate
   J_{21} = release of phosphate via condensation
   J_{23} = phytoplankton uptake of DOP
   J_{24} = bacteria uptake of DOP
   J_{32} = phytoplankton excretion of DOP
   J_{35} = zooplankton grazing of phytoplankton
   J_{37} = phytoplankton death
   J_{45} = zooplankton grazing of bacteria
   J_{47} = bacteria death
   J_{51} = zooplankton excretion of DIP
   J_{57} = zooplankton death
   J_{61} = polyphosphate utilization
   J_{71} = autolysis of DIP
   J_{79} = loss of DET to sediments
   J_{84} = loss of precipitated DIP to sediments

greater the percentage of excretion. Phytoplankton, bacteria, and zooplankton die (J_{37}, J_{47}, J_{57}), forming the detrital phosphorus pool (Q_7). Within several hours after death, 25 to 75% of cell phosphate may be released through autolysis (J_{71}). The remaining cell phosphates, mostly slow-degrading nucleic acids, sediment out (J_{79}). Depending on pH, iron, and phosphate concentrations, some phosphate is removed through physical complexation and precipitation processes (J_{18}, J_{19}). Other ligands, such as calcium, magnesium, and aluminum, may also enter these reactions. Light and temperature affect biotic rates and thus the phosphorus demands of the organisms. In some lakes allochthonous inputs, sediment recharge, and a littoral macrofaunal community may be important in the phosphorus cycle, but these were shown to be insignificant during the summer in Castle Lake (Richey, 1974).

The processes or pools indicated in Figure 1 change over depth and time. Translating Figure 1 into a field program and subsequently into computer models required the measurement of the different Q_i, J_{ij}, and k_{ij} values and the different factors that influence them. Phosphorus is a particularly ephemeral substance to work with, as it is present in a number of different forms at extremely low concentrations and it is quickly recycled. Thus it was most important to develop sensitive techniques to monitor simultaneously the rates of change of the phosphorus system to work out a meaningful, testable model. Perturbing any one of these fractions or flows, either in the model or experimentally in a lake, provides insight into the response of the system to some sudden stress, such as pollution.

Some of the pools and flows outlined in Figure 1 are very difficult to assess experimentally. There is doubt concerning the accuracy of the dissolved inorganic phosphate (DIP) determination (Rigler, 1973), and it is awkward to separate the bacteria, phytoplankton, and zooplankton phosphorus pools from the measured total particulate fraction. In addition, the individual flows, J_{13}, J_{14}, J_{18}, J_{37}, J_{47}, and J_{57}, are difficult to measure directly. If a quantitative model of the phosphorus cycle is ever to be understood, however, some approximations must be made. The primary goal here was not only to provide direct measurements, but also to outline an initial model of the phosphorus cycle and the assumptions that must be made. Such models, if validated against available field data, can be used to estimate the magnitude of those parameters difficult to measure and to verify model assumptions. The "next generation" of models would grow from a new field program, suggested by the successes and failures of the initial model.

STUDY SITE AND FIELD MEASUREMENTS

Castle Lake is a mesotrophic, glacial cirque lake at an elevation of 1708 m in the Klamath Mountains of northern California (T.39N, R.5W, S13). The lake has a surface area of 19.7 hectares and is divided into a shallow end over a terminal moraine, with an average depth of about 4 m and a deep end (35 m) off the cirque face.

During the summer there is virtually no rain, inflows are minimal, and there is no significant littoral community. Samples were taken from a central station located in the deep end of the lake every 5 days from August 14 to September 18, 1972 at 3.0, 7.5, 12.5, 17.5, 22.5, 30.0, and 32.5 m, and at varying depths and time intervals from April 1972 through February 1973. The model is representative of this pelagic water column.

Phosphate, dissolved organic phosphorus, and total particulate phosphorus pools were measured by chemical techniques. Phosphorus contained in the phytoplankton, zooplankton, and bacteria pools were measured as carbon and converted to P from known ratios of carbon to phosphorus. Phosphate fluxes were measured with a radioactive tracer and enzyme techniques.

The measurement of primary productivity, a possible driving force of phosphate flux, was made in situ using the ^{14}C method of Steemann Nielsen (1952), with the Goldman (1963) modifications as applied to Castle Lake (Goldman, 1969).

A series of bioassay experiments were conducted to assess the importance of different nutrients as potential limiting factors. Phosphorus and sulfur (previously shown to be limiting on occasion in Castle Lake; Goldman, 1964) were added to enclosed flasks with ^{14}C, and the changes in ^{14}C activity were monitored (Goldman 1960). Although the biota of a lake will usually change with a prolonged nutrient addition, this short-term method is useful as a relative indicator of immediate nutrient limitation. Details of all methods are provided in Richey (1974).

RESULTS OF FIELD MEASUREMENTS AND CARBON-PHOSPHORUS CONVERSIONS

This analysis emphasizes the results from one 1-month period—August 19 through September 18, 1972, a period when we made a concentrated effort to measure all important biotic components of the Castle Lake pelagic community. A detailed description of the phosphorus phytoplankton, zooplankton, and bacterial communities of Castle Lake during this period is provided by Richey (1974), Williams (1973), and Jassby (1973). Table 1 summarizes the more important pools of phosphorus during this period. There was no change in either total phosphorus or total dissolved phosphorus during this time, so any changes must have been a result of uptake or release of phosphorus within the pelagic zone.

Table 1   Chemical and Stoichiometric Measurements for August 19, 1972.[a]

Depth (m)	Phosphorous Fraction						
	SRP	TOTP	TDP	Q_3	Q_4	Q_5	PP
3.0	0.1	3.0	2.2	4.8	–	2.3	7.1
7.5	0.1	1.6	3.6	4.3	–	1.8	6.1
12.5	0.1	0.8	0.9	2.3	0.3	1.4	4.5
17.5	0.1	2.0	1.5	1.3	0.0	1.1	2.4
22.5	0.1	1.8	1.1	1.4	0.5	0.7	2.6
30.0	0.1	5.5	1.9	–	–	–	–
32.5	0.1	8.4	3.8	–	–	–	–

[a] SRP $\pm$ 0.1, TOTP and TDP $\pm$ 1.1 μg liter^{-1}; Q_3, Q_4, Q_5 $\pm$ 40% ($p < 0.05$).

There was a distinct relation between the primary production of Castle Lake and the rate of phosphorus exchange. Phosphate uptake ranged from 0.03 to 0.12 μg liter^{-1} hr^{-1} (Table 2), and most of the error in the calculation results from the uncertainty of the SRP ($\cong$ dissolved inorganic phosphorus) measurement. There was no significant variation in phosphate uptake over depth on any one day in Castle Lake, although pronounced vertical variations have been demonstrated in Lake Tahoe (M. Perkins, unpublished data) and in Lake Washington (Richey et al., 1975).

Table 2   Experimental Determinations of Phosphate Uptake ($J_{13} + J_{14}$).[a]

Depth (m)	Aug. 19	Aug. 24	Sept. 3	Sept. 8	Sept. 13	Sept. 18
3.0	0.124	0.048	0.034	0.045	0.031	0.054
7.5	0.101	0.048	0.053	0.039	0.053	0.043
12.5	0.103	0.033	0.035	0.046	0.036	0.035
17.5	0.109	0.041	0.036	0.051	0.035	0.030
22.5	0.106	0.050	0.043	0.056	0.038	0.043
30.0	0.110	0.097	0.107	0.039	0.051	0.063

[a] In μg liter^{-1} hr^{-1}. Average error = 102%, August 19 to September 18, 1972.

Because of the uncertain kinetics of DOP evolution and utilization, it was not possible to quantify directly the DOP excreted. However, relative results, useful for spatial and temporal comparisons, could be measured and are presented in Table 3. The evolution of $DO^{32}P$ showed a pronounced vertical variation, with high counts at the surface on August 18 and 24 and September 13 and 18. The results of a compartmental analysis experiment showed that DOP was not an important source of phosphorus to the phytoplankton and bacteria, at least at the time of the experiment. It should also be noted that the greatest uptake of $[^{32}P]PO_4$ and release and uptake of $[^{32}P]DOP$ occurred on August 19, which was also the day of the greatest photosynthetic activity (see following discussion). There were no significant differences in either carbon or phosphorus flux among any of the other days during this period.

Bioassay results showed that the addition of phosphorus to Castle Lake at 3 and 12.5 m was rarely stimulatory and on occasion even inhibitory (Richey, 1974). Bioassays conducted by A. Jassby on Castle Lake in 1971 (unpublished data) showed the same pattern. Of course, the bioassays discussed here measured only short-term changes in the rate of carbon fixation. Population or biomass could change under the long-term influence of an altered nutrient regime. Also, the true picture of nutrient limitation lies in the spectrum of a variety of complex nutrient interactions and only rarely as one single nutrient, such as phosphorus. However, we can assume for the purposes of the present model that phosphorus is not limiting.

Primary production rates for each sampling day were similar and had a bimodal profile with depth (Figure 2, observed). A maximum production of 3 to 5 mg C m^{-3} hr^{-1} at 3 or 5 m was seen, with a second peak of 1 to 3 mg C m^{-3} hr^{-1} at 17.5 to 20 m. Total daily areal production of 662 mg C m^{-2} day^{-1} on August 19 was significantly greater ($p < 0.05$) than the production

Table 3 $DO^{32}P$ Evolution.[a]

Depth (m)	Aug. 19	Aug. 24	Sept. 3	Sept. 8	Sept. 13	Sept. 18
3.0	434,669	164,706	30,151	5,366	229,634	204,623
7.5	160,759	55,615	55,440	0	91,584	74,612
12.5	111,929	36,124	44,954	3,701	102,492	46,723
17.5	93,969	34,592	80,266	6,004	50,701	23,130
22.5	79,342	21,691	89,148	18,125	22,904	12,788
30.0	65,340	19,882	–	–	29,904	10,948

[a] In distintegrations per minute (dpm). Average error = 21%, August 19 to September 18, 1972.

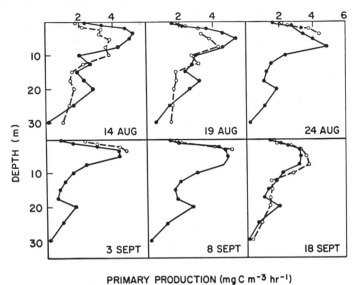

PRIMARY PRODUCTION (mg C m^{-3} hr^{-1})

Figure 2 Observed (•—•) versus predicted (Eq. 1, ○—-○) primary production. Absence of (•—-•) indicates model failure.

on the other sampling days, while none of the other days differed significantly among themselves.

DEVELOPMENT OF MODEL EQUATIONS

The equations developed below may be considered as a set of hypotheses concerning how the phosphorus cycle functions. As the experimental results showed that phosphorus flux appears to be dependent on photosynthesis in Castle Lake and not vice versa, a model of photosynthesis must be included in a model of the phosphorus cycle.

There have been a variety of photosynthesis models reported in the literature, as reviewed by Patten (1968). The model used here (Table 4, Eq. 1) was developed to predict photosynthesis per unit volume as a function of phytoplankton carbon and light.

The peak in primary production (Figure 2) is not at the surface, due to light inhibition in the top several meters. Goldman (1963, 1969) reviewed some of the mechanisms and reported the light intensities that caused inhibition. An inhibition term was derived as a function of uv + ir light by normalizing maximum, uninhibited photosynthesis to 1 and taking inhibition as a function of available uv and ir (Table 4, Eq. 1b).

Table 4 Phosphorus Model Equations.[a]

Eq.		Process	Curve Form
1	$\dfrac{dC}{dt} = (F\gamma)^{1/2}\,\rho\alpha P_{\text{mod}}$	Carbon flux (mg C m^{-3} hr^{-1}), F = phytoplankton biomass	—
1a	$\gamma = \exp(0.685 I_z + 0.4)$	Increase of photosynthetic efficiency with decreased light (i.e., with depth)	Exponential
1b	$\rho = I - \dfrac{\ln[(uv + ir)_z + 5]}{[\ln(uv + ir)_0 + 5](1.54)}$	Light inhibition as a function of uv (ultra violet) and (infrared) ir at surface (0) and depth (z)	Exponential
1c	$\alpha = \dfrac{I_z/I_k}{[1 + (I_z/I_k)^2]^{1/2}}$	Light saturation, I_z = light at depth z, I_k = light at onset of saturation	Hyperbola
1d	$P_{\text{mod}} = \dfrac{vQ_1}{k_m + Q_1}$	Michaelis-Menten expression for phosphate limitation (Q_1); k_m = half-saturation, v = maximum flux	Hyperbola
2	$J_{13} = \dfrac{0.025\,dC}{dt}$	Phosphate required to support photosynthesis	—
3	$J_{14} = Q_4\,\dfrac{2.46S\,\exp(-0.76/T)}{24(1.17 + S)}$	Phosphate required to support bacteria; S = concentration of dissolved organic carbon, T = temperature	—

4	$J_{15} = Q_1 \left[\sum\limits_{k=1}^{2} \sum\limits_{j=1}^{5} 0.51 L_{kj}^2 (0.44 + 0.05T) + \sum\limits_{j=1}^{5} 0.0034 L_{3j}^2 T \right]$	Zooplankton grazing (see text for description) —
5	$J_{51} = 0.0286 Q_{5j} W^{-0.383} \exp(0.0387T + 10^{-5}C - 3.34P)$	Zooplankton excretion of phosphate (see text for description) —
6	$J_{37} = \dfrac{0.005 Q_3}{t}$	Phytoplankton death per time t —
7	$J_{47} = 0.7 J_{14}$	Bacteria death —
8	$J_{57} = \dfrac{0.005 Q_5}{t}$	Zooplankton death —
9	$J_{71} = \dfrac{0.75 Q_7}{t}$	Phosphate regeneration through autolysis —

[a] J_{ij} expressed in μg liter^{-1} hr^{-1}.

The second, deeper peak in primary production, at the vastly reduced light intensity, implies that the deep-water population has adapted its enzyme systems to become more efficient at utilizing light. Indeed, Goldman (1969) showed that during the summer of 1968 in Castle Lake, efficiency, defined as $mg\,C\,m^{-3}$ fixed per unit of light, increased greatly with depth. This phenomenon is described by Eq. 1a (Table 4) by fitting the 1968 data.

The light-saturation equation (Table 4, Eq. 1c) is the same as that developed by Talling (1957), Vollenweider (1965), and Fee (1969). The effect of a single limiting nutrient (Table 4, Eq. 1d) is expressed through the familiar Michaelis-Menten formulation. Multiplication of Eq. 1 by a P/C ratio of 0.025 yields the amount of phosphate required to support photosynthesis (Table 4, Eq. 2) assuming that J_{23} is negligible.

The phosphorus uptake required to support bacterial growth was calculated by modifying the bacteria generation time equation of Jassby (1973), again assuming that J_{24} is negligible. The phosphorus demand is a function of the number of doublings per day (Table 4, Eq. 3).

Williams (1973) derived feeding rate equations for the three dominant zooplankters in Castle Lake. For *Daphnia rosea* and *Holopedium gibberum*, the filtering rate in milliliters per individual per hour is given by 0.51 L^2 (0.44 + 0.05T), and for *Diaptomus novamexicanus* by 0.0034 $L^2 T$, where L = body length in millimeters and T = temperature in degrees centigrade. The amount of phytoplankton and bacteria phosphorus removed by the zooplankton is the water volume filtered by the zooplankton times the particulate phosphorus concentration. This model ignores size preference and assumes that all phytoplankton and bacteria are available for grazing. The total quantity of phytoplankton and bacteria phosphorus grazed is given by Eq. 4 (Table 4, where $i = 3$ is phytoplankton and $i = 4$ is bacteria grazed; $k = 1$ is Daphnia, $k = 2$ is Holopedium, and $k = 3$ is Diaptomus; and $j = 1, 5$ are size classes of zooplankton).

The model for zooplankton excretion reported by Peters and Rigler (1973) appears to account for the ranges of excretion reported in the literature, so it was used here. They describe the rate of phosphorus release, with this author's correction for the phosphorus content of the zooplankton population, as given in Eq. 5 (Table 4), where T = temperature in degrees centigrade, C = cell food concentration (cells per milliliter), P = food phosphorus concentration, W = dry weight (milligrams) of the individual zooplankters, and Q_{5_f} = the sum by individual of the zooplankton phosphorus mass.

There are several approaches to determining the size of the detrital phosphorus pool (Q_7). The direct approach is to subtract the known fractions from the total:

$$Q_7 = \sum_{i=3}^{8} Q_i - \left(\sum_{i=3}^{6} Q_i + Q_8 \right)$$

This would give only the residual phosphorus, mostly nucleic acids, after autolysis had liberated the more labile fractions from the cell shortly after death. Therefore, the rate of accumulation of this fraction would be J_{79}, the contribution of the detrital pool to the sediments. Sensitivity analysis showed that this fraction was insignificant; thus it will be ignored. The total detrital phosphorus is actually a function of the death rates of the phytoplankton, bacteria, and zooplankton, as autolysis releases much of the cellular phosphorus very quickly in an inorganic form available for assimilation. In fact, Jassby (1973) demonstrated that the death rate of bacteria was almost as great as the growth rate; thus a large fraction of bacterial phosphorus might be made available through death. The death rates were measured as about 10% of the population per day for the plankton and 70% of the bacteria per day. Regeneration from the detrital pool, P_7, was estimated as the autolysis of the inputs to that pool at 0.75 hr^{-1}.

As has been discussed, physical complexation and precipitation processes affect the concentration and biotic availability of the different phosphorus species. It is assumed here that the protonation of the phosphate ions does not affect their availability. The equations of Table 5 calculate the distribution, complexation, and precipitation of the different P species as a function of iron concentration and pH (after Stumm and Morgan 1970).

The field research indicated that DOP and its associated fluxes and enzymes, polyphosphate, sedimentation, and eddy diffusion did not immediately affect short-term cycling; thus they were not included in the model.

The processes we have considered, as represented by the equations of Table 4, were translated into FORTRAN and run on a digital computer using a time step of 1 hr.

Table 5 Model Equations of Distribution, Complexation, and Precipitation of Phosphate Species as a Function of pH and Iron.[a]

$$[PO_4^{3-}] = P_T/(1 + [H^+]/10^{-12.3} + [H^+]/10^{-19.5} + [H^+]^3/10^{-21.7})$$
$$[HPO_4^{2-}] = P_T/(1 + (10^{-12.3}/[H^+] + [H^+]/10^{-7.2} + [H^+]^2/10^{-9.4})$$
$$[H_2PO_4^-] = P_T/(1 + [H^+]/10^{-2.2} + 10^{-7.2}/[H^+] + 10^{-19.5}/[H^+]^2)$$
$$[H_3PO_4] = P_T/(1 + [H^+]/10^{-2.2} + [H^+]^2/10^{-9.4} + [H^+]^3/10^{-21.7})$$
$$HPO_4S = 10^{-11}[H^+]/[Fe^{3+}]$$
$$HPO_4P = HPO_4S/HPO_4^{2-}$$
$$[DIP] = 3.1 \times 10^7(P_T - [HPO_4^{2-}](1 - HPO_4S[HPO_4^{2-}])$$

[a] P_T = total species, DIP = phosphate left in solution. HPO_4S = HPO_4 removed, HPO_4P = percentage removed. Brackets indicate concentrations.

COMPARISON OF OBSERVED AND PREDICTED RESULTS

The model tests a set of postulated relationships concerning the functioning of the phosphorus cycle in Castle Lake. A comparison of model predictions to field measurements is one check on the validity of the model assumptions. If the data match, the assumptions made about the ecosystem and the measurement of the important parameters may be correct, whereas a large discrepancy indicates a need for further understanding. The outcome of the primary productivity submodel is shown along with the field data (Figure 2). On August 14 and 19 and September 19, the agreement throughout the water column is quite close, within the error of the field analytical methods. On the remaining days, the agreement is close in the upper 3 to 5 m, but then the model failed by giving values far greater than the observed rates of carbon fixation. Failure is probably due to an overestimate in the field of viable phytoplankton (Richey 1974). This model suggests the necessity of refining our methods for separating photosynthetically-active plankton from non-active plankton.

Figure 3 compares the observed rate of phosphate uptake with the model calculations of uptake ($J_{13} + J_{14}$) and regeneration ($J_{51} + J_{71}$). Agreement

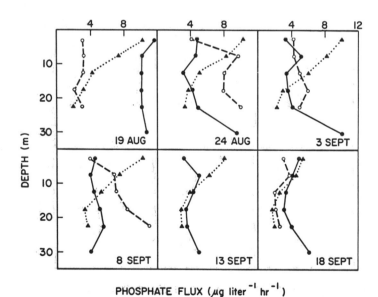

PHOSPHATE FLUX (μg liter^{-1} hr^{-1})

Figure 3   Observed phosphate uptake (•——•), predicted phosphate uptake ($J_{13} + J_{14}$, ○———○) and model phosphate regeneration ($J_{51} + J_{71}$, △ · · · △). Absence of (•———•) indicates model failure.

between observed and predicted fluxes is quite good for September 3 and September 18 within the limits of error of the field measurements, but not so good on the other days. Since the model and the field data did not always agree, additional processes not accounted for in the model may have been occurring, or pool size determinations may not have been accurate. It may be implied, however, that because uptake predicted by the model and observed in the field are generally close and are balanced in turn by regeneration, the general procedure outlined here may be a valid first approximation to understanding some of the aspects of the phosphorus cycle. It also implies that the system was in a steady state at the time of study. With this justification, a steady-state solution to the system of equations in Table 4 provides insight into the behavior of some of the other terms of the model.

Under steady-state conditions, the net flows into and out of each compartment must balance to give an equilibrium solution. This solution was provided by taking representative values as discussed above and solving the model equations for the different Q_i values (Table 6). These calculations show that the mechanisms for the regeneration of phosphorus are sufficient to supply the community, without DOP utilization. Grazing of bacterial phosphorus by zooplankton results in a minor depletion of bacteria, but bacterial autolysis is considerable. If Jassby's (1973) hypothesis that bacteria death is almost equal to bacteria growth is correct and if autolysis is rapid, then the phosphorus released by bacteria through autolysis is almost enough to sustain bacterial growth. The turnover of zooplankton phosphorus via excretion is significant enough to supply the demands of phytoplankton under steady-state conditions. Thus within the limits of field observations, it appears that the exclusion of DOP from this model does not appear to have caused serious error.

Table 6   Steady-State Solution to Equations 1 to 9.

Pool (μg liter^{-1})		Flux (μg liter^{-1} hr^{-1})		Rate Constant (hr^{-1})	
Q_1	0.10	J_{13}	0.032	r_{13}	0.319
Q_3	1.65	J_{14}	0.005	r_{14}	0.050
Q_4	0.11	J_{35}	0.025	r_{35}	0.015
Q_5	0.24	J_{45}	0.002	r_{45}	0.015
Q_7	0.01	J_{51}	0.025	r_{51}	0.104
–	–	J_{37}	0.007	r_{37}	0.004
–	–	J_{47}	0.004	r_{47}	0.038
–	–	J_{57}	0.001	r_{57}	0.004
–	–	J_{71}	0.012	r_{71}	0.840

ADDITION OF PHOSPHATE IN THE MODEL

It might be interesting to see the result of a sudden phosphorus input to this steady-state phosphorus cycling system. For example, what would have been the result if a plane loaded with phosphate fertilizer had crashed at 3 m in the lake on August 19, 1972?

The phosphorus model equations were expressed in the explicit difference form

$$Q_{i,t+1} = Q_{i,t} + \left(\sum J_{ji} - \sum J_{ij}\right)_t \Delta t$$

and solved. It was assumed that the fertilizer was instantly mixed to a concentration of 10 μg liter^{-1}, as orthophosphate. At a pH of 7 and an Fe^{3+} concentration of $10^{-6.7}$M (a typical value for Castle Lake), over 50% of the addition was precipitated out, leaving only 4.1 μg liter^{-1} as phosphate (Q_1). Surprisingly, the model results indicate that the additional phosphorus remains in the Q_1 pool and that none appeared elsewhere; However, none of the other compartments are limited by Q_1, and thus there should be no change in the flows into and out of these compartments. Where the addition does become evident is in the difference in turnover rates. The rate constant, k_{13}, went from 0.314 hr^{-1} to 0.008 hr^{-1}, and k_{14} went from 0.015 hr^{-1} to 0.001 hr^{-1}. Thus there is no direct relation between the amount of a solute present and its rate of utilization. This further supports the concept that knowing the mere ambient concentration of a nutrient may tell one little about the importance of that nutrient in the system.

To test what might happen to the phosphorus cycle in Castle Lake if phosphate were added when phosphorus was limiting, the following scenario was constructed. Assume in Eq. 1 (Table 5) that $v = 2$ and $K_m = 0.1$. Then the addition of phosphate should result in an increased demand for phosphorus through the simulation of photosynthesis.

Figure 4 shows the results of the addition of 10 μg liter^{-1} phosphate to the phosphorus-limited system. Again, only 4 μg liter^{-1} of Q_1 is left after precipitation with iron. After 48 h, the Q_1 pool is reduced, while the phytoplankton phosphorus (Q_3) increases. Zooplankton (Q_5) increases, presumably because of the increased phytoplankton. Detrital phosphorus (Q_7) also increases. Bacterial phosphorus (Q_4), however, does not change. Zooplankton does not increase enough to exert a significantly increased grazing pressure on the bacteria. The formulation for bacteria uptake depends only on the concentration of dissolved organic carbon and temperature; thus increased phosphorus would not directly affect the bacteria. In real life the increased phytoplankton population would probably excrete more carbon as extracellular products of photosynthesis or as dissolved organic phosphorus, which would increase the bacterial activity. This is another argument for

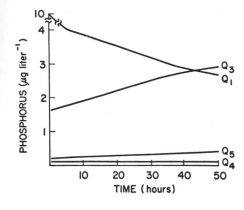

Figure 4 Response of phosphorus model under phosphate-limiting conditions to plane crash of fertilizer; phosphate (Q_1), phytoplankton (Q_3), bacteria (Q_4), zooplankton (Q_5).

further research into the role of excreted organic products by phytoplankton. This model predicts only those changes that might occur shortly after the nutrient addition. In nature the populations would change and a complete new system would evolve (cf. Schindler et al., 1973).

Approaching the study of phosphorus dynamics in Castle Lake with a modeling perspective served several purposes:

1   The formulation of an understanding of the phosphorus cycle as a whole was aided by a conceptual model that indicated what parameters ought to be measured in the field.
2   Comparison of the field data to model output served as a check on the validity of the model assumptions and thus on the understanding of the system as expressed through the equations.
3   Steady-state solutions to the equations allowed estimates of parameters difficult to measure experimentally. The field data and the model suggest that the regeneration of phosphate is adequately accounted for by zoo-plankton excretion. Autolysis from bacteria also may be a significant source of phosphate. Physical complexation and precipitation processes may remove considerable amounts of phosphate from solution. Dissolved organic phosphorus may be an important source of phosphate in some systems, but in Castle Lake there appear to be enough alternative sources of phosphorus to maintain the population in a non-phosphorus-limited state, even at extremely low ambient nutrient levels. This highlights the importance of obtaining nutrient flux rates as well as quantity to assess the importance of a nutrient to the system.
4   The model provided insight into what might happen to Castle Lake if a pollution stress occurred, under both phosphorus-limiting and non-limiting conditions.

Further model manipulation would serve little purpose, as the data are not available to verify the results. Instead, the model highlights a number of problems that need further investigation, before an adequate dynamic model of the phosphorus cycle is completed: (1) Accurate and sensitive chemical analyses of the different phosphorus pools are needed, (2) the partitioning of phosphorus uptake between phytoplankton and bacteria must be resolved, (3) the role of dissolved organic phosphorus, in particular, needs careful study, (4) accurate determination of death rates and the subsequent nutrient release is important, (5) zooplankton excretion needs further work if the hypothesis that it is the main source of phosphate renewal is to be accepted, and (6) also of great importance is the necessity to determine properly through a variety of bioassay techniques the degree of limitation in a lake of phosphate and phosphate plus other nutrients.

This initial phosphorus model has provided a base for a continued "second-generation" field and modeling program in lakes of the Lake Washington (Seattle, Washington) drainage basin, where phosphate is limiting. Current research is concentrating on the six main problem areas identified by this study (Richey et al., 1975; Richey, in preparation; Wissmar and Richey, in preparation). The agreement between the model and field measurements suggests that the interaction between models and field experimentation may provide a powerful tool for the study of aquatic nutrients. These methods are applicable to the study of whole lake ecosystems and could lead to models of the eutrophication processes. Such models would be invaluable tools in water quality management.

ACKNOWLEDGMENTS

Research was funded by National Science Foundation grants GB–19052 and GB–35371 to C. R. Goldman and GB–20963 and GB–36810X1 to the Coniferous Forest Biome, Ecosystem Analysis Studies, International Biological Program. Drs. C. R. Goldman, D. W. Schindler, C. A. Hall, F. Taub, and E. P. Richey made valuable comments on the manuscript, although I assume responsibility for any remaining errors or incorrect interpretations. P. O. Neame, F. Sanders, N. Williams, and A. Jassby provided field support.

REFERENCES

Dugdale, R. C. 1967. Nutrient limitation in the sea: Dynamics, identification and significance. *Limnol. Oceanogr.* **12**: 685–695.

Fee, E. J. 1969. A numerical model for the estimation of photosynthesis production, integrated over depth and time in natural waters. *Limnol. Oceanogr.* **14**: 90.

Goldman, C. R. 1960. Molybdenum as a factor limiting primary productivity in Castle Lake. *Science* **132**: 1016–1017.

Goldman, C. R. 1963. The measurement of primary productivity and limiting factors in fresh water with carbon-14. In M. S. Doty (Ed.), *Proc. Conf. Prin. Prod. Meas., Mar. and Fresh Water*. U.S.A.E.C. TID-7633. Pp. 103–116.

Goldman, C. R. 1964. Primary productivity and micronutrient limiting factors in some North American and New Zealand lakes. *Int. Ver. Theor. Angew. Limnol. Verh.* **15**: 365–374.

Goldman, C. R. 1969. Photosynthetic efficiency and diversity of a natural phytoplankton population in Castle Lake, California. *Proc. IBP/PP Tech. Meet., Trebon, 14–21 Sept. 1969.* Pp. 507–517.

Jassby, A. 1973. The ecology of bacteria in the hypolimnion of Castle Lake, California. Ph.D. thesis, Univ. Calif., Davis. 186 pp.

Lee, G. F. 1973. Role of phosphorus in eutrophication and diffuse source control. *Water Res.* **7**: 111–128.

Patten, B. C. 1968. Mathematical models of plankton production. *Int. Rev. ges. Hydrobiol.* **53**: 357 408.

Peters, R. H., and F. H. Rigler. 1973. Phosphorus release by *Daphnia*. *Limnol. Oceanogr.* **18**: 821–839.

Richey, J. E. 1974. Phosphorus dynamics in Castle Lake, California. Ph.D. thesis, University of California, Davis. 160 pp.

Richey, J. E., A. H. Devol, and M. A. Perkins. 1975. Diel phosphate flux in Lake Washington, U.S.A. *Verh. Internat. Verein. Limnol.* **19**: 222–228.

Rigler, F. 1973. A dynamic view of the phosphorus cycle in lakes. In E. Griffith, A. Beeton, J. Spencer, and D. Mitchell (Eds.), *Environment phosphorus handbook*. Wiley, New York. Pp. 539–568.

Schindler, D. W., H. Kling, R. V. Schmidt, J. Protopowich, V. E. Frost, R. A. Reid, and M. Capel. 1973. Eutrophication of lake 227 by addition of phosphate and nitrate: The second, third and fourth years of enrichment, 1970, 1971, and 1972. *J. Fish. Res. Board Can.* **30**: 1415–1440.

Steemann-Nielsen, E. 1952. The use of radioactive carbon (C14) for measuring primary production in the sea. *J. Cons. Perm. Int. Explor. Mer.* **18**: 117 140.

Stumm, W., and J. Morgan. 1970. *Aquatic chemistry*. Wiley-Interscience, New York. 583 pp.

Talling, J. F. 1957. Photosynthetic characteristics of some freshwater plankton diatoms in relation to underwater radiation. *New Phytol.* **56**: 29–50.

Vollenweider, R. A. 1965. Calculation of photosynthesis-depth curves and some implications regarding day rate estimates in primary production measurements. In C. R. Goldman (Ed.), *Primary productivity in aquatic environments*. Mem. Ist. Ital. Idrobiol. 18 Suppl., Univ. California Press, Berkeley. Pp. 435 437.

Williams, N. 1973. A mathematical model of plankton dynamics in Castle Lake, California. Ph.D. thesis, University of California, Davis. (In preparation.)

REVIEW REFERENCES

Atkins, W. R. G. 1923. The phosphate content of fresh and salt waters in its relationship to the growth of algal plankton. *J. Mar. Biol. Assoc. U.K.* **13**: 119–150.

Baudouin, M. F., and O. Ravera. 1972. Weight, size and chemical composition of some freshwater zooplankters: *Daphnia hyalina* (Leydig). *Limnol. Oceanogr.* **17**: 645–649.

Beers, Y. 1953. *Introduction to the theory of error.* Addison-Wesley, Reading, Mass. 167 pp.

Beers, J. R. 1966. Studies on the chemical composition of the major zooplankton groups in the Sargasso Sea off Bermuda. *Limnol. Oceanogr.* **11**: 520–528.

Brown, W. 1973. Solubilities of phosphates and other sparingly soluble compounds. In E. Griffith, A. Beeton, J. Spencer, and D. Mitchell (Eds.), *Environment phosphorus handbook.* Wiley, New York. Pp. 205–239.

Carpenter, E. J. 1970. Phosphorus requirements of two planktonic diatoms in steady-state culture. *J. Phycol.* **6**: 28–30.

Chamberlin, W., and J. Shapiro. 1973. Phosphate measurements in natural waters: A critique. In E. Griffith, A. Beeton, J. Spencer, and D. Mitchell (Eds.), *Environmental phosporus handbook.* Wiley, New York. Pp. 355–366.

Denige, G. 1920. Réaction de coloration extrêmement sensibles des phosphates et des arsenates. *Cr. Acad. Sci.* **171**: 802–804.

Fuhs, G. W. 1969. Phosphorus content and the rate of growth in the diatoms *Cyclotella nana* and *Thalassiosira fluviatilis. J. Phycol.* **5**: 312–321.

Gardiner, A. C. 1937. Phosphate production by planktonic animals. *J. Cons., Cons. Perm. Int. Explor. Mer.* **12**: 144–146.

Golterman, H. 1964. Mineralization of algae under sterile conditions by bacterial breakdown. *Int. Ver. Theor. Angew. Limnol. Verh.* **15**: 544–548.

Golterman, H. 1973. Natural phosphate sources in relation to phosphate budgets: A contribubution to the understanding of eutrophication. *Water Res.* **7**: 3–17.

Harold, F. M. 1966. Inorganic polyphosphates in biology: Structure, metabolism, and function. *Bacteriol. Rev.* **30**: 772–788.

Harrison, M. T., R. E. Pacha, and R. Y. Morita. 1972. Solubilization of inorganic phosphates by bacteria isolated from Upper Klamath Lake sediment. *Limnol. Oceanogr.* **17**: 50–57.

Jagendorf, A. T. 1973. The role of phosphate in photosynthesis. In E. Griffith, A. Beeton, J. Spencer, and D. Mitchell (Eds.), *Environment phosphorus handbook.* Wiley, New York. Pp. 381–392.

Johannes, R. E. 1964. Phosphorus excretion and body size in marine animals, microzooplankton and nutrient regeneration. *Science* **146**: 923–924.

Johannes, R. E. 1968. Nutrient regeneration in lakes and streams. In M. Droop and E. Wood (Eds.), *Advances in the microbiology of the sea.* Academic, London and New York. Pp. 203–213.

Keck, K., and H. Stich. 1957. The widespread occurrence of polyphosphate in lower plants. *Ann. Bot.* **21**: 611–622.

Kuenzler, E. J. 1965. Glucose-6-phosphate utilization by marine algae. *J. Phycol.* **1**: 156–164.

Kuenzler, E. J. 1970. Dissolved organic phosphorus excretion by marine phytoplankton. *J. Phycol.* **6**: 7–13.

Kuhl, A. 1962. Inorganic phosphorus uptake and metabolism. In R. Lewis (Ed.), *Physiology and biochemistry of algae.* Academic, New York. Pp. 211–229.

Kuhl, A. 1968. Phosphate metabolism of green algae. In D. F. Jackson (Ed.), *Algae, man and the environment.* Syracuse Univ. Press, Syracuse, N.Y. Pp. 37–52.

Lean, D. R. S. 1973a. Phosphorus dynamics in lake water. *Science* **179**: 678–680.

Lean, D. R. S. 1973b. Movements of phosphorus between its biologically important forms in lake water. *J. Fish. Res. Board Can.* **30**: 1525–1530.

Mortimer, C. H. 1941. The exchange of dissolved substances between mud and water in lakes. Vol. I, *J. Ecol.* **29**: 280–329.

Mullin, M. M., P. R. Sloan, and R. W. Eppley. 1966. Relationship between carbon content, cell carbon and area in phytoplankton. *Limnol. Oceanogr.* **11**: 307–311.

Otsuki, A., and R. G. Wetzel. 1972. Coprecipitation of phosphate with carbonates in a marl lake. *Limnol. Oceanogr.* **17**: 763–767.

Peters, R., and D. R. S. Lean. 1973. The characterization of soluble phosphorus released by limnetic zooplankton. *Limnol. Oceanogr.* **18**: 270–279.

Phillips, J. E. 1964. The ecological role of phosphorus in waters with special reference to micro-organisms. In H. Haukelekian and N. Dondero (Eds.), *Principles and applications in aquatic microbiology*. Wiley, New York. Pp. 61–81.

Pomeroy, L. R. 1963. Experimental studies of the turnover of phosphate in marine ecosystems. In V. Schultz and A. Klement (Eds.), *Radioecology*. Reinhold Publ. Corp., New York. Pp. 163–166.

Rabinowitch, E., and Govindjee. 1969. *Photosynthesis*. Wiley, New York. 273 pp.

Redfield, A. C., B. H. Ketchum, and F. A. Richards. 1963. The influence of organisms on the composition of sea water. In M. N. Hill (Ed.), *The Sea*, Vol. II. Interscience Publ., New York. Pp. 26–77.

Reichardt, W., J. Overbeck, and L. Steubing. 1967. Free dissolved enzymes in lake waters. *Nature* **216**: 1345–1347.

Rhee, G. 1972. Competition between an alga and an aquatic bacterium for phosphate. *Limnol. Oceanogr.* **17**: 505–514.

Riggs, D. S. 1963. *The mathematical approach to physiological problems*. Williams and Wilkens, Baltimore. 445 pp.

Rigler, F. 1956. A tracer study of the phosphorus cycle in lake water. *Ecology* **37**: 550–562.

Rigler, F. 1961. The uptake and release of inorganic phosphorus by *Daphnia magna* Straus. *Limnol. Oceanogr.* **6**: 165–174.

Scharpa, L. 1973. Transformations of naturally occurring organophosphorus compounds in the environment. In E. Griffith, A. Beeton, J. Spencer, and D. Mitchell (Eds.), *Environmental phosphorus handbook*. Wiley, New York. Pp. 393–412.

Soeder, C., H. Muller, H. Payer, and H. Schulle. 1971. Mineral nutrition of planktonic algae; some considerations, some experiments. *Mitt. Int. Verein. Limnol.* **19**: 39–58.

Solomon, A. K. 1960. Compartmental methods of kinetic analysis. In C. L. Comar and F. Brunner (Eds.), *Mineral metabolism: An advanced treatise* Vol. I (A) Academic, New York. Pp. 119–162.

Sommer, J. R., and J. J. Blum. 1965. Lytochemical localization of acid phosphatases in *Euglena gracilis*. *J. Cell. Biol.* **24**: 235–254.

Strickland, J. D. H., and T. R. Parsons. 1968. A practical handbook of seawater analysis. *Bull. Fish. Res. Board Can.* No. 167. Ottawa. 311 pp.

Van Wazer, J. 1973. The compounds of phosphorus. In E. Griffith, A. Beeton, J. Spencer, and D. Mitchell (Eds.), *Environmental phosphorus handbook*. Wiley, New York. Pp. 169–178.

Vollenweider, R. A. 1970. *Water management research. Scientific fundamentals of the eutrophication of lakes and flowing waters with particular reference to nitrogen and phosphorus as factors in eutrophication*. Tech. Rep. to the Organization for Economic Cooperation & Development, Committee for Research Cooperation, Paris. 159 pp.

12
A Model of a
Thermal Spring Food Chain

RICHARD G. WIEGERT

Ecological models range from simple, highly aggregated, box-and-arrow flow diagrams to models representing the complexities of behavior exhibited by individual organisms. The former may not represent all important mechanisms of interaction within the system and thus may be little more than quick and convenient data summaries. The latter demand so much information on interactions that they can be implemented, if at all, only at the population level.

Within this range one may identify an intermediate class of model: one in which the aim is not only to simulate faithfully an existing (measured) set of data, but, in addition, to provide some explanation of the system as well as some predictive capability. This class of model, in short, can be (1) constructed at reasonable cost, (2) validated in some way, and (3) used to investigate the operation of a specified ecosystem. In this chapter I tell the story of the development of such a model, one describing the simple algal-fly-predator community found in the thermal outflows of Yellowstone National Park. Along the way, I touch on a few of the pitfalls as well as the advantages of model building in ecology and use the example of this simple thermal system to describe a model structure general enough to serve as the theoretical framework for any population or community of organisms. Finally, I discuss my experiences in constructing models of the same system but different degrees of aggregation.

The simple algal-arthropod food chains of Yellowstone are ideal for modeling because:

1   They are isolated natural ecosystems that have evolved considerable trophic diversity, yet the same families, genera, and even species are found in other thermal springs throughout the world.
2   They are not so large as to preclude or make difficult a total system study, yet they are large enough to permit adequate sampling without undue perturbation of the system.
3   The many analogous ecosystems that exist within any given thermal area and throughout the world permit corroboration of general conclusions by other investigators.
4   They are easy to manipulate, so field experiments to test model predictions are possible.
5   The anabolic and catabolic processes occurring in these heated systems are generally very rapid, in many instances reducing the time necessary for succession experiments in the field to a few weeks.

The Thermal Community

The main study area is a small meadow next to the Firehole Lake Drive in the Lower Geyser Basin of Yellowstone National Park. This meadow contains a number of small thermal outflows, differing in flow rate and ranging in temperature from 43 to 90°C. The hot pools in this meadow are all small and have no official names. Because of the chance "discovery" of this study area in early 1968 in company with M. L. and T. D. Brock, I call it Serendipity Springs.

We initiated studies of the biological communities in the effluents of these springs, where temperature ranged from 50°C to slightly above ambient. Within this temperature range animals can exist in the community given certain conditions of spatial and temporal heterogeneity.

Because of the variable and ephemeral nature of many natural flows, "artificial" communities were grown on board "troughs" with a controlled source of water (Frontpiece 3). The water is close to chemical neutrality and contains large amounts of dissolved minerals. Inflow temperature of the various board troughs ranges from 43 to 56°C, but varies seasonally on a given trough by only a degree or two.

Immediately downstream from the inflow a mucilaginous (gelatin-like) mat of filamentous algae and bacteria develops. Beginning with a bare substrate, this algal-bacterial mat grows to a thickness of 1 to 3 cm within 2 to 3 months. During this time some areas grow faster than others, causing disruption of the original laminar water flow and producing temperature heterogeneity in the mat, hotter in flowing areas than in stagnant parts of the mat. If the maximum diel temperature is below 40°C, the eggs, larvae, and pupae of brine flies (Ephydridae) are able to survive and graze on the algae, but the blue-green algae do not grow in this stagnant cool condition. Thus successional growth quickly produces a mosaic of (1) hot flowing areas with no flies and growing algae; (2) hot stagnant areas with adult flies, eggs, and small larvae on the surface but little available habitat for survival of the 3rd larval stage (instar); and (3) cool stagnant areas with no algal growth but high densities of all life history stages of the fly.

THE MODELING PROCESS

General Description

Preparing a list of the significant biotic and abiotic components is the first step in modeling any system. This includes a thorough sampling of the species and a description of the kind and quality of soil and water, along with ranges in solar radiation, temperature, pH, and so forth. At this preliminary stage the modeler begins to formulate a list of desired components or state variables.

Once a tentative list has been drawn up, the modeler may begin to ask which components are connected, that is, interact with each other in some significant way. This web of interconnections is the trophic (food) web representing the pathways of material or energy flow through the community. If each chain in the trophic web were followed upward from the source, steps or levels could be defined at each point that material passed through a living population. Gathering together all flows at each step, we could construct trophic levels, for example, plants (autotrophs), first-level consumers, second-level consumers, and so forth. Trophic levels can thus be constructed from a knowledge of the species (or functional group) food web, but the reverse is not possible.

In these first two stages of model building some difficult decisions have to be made concerning the degree to which species will be combined into multi-species model components and how these combinations are to be made. These decisions are difficult because each species (and in most cases each life history stage) has different density-control mechanisms and different responses of variable parameter values (e.g., weight-specific respiration and ingestion) to change in density, available resources, and abiotic factors. The end result, even in the simplest of natural ecosystems, is always a compromise with the reality of extreme taxonomic and structural complexity. One of the major objectives of the research undertaken on these thermal spring communities was to formulate some guidelines for such lumping. I take up this topic again at the end of the chapter.

Trophic Relationships

Three species of dolichopodid fly, two species of water mite, at least two species of bird, two species of wolf spider, one species of tiger beetle, and two or more species of parasitic wasp prey on or parasitize the brine fly.

Although this list represents a small total community species component compared to most natural communities, a food web detailing all of their interactions would still be a formidable array from which to construct a model. Such a total model is in fact the overall goal of this ecological study begun in Yellowstone in 1968. However, the principles of developing and using a predictive model of an ecosystem can be demonstrated by considering a simpler subset of these component populations. This is my objective here.

All of the autotrophic blue-green algae except one (relatively rare at the midtemperatures, 40 to 56°C) are filamentous forms with similar growth properties. Almost all of them also fix nitrogen. One species, *Mastigocladus laminosus*, constitutes more than 75% of the total biomass. In the model, the blue-green algae, the autotrophs of the system, constitute a single functional group because all are eaten by the brine fly and they all respond similarly to regulatory mechanisms.

Of the three species of brine fly only one, *Paracoenia turbida*, was abundant in the particular communities modeled.

The filamentous and unicellular bacteria, for the most part saprophages, are dependent on the algae for food, and on egestion and mortality by the grazing flies. Because of its passive dependence, this group is not included in the model described in this chapter.

Predation and parasitism on the flies is included, but only as an energy drain on the flies. The functional dynamics of the individual predator and parasite populations are not incorporated in this model in the interests of simplicity of illustration. Predation and parasitism in these systems are known to play a passive role in these systems except under certain conditions of increased availability of algae to flies. Under normal conditions, the algal-fly-interaction stabilizes nicely with intraspecific competition for space among fly larvae acting as the major population control mechanism. Considerable condensation of species is made in the food web (Figure 1), although certain aspects are expanded, for example, the recognition of three types of algal mat based on temperature-flow characteristics and the treatment of individual life history stages of the brine fly as separate components or state variables. Furthermore, time delays in development are incorporated within the model. This wealth of detail was included in the model to permit later examination of the effects of progressive lumping together or condensation.

Construction of a food web requires qualitative knowledge of interconnections within the system that can be obtained only by direct observation or experiment. Assigning the role of autotroph to the filamentous algae was a simple first step.

Brine flies were studied by direct observation of their feeding and by the use of radioisotopes of carbon. Either carbonate or glucose tagged with ^{14}C was introduced into an intact algal-bacterial mat. Carbonate was incorporated only by the autotrophic algal cells, glucose only by the heterotrophic bacteria. Brine flies (both larvae and adults) were offered radioactively tagged algal-bacterial mat. Subsequent measurements of the radioactivity in the tissue of brine fly larvae and adults showed both capable of assimilating the organic materials of algae and bacteria. However, because no selective feeding on either component was observed and because the biomass of the mat was more than 95% algal material, the latter must be considered as the principal food of the brine flies.

The separation of the various life history stages of the brine fly in the food web diagram was deemed necessary because of the specific nature of the different sources of predation. Dolichopodid flies were observed to feed upon both eggs and one- to two-instar larvae, but the larger third instar larvae, pupa, and adults were unmolested by these predatory flies. In contrast, brine flies lose material and energy from the adults to the larvae of the red water mite *Partnuniella* and from the pupae to the small parasitic pteromalid wasp, *Eurolepis.* Spiders and shore birds (spotted sandpiper) take both adult flies and third instar larvae.

Equations of Interaction: Developing the General Equation

With a list of components and a diagram of their pathways of interconnection in hand, the modeler next specifies the mechanisms of interaction. This is a difficult step and requires considerable ecological knowledge. The value of the model as a tool with which to make predictions or formulate theory will depend on the care and thoroughness with which these mechanisms are specified. The interactions may first be described in words; then these verbal constructions are transformed into suitable mathematical form. The thermal spring model began with a consideration of the most basic facts about population growth and regulation. These facts were incorporated into a flux equation employing a maximum specific rate of ingestion together with feedback control terms that may take an infinite variety of functional forms; the particular function depends on the type of ecological interaction involved.

For example, some commonly observed attributes of living populations are:

1   They are capable of autocatylytic growth and, when provided with optimum amounts of space and material resources, will increase exponentially at some genetically-fixed maximum specific rate. This upper bound rate of growth is represented by

$$\frac{dQ}{dt} = Q(\tau - \lambda) \tag{1}$$

where Q = standing crop of the population
τ = the maximum specific rate of ingestion
λ = the minimum possible rate of physiological loss, that is, respiration plus excretion and mortality (an egestion loss will be incorporated later).

The difference, $\tau - \lambda$, represents the maximum specific rate of growth under optimum conditions (r_{max}).

2   A population cannot continue positive exponential growth for an unlimited time. In practice organisms seldom if ever achieve their maximum potential growth. Feedback controls, mortality, and other-than-minimum physiological losses combine to control growth. Populations in the field

exhibit a steady state density or fluctuate about some long-term mean. In either case, a time, t, can be found such that growth is zero.

$$\frac{dQ}{dt} = 0 \tag{2}$$

3   When deprived of any material resources, a population declines in accordance with losses. If these losses are restricted to the minimum imposed by physiological considerations, then

$$\frac{dQ}{dt} = -\lambda Q \tag{3}$$

From these observations, plus the trivial but necessary requirement that ingestion can never be negative, a general equation of population growth can be constructed.

$$\frac{dQ}{dt} = Q[(\tau - \tau f)_+ - \lambda] \tag{4}$$

where f = a dimensionless feedback control term determining the actual rate of ingestion as a function of population crowding.

The subscript plus $(\ )_+$ notation specifies that the term within parentheses, represented by a dot below, must always be greater than or equal to zero

$$(\cdot)_+ = \begin{cases} 0 & \text{if } (\cdot) \leq 0 \\ (\cdot) & \text{if } (\cdot) > 0 \end{cases}$$

Equation 4 now specifies the realized rate of ingestion and the minimum physiological losses by a population. If the form of f, the function specifying the control on ingestion, were known, such an equation could be employed in modeling the behavior of a population in the context of an ecosystem. Unfortunately, a single feedback control function is seldom adequate to represent even the simplest of real world situations. For example, a population is usually limited or controlled (simultaneously or at different times) by scarcity of at least two very different kinds of resources. Scarcity of one or more material resources may cause realized ingestion to be less than the maximum value. Alternatively, intraspecific competition for space may be a factor in holding ingestion to less than optimum levels. Generally, we can distinguish these as material (renewable) resources versus nonmaterial, spatial (nonrenewable) resources. Separation of these two kinds of feedback control permits a further useful generalization of Eq. 4.

$$\frac{dQ}{dt} = Q[\tau - \tau f - \{\tau - \lambda\}f']_+ - Q\lambda \tag{5}$$

where f = the feedback control of ingestion determined by scarcity of material resources

f' = the feedback control of ingestion determined by crowding and competition for space (nonrenewable resources). Note that when f' is unity, ingestion just balances minimum losses, that is, growth would be zero no matter the state of the material resources.

Equation 5 satisfies the observations made earlier. Whenever both space and material resources are optimum, the feedback terms, f and f', will be zero and ingestion will be maximized. Possibilities exist for any combination of feedback control to reduce ingestion to the point where it is equal to losses. Whenever material resources are totally unavailable, f will equal 1 and the population will decline at rate λ.

There remain two difficulties in the way of adapting Eq. 5 to models of ecological systems. First, such models consider the flow of materials or energy between a series of interconnected and interacting compartments. The ingestion of each compartment comprises the only gain. Losses then are the sum of "ingestions" by all other compartments receiving input from the compartment of interest. Thus the focus is on the ingestion; growth or decline of the compartment is arrived at by summing all gains and losses.

A second problem is the failure to represent the assimilation efficiency in Eq. 5. For example, if space was so severely limiting as to cause f' to equal 1, ingestion would be just sufficient to balance physiological losses. But in most populations, a fraction of all material egested is passed through the gut and lost. Thus in order for *assimilated* as opposed to ingested material to equal physiological losses, some extra must be taken in to compensate for losses to egestion.

The first problem is easily dealt with by writing the equation only for ingestion. Thus each flow in the model is separately expressed, and the dynamics of each component of the ecosystem can be represented by an equation that sums the gains and losses.

The second difficulty is eliminated by simply dividing the specific rates of physiological loss by the proportion assimilated (usually more conveniently expressed as one minus the proportion egested). Combining these two changes, we can write the equation for the ingestion by any component, j, feeding on a resource compartment, i. Subscript j identifies a term as associated solely with component j, while subscript ij identifies the interaction or exchange between i and j.

$$ J_{ij} = Q_j \left[\tau_{ij} - \tau_{ij} f_{ij} - \left\{ \tau_{ij} - \frac{\lambda_j}{1 - \varepsilon_{ij}} \right\} f_j \right]_+ \qquad (6) $$

where J_{ij} = flux into compartment Q_j

ε_{ij} = proportion of ingestion that is egested.

The terms, τ_{ij}, λ_{ij}, f_i, and f_j, correspond to the terms, τ, λ, f, and f', described earlier. When necessary the λ_j parameter can be divided into its component respiration and excretion terms, that is, ρ_j, η_j, and μ_j, respectively.

Thus a predictive model of any ecological system is constructed by ascertaining the specific feedback functions involved, evaluating the parameters, and determining the role played by spatial and temporal heterogeneity in each interaction equation. Let's see how this works in setting up a model of the algal-fly food chain of the thermal community.

THE THERMAL MODEL

General Description

Designating the flux of energy between each of the components (including the surroundings, component zero) shown in Figure 1 by J_{01}, J_{12}, and so forth, a set of equations of change is written

$$dQ_1/dt = J_{01} - J_{10} - J_{12} - J_{14} - J_{15} \tag{7}$$
$$dQ_2/dt = J_{12} + J_{62} - J_{20} - J_{23} \tag{8}$$
$$dQ_3/dt = J_{23} - J_{30} - J_{34} \tag{9}$$
$$dQ_4/dt = J_{14} + J_{34} - J_{40} - J_{45} \tag{10}$$
$$dQ_5/dt = J_{15} + J_{45} - J_{50} - J_{56} \tag{11}$$
$$dQ_6/dt = J_{56} - J_{60} - J_{62} \tag{12}$$

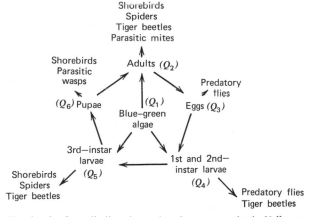

Figure 1 Food web of an alkaline thermal spring community in Yellowstone Park.

This set of equations reduces the information in Figure 1 to a form representing a balanced flow of energy. The next step in modeling these components is to consider the functional form of each of the above 15 fluxes, including the feedback control terms and the techniques for incorporating spatial and temporal heterogeneity where necessary.

Algae

A series of in situ measurements of gross production, net production, and chlorophyll content of algal cores was made on the larger pair of board troughs (Frontpiece 3b), and the results were compared with light intensity and mat temperature. Neither of these abiotic variables was correlated significantly with gross production or respiration. The algal-bacterial mat as a whole was light saturated at low light levels ($\approx 10\%$ of full sunlight) and acclimated to the temperature at which it had developed.

However, there was a strong relationship between density of the mat (expressed as g ash-free dry wt per m^{-2}) and the weight-specific gross photosynthesis rate. For a given flow, the maximum weight-specific rate of gross photosynthesis decreased with the 0.29 power of biomass increase, modified by the concentration of free CO_2 and the day length. Both of these were incorporated into an equation of photosynthesis as simple multiplicative factors: the former on the basis of a simple diffusion of the CO_2 across the cell wall proportional to concentration, the latter on the basis that the mat was light saturated during virtually the entire daylight period. Thus if the algal component is represented by Q_1 (g ash-free dry wt m^{-2}) and the weight specific gross photosynthesis rate by τ_{01},

$$\tau_{01} = 0.04\beta_2\beta_1Q_1^{-0.29} \tag{13}$$

where $\beta_1 = $ ppm $CO_2 \times 10^{-1}$
$\beta_2 = $ daylength in hours. (The summer value of 14 was used in the model).

Then having determined the manner in which free CO_2 and day length interact to determine gross rate of photosynthesis (τ_{01}), we next use this rate to construct an equation incorporating the necessary feedback controls. The general form of this equation is that of Eq. 6. However, the term, $\tau_{ij}f_i$, is unnecessary, because free CO_2, the material resource in this instance, is already used to determine τ_{01}. Thus only the space-related feedback control term of Eq. 6 was used, and we have

$$J_{01} = \lambda_{01} + Q_1[\tau_{01} - (\tau_{01} - \rho_1)f_1]_+$$

where λ_{01} = input of algae in source water (kcal m^{-2} day^{-1})
$\quad\quad \rho_1$ = respiration loss (kcal kcal^{-1} day^{-1})
and

$$f_1 = \left[\frac{1 - \alpha_{11}/Q_1}{1 - \alpha_{11}/\gamma_{11}}\right]_+ \tag{14}$$

where $\quad \gamma$ = the maximum density of Q possible in absence of grazing under the given flow regime
$\quad\quad \alpha_{11}$ = the algal threshold response density above which the rate of gross photosynthesis is reduced below the maximum value (τ_{01}) due to CO_2 depletion.

The form of the feedback control function (f_1) represents a hypothesis of how a steadily increasing mat thickness would affect the concentration of free CO_2 in the interstitial water immediately adjacent to the cell. Because the mat is a reasonably compact structure, turbulent flow does not penetrate far, and the supply of CO_2 to these deeper layers should follow Ficke's Law of diffusion. Thus the effect of adding another increment of algae to the mat is inversely proportional to the thickness already present, hence the form of Eq. 14. Adding a millimeter to the thickness of the mat when the original mat thickness is small has a greater depressing effect on photosynthesis than adding the same increment when the mat is thick.

The threshold thickness (α_{11}), expressed as kcal m^{-2}, was measured by inserting micro-pH probes into the mat (1 to 2 mm) until the pH began to change, indicating the point where turbulent replenishment of interstitial water stopped.

The maximum or equilibrium density was obtained from the maximum density of mat extrapolated from the largest single core sample taken in the field. Note that feedback repression of gross photosynthesis begins as soon as the thickness of the mat exceeds the α_{11} value.

Loss of energy from the algal mat was caused both by washout (λ_{10}) and by respiratory energy loss (ρ_{10}). Washout was measured by periodic samples taken from the outflow of the troughs. The specific rate of respiratory energy loss was obtained from measurement of oxygen consumption of algal cores in dark bottles. The flux of energy from the algal-bacterial mat to the surroundings was

$$J_{10} = (\lambda_{10} + \rho_1)Q_1 \tag{15}$$

Spatial heterogeneity of the mat was caused by shifts in current due to differential mat growth. The result was the creation of distinct habitats on the mat. These were designated hot flowing (water covers algae, no flies), hot stagnant (no surface flow, adult flies feed but no larvae), and cool stagnant (no flow, adult flies, eggs, larvae). Algal growth was maximum in the hot-flowing condition, and fly growth was maximum and algal growth zero in the cool-stagnant areas. This spatial heterogeneity had to be incorporated in the model by setting up separate *substate variables* for each of the three conditions and specifying the proportions of algal mat in each. Then from field measurements of the probability of a given area remaining in any given state, transfers were effected each day. These transfer rates were additionally made probabilistic, with standard deviation equal to 0.25 of the mean so that the model now had a stochastic element.

Transfers of algal condition proceeded in the direction hot flowing (β_{12}) to hot stagnant (β_{23}) to cool stagnant and back to hot flowing (β_{31}), resulting in considerable simulated mortality of fly larvae and eggs.

Adult Brine Flies

Ingestion by adult brine flies was modeled according to Eq. 6, using controls representing both scarcity of material resources and of space. The equation describing the ingestion flux was

$$J_{12} = Q_2 \left[\tau_{12} - \tau_{12}f_{12} - \left(\tau_{12} - \frac{\rho_2}{1 - \varepsilon_{12}} \right) f_2 \right]_+ \tag{16}$$

with

$$f_{12} = \left[\frac{\alpha_{12} - Q_1}{\alpha_{12}} \right]_+$$

and

$$f_2 = \left[\frac{Q_2/\beta_{22} - \alpha_{22}}{\gamma_{22} - \alpha_{22}} \right]_+$$

where P_2 = the rate of respiratory energy loss.

The parameter, β_{22}, or adult fly fractional concentration factor was necessary because part of the mat surface at any given time is covered with flowing water and is thus unavailable to adult *Paracoenia* flies. Thus the real or ecological density is the density over the whole mat divided by the fractional portion of mat available to the flies.

Both feedback terms use a simple linear relationship between an increment of change in the resource or recipient density and the resulting effect on rate of ingestion. This involves the fewest assumptions about the nature of feedback control and thus is the equation of choice in the absence of data suggesting a more complicated relationship.

The threshold response to resource scarcity (α_{12}) was based on both field and laboratory observations of adult flies feeding on thin films of algae. No refuge for algae exists; all algae can be eaten. Thus γ_{12} was not included in the feedback term.

Crowding or competition for space is seldom a factor in controlling ingestion by the flies. They are often so crowded on favored oviposition sites as to virtually cover the algal surface. Thus the maximum density (γ_{22}) was computed as the maximum number of flies that could be packed into a single layer on the surface. No significant interference with oviposition was observed until flies reached approximately 90% of this maximum, thus providing a value for α_{22}.

Direct measurements of the maximum rate of ingestion by adult flies would have been difficult if not impossible under field conditions. Instead, this rate (τ_{12}) was computed as the sum of respiratory energy loss (ρ_2) plus the maximum rate of oviposition (τ_{23}), all divided by the fraction assimilated $(1 - \varepsilon_{12})$.

Respiration was measured in the laboratory at the average summer temperatures in the field during daytime (25°C), the only period during which the flies are active.

Because of the relatively large proportion of the mat surface available to feeding by adult *Paracoenia*, neither resource, material or space, is usually an important factor limiting ingestion by adult flies during model runs that simulated the normal state of the algal-fly system.

Losses from the adult brine fly compartment consist of oviposition and of combined egestion, respiration, and mortality (J_{23} and J_{20}, Eq. 8). Simulation of oviposition employed the equation

$$J_{23} = \frac{J_{12}\tau_{23}}{\tau_{12}} \tag{17}$$

This equation contains a maximum rate of oviposition (τ_{23}) obtained from flies held under assumed optimum conditions in the field. Note under less than optimum conditions, that is, when J_{12} is less than maximum, the oviposition rate is less than the maximum possible $Q_2\tau_{23}$ (from Eq. 16, maximum J_{12} equals $Q_2\tau_{12}$).

Losses to egestion, respiration, and nonpredatory mortality are represented very simply as an egestion fraction multiplied by the total ingested plus the sum of the specific rates of respiration and nonpredatory mortality multiplied by the adult fly standing stock

$$J_{20} = J_{12}\,\varepsilon_2 + (\rho_2 + \mu_2)Q_2 \tag{18}$$

Predatory mortality was imposed as an external loss under control of the modeler.

Eggs

The number of eggs in the system at any one time responds to input from oviposition (Eq. 17) and losses to hatching, mortality, and respiration.

The average time to hatching was 2 days, necessitating a time delay in the equation simulating transfer from eggs to first-instar larvae. The equation employs a vector whose dimension corresponds to the length of the hatching period in days. Oviposition is deposited in the first age class of the array. Each day the content of the last age class of the array is transferred to the first age class of the first-instar larval compartment. Total egg standing crop at any time is the sum of all age classes in the vector. Thus for eggs the hatching period (β_{30}) equals 2 and $Q_3 = \sum_{i=1}^{2} \mathbf{Q}_{3_i}$, where $\mathbf{Q}$ represents a vector and

$$J_{34} = Q_{3_2} \tag{19}$$

where Q_{3_2} represents the last age class of the vector.

Nonpredatory mortality of eggs is caused by oviposition into areas too hot for survival and by the shifting flow of hot water.

Measurements of both of these factors are combined into one nonpredatory mortality rate, and this together with a respiration rate is used to compute losses in a manner similar to that for adult flies.

$$J_{30} = Q_3(\mu_3 + \rho_3) \tag{20}$$

Larvae

In the model two compartments, Q_4 and Q_5, represent first- and second-instar larvae and third-instar larvae, respectively. The separation of the larval components is done to better represent predatory mortality, since some predators prey on only one or the other of these two groups.

Larvae require mat temperatures of 40°C or less for survival and growth (35°C is optimum). Otherwise, they have many of the same requirements as adults, and the equations simulating their ingestion and losses follow directly from Eq. 16 with time delays to simulate larval development constructed as in Eq. 18. Development time of first- to second-instar stages is 2 days, that of third-instar larvae is 4 days.

The equations for fluxes of energy into and out of the larval compartments are

$$J_{14} = Q_4\left[\tau_{14} - \tau_{14}f_{14} - \left[\tau_{14} - \frac{\rho_4}{1 - \varepsilon_4}\right]f_4\right]_+ \tag{21}$$

$$J_{15} = Q_5\left[\tau_{15} - \tau_{15}f_{15} - \left[\tau_{15} - \frac{\rho_5}{1 - \varepsilon_5}\right]f_5\right]_+ \tag{22}$$

$$J_{40} = \varepsilon_4 Q_{14} + (\rho_4 + \mu_4) Q_4 \tag{23}$$

$$J_{50} = \varepsilon_5 Q_{15} + (\rho_5 + \mu_5) Q_5 \tag{24}$$

$$J_{45} = (1 - f_{14} - f_4)_+ Q_{42} \tag{25}$$

$$J_{56} = (1 - f_{15} - f_5)_+ Q_{54} \tag{26}$$

Note: (1) The larval non-predatory mortality rates, μ_4 and μ_5, are in this instance variables that increase with the degree of stress placed on the population due to the crowding control factors, f_4 and f_5. (2) The transfer of larvae from Q_4 to Q_5 or from Q_5 to Q_6 (pupae) in Eqs. 25 to 26 is also controlled by the degree of stress placed on the population both by control due to scarcity of resources and by competition for space (factors f_{14}, f_{15}, f_4, and f_5). In other words, growth and pupation are both affected by scarcity of resources or crowding or both.

Pupae

Simulation of the pupal stage, because it does not feed or reproduce, is straightforward and simple. Losses consist of respiration and mortality plus emergence of adult flies.

The vector form was needed to model the latter pathway because of the length of time in the pupal state (average of 6 days).

Transfer equations are:

$$J_{60} = Q_6(\rho_6 + \mu_6) \tag{25}$$

$$J_{62} - Q_{6_6} \tag{26}$$

SIMULATION

Simulation is the major reason the algal-fly model was constructed. By simulation I mean the simultaneous solution of the set of Eqs. 7 to 12 representing the ecological interactions between compartments. Such computation can accomplish a number of objectives. Among these are (1) validation, that is, comparing predictions of the dynamic behavior of the system with measured values; (2) prediction of predator-prey interactions, that is, examining the effects of certain predator drains on the steady or cyclic behavior of the algae and the flies; (3) development of rules for condensing or lumping state variables to make the model "simpler." This later objective is extremely important to the modeler, because most communities require extensive lumping of species right from the beginning. Few total community models can attempt the detail that is being incorporated in the thermal springs construction.

Flow Chart

The flux equations (Eqs. 16 to 20) form the basis of a detailed Fortran-IV computer model of the algal-fly components of a thermal community, with

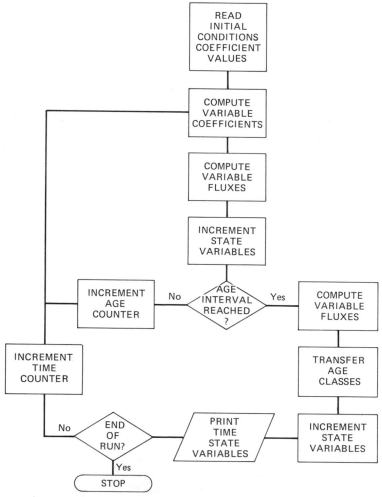

Figure 2 Flow chart of a thermal algal-brine fly computer model.

provisions for examining the effects of varying mortality imposed on different life history stages of the fly as well as for representing the spatial heterogeneity caused by temperature and flow irregularities.

Steps in the computation can be followed on the flow chart (Fig. 2). First the initial conditions are set, then those coefficients that change with every iteration (i.e., the variable coefficients) are calculated. One of these is the rate of photosynthesis, dependent on the standing crop of algae. Mortality rates of larvae also need to be evaluated each iteration, since they are increased whenever the feedback control terms are greater than zero.

The variable coefficients are then used to compute all variable fluxes. Next the state variables are incremented. I chose an integration interval of 0.1 day and a simple iterative addition (Euler) method of integration. This choice of interval and integration method provided reasonable accuracy at low cost. (Many more sophisticated and accurate methods of numerical integration are available and on occasion may be required for models of ecological systems. However, for systems showing periodic fluctuations around some mean value, the slight increase in accuracy is seldom worth the considerable increase in computation time. For example, some tests made on a model of this type constructed to simulate the dynamics of a coastal salt marsh community showed that the Euler method could virtually equal the precision of a numerical method called the fourth-order Runge-Kutte at less than one third the computational cost.) The area of hot stagnant and cool stagnant algal mat was controlled by three variables ($\beta_1 - \beta_3$, Table 1), each randomly distributed with the mean equal to the measured probability of a piece of algal mat remaining in a given condition. This aspect contributes much to the success of the model as an accurate simulator.

When ten iterations have been performed (enough to equal 1 day at 0.1 day per iteration), control is transferred to the portion of the program that calculates those fluxes that vary daily. Then the age class transfers and the values of the state variables are printed.

Initial Conditions

The initial values of parameter and state variables used in all simulation runs (Table 1) are determined from field and laboratory observations and experiments. An initial algal energy density of 1.0 kcal m^{-2} was used together with a fly density of 0.005 kcal m^{-2}. The latter value is the equivalent of one adult fly per m^2, a density commonly observed as a result of movement of flies from nearby mats. The initial algal density represents the normal accumulation of algae downstream from a mat already established. All simulations use a 14-hr day without incorporation of day/night temperature fluctuations.

Table 1 **Initial Parameter Values Used in the Thermal Spring Simulation Model.**

Symbol	Description	Units	Value
Transfer Coefficients			
τ_{01}	Gross photosynthesis rate	kcal kcal^{-1} day^{-1}	[a]
ρ_1	Algal respiration rate	kcal kcal^{-1} day^{-1}	0.031
λ_{10}	Algal emigration rate (including DOM)	kcal kcal^{-1} day^{-1}	0.007
τ_{12}	Adult fly ingestion	kcal kcal^{-1} day^{-1}	0.751
τ_{23}	Oviposition	kcal kcal^{-1} day^{-1}	0.056
ρ_2	Adult fly respiration	kcal kcal^{-1} day^{-1}	0.470
μ_2	Adult fly nonpredatory mortality	kcal kcal^{-1} day^{-1}	0.148
ρ_3	Egg respiration	kcal kcal^{-1} day^{-1}	0.060
μ_3	Egg nonpredatory mortality	kcal kcal^{-1} day^{-1}	0.215
τ_{14}	First-, second-instar ingestion	kcal kcal^{-1} day^{-1}	2.940
ρ_4	First-, second-instar respiration	kcal kcal^{-1} day^{-1}	0.360
μ_4	First-, second-instar nonpredatory mortality	kcal kcal^{-1} day^{-1}	0.221
τ_{15}	Third-instar ingestion	kcal kcal^{-1} day^{-1}	1.456
ρ_5	Third-instar respiration	kcal kcal^{-1} day^{-1}	0.330
μ_5	Third-instar nonpredatory mortality	kcal kcal^{-1} day^{-1}	0.122[b]
ρ_6	Pupal respiration	kcal kcal^{-1} day^{-1}	0.030
μ_6	Pupal nonpredatory mortality	kcal kcal^{-1} day^{-1}	0.056
Limit-Egestion Proportions			
α_{11}	Algae, self-control threshold density	kcal m^{-2}	100
α_{22}	Adult fly, self-control threshold density	kcal m^{-2}	284
α_{44}	First-, second-instar self-control threshold density	kcal m^{-2}	100
α_{55}	Third-instar self-control threshold density	kcal m^{-2}	50
α_{12}	Adult fly, resource-control threshold density	kcal m^{-2}	50
α_{14}	First-, second-instar resource-control threshold density	kcal m^{-2}	50
α_{15}	Third-instar resource-control threshold density	kcal m^{-2}	50
ε_2	Adult fly, egestion proportion	none	0.300
ε_4	First-, second-instar egestion proportion	none	0.300
ε_5	Third-instar egestion proportion	none	0.300
Maximum Standing Crop			
γ_{11}	Algae	kcal m^{-2}	1600
γ_{22}	Adult flies	kcal m^{-2}	315
γ_{44}	First-, second-instar larvae	kcal m^{-2}	150
γ_{55}	Third-instar larvae	kcal m^{-2}	150

Table 1 (*Continued*)

Symbol	Description	Units	Value
Algal Factors			
β_{11}	Dissolved CO_2	10^{-1} ppm	4.9
β_{12}	Day length	hours	14
β_{13}	Proportion of area hot flowing	dimensionless	0.25
β_{14}	Proportion of area hot stagnant	dimensionless	0.70
β_{15}	Proportion of area cool stagnant	dimensionless	0.05
Crowding Factors			
β_{21}	Adult flies	dimensionless	0.75
β_{41}	First-, second-instar larvae	dimensionless	0.05
β_{51}	Third-instar larvae	dimensionless	0.05
Development Times			
β_{30}	Eggs (to hatching)	days	2
β_{40}	First-, second-instar larvae	days	2
β_{50}	Third-instar larvae	days	4
β_{60}	Pupae	days	6

[a] Gross photosynthesis rate is a variable, recomputed each interval using Eq. 12.
[b] Third-instar nonpredatory mortality is a variable, increased from the base rate (0.122) by twice the amount of the self-control feedback term ($f_{3,3}$).

Model Validation

Tests of the predictive accuracy of the model take three forms. First, the successional growth of the algal mat alone without fly grazing is simulated proceeding from a bare substrate to the mature mat. The growth curve predicted by the model for varied free CO_2 concentrations compares favorably with curves constructed from replicated field measurements of successional development of the algal mat (Figure 3). Second, the growth of the total fly population was simulated under optimum conditions, that is, with no controls on fly density. From this curve (once the age distribution stabilized) a specific growth rate of 0.25 flies per fly day was computed, a value identical to the maximum specific growth rate (r_{max}) computed from my unpublished brine-fly life-table data. Third, the model was run with provision for 5% of the area of the mature mat available for use by fly larvae, 75% of the algal mat

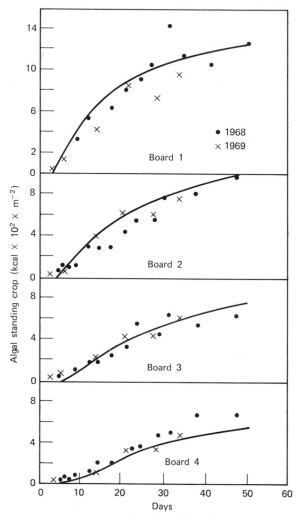

Figure 3 Successional increase in standing crop of algae through time on the four 1×2 m experimental board troughs. The data are from replicate experiments in 1968 and 1969. The lines are predictions of the model.

surface available to adult flies and normal predation losses (Table 1). The resulting predictions of steady-state algal and fly standing crops were close to the values from field measurements made on the mature mats in the field (Table 2). The success of these three validation comparisons indicated that the model might be useful for simulating the dynamics of this system under other conditions of algal mat availability and predation losses.

Table 2   Predicted and Measured Steady-State Standing Crops of Algae and Flies in a Thermal Ecosystem.[a]

	Standing Crop (kcal m^{-2})		Percentage of Total Flies in Each Age Class				
	Algae	Flies	Eggs	Larvae 1–2	Larvae 3	Pupae	Adults
Measured in field	889	7.3	2.1	5.1	42.1	33.6	17.3
Predicted from model	1030	6.0	1.6	3.7	40.7	34.0	20.0

[a] Percent habitat available to flies in the field ranged from 4 to 6. Prediction percentage taken from Day 398 in each age class.

Model Condensation—Effects of Perturbation

Predatory mortality in the algal-fly model is introduced as an external drain of energy from the fly compartments. The level applied to the algal-fly system in this instance is equivalent to the predator losses borne by each fly life history stage under "normal" conditions, that is, 5% of the algal area in the cool-stagnant condition and 70% in the hot-stagnant condition the remainder in the hot-flowing condition.

Different levels of condensation (lumping) were employed. Model 1 is the most detailed, validated model discussed in the preceding sections. In model 2, a crude averaging procedure replaced the detailed modeling of the spatial heterogeneity of algae. In model 3, the second model was further simplified by removing the vector system of time delays. Model 4 continued the condensation by lumping the three compartments, eggs, first to second instar, and third instar into one compartment. This lumping process reached the ultimate limit in model 5 with only two separate components, algae and flies.

Each of these models was run for 200 days, at which time all components exhibited either a "steady-state" or a constant long-term mean value, that is, no upward or downward trend. At this time drain due to predation was applied and the simulations continued for another 200 days. Expressing the results as a percentage change in the compartment value on Day 199 compared to that on Day 399, models 1 and 2 showed significant effects of predation on each of the various life history stages, but differed from each other very little (Table 3). Model 3, lacking realistic time delays, predicted results very different in some instances from the predictions of the first two models. Model 4, with partial lumping of the life history stages, predicted complete extinction of the flies! And model 5, simplest of all, although it could say

Table 3   Predicted Response of Algae and Brine Flies to Simulated Predation begun on Day 200. Values Represent the Percentage Change in Standing Crop from Day 199 to Day 399.

	Algae	Flies					
		Eggs	1–2-Larvae	3-Larvae	Pupae	Adults	Total Flies
Model 1	+3.4	−23.4	−1.4	−28.6	−17.9	−20.2	−22.7
2	+3.1	−25.6	+1.5	−36.2	−15.1	−46.6	−26.0
3	+0.6	−13.3	+48.8	−14.5	−5.8	−12.2	−9.3
4	+18.0	−100	−100	−100	−100	−100	−100
5	+1.2	—	—	—	—	—	−19.4

nothing about the various life history stages, predicted a total fly population reasonably close to the densities of models 1 and 2.

The moral of this story seems to be; condensation of the model with respect to life history stages should be total, that is, if you cannot afford to model the system in considerable detail, opt for the crude lumped model with appropriate averaged constraints. For example, models 1 and 2 look in detail at the various life history stages of a species, separately modeling growth rates and time delays. Model 5 on the other hand, incorporates all of these criteria into the maximum specific rate of growth (r_{max}) or more appropriately the maximum specific rate of ingestion (τ_{max}).

This suggests another tentative rule. When lumping two or more species into a single compartment, place things together that have similar maximum growth rates. Otherwise, the behavior of the lumped compartment under a predation drain may bear little resemblance to the real behavior of the species.

Model Condensation—Effects of Predation

The same models (1 to 5) were run for 400 days with the "normal" predation drain begun on Day 200. Only this time a number of runs of each model was made, with the proportion of algae in the cool, stagnant "available" condition increased progressively from 5% to 25, 50, 75, and 95%. In nature this is equivalent to progressively diminishing the flow of hot water to a patch of algae, an experiment that is continually being made in Yellowstone Park by fluctuations in natural flows and disturbance to channels. If the cooled patch remains small enough and close enough to a hot-water flow to stay warm, at least during the day, the results predicted by model 1 (Figure 4) appear adequate, although quantitative data bearing on this point are lacking at present. The relative efficacy of the various condensed models as predictors

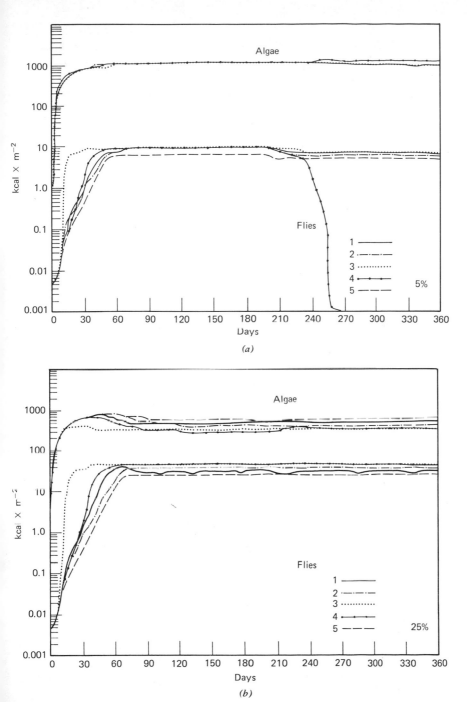

Figure 4 Simulations of the dynamics of the algal fly system with models 1 to 5 under the following conditions of algal availability to fly larvae (*a*) 5%, (*b*) 25%, (*c*) 50%, (*d*) 75%, and (*e*) 95%.

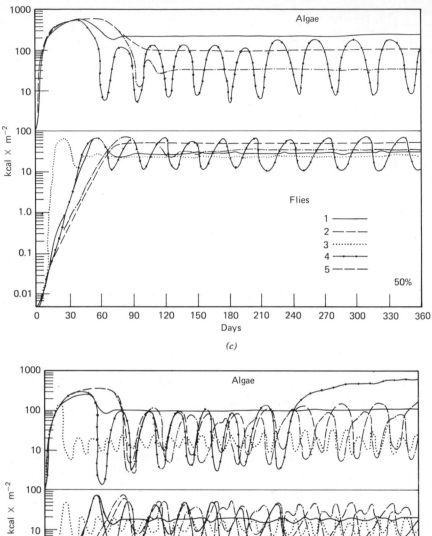

Figure 4 (*Continued*)

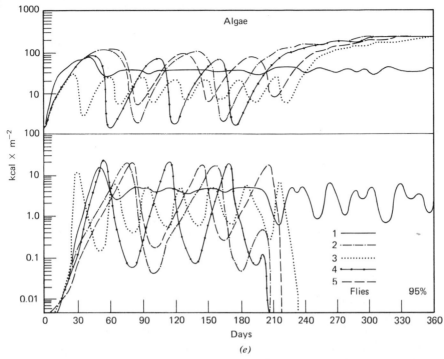

Figure 4 (*Continued*)

of the situation described by model 1 can be judged from a quick perusal of Figure 4. In general, they fail most markedly as the magnitude of the perturbation (the difference from the "normal" 5%) increases.

All the reasons for this are not clear, but the major failure is caused by the abnormally violent oscillations of flies predicted by models 2 to 5, carrying this group to such a low level that the constant predation drain renders them extinct.

Of course, the predator drain in a more complex and realistic model would not remain constant. And model 5, the crudest of the five, again seems to compare with model 1 about as well as any other. This suggests that a high degree of model condensation is possible provided the ecology of the species being lumped is carefully studied, at least to the point of providing a good estimate of r_{max} and the factors that control the realized rate of growth. Failing that, I would choose the path of lumping several closely allied species together, not into a single compartment, but rather in separate life-history stages. This may often prevent unnaturally violent oscillations in the predicted behavior while holding the model to a reasonable degree of complexity and number of compartments.

CONCLUSION

The model of thermal spring communities was developed to explain the dynamics of a relatively simple trophic system where considerable ecological sophistication was possible. But the basic framework of the equations should be generally applicable. The transfer of matter-energy to any living component of the model comprises a maximum, genetically determined rate of ingestion and various threshold densities that determine the switching on and off of the mechanisms of control. The functional form of the latter may be varied infinitely to conform to the biological realities of any situation. Thus the model structure employed in defining the algal-brine fly food chain in the Yellowstone thermal spring should, in principle, be applicable to other ecosystems. Particularly challenging in this respect are large, complex heterogenous systems of use to (and often endangered by) man. As a first step in testing this view, I have begun applying some of the lessons learned in the thermal spring work to the construction of an ecosystem model of those coastal Georgia salt marsh communities dominated by cord grass (*Spartina alterniflora*).

The first results of this effort are encouraging. The early versions of the marsh model provide much insight into the way the estuarine system operates and to the possible sensitivities of various biotic and physical parameters. In addition, the models have pointed to some serious gaps in our knowledge of processes in the field. Further work aimed at bringing these marsh models to a level of detail comparable to the thermal spring model should show whether this approach truly provides a general theoretical structure for predictive models of ecosystems.

ACKNOWLEDGMENTS

The work upon which this chapter is based was supported by research grants NSF GB7683 and GB21255. I am deeply indebted to the many colleagues and students who have worked with me in Yellowstone during the past seven summers.

REFERENCES

Brock, T. D. 1970. High temperature systems. *Ann. Rev. Ecol. Syst.* **1**: 191–220.

Collins, N. C. Population biology of a brine fly (Diptera: Ephydridae) in the presence of abundant algae food. *Ecology* **56**: 1139–1148.

Fraleigh, P. C. and R. G. Wiegert. A model explaining successional change in thermal blue-green algae. *Ecology* **56**: 656–664.

Kuenzel, W. J. and R. G. Wiegert. 1973. Energetics of a Spotted Sandpiper feeding on brine fly larvae (*Paracoenia*; Diptera; Ephydridae) in a thermal spring community. *Wilson Bull.* **85**: 473–476.

Mitchell, R. 1974. The evolution of thermophily in hot springs. *Quart. Rev. Biol.* **49**: 229–242.

Mitchell, R. and B. L. Redmond. 1974. Fine structure and respiration of the eggs of two ephydrid flies (Diptera: Ephydridae). *Trans. Amer. Micros. Soc.* **93**(1): 113–118.

Tuxen, S. L. 1944. The hot springs, their animal communities and their zoogeographical significance. In *The Zoology of Iceland* 1: Part 2, Munksgaard, Copenhagen. 206 pp.

Wiegert, R. G. 1973. A general ecological model and its use in simulating algal-fly energetics in a thermal spring community. In P. W. Geier, L. R. Clark, D. J. Anderson, and H. A. Nox (Eds.), *Insects: Studies in population management*, Occasional Papers. Ecol. Soc. Austr., Canberra, Vol. 1, pp. 85–102.

Wiegert, R. G. 1975. Simulation modeling of the algal-fly components of a thermal ecosystem: Effects of spatial heterogeneity, time delays and model condensation. In B. C. Patten (Ed.), *Systems analysis and simulation in ecology*, Vol. III, Academic, New York.

Wiegert, R. G. and P. C. Fraleigh. 1972. Ecology of Yellowstone thermal effluent systems: Net primary production and species diversity of a successional blue-green algal mat. *Limnol. Oceanogr.* **17**: 215–228.

Wiegert, R. G. and R. Mitchell. 1973. Ecology of Yellowstone thermal effluent systems: Intersects of blue-green algae, grazing flies (*Paracoenia*, Ephydridae) and water mites (*Partnuniella*, Hydrachnellae) *Hydrobiologia* **41**: 251–271.

Wiegert, R. G., R. R. Christian, J. L. Gallagher, J. R. Hall, R. D. H. Jones, and R. L. Wetzel. 1975 A Preliminary Ecosystem Model of Coastal Georgia *Spartina* Marsh. *Estuarine Research*, *Vol. I. Chemistry, biology and the estuarine system.* Academic, New York. Pp. 583–602.

PROLOGUE TO
SECTION III

In 1969 the U.S. Congress created the National Environmental Protection Act (Chapter 8). As a result of this law, an "environmental impact statement" is required for many projects that are expected to have significant impact on the environment. This law has spawned an enormous amount of effort aimed at understanding the effects of potential environmental disruptions.

Models may be useful in this assessment process, particularly when impacts are large and complicated. The following chapters present a number of examples of the use of models in determining environmental impact. However, the distinction between the assessment models presented here and the optimization models of the next section is not always clear. Assessment models, at least in theory, are not designed to determine the best course of action but rather to predict the consequences of a particular course of action. Results of such models can be used along with other criteria in decision making. For example, Chapter 13 attempts to predict the consequences of a continuation of present patterns of waste disposal on the environmental health of the Baltic Sea. It is up to the governments of the Baltic countries to weigh these consequences against the cost of disposing wastes in some other manner. In addition, this model represents an assessment of the research project itself. One conclusion from their work is that it is better to develop models relatively early in the process of research and then use the models to guide research effort.

Chapter 14 is an example of one increasingly common uses of models—to assess the impact of the operation of electric power plants on aquatic environments. In addition to developing the model, this chapter considers the effects of the model on public decision making. This and related models have been important in influencing large-scale environmental decisions, in large part due to the environmental litigation associated with the models. Again, the models were not meant per se to proscribe a course of action, but to serve as one input into the decision-making process.

In Chapter 15 we consider a model that could *not* be made to work, mainly due to an insufficient data base. Nevertheless, this model was valuable for a number of reasons given in the chapter. The paper also demonstrates that models can be devised, if not completed, for large systems as well as smaller ones. In Chapter 17 Brown very ambitiously attempts to build an objective model of a complicated social process—a nation at war and peace. It suggests that there were important questions that were not asked concerning the Vietnam war—before the war was undertaken—that might have been addressed easily with models. We think this chapter demonstrates very nicely Botkin's perceptive statement that models help you to understand the implications of your assumptions. The computer cannot always give you the right answers, but it is useful in exploring what the consequences of a decision might be within the limits of the information at your disposal.

Finally, there are two more chapters that deal with a popular subject in this book—estuaries. First Day, Hopkinson, and Loesch continue their account of the vast and productive Louisiana Marshes with two models that examine the consequences of man's activities in the coastal zone. The models raise questions concerning cumulative impact of management activities on the Mississippi River and the coastal marshes. These Louisiana models and the research behind them also have been important in influencing the decision-making process. Finally, Kelly and Spofford have combined an ecological and an economic model of the Delaware River estuary to explore the environmental consequences of different effluent discharges and discharge regulations on the different organisms living in the Delaware river. It turns out that "pollution" isn't always bad for all organisms, and that such models could greatly decrease the cost of cleaning up a river.

III
CASE HISTORIES
Assessing Environmental Impact

13
Baltic Ecosystem Modeling

BENGT-OWE JANSSON
FREDRIK WULFF

The Baltic Sea has been an important resource to its neighboring nations for many centuries. For example, the herring fisheries of the Baltic were an important factor in the development of the commerce and hence culture of the adjacent medieval cities. However, the increasing urbanization and industrialization of the seven bordering countries (Denmark, E. Germany, Finland, Poland, Sweden, U.S.S.R., and W. Germany) is stressing the Baltic's capacity for fishing, recreation, transportation, fresh water supplies, and even for assimilating the different wastes from these nations. Since many complicated factors interact to produce pollution in the Baltic, an optimal use of these resources on a long-term basis requires the creation and simulation of management models, either as a complex set of conceptual models or the computer models we present here. The Baltic Sea does have several features making it suitable for modeling: the inputs and outputs are relatively easy to determine and measure, the fauna and flora in this brackish water area have a low diversity compared to the oceans, and the main characteristics of the sea are well known due to the physicochemical and biological research undertaken during recent decades.

One particularly critical problem and the one we focus on here is the increasing eutrophication of the Baltic. Algal growth, especially that of the sessile green algae, has been stimulated by the addition of nutrients in sewage and industrial waste. A filamentous cover is spreading on the rocks out in the archipelagos far from the principle sources of pollutants. The excessive plant growth also has increased the frequency of stagnation periods and contributed to formation of hydrogen sulfide in the deeper, oxygen-free parts due to organic decomposition by anaerobic bacteria. This large-scale decline in water quality is similar in many respects to the Great Lakes eutrophication problem in Canada and the U.S.A.

Figure 1 gives the approximate distribution of hydrogen sulfide in the deepest Baltic waters. The presence of this poisonous gas not only greatly reduced the bottom fauna and hence the food for demersal fish, but also changed the migration pattern of several species of fish such as the cod, by creating large areas that the fish will not enter. In the presence of hydrogen sulfide, inorganic nutrients such as phosphorus are released from the sediment and carried to the surface layer, where they stimulate the growth of phytoplankton. The sedimentation of this increased algae will increase the already heavy input of organic material to the deep water, using up more oxygen and continuing a vicious circle. In order to understand the reactions of this system to natural and man-made impacts, the Swedish Natural Science Research Council sponsored a basic research project, "Dynamics and energy flows in the Baltic ecosystem," carried out by the Askö Laboratory, University of Stockholm.

The following is a short description of the present status and difficulties

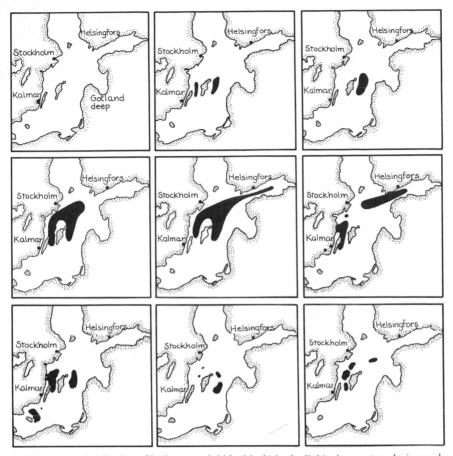

Figure 1 The distribution of hydrogen sulphide (black) in the Baltic deep waters during each year for 1965 to 1973. The maximum distribution during 1969 is broken by a salt-water inflow from the North Sea during the winter of 1969 to 1970. A new stagnation follows with the formation of hydrogen sulphide in the deeps east of Gotland, but is broken again by a new salt-water inflow during 1971 to 1972 (Fonselius, 1969, in Lindqvist, 1973).

of this project. We start with a presentation of the characteristics of the Baltic Sea as an ecosystem.

THE BALTIC SEA ECOSYSTEM

The Baltic Sea is a fairly large water body of about 365,000 km^2 with a mean depth of only 60 m, connected to the North Sea through two narrow straits of 17 and 8 m depth. It might be described as a stratified fiord with a rapid decrease in salinity near the mouth and low and fairly stable salinities in the interior surface waters. The fauna and flora consist mainly of a few euryhaline species (Figure 2). The diversity of the macrofauna drops from about 1500 species in the Skagerack to about 50 in the Bothnian Bay. Although the micro- and meiofauna show a less drastic reduction in numbers of species, the Baltic brackishwater system still comprises a fairly simple biotic component compared to the intricate marine systems of the world.

The primary production of the open Baltic shows the usual annual pattern of the temperate latitudes, that is, a spring bloom of dinoflagellates and

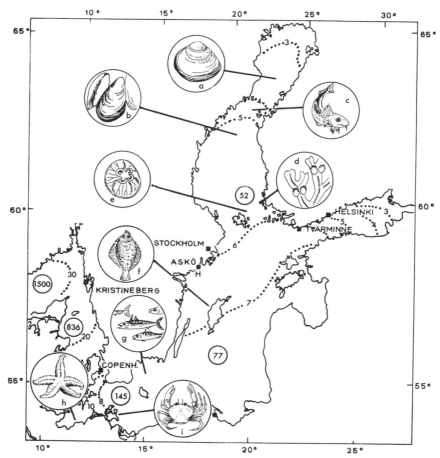

Figure 2   The Baltic Sea. Dotted lines are surface-water isohalines. Numbers within circles are number of macrofauna species. *a* to *i* are distribution limits for some common marine species: (*a*) *Macoma baltica*, (*b*) *Mytilus edulis*, (*c*) cod, (*d*) *Fucus vesiculosus*, (*e*) *Aurelia aurita*, (*f*) plaice, (*g*) mackerel, (*h*) *Asterias rubens*, (*i*) *Carcinus meanas* (modified from Jansson, 1972).

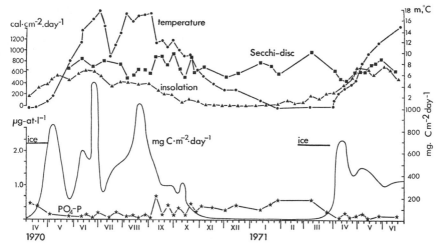

Figure 3 Annual variation in primary production (mg C m^{-2} day^{-1}), nutrients (PO$_4$–P in μg at l^{-1}), surface temperature, insolation, and Secchi-disc readings at Station 5, Landsort 1970 to 1971 (Jansson, 1972).

diatoms when the ice breaks up, followed by several peaks during the summer when the blue-greens dominate (Figure 3). Although discrepancies occur between the few estimations of the pelagic primary production in the Baltic Sea, the total annual production is about 75 to 100 g C m^{-2} as measured by the ^{14}C technique. This is higher than the open sea but not nearly as great as many inland waters.

The benthic secondary production of the open Baltic is concentrated in the vast soft bottoms above the halocline (i.e., the primary salinity-caused density layer at about 60 m). The macrofauna is dominated by a few species of amphipods, the clam *Macoma baltica*, and some polychaetes. The meiofauna is surprisingly abundant with biomass as great as many rich marine systems. Bacterial activity is intense, and microbes are the dominant organisms below the halocline.

The coastlines of the northern and central Baltic Sea are dominated by rocky archipelagos. The thousands of islands and skerries found there, along with their biotic communities of green, red, and brown algae, constitute a huge "filtering apparatus" with an active membrane of algae that act as both mechanical and biological filters by physically retaining coarser particles and by taking up the nitrogen and phosphorus of the bypassing water. In a sense, these archipelagos help to chemically "buffer" the main Baltic by trapping river-borne nutrients. The southern coasts are shallow, straight, and sandy with less ability to trap the nutrients discharged from rivers.

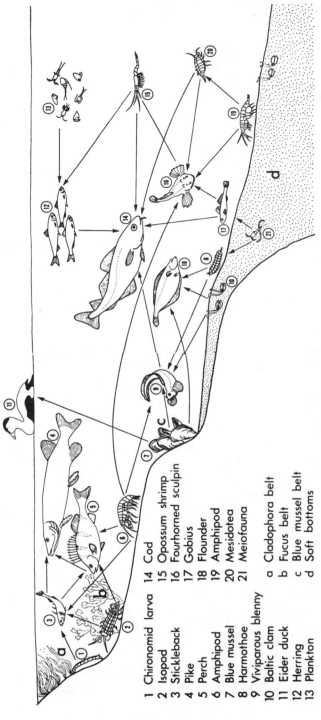

Figure 4 Some fundamental food chains in the Northern Baltic proper (Jansson, 1972).

1 Chironomid larva
2 Isopod
3 Stickleback
4 Pike
5 Perch
6 Amphipod
7 Blue mussel
8 Harmothoe
9 Viviparous blenny
10 Baltic clam
11 Eider duck
12 Herring
13 Plankton
14 Cod
15 Opossum shrimp
16 Fourhorned sculpin
17 Gobius
18 Flounder
19 Amphipod
20 Mesidotea
21 Meiofauna

a Cladophora belt
b Fucus belt
c Blue mussel belt
d Soft bottoms

328

The achipelago is the most complicated coastal system of the Baltic. The green (Cladophora), brown (Fucus), and red algal belts constitute important habitats for fish and fish food and contribute greatly to the pool of dissolved and particulate organic matter of the water and sediment bottoms. The sharp seasonal pulses of the Baltic is characteristic of many detritus-based systems. In Spring the shallow waters of the archipelagos are warmed rapidly, and a rich plankton soup is cooked with nutrients furnished from the winter decomposition of last year's biological communities and from land runoff. Fish from the open sea, such as herring and butt, now arrive for this "spring lunch," and their spawning activity is great in bays and straits. An example of the most common organisms of the archipelagos and their place in the food web of the coast is shown in Figure 4.

The algal belts of the coast, the primary and secondary producers of the open sea, the water, and the sediments all work together in a huge system. Solar radiation, sedimentation, upwelling, and vertical and horizontal migration of fish and plankton are the forcing functions that drive the flows and transports of energy and matter through the vast systems of plants, animals, and dissolved inorganic and organic pools (Figure 5).

In this complicated pattern the physical processes play a critical role. The main hydrographic feature of the system is the salinity stratification, with the stable primary halocline occurring at about 60 m depth. The salty (ca. 17‰ salinity) and dense North Sea water flows along the bottom of the deeper strait into the Southern Baltic, filling the depressions with oxygenated water, and replacing the old stagnant bottom water. The inflowing water from the North Sea follows the eastern coasts of the Baltic, flowing to the north while mixing slightly with the surface layer. This water turns counterclockwise around Gothland and returns to the south along the Swedish east coast together with the main flow of Baltic surface water from runoff. The boundary between the surface and bottom water is very sharp, and while the surface water is well aerated except for short periods of temperature stratification, the oxygen level of the bottom water is very low due to respiration by aerobic organisms, especially bacteria. The higher salinity and the lower oxygen concentration of the deeper waters, and the large amounts of phosphate released in the absence of oxygen and presence of hydrogen sulfide is shown in Figure 6. The North Sea water mixes with the Baltic more or less continuously in small quantities (a few km^3 per day). However, intermittent pulses of great masses of salty, oxygenated water forced in by changes in the macroclimate are much more important. For example, during a few weeks in 1951, 200 km^3 of salt water flowed into the Baltic. Owing to the different densities of the inflowing water, secondary and tertiary haloclines may be formed.

Many inflows of North Sea water have occurred during the last decades and have registered as pulses of oxygen concentrations in the bottom water.

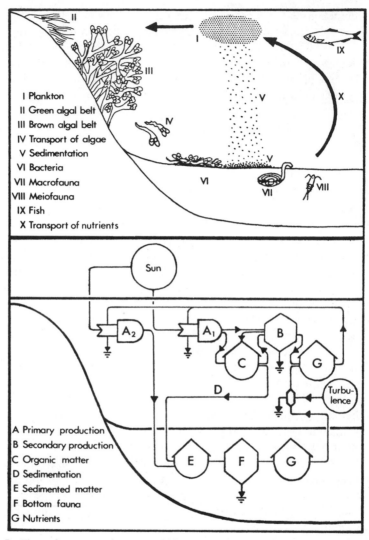

Figure 5 Flow of energy and matter within a coastal area. (Above) A naturalist's picture. (Below) The same system expressed in energy language.

This increased frequency has caused a vertical displacement of the primary halocline since the beginning of the century from 80 to the present 60 m, representing a 200-km^3 increase in volume of the bottom water.

The heavy stratification, the enclosed basin, and the absence of flushing tides in the Baltic Sea makes it an effective trap for organic matter and persistent substances. The mean residence time of Baltic water has been

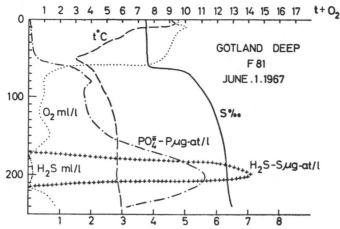

Figure 6 Hydrographical profile in the Gotland Deep (east of Gotland), showing the primary halocline at 60 m, the low oxygen concentration in the bottom water, and the coinciding maxima of PO_4–P and H_2S at 200 m depth (Fonselius, 1969).

calculated by various authors as 25 to 40 years. Though geologically a fairly young system, the Baltic Sea had attained a reasonable biotic stability and a fairly good ratio between production and decomposition, until man, with the power of fossil fuel, switched the Baltic to an anoxic and poisonous state. Recent estimations show that 17,500 metric tons of total phosphorus and 58,500 tons of total nitrogen are discharged into the Baltic basin each year, of which 80 and 50%, respectively, are caused by man's activities. The total input of organic matter per year has been estimated as 5.8×10^6 tons of carbon. High concentrations of DDT, PCB, and mercury accumulate in the foodchains and many organisms show values of these toxins ten times higher than corresponding populations along the Swedish west coast. How much can the Baltic ecosystem stand? How far can we go with organic loading? We need good models of the different parts of the Baltic Sea to optimize man's use of the Baltic system. A first attempt is presented here.

THE ASKO PROJECT

The annual budget offered to us by the Swedish Natural Science Research Council amounts to about one and a half million Swedish crowns ($400,000). This pays for nine full-time scientists and our field station (the Asko Laboratory) with personnel, vessels, and equipment. Intimately incorporated with this are the monitoring and modeling sections, comprised of five scientists and sponsored by the Swedish Environment Protection Board.

Though a big team by Swedish standards, it is certainly too small for the size of the research task. The purpose of our project is to define the main subsystems of the Northern Baltic, show their degree of interdependence and role in the total system, and determine how this total system behaves under the influence of wastes from land and nutrients from upwelling. This calls for an effective organization of the different basic research tasks and the monitoring studies. After the first 2 years we are still fumbling around after the perfect cast, and important areas are still uncovered. Our work proceeds along three parallel lines:

1 Rough modeling of the important total system to define the proper connections between the different subsystems.
2 Detailed modeling of subsystems down to at least functional groups of organisms.
3 Intensive and extensive field work to fill in the most important information needed for the models.

After 10 years of field and laboratory studies in the area prior to the start of this project we have a fairly good qualitative knowledge of the main components of the system, the annual biomass variations of different trophic levels, and the effects of "microcatastrophies" by wind and weather upon the various subsystems. When trying to make a model, however, we find a great lack of quantitative data and a poor knowledge of the rates of important flows. We have made several preliminary simulations: a total Baltic model (Sjöberg et al., 1972) and the Baltic hypolimnion (B.-O. Jansson, 1976), both representing a first level of rough modeling, and the Cladophora belt (A.-M. Jansson, 1974, 1975), constituting a more detailed modeling of a subsystem. The first one is presented here in more detail.

Our model building starts with the translation of the appropriate set of connected physical, chemical, and biological processes to H. T. Odum's energy-circuit language (Chapter 2). This is a symbolic representation of differential equations where forcing functions, such as solar radiation and turbulence, are connected to state variables, such as producers, consumers, or dead organic matter, by flows of energy and matter. In this way we have constructed diagrammatic models for the different subsystems that we now consider important for the behavior of the total system. At the Askö Laboratory, research groups are actively accumulating additional data on the algal belts, the pelagic zone, and the soft bottoms. A more detailed description of both the Baltic ecosystem and the Askö project is found in Jansson (1972). Recently a physical oceanographer has joined the project, exploring the main hydrodynamical features of our investigation area, such as upwelling and downwelling processes.

The initial versions of these submodels were very complex and "impressive," and reflected the total information collected by the individual scientists. It soon became apparent that these complete submodels were too complex to handle when hooked together to form the complete model. In the following pages, the model of the pelagic system (Figure 7) will exemplify the more detailed submodels, whereas in the model of the total Baltic (Figure 8) the details of the submodels are sacrificed for the benefit of totality. Here the dependent variables of the pelagic models, for example, are reduced to three storages only. Later models of the complete system will incorporate more of the details of the submodels, but only if they are of importance to the behavior of the total system.

For the pelagic systems of the Baltic our main questions are:

1 How much of the organic matter produced by plants is channeled into higher trophic levels, thus serving as a potential food source for fish?
2 How much is leaving the system as sedimenting organic matter, affecting heterotrophic activity and oxygen consumption of the sediments?
3 How does this system respond to the annual variation of insolation and temperature normally found from year to year and to increased inputs of nutrients in the forms of nitrogen and phosphorus?

For predictions of primary production alone, simplified empirical models such as those recently developed by Fee (1973) may be sufficient. However, we are not interested only in predicting "primary production" (as measured by ^{14}C uptake), but also in how the organic matter is utilized during different environmental conditions. The objectives of our pelagic study have to a large extent motivated the compartmentalization used in the model shown in Figure 7. But as it is not always possible to measure all the information needed, the model parameters are also chosen according to the available research methods. For example, the phytoplankton is subdivided into three functional units: diatoms ($Q1$), bluegreens ($Q2$), and nannoplankton ($Q3$), a compartmentalization motivated by their differences in sinking rates, turnover rates, and nutritional value for the zooplankton. The different phytoplankton groups also have different responses to water temperature, insolation, and to phosphate ($Q7$) and inorganic nitrogen ($Q8$), as well as ability to use atmospheric N_2 as a nitrogen source (N). The zooplankton ($Q4$) are treated as one compartment, because as yet we have no indication that the different species have different modifying effects on the system with respect to the questions we are asking. For the same reason, the decomposers ($Q5$) are treated as one unit. Annual variation in insolation (I), water temperature (T), and inputs of nutrients from land or other subsystems are regarded as

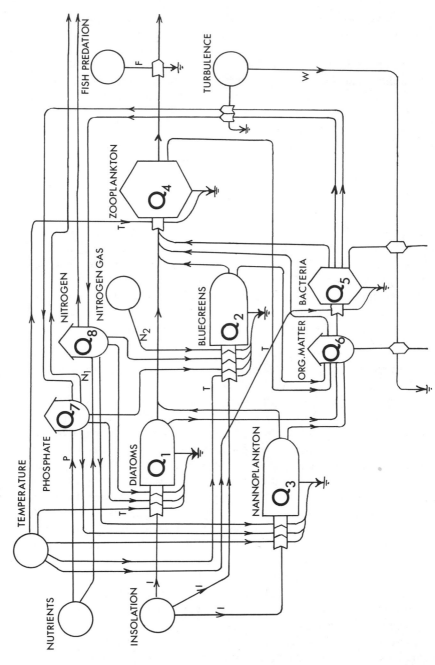

Figure 7 A model of the pelagic subsystem for the simulation of the phytoplankton succession. The primary halocline is represented by the two-way gates at the bottom of the picture and oxygen is omitted as it is regarded as nonlimiting in the epilimnion.

334

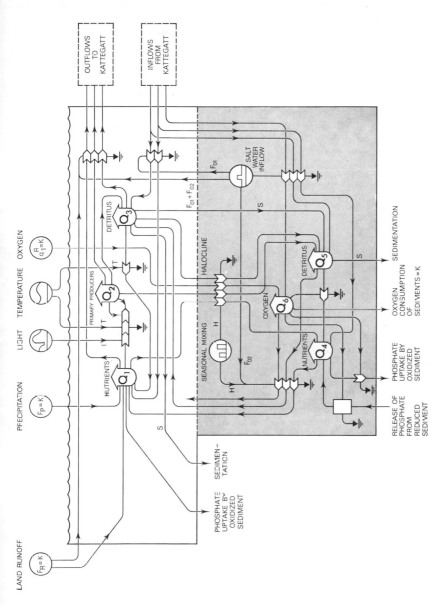

Figure 8 Model of the major processes of the whole Baltic, expressed in energy-circuit language (Sjöberg et al., 1972).

335

the main forcing functions on the system. The model has very limited hydro-dynamical information, reflecting both our poor knowledge of these processes and our limited capacity to measure them. However, "water turbulence" (W) is included in the model, and it affects nutrient recycling, sedimentation, and exchanges with other subsystems of the materials considered in the model. In the field it is measured from temperature stratification patterns of the water column as a semiquantitative stability term and is used in the mathematical formulations of the model as a forcing function varying between zero and one. The spatial limits in the model are determined vertically by the primary halocline and horizontally by distances from the shore and the geography of the Baltic. Definition of these horizontal boundaries is made from plankton composition and biomass found from surveys.

The pelagic fishes initially are included in the model as a forcing function (F) rather than a state variable due to their seasonal occurrence. The fish populations are studied separately as they migrate between different subsystems taking advantage of the seasonal peaks of available food. Therefore, they are not entirely dependent on the annual production in any one separate system.

We measure dead organic matter ($Q6$) as suspended and dissolved organic nitrogen and phosphorus, and we have to disregard the role of other organic fractions due to our lack of the sophisticated equipment needed to measure them.

MODIFICATIONS IN SAMPLING PROCEDURES RESULTING FROM OUR MODELING

We have found that many of the standard techniques used in marine ecological research produced inadequate and scanty data that was insufficient for our system approach. Therefore, we continuously are modifying our investigations, in large part because of "holes" in our data that became apparent only after we had developed our preliminary models. Some methods have been abandoned or greatly modified already, and new techniques are tried constantly in order to collect as much relevant data as possible in relation to the resources available to us.

It is essential that field measurements be made on as many parameters as possible during the same state of the natural dynamic system. Therefore, the simultaneous use of a variety of techniques is needed. A quantitative estimate of the whole plankton community is difficult to obtain in one sample due to a difference in abundance and size of the organisms. We now filter large quantities of water by shipboard pumping and simultaneously analyze the chemical content of the different size fractions, both as a complement to the traditional direct counts and classification of the organisms in subsamples. The different plankton groups can be sampled and analyzed separately relatively easily by this technique because of the distinct blooms, with one species dominating, that normally occur.

As the models can be regarded as budgets of matter and/or energy it is essential to measure the total content of each parameter in the ecosystem studied. Therefore, it is important that all variables can be expressed in the same unit, either directly or by using conversion factors. Due to the different quantitative techniques used, biomass estimates are initially given as specimen counts and chlorophyll, nitrogen, and phosphorus as concentrations. Values are recalculated as matter and energy equivalents when used in the model. Estimates of total quantities of matter made by chemical analysis of mixed samples are a check against the summation of the different components. We have found that it is necessary to make frequent independent checks of many of the conversion factors used in order not to introduce large errors in the model parameters.

By using a coarse division of space into a few boxes during modeling, it is sometimes possible to use integrating sampling techniques such as vertical hauls and continuous-flow analysis that greatly simplify the field work. For example, we take samples for phytoplankton, chlorophyll, and nutrients for the whole water column by using a plastic tube that is continuously filled with water as it descends. Similarly, integrated samples are obtained for an entire water column by using vertical net hauls for zooplankton, since they are considered as only one state variable in the model regardless of their depth, at least down to the halocline.

Even when using a coarse spatial separation in our pelagic study, a large number of vertically integrated samples still must be taken in order to account for horizontal and vertical distribution patterns. Therefore, distribution of chlorophyll a determined by in vivo fluorescence in a flow-through cuvette (Lorenzen, 1966) is now used as a complement to our discrete sampling program.

Patchiness, however, is not only a sampling problem. The box model assumption that the variables are evenly distributed in the system disregards important structural properties of ecological systems. The inhomogeneities found in the distribution of different species are caused not only by physical and chemical gradients but also by interactions between the species. Even if it is sometimes possible to measure and in multidimensional models describe these processes, the complexity of the model is increased enormously and the overall view easily lost. We are, therefore, trying to avoid such models in our initial studies of the subsystems.

Estimates on concentrations need to be supplemented by rate measurements relevant to the field conditions. Feeding rates and metabolic activity of specimens held in the laboratory are different from those found in a natural, mixed-species community.

In a dynamic natural system the same ecological habitat may be occupied by a succession of different species as response and adaptation to environmental changes occurs. As a result, single species' observations concerning, for example, optimal environmental conditions and adaptational capacities may not be applicable for a compartment in the model, since this was designed with "built-in" succession of species. Therefore, we use short-term enclosure experiments using plastic bags of entire pelagic communities in order to determine rate functions applicable for whole systems as well as for larger compartments.

A MODEL OF THE TOTAL BALTIC

In the first stage of the research program it was of prime interest to gain a better understanding of those processes and subsystems that are most important in the behavior of the whole Baltic system. Therefore, we first concentrated our simulation efforts on a model that aimed to describe basic and characteristic features of the Baltic as a whole, although superior data were available for some of the subsystems.

A model capable of describing primarily the oxygen and phosphorus balance in the Baltic may give boundary conditions for the development of the different subsystems during various environmental conditions. The specific hydrochemical conditions that characterize the Baltic govern the overall material balances and determine the possible energy pathways within the ecosystem. Therefore, the pronounced gradients and discontinuities, especially the vertical separation of water masses of different salinities, are conditions of prime interest. By using the same spatial separations and physical limits in our model as in the static box models developed earlier by Fonselius (1969, 1972) and Bolin (1972), their same field data could be used, and direct comparisons of the results were possible. They both divided the Baltic into "boxes" or compartments representing the water volumes above and below the primary halocline, including both the Baltic proper and the Gulf of Finland. Thus the flows from the coastal zones, the northern area between Sweden and Finland (Gulf of Bothnia) and the waters outside the sounds between Sweden and Denmark (Kattegatt), were considered only as "forcing functions" or inputs to the system. The influences of flows and processes upon the variables considered in the model were expressed as a set of ordinary, time-dependent differential equations derived from a continuity equation. Contrary to the approach taken by Fonselius and Bolin, we created a time-dependent model capable of describing the system in a nonsteady state, making it possible to study the effects of the different perturbations to which the Baltic is exposed, both naturally and by man. The final version of the model is shown in Figure 8, expressed in the energy-circuit language. Six variables are considered in the model: nutrients ($Q1$), primary producers ($Q2$), and detritus ($Q3$) in the water above the halocline and nutrients ($Q4$), detritus ($Q5$), and oxygen ($Q6$) below the halocline. The oxygen variable, expressed in ml liter^{-1}, represents hydrogen sulfide when it has a negative sign. The annual variations of the main forcing functions described earlier are shown in Figure 8 with a graphic representation of their seasonal nature. A two-step function generates the seasonal mixing of water and water-transported substances through the halocline that occurs mainly in the shallow coastal zones during spring and autumn. The salt-water inflow above and below the halocline is considered as a constant force, compensating the outflow of less saline surface water. A step function is indicated in the symbol and represents the occasional, sudden inflow of high-saline water that penetrates into the deep basins of the Baltic, causing an upwelling of nutrient-rich bottom water.

In the model nutrients are added to the surface waters by inflowing water from the Kattegat, by land runoff and by precipitation. Nutrients leave the system with the outflow of water to the Kattegatt, by sedimentation (S) of suspended organic particles and by uptake of phosphate by oxidized sediments. The annual variation of light (I) and temperature (T) are the somewhat modified sine-wave forcing functions that regulate biotic activity and hence determine the build-up of organic matter and the recycling of nutrients within the surface water layer. The suspended organic matter that sinks down through the halocline is decomposed primarily by aerobes (i.e., bacteria) that in the deep basins have to depend on the limited oxygen supply brought in by the intermittent salt-water inflows. The box-shaped symbol indicates a function describing various bottom processes including the anaerobic decomposition of organic matter, the formation of hydrogen sulfide, and the release of phosphate from reduced sediments in relation to mean oxygen concentration and bottom-volume relation of the deep basins. A detailed description of the model and the simulations carried out is given by Sjöberg et al. (1972).

Initial values for the variables and constants were calculated by assuming a steady state of the system using mean annual values, and the simulations were made on a PDP-7 digital computer at the Division of Automatic Control, Royal Institute of Technology, Stockholm. These initial values were based on the data compiled by Fonselius (1969), representing conditions during 1968. However, the simulations of the model did not give a realistic state of equilibrium when these values were assumed to represent a steady state. As the Baltic has not even been near a steady state during the last decades, we felt obliged to set up the model representing the Baltic 30 to 40 years ago (i.e., in an aerobic steady state). The long-term recording of measurements carried out in the Baltic made it possible for us to extrapolate today's values backwards in time. From this hypothetical steady state, realistic perturbations should give a model that represents the Baltic of today and give indications of the future development of the system. A comparison between the amounts of phosphorus and oxygen of the Baltic today given by Bolin (1971) and Fonselius (1969) show a good agreement with the model as indicated in Figure 9. However, it is essential to remember that a model always represents an abstraction of the true ecosystem. The results of the simulations will explain the properties of the model and may not be directly applicable to the real system. The features of the model must be understood and then compared with the true system for a verification of the results.

A feature observed in all the different kinds of perturbations was that new steady state conditions were reached only after a considerable time period. This is due to relatively small inflows and outflows of phosphorus compared with the amounts within the system. An increase in the annual discharge of phosphate from 7100 to 20,000 tons gave a slow but dramatic change in the system, and after about 50 years the system was close to a new steady state. The total amount of phosphorus in the system had increased by about 50%, compared with the initial steady state.

When the phosphorus was increased the annual downfall of detritus through the halocline increased by more than 60%, and as a result the oxygen consumption below the halocline increased by more than 50%. This resulted in a lowering of the mean oxygen concentration below the halocline from 2.6 to 0.76 ml per liter, which means a complete oxygen deficiency in most of the deeper basins of the Baltic. This in turn increased the release of phosphorus from the sediments because phosphorus is much more soluble under anaerobic conditions. Thus the increase of phosphorus outputs to values that are representative of the real situation of today caused anoxic conditions in the model, just as in the Baltic, without any alternations of the hydrodynamic situation.

When all the forcing functions on the system were changed simultaneously to represent today's values, an anaerobic, steady-state situation was obtained after 50 years. This system was perturbed in several ways comparable with the situation of the Baltic today in order to study the relative importance of various processes for the whole system. We found that the water-exchange processes between the Baltic and the Kattegatt are critical to the behavior of the system. Changes in the stability of the halocline as well as the fall of detritus through the halocline and the sedimentation rates were also important. The effects of the increase in the release of phosphate from the sediments

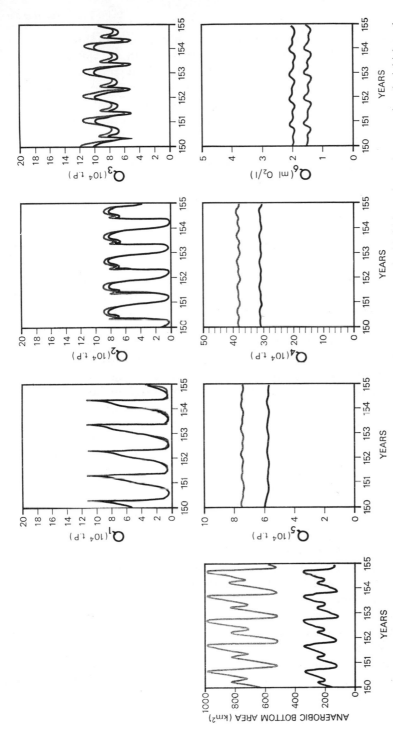

Figure 9 The new steady state obtained when the release of phosphate per unit bottom area was doubled (shown in grey) compared to the initial steady state. The diagrams show the years, 150 to 155, of the simulations (Sjöberg et al., 1972).

341

because of anaerobic conditions are shown in Figure 9. In this case, the release per bottom area was doubled compared with the initial anaerobic steady state which is shown in the same figure. At the new steady state the anaerobic bottom area had increased more than three times, and the annual release of phosphate from anaerobic sediments had increased from 1000 to 6600 tons. When the deep system was perturbed by a large sudden salt-water inflow, this feedback effect was further accentuated, as the resulting upwelling of nutrients caused by the salt-water inflows increased primary production and also the fall of detritus through the halocline. The amount of oxygen brought in by the new salt water was small compared with the increased oxygen demand, and, therefore, the anoxic bottom areas were further expanded. These features of the model emphasize the nature of the positive feedbacks resulting from nutrient additions.

The simulations of this model demonstrated the importance of gaining a better knowledge of the Baltic as a hydrodynamic system. The water exchange through the Danish sounds and the water exchange processes within the Baltic to a large extent determine the time constants and the material balances of the system. Direct measurements of these processes are needed in order to understand the extent and rate of alterations in the biological and chemical processes in the different subsystems. Therefore, we think the model is justified if only for determining the direction of future research.

Experience from other programs aimed at studying whole ecosystems indicates that a massive data collection not coupled from the beginning with simulations is unlikely to produce a final model worth the efforts in the field and laboratory. Therefore, we have made actual simulations of a simplified version of the pelagic model described above even before we had an adequate data base. At present the results are of limited value since we still have to rely heavily on literature data from many divergent sources. Frankly, we have had the same somewhat depressing experience that most ecologists have had trying to find data for their particular model: for most of the variables and rates there exist a large range of possible values. When the model did not behave reasonably with the initial parameter set, it was possible to find another set within the possible ranges so that the model fitted the observed variables! However, our initial simulation of the system has had many advantages. Engineers and biologists had to come together and have intimate discussions about how to transform the real world into mathematical formulations; thus many of the difficulties in future simulations of the model using our own data have been avoided. It also has been possible to make some initial tests of the model system's behavior in order to test the validity of alternative mathematical formulations of ecological processes, the effects of feedback loops, and so forth, by means of the techniques developed in the engineering sciences.

A LOOK INTO THE FUTURE

Parallel to the modeling of the subsystems of our investigation area, we are developing close collaboration with scientists in the other Baltic countries. A similar project is carried out in Kiel, West Germany at the Institute for Meereskunde, using the Odum language and the same simulation techniques. We also exchange scientists with the U.S.S.R., which has an outstanding team of Baltic hydrodynamicists in Tallinn. The informal organizations, *the Baltic Oceanographers* and *the Baltic Marine Biologists*, carry out joint measurements and intercalibrations. As members of a working group within ICES/SCOR we have contributed to a general plan of modeling the Baltic. In all these efforts the ecosystem approach and the energy circuit language have proved to be invaluable tools for understanding and communication between the disciplines, for the understanding and assessment of the various mechanisms, and as working schemes for both scientifically and geographically heterogeneous teams.

As a study area the Baltic Sea offers all types of environmental problems. We have reported on our beginning ecological models. Future environmental models must be merged with information from sociology, economics, and politics for a proper management of the Baltic.

REFERENCES

Bolin, B. 1972. Model studies of the Baltic Sea. *Ambio Spec. Rep.* **1**: 115–119.

Fee. E. J. 1973. Modelling primary production in water bodies: A numerical approach that allows vertical inhomogeneities. *J. Fish Res. Board Can.* **30**: 1469–1473.

Fonselius, S. H. 1969. Hydrography of the Baltic deep basins III. Fishery Board of Sweden, *Series Hydrography*, 23. 97 pp.

Fonselius, S. H. 1972. On biogenic elements and organic matter in the Baltic Ambio Spec. Rep. **1**: 29–36.

Jansson, A. M. 1974. Community structure, modelling and simulation of the Cladophora ecosystem in the Baltic Sea. *Contr. Askö Lab.* Univ. Stockholm, Sweden, No. 5, 130 pp.

Jansson, A. M. 1975. System analysis and simulation of the green algal belt (Cladophora) in the Baltic. *Havsforskningsinst. Skr.* No. 239: 240 247.

Jansson, B.-O. 1972. Ecosystems approach to the Baltic problem *NFR Bull.* No 16. 82 pp.

Jansson, B.-O. 1976. Modeling of Baltic ecosystems. *Ambio Spec. Rep.* No. 4. (in press.)

Lorenzen, G. J. 1966. A method for the continuous measurement of in vivo chlorophyll concentration. *Deep-Sea Res.* **13**: 223–227.

Sjöberg, S., F. Wulff, and P. Wåhlström. 1972. Computer simulations of hydrochemical and biological processes in the Baltic. *Contr. Askö Lab.* **1**, 90 pp.

Lindqvist, A. 1973. Arsberättelse *Medd. Havsfiskelab.* No. 145 (in Swedish). Lysekil, Sweden.

14
Models and the Decision Making Process: The Hudson River Power Plant Case

CHARLES A. S. HALL

E stuarine environments are used extensively by both fishes and man. The large volume of flowing water, biotic richness, and ready access to both sea and rivers have made estuaries attractive to fishes for feeding and spawning and attractive to man for food, shipping, and industry. In the recent past this nation's electric power generating capacity has increased by about 10% per year. Estuaries have been a convenient and economical location for siting new power plants, since about half of our nation's largest cities are located on or near estuaries. To what extent is this use of an estuary compatible with estuarine fisheries?

I address this question for the Hudson River, an important spawning grounds for the striped bass, an important commercial and sport fish. Computer models were an important part of the analysis of the potential effects of the Hudson River power plants on the striped bass and in the many complicated interactions that took place as model builders of different persuasions interacted within the decision-making process.

To do this I first introduce the biotic significance of estuaries, then develop the life history of the Hudson River striped bass, tell briefly how our knowledge of the striped bass, the Hudson River, and its power plants were translated into a computer model, and finally and most importantly discuss how the computer results influenced the industrial and political decisions that were made in the Hudson region.

LIFE-HISTORY PATTERNS OF ESTUARINE-DEPENDENT ORGANISMS

The chief biotic characteristic of estuaries is their very high productivity. Measurements of the natural biotic energy fixed and used in these places indicate that, in general, estuaries are among the richest natural type of ecosystem found anywhere in the world. These conditions have made estuaries

An earlier version of this paper was published in Ecosystem Analysis and Prediction, 1974. S. A. Levin, (Ed.) Soc. Industr. Appl. Math., Philadelphia.

I was involved in the construction of the "first generation" of the Hudson River entrainment models with other scientists, especially Phil Goodyear, while at Oak Ridge National Laboratory. I have undertaken additional modeling studies of the Hudson problem while at Brookhaven National Laboratory, Cornell, and MBL. The conclusions and opinions represented in this paper are solely my own and are in no way an official view of either National Laboratory or any other institution. I thank Leslie Lasker for programming assistance.

As reported in the text the different National Laboratory models and several others tend to give results more or less similar to the results of our earliest model. There are other models prepared by consultants to the principal utility involved that give results different in some important respects from those reported here. Those who may wish a different view on this controversial topic might consult Wallace (1975). I do not agree at all with the conclusions of that paper.

rich feeding grounds for fishes, and it is also here that man takes his greatest harvest of sea foods (Smith, 1966; Teal and Teal, 1969; Douglass and Stroud, 1971; Woodwell et al., 1973; Day et al., 1973). Five of the six most important commercial fish species in the United States are considered "estuarine dependent," as are most of our salt-water sport fishes.

The utilization of estuaries by marine organisms is not random. Many species have selected through evolution behavioral, morphological, and physiological adaptations that optimize the use of estuarine richness during the juvenile stages of organisms by the timing of reproduction and migration patterns. What are these patterns? In general, estuarine-dependent organisms utilize estuaries as follows: Adults migrate from their normal feeding regions, which often are far from estuaries, and spawn so that eggs or larvae will drift into estuaries. The young hatch in or on their way to estuaries and upon arrival exhibit various behavioral and morphological characteristics that tend to keep them there. For example, young finfish maintain their position

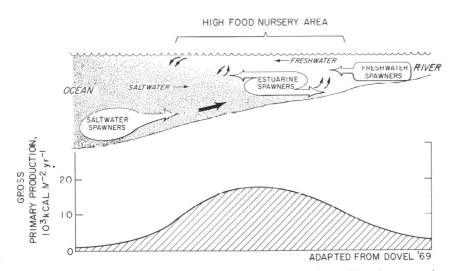

Figure 1 A representation of several important physical and biotic conditions in an estuarine environment. In an estuary with large fresh-water inflows, such as the Hudson River, the less-dense fresh water flows seaward above the salt water. The salt water, in turn, flows *landward* below the fresh water to replace other salt water that is entrained (drawn into) the surface fresh water. In the Hudson River the salt wedge interface (which is gradual, not sharp, due to mixing) can be found from Manhatten to Poughkeepsie, depending on the amount of fresh-water discharge, and it moves landward and seaward with the tides. The shoreward flowing salt water can be detected many tens of kilometers to sea. The intensity of photosynthesis along the river-sea gradient is represented as the top graph on the figure and is greatest in the vicinity of the salt wedge. Different strategies that fish have evolved to exploit the riches of estuaries for juvenile stages are represented by the three types of "spawners."

in estuaries by swimming, by day-night vertical movements, or by settling on the bottom. Shellfish often attach themselves to substrates by various claws and shrouds. The young spend from a few weeks to 1 or 2 years in estuaries, and may later drift with currents or swim to adult feeding regions.

The salt-wedge circulation pattern (Figure 1), characteristic of estuaries with large freshwater inflows, consists of fresh water moving seaward near the surface and salt water moving into the estuary on the bottom. The deep salt water moves landward to replace other salt water that mixes with surface water and is carried to sea. The inward-flowing salt water that supplies the salt wedge may be found from many tens to hundreds of kilometers offshore, and the boundaries of the salt wedge shift landward and seaward in response to tides and seasonal variations in freshwater input.

The bidirectional flow of water in estuaries has allowed the development of three major patterns by which organisms use estuaries for reproduction and juvenile feeding. These three patterns are (1) ocean spawning, followed by inmigration of the larvae in the landward-moving, deeper salt water, (2) estuarine spawning in which the larvae do not move appreciably, and (3) river spawning, followed by downstream drift or swimming of larvae or juveniles. The commercial shrimps of the Gulf of Mexico (genus *Penaeus*) are examples of a group of organisms that use the first pattern, various *Fundulus* or killifishes are examples of the second, and salmon and the striped bass (*Morone saxatilis*) are examples of the third. These general relations are presented diagrammatically in Figure 1.

THE STRIPED BASS

The striped bass, an example of a fish that spawns above estuaries, is a large, migratory, anadromous* fish that spends much of its life cycle directly in or near estuarine water. They rarely are found at distances of greater than 10 km from the shore and normally are found much closer. The original range was principally from North Carolina to Massachusetts; however, the species has been successfully introduced to the Pacific Coast of the United States and even several freshwater lakes. The only important spawning grounds for migratory stripers on the East Coast are the Hudson River and the rivers flowing into Chesapeake Bay. The fish is an important commercial species with landings of about 1.6 million pounds from New York State waters alone. Sports fishermen catch an additional 25 million or so pounds from New York waters, and stripers, or rock as they often are called, are an important part of the tourist economies of such regions as eastern Long Island and Cape

* Spawns in freshwater and lives in salt.

Cod. Many sports fishermen are extremely enthusiastic and vocal about stripers.

Talbot (1956) has reviewed the life history of the striped bass and documented the crucial role estuarine ecosystems play in their life history. The following account of the life history pattern of the striped bass in the Hudson River (New York) is based on the work of Talbot and others, as well as analyses that we developed at Oak Ridge National Laboratory as part of environmental-impact studies for Hudson River nuclear plants. The essential features of the life history of the Hudson River Striped Bass are the following: the fish are spawned in May and June above the salt water principally in the Saugerties to Cornwall sections (Figure 2). The eggs and larvae appear to spend some days or weeks very near the bottom of the river, and during this time they do not drift very much as the bottom currents are not strong. However, by the age of 2 weeks or so the young bass are found throughout the water column, and they drift downstream with the water currents as they develop. As the fish become strong enough to swim feebly they begin to exhibit a diurnal pattern of migration, that is, they tend to swim to the bottom during the daytime and to the surface at night. When they drift downstream into the influence of the salt wedge the young fish are swept seaward during the night when they are in the relatively fresh surface water and upstream during the day when they are found nearer the bottom. Once the fish are about 6 weeks old they are able to maintain their position by swimming against the current, and at this time the majority of the fish are found in shallow regions in Haverstraw Bay and the Tappen Zee (these are the broad bays found from just below Peekskill downstream to the New Jersey state line) or along the shoreline.

The diurnal migration pattern of the 2- to 6-week-old fish appears to be a critical factor in the life history of the striped bass. Without this pattern the young fish would be too weak to swim against the seaward flowing surface waters, and they could be swept to sea if they remained in the main currents. If they remained entirely on the bottom they would be moved upstream above their "nursery grounds." Since the salt wedge is normally associated with the biotically richest regions of estuaries (Figure 1) any fish that exhibited a pattern of diurnal vertical migration would be concentrated in the region of highest available food supplies. Many other anadromous fish also show this pattern, and it seems clear that there has been natural selection to develop this behavioral pattern in the young of many estuarine fish species.

However, there is another implication of the diurnal migration pattern that is important for any attempts to understand the effects of the power plants on the fish. Most of the striped bass are spawned within or above the Peekskill section; yet most of the juveniles are found below the Peekskill section by mid to late summer. This means that the majority of the striped

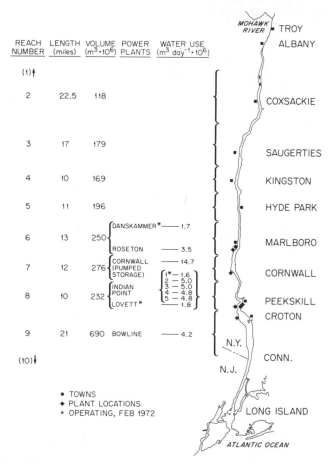

Figure 2   The Hudson River, the divisions used for analysis (named for towns), the approximate volume of each section, the power plants projected for each section as per 1972, and the quantity of water used by each power plant. As a general rule Striped Bass are spawned in or above the Cornwall section and grow up in the Cornwall section or below, particularly in the vicinity of Haverstraw Bay (the wide bay in the northern part of the Croton section).

bass that survive until the summer must pass by the series of power plants that have been built or that may be built in or between the Cornwall and Croton sections (Figure 2). In addition, the salt wedge is normally found in the vicinity of the Indian Point power plants during the month of June, when the young fish are drifting downstream. Any fish that exhibited a diurnal pattern of migration would be swept back and forth on successive days and nights as it was drifting alternately with the surface and bottom currents. This effect could increase the potential for the power plants to interact with the young stripers.

A POPULATION MODEL OF YOUNG HUDSON RIVER STRIPED BASS

Some early probability studies done by staff members at Oak Ridge National Laboratory indicated that the effects of the Indian Point power plants on the young striped bass might be considerably greater than had previously been thought likely. However, it soon became apparent that probability equations were insufficient to explore the complicated potential impact of a series of power plants built at different locations on the river and operating according to different series of assumptions. It had become obvious that some sort of computer model was necessary to handle this complexity. I was asked to work with the Oak Ridge group for about 4 months since I had modeling experience (rarer amongst ecologists then compared to now!) and because I had experience with estuarine fish. The model that was developed by Phillip Goodyear and myself during the winter of 1971–1972 was the first of many subsequent, and often more sophisticated, models to come, each of which attempted to investigate what the operation of a series of power plants on the coastal river would do to the striped bass population.

The model that we developed and most of the subsequent ones is based on river flow, temperature, and quantitative spawning patterns as the most important forcing functions (Table 1; Figure 3). The model used a series of state variables to represent the population of striped bass subdivided by location and age. In practice, FSC(K,L) represents the total population (standing crop) of stripers in the Kth section of the river (e.g., $K = 2$ for the Coxsackie section, $K = 3$ for the Saugerties section, etc.). Subscript L represents the age class of the fish ($L = 1$ for fish born before May 21st, $L = 2$ represents fish born between May 21 and 28, etc.). Fish were added to each section by spawning (according to the number of eggs that had been measured in the Hudson) and by migration in with the water currents from upstream sections. The latter was derived from data on the river morphology, hydrology, and records of water flow. Fish were removed from each section in the model by migration out, (determined in the same fashion as for migration in) by natural mortality (estimated from field samples), and where appropriate by the action of power plants.

Since the effects of tide were very important the state variables were solved four times per day, during which adjustments in state variables were made for tidal effects and for the proportion of the population that was estimated to be in the upper and the lower portion of each section. In all cases, field data was used to try to derive estimates of the important parameters that were as

Table 1   The Basic Mathematical Structure of the Computer Model Developed by Hall and Others for the Estimation of Entrainment-Induced Mortality on Hudson River Striped Bass.

I. These equations are used to simulate the striped bass population in the Hudson River as a function of known and implied attributes of their life history and to compare such simulation results with
(1) field observations of the distribution of young fish and
(2) simulations of the numbers and distribution of fish with the operation of various possible configurations of electric power plants in operation along the river.
The principal state variable, representing the standing crop of the striped bass population by river section and time, is solved with a time step of four per day representing two ebb and two flow tides. Different assumptions used during conceptual sensitivity analyses are included as options below.

II. Fish standing Crop, FSC, was subdivided according to river segment (K) and age of fish at weekly intervals (L). PFSC(K,L) represents the same state variable at the end of the previous time step. (FORTRAN statements are in capitals). Hence:

FSC(K,L) = PFSC(K,L) ± Births, natural mortality, movements, and,
where appropriate, entrainment-induced mortality

Subscript K is used in the most aggregated runs to represent nine river segments according to Figures 2 and 3, for example, K = 2 for Cocksackie, ... K = 9 for Croton. In less aggregated runs: K = 1, 10 for the lower half of each of these sections, and K = 11 to 20 for the upper half of each segment. In still less aggregated runs K was used to represent across-river subdivisions as well as vertical and longitudinal subdivisions.
The subscript L is used to represent the age of fish by week. Thus FSC(8,3) is a fish presently located in the Peekskill section that was born (possibly in another section) during the first week of June.

III. Births: Newly hatched fish were added to fish standing crops according to the distribution of egg concentrations observed for various years during field studies, corrected for expected hatching success and egg development times.

BIRTHS = [ECONC(K,L)*EGGHCH]/DEVT

IV. Natural mortality per time: Assumed in different runs to be either a constant proportion of the population (D1) or an exponentially decreasing rate representing less mortality with older fish. D1 and D2 were derived from literature rates, or from field data for the Hudson River. M is the week number of the model run, and L is as before. T is time in quarter days, and E is the natural log.

NMRATE(M,L) = D1*FSC(K,L)
or NMRATE(M,L) = D1*(E**D2*T)*FSC(K,L)

Table 1 (*Continued*)

V. Migration. This represents the movement of fish out (downstream) of each segment per tide. In the earliest runs this was made a function of the net downstream flow of water (QDDIS), the river volume of each segment, and the number of fish in that segment:

$$OUTFMG(K,L) = (QDDIS/RVSGVL(K))*PFSC(K,L)$$

Movement into the next section downstream (K + 1) was made equal to the downstream movement from the next upstream segment:

$$INFMIG(K + 1) = OUTFMG(K)$$

Later models solved in and out migration for top and bottom segments, included bidirectional tidal effects for both vertical and temporal patterns, included latitudinal distribution-of-fish patterns and/or delayed movements (and the initiation of entrainment) for 2 weeks to represent possible behavioral characteristics of young fish. As a general rule passive drift (and entrainment) of young bass was stopped in the model when the fish were 5 or 6 weeks old.

VI. Entrainment. Entrainment mortality was made a function of the volume of water withdrawn per tide by the power plant(s) "running" during a given computer run in each river segment, the expected proportion of fish entrained that were assumed to die during passage through the power plant intake and condensor coils, the total volume of water in that river section (for various configurations of K as in section II), and the number of fish in that section. In other words, the power plants "eat" a fraction of all fish in a given section according to the ratio of water withdrawn to all water in that section:

$$ENTR(K,L) = (VOLPP(K)/RVSGVL(K))*PMORT*PFSC(K,L)$$

For example, in mid-June, 1973, there were about 92 million young striped Bass in the Cornwall section of the Hudson River (which contains about 347 million cubic meters of water). If the Storm King pumped storage power plant had been fully operational at that time it would have been extracting some 13 million cubic meters of water per day, pumping it some 3 km to the top of Storm King Mountain, and letting it run back down again later that day. Testimony of Morgan (1974) has suggested that the mortality factor for young striped bass subject to the 6-km trip, and two passages through the turbine blades would be essentially 100% (or 1.0). Thus according to our entrainment formula the number of young striped bass killed per tide (i.e., 0.25 day) in this section would be:

$$ENTR(K,L) = ((0.25*13E6)/347.OE6)*1.0*92E6$$
$$= 861,670 \text{ fish killed per 6-hour tide (on average)}$$

(The E6 is FORTRAN shorthand for one million). In the computer this equation is solved four times per day (modified for pumping and nonpumping times) for

353

Table 1 (*Continued*)

the 4 to 6 weeks that the young fish are susceptible to entrainment. At the same time, of course, population estimates are readjusted for births, natural mortality, and migration. This analysis could be modified to take account of, for example, different concentrations of fish in different parts of the river (e.g., in some years the fish tend to be concentrated on the side of the river where the most water will be withdrawn, and on other years they are concentrated in other parts of the river. However, until we really know the shape of the "intake plume" further refinement may be spurious.)

VII. Overall basic structure of model
   A. Set initial conditions, read in water volumes of each section, fresh water discharge (at monthly or weekly intervals), water used by power plants
   B. Solve population equations over time:

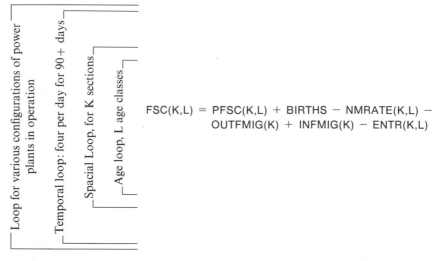

$$FSC(K,L) = PFSC(K,L) + BIRTHS - NMRATE(K,L) - OUTFMIG(K) + INFMIG(K) - ENTR(K,L)$$

   C. Print summaries of fish concentrations by river section.
      Compare runs with and without power plants for fish surviving in late July (or other times).
      Investigate possible density compensation effects.
      Print results.
      Plot results of field survey data, fish population estimates by section with no power plants, fish population estimates with different configurations of power plants in operation.
   D. End program.

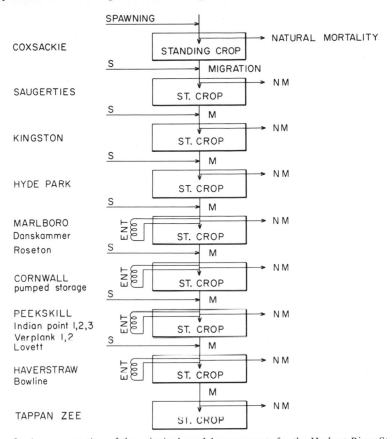

Figure 3 A representation of the principal model components for the Hudson River Striped Bass population model. The principal state variable is the fish population in each section. Fish are added to each section by spawning and by immigration from adjoining sections. Fish are subtracted from each section by mortality, outmigration, and entrainment mortality from the power plants.

realistic as possible. The population estimates were solved iteratively for various periods of from 2 to 6 months, and the results of computer predictions of striped bass population numbers by river segment were plotted on a computer-driven plotter as well as the population estimates derived from field sampling (Figure 4). Since the population estimates of the larvae were not used in the computer model they served as one validity check on the accuracy of the model.

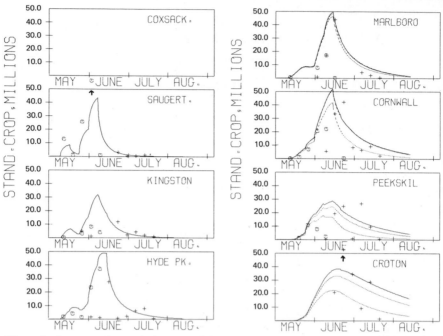

Figure 4 A more-or-less typical output of the simulation program. The eight boxes represent the eight principal sections of the river used for analysis from upstream (Coxsackie) to downstream (Croton). The circles represent the number of eggs found in each section week by week based on 1967 data. The crosses represent the number of larvae and juvenile fish found in each section based, again, on the 1967 data. The total population measured in each section would be the summation of eggs and larvae-juveniles. Since the fingerlings become strongly net shy after a month or so the population estimates after July 1 should be considered minimal estimates. The solid lines in each section represent the population estimate of the computer model without any power plants running. The upper dotted line represents the population estimates with four power plants in operation, and the lower dotted line represents population estimates with all ten power plants in operation.

MODELING POWER PLANTS

Power plants in the Hudson River may affect fish in at least four ways: by thermal, chemical, impingement, and entrainment processes. The direct addition of heat to the river does not appear to be important relative to other effects (AEC, 1973). Chemical toxins periodically added to the cooling water intake to clear fouling organisms as well as erosion of metal cooling coils may damage fish near the power plant outfall (AEC, 1973; Clark and Brownell,

1974). Impingment (e.g., impaction) of adult fish on the trash screens located in the intakes of the power plants may kill hundreds to hundreds of thousands of fish per day per power plant, especially during the winter when fish cannot swim against even weak currents. However, the most important problem appeared to be *entrainment*, that is, the pulling of eggs and larvae through the cooling coils with the cooling water. Figure 2 includes estimates of the quantity of cooling water that would be used at full power output for each of the power plants that were operational or proposed in January 1972. One way to visualize what this quantity of water means relative to the total river volume is that these plants would use the equivalent of the water contained in 3 km of the mid-Hudson every day, or all of the water found in the Marlboro, Cornwall, and Peekskill sections in about 18 days. Thus a significant potential impact exists since the young stripers are largely planktonic (i.e., they drift with the water currents) during about a month of their early life.

In order to assess what the entrainment effects might be on the young striped bass the simplest and most reasonable adjustments to the population model given above were made in order to predict what the effects of building these power plants might be on the young striped bass. Since the fish were planktonic and appeared to go wherever the water went fish were "killed" in the computer (i.e., the quantity of fish in the appropriate state variable were reduced by some amount) as a function of the mean concentration of fish in each section, the quantity of water used by the power plants, and the expected mortality of the fish as they were drawn through the power plants. As time went on many modifications in this basic design were made to reflect increasing knowledge of the fish and the river, but the above parameters remain the most important in assessing the impact of the plants on the fish.

The results of this computer model indicated that if all ten power plants are constructed with once-through cooling there would be major effects on young striped bass. The results of a typical run of one of our earlier models is shown in Figure 4. The predicted reduction in the population levels of young stripers due to entrainment mortality are shown for the same conditions as the population simulation mentioned above. The solid line represents, as indicated before, a simulation of the population without any power plants in operation. The upper dotted lines in Figure 4 represent the estimated population of striped bass remaining in each section with the Danskammer, Lovett, Indian Point 1, and Indian Point 2 power plants operating—conditions existing during the spring of 1973. The lower dotted lines represent population estimates for each segment if all ten plants projected in 1972 were actually built with once-through cooling. The first situation represents about 20% mortality of the fish during their first four months or so; the second condition (i.e., all ten plants running) would reduce the population of four-month-old fish by about 70 to 80%. This particular run was made with the assumptions that each river section was well mixed and that all of the fish that went through the plant would die.

Since the estimated impact of the power plants on the fish is very much a function of the assumptions that go into model building, we and others at Oak Ridge have done extensive sensitivity analysis on both (1) the model *structure* and (2) the *coefficients* used. The series of other computer models that grew out of our earliest version examined the sensitivity of the power-plant-induced mortality to differences in importance of migration within the salt wedge, different assumptions about vertical and horizontal variations in fish density, and different mortality factors for the organisms entrained. These results are given in detail elsewhere (e.g., AEC, 1975) but are broadly similar to those presented in Figure 4.

Runs were also made using what we considered to be the maximum possible density compensation during the first 3 months. This is based on the assumption that the mortality rate is a function of population density, not age of fish. When entrainment mortality reduced the population in a computer run, the natural mortality was also reduced to the mortality level of the natural population at that same concentration. This assumption of maximum density dependence during the first three months reduced the effect of plants by 20 to 30%, that is, to 50 to 60%.

A major impression we have of the results of the many different models and the many different assumptions used for each is that the computer-derived estimates of cumulative impact of the power plants on the Hudson River striped bass during their first 6 months or so has not been terribly sensitive (e.g., ±25%) to the different sets of assumptions that have been used in the models. In other words, all of the different assumptions that have been built into the model with time have not drastically altered the estimates of the magnitude of entrainment of the bass. In many cases, changing assumptions decreases the impact of some plants but increases the impact of others. At least four different models made by AEC staff or consultants gave similar results (Clark, 1974; Wallace, 1975). [However, models prepared by consultants to the utilities have given different results.* (Wallace, 1975)]

---

* Some of the principal assumptions that have been used by the consultants to the electric utilities that would cause their estimates of mortality to be lower than ours include:

a   The use of tidal velocity rather than net downstream velocity or behavioral movements to remove fish from the vicinity of power plants. This is erroneous and is no longer used by anyone.
b   The use of "on-line" rather than "predicted-on-line" times for plant pumpage. The Hudson generating plants have in many cases not operated for as much time per month as was

What then would be the effect of this additional juvenile mortality on the striped bass fisheries? The answer to this is less subject to analysis than were the entrainment mortalities and is strongly conditional on the relative importance of density-dependent versus density-independent factors for post-juvenile fish. It may be that in the case of striped bass density-independent

---

called for in their design due to operating problems. It is not certain what percentage of the design time they will be operating in future years. This has been used in the assessment (not technically a model) of Texas Instruments (1975).

c The use of mortality factors less than one. All of the young fish that go through the plant may not in fact die. Mortality factors of from 0.5 to 1.0 were incorporated in our original models during sensitivity analysis, with obvious effects on plant-induced mortality. This would seem to be a simple question to answer, but it has in fact been elusive, due both to the difficulty in getting such samples and seemingly poorly designed sampling schedules. At this time consultants to the utilities are gathering such information. Past studies have indicated that mortality is probably very high (e.g., Marcy, 1971).

d Different handling of spatial distribution patterns. This is a real problem, and again the needed data is being gathered and interpreted by consultants to the utilities. (No environmental organization could possibly finance such an undertaking, as it costs millions of dollars per year.) The spatial distribution patterns are important because the different power plants draw water from different parts of the river channel. For example, the Indian Point plants draw their cooling water from the top 10 m on the east side of the river, and the Storm King Plant would have an intake near the west shore. Data obtained for some years have shown that the fish are more abundant along the west shore, for other years along the east shore, and data for still others indicate that fish are found mostly in the middle. Thus the assumption that the fish are randomly distributed across the river during an average year seemed valid as a first approximation. However, the later models prepared by various groups have tended to use much more complicated patterns for the spatial distribution of fish. Such spatial patterning may make the fish more or less susceptible to entrainment depending on the spatial relation of the fish to the power plant intakes. Longitudinal patterns likewise may be important.

e Different assumptions about the ability of young fish to actively avoid moving with the intake water into the power plants.

f Different interpretations of longitudinal distribution of juvenile fish. My interpretation of the data summaries supplied by the consultant companies pertaining to spatial and temporal distribution patterns of young striped bass in the river in 1973 is that our original assumptions about how the young fish were distributed in the river were essentially correct. Eggs are found most abundantly in the sections at or above Cornwall, and the majority of young fish are found at or below the Indian Point and Croton sections by mid June to early July. However there is still a remarkably large amount of variance in the data, and one impressive finding of the recent, very extensive (and expensive) studies is that the data does not appear any "smoother" than earlier less-extensive surveys. In other words, there is still enormous difference in the estimated fish population estimates for each respective river stretch from one 2-week period to the next. The variance in the fish population estimates seen, for example, in the 1967 data plotted in Figure 4, Croton section, is also found in a recent summary report by one of the consultant companies (Texas Instruments, 1975). It is not unlikely that the difficulty in getting really solid fish estimates in the Hudson River will defy our best scientific efforts.

factors dominate, at least for the strong year classes that dominate fisheries. One line of argument is that striped bass fisheries show pronounced year-class dominance. For example, the 1934 and 1957 year classes dominated the fisheries in the 2- to 5-year period after their birth. In other words, the fish populations seem more dependent on favorable environmental conditions than on lack of competition for their survival. In addition, the existence of a congeneric competitor that is less susceptible to entrainment (white perch) would help maintain a high level of competition, regardless of the density of striped bass. Therefore, a case can be made that the loss to the adult fisheries would be roughly proportional to the loss of young fish and that losses could become greater over the years as fewer spawners return each year. However, this point is certainly open for question (Texas Instruments, 1975). There has been a considerable amount of effort expended on answering this, and to date the results are still conflicting.

In the meantime, there has been some decline reported in the striped bass catch in New York waters (Striped Bass Fund, Inc., no date; Texas Instruments, 1975). I would not be so rash as to attribute this entirely to the power plants presently operating, but our model does predict the direction of the catch statistics. Since seven of the original ten plants are built and operating with once-through cooling, we may have a nice opportunity to test our model output—with the one difficulty that since the bass naturally go through wide population fluctuations there is no way to get a complete check on the model's predictions.

DECISION-MAKING RESULTS BASED ON THE MODEL RESULTS

The Atomic Energy Commission regulatory agency was justifiably concerned about the effects that the atomic plants under their jurisdiction would have on the fish. The agency recommended that all future Hudson River plants be equipped with wet cooling towers and that cooling towers would have to be backfitted on existing plants. Cooling towers withdraw only about 4% as much water as once-through cooling. However, cooling towers have other adverse environmental features and require substantial quantities of energy to run (Chapter 21).

In early 1974 a review panel composed of two physicists and a lawyer rejected our estimates of fish mortality, even though our model had been prepared by about a dozen ecologists, engineers, hydrologists, and operations researchers over more than 2 years. One important aspect of the fish's life history was dismissed by fiat. I confess that I am perplexed as to their scientific rationale. The legal decisions based on this are currently in flux.

Consolidated Edison (Con Ed), the major utility involved, has dropped plans for Verplank 1 and 2 (Indian Point 4 and 5), perhaps due in part to these studies but no doubt additionally influenced by operating difficulties of the earlier nuclear plants. However, other utilities, in association with Con Ed, have constructed and begun operation of the other large fossil fuel plants (Bowline and Roseton). Con Ed is proceeding with plans for Storm King (Cornwall)* and perhaps another fossil fuel plant (Ossining) on the Tappan Zee. These plants, of course, are not subject to the restrictions of AEC licensing and are not equipped with cooling towers. All of these plants, particularly Storm King, are subject to a variety of other legal restraints and battles, and which "side" is "winning" appears to change month by month. Other large power plants are being constructed on the striper's nursery grounds in the Chesapeake.

Recently the AEC has begun a program of research aimed at eventually placing nuclear power plants off the shore of New York-New Jersey. This decision, like the cooling towers, will require enormous amounts of energy to implement and will have other environmental impacts. If these decisions were based in part on the results of our modeling efforts it is not at all clear to us that the best social/environmental decisions have been made. In general, we were excluded from partaking in many of the decisions based, in part, on our model results.

---

* The Storm King (or Cornwall) project is a special case deserving separate attention. The plant does not generate electricity, but instead uses electricity generated elsewhere at night (when demand is low) to pump water to the top of Storm King Mountain through a 3-km tunnel. Electricity would be generated when demand was high by running the water back down the mountain and through turbines.

Since the pumped-storage project uses considerably more water per kwh of electricity pumped/stored than does a generating plant to make a kilowatt of electricity and since the pumped storage plant still requires a generating plant to make the electricity used to store the water (at about 70% efficiency), some six times more fish would be killed per kilowatt delivered to consumers than if it were not diverted through the Storm King project first (Hall, 1973).

However, many of us think that the most important issue in the Storm King case has to do with a "net energy" assessment of the project (Chapters 7 and 21). The Storm King project does not generate electricity; it merely stores it as elevated water. Nearly 1.4 cal of electricity must be used to get 1.0 cal back (at a more desirable time of day), not counting the huge initial energy capital investment to build the project. Since a large part of the baseload of the Con Ed system is oil fired, the existence of the Cornwall project would mean that about 50,000 barrels of oil per day would be burned simply to overcome friction in the turbines and rock tunnels of the project. We think that studies such as that in Chapter 21 should be undertaken before the project is built, and that making the price of electricity relatively more expensive during periods of high demand should be studied as a possible means of accomplishing the same objectives (meeting peak demands) with less waste of industrial energy and less environmental impact.

VARIOUS MODEL OUTPUTS AND VARIOUS LOYALTIES OF MODELERS

It is well known that a model can give you almost whatever results you wish, depending upon what sort of assumptions you want to make and what numbers you choose as representative for your inputs and coefficients. This, of course, was obvious to us at the beginning of this project. What was not so obvious was how the results of at least the early calculations and models appeared closely correlated with the personal convictions of the modelers. Our earliest models, created by a couple of ecologists who like to fish and who are, in addition, suspicious of continued industrial expansion, predicted dire effects on the fish. Early results of consultants to the power companies predicted negligible effects on the fish. However, with time the more obvious fallacies have been weeded out, and the results of the modelers of different persuasions have begun to converge, at least with respect to the number of fish entrained. Our estimates of mortality have dropped a little, those of the consultants have risen some. I consider this a positive sign.

We used what we considered median assumptions in our models where parameters were not known precisely, but we have been impressed by the efforts that the consultants to the utilities have taken to demonstrate that one extreme or another of a poorly known parameter (whichever would give the smallest effect on the fish) was the proper one to use. We suppose that they thought the same of us. We are impressed at how "objective" science, in this case, seemed very much a function of one's personal convictions pertaining to other matters. This of course is the case in many environmental matters.

In the Hudson River case we have found that models were critically important in forcing different sides to define their processes precisely. Some early erroneous assumptions were eliminated in constructing the models, because the assumptions made no sense when translated into FORTRAN. Although we do not yet know for sure, it is a possibility that modeling may assist in the merging of scientific and legal "truth" (Chapter 8) in the Hudson River court proceedings.

Our goal was an objective model of the river, the fish, and the power plants. In theory this would be entered into a larger framework of decision making in which the fish would be weighed against the pluses and minuses of more electricity and other matters. The fish per se probably would not count for much in the overall assessment, although the tourist industries directly or indirectly dependent upon the striped bass are worth some hundreds of millions of dollars per year. We would suspect that more people have watched stripers caught on (the electricity consuming) television than have caught the fish themselves. Yet large, expensive decisions are being made on the basis of the fish's supposed welfare, and they will probably continue to be made on that basis. In this case the analytical and decision-making processes appear to have been channeled by tractability, not necessarily importance. The fish can be modeled easily; social values cannot. Yet the fish issue gives a lever to conservationists who oppose industrial expansion in general, and the specific issue is more easily fought than the vaguer but potentially more important issues of fuel depletion, regional air pollution, problems of possible nuclear accidents, and the other industrial expansion made possible by the availability of electricity. Is this good? Are we sufficiently affluent to preserve relict pieces of wilderness along our shores? Can we afford not to? I don't know.

ACKNOWLEDGMENTS

I thank John Day, Thurman Grove, and Simon Levin for helpful comments.

REFERENCES

Clark, J. 1974. Testimony before U.S. House Subcommittee on Merchant Marine and Fisheries. Feb., 1974.

Clark, J., and W. Brownell. 1973. *Electric power plants in the coastal zone*, The Striped Bass Fund and The American Littoral Society, Highlands, N. J.

Day, J. W., W. G. Smith, P. R. Wagner, and W. C. Stowe. 1973. Community structure and carbon budget of a salt marsh and shallow bay estuarine system in Louisiana. Cent. Wetland Resour. *Publ.* LSU-73-04.

Douglass, P. A. and R. H. Stroud (Eds.). 1971. *A symposium on the biological significance of estuaries*. Sport Fishing Institute, Washington, D.C.

Hall, C. A. S. 1973. Testimony before U.S. House Subcommittee on Merchant Marine and Fisheries. Feb., 1974.

Hall, C. A. S. 1974. Models and the decision making process. In S. Levin (Ed.) *Ecosystem analysis and prediction*. Soc. Ind. and Appl. Math., Philadelphia.

Koo, T. S. Y. 1970. The striped bass fishery in the Atlantic States. *Chesapeake Sci.* 11: 73–93.

Marcy, B. T., Jr. 1971. Survival of young fish in the discharge canal of a nuclear power plant. *J. Fish Res. Board Can.* 28: 1057–1060.

Rathjen, W. B., and L. C. Miller. 1957. Aspects of the life history of the striped bass [*Roccus* (= *Morone*) *saxatilis*] in the Hudson River. *N.Y. Fish Game J.* 4: 43–60.

Smith, R. F. 1966. (Ed.) *A symposium on estuarine fisheries*. Spec. Publ. No. 3 Amer. Fisheries Soc.

Striped Bass Fund, Inc. (no date). *We're concerned . . . you should be too*. Leaflet. 4 pp.

Talbot, G. B. 1966. Estuarine environmental requirements and limiting factors for Striped Bass. In R. F. Smith (Ed.), A *symposium on Estuarine Fisheries*. Spec. Publ. No. 3 Amer. Fisheries Soc.

Teal, J. M., and M. Teal. 1969. Life and death of the salt marsh. Ballantine, New York.

Texas Instruments, Inc. 1975. *First annual report for the multi-plant impact study of the Hudson River Estuary*. Texas Instruments, Inc., Dallas.

U.S. Atomic Energy Commission. 1973. *Final environmental impact statement, Indian Point 2.* U.S. Atomic Energy Commission, Washington, D.C.

Van Winkle, W., B. W. Rust, C. P. Goodyear, S. R. Blum, and P. Thall. 1974. *A striped bass population model and computer programs.* Preliminary report available from Environmental Sciences Division, Oak Ridge National Laboratory, Oak Ridge, Tennessee 37830.

Wallace, D. N. 1975. A critical analysis of the biological assumptions of Hudson River striped bass models and field survey data. *Trans. Amer. Fish. Soc.,* 1975: 710–717.

Woodwell, G. M., P. H. Rich, and C. A. S. Hall. 1973. Carbon in estuaries. In G. M. Woodwell and E. Pecan (Eds.), *Carbon in the biosphere.* Nat. Tech. Inf. Center., Arlington Va.

15
A Simulation that Failed: The Biospheric Productivity Model

DANIEL E. WARTENBERG
CHARLES A. S. HALL

"Holmes," I cried, "this is impossible."
*"Admirable!" he said. "A most illuminating remark. It is impossible as
I state it, and therefore I must in some respect have stated it wrong..."*

Sir Arthur Conan Doyle
The Adventure of the Priory School

The rate of global and regional primary production, frequently taken for granted in our supermarket society, is actually the single most important factor determining the welfare and happiness of most of the Earth's people. Yet over recorded times regional, and perhaps global, primary production has changed considerably, and these changes have been of enormous consequence to the people living on our planet (e.g., Borgstrom, 1969; Ladurie, 1971; Bryson, 1974; Winkless and Browning, 1975). It is a sobering experience to examine the graphs of the population of many nations over their recorded history—for example, the population of Egypt in Biblical times was about the same as it is today, but it also has been less and frequently much less between then and now. The same can be said for most other old civilizations—the human populations have tended to decline about as frequently as increase. The population "explosion" that has concerned us so much of late is a rather common experience of mankind. However, population declines also have been about equally common over the past several thousand years.

Why have population growth patterns changed so much over human history? There are many reasons, of course, but high on the list of causative factors has been famine and/or other failures of biotic life-support systems. Famine may kill people directly through starvation, but indirectly as well through the lowering of resistance to disease. And perhaps more important, hungry people often have satisfied their hunger, or at least tried to, at the expense of their neighbors. Thus a consideration of future patterns of primary production of the earth is important.

This paper describes a preliminary attempt to build a model that might help to predict the future primary productivity of the earth. This model emphasizes some artifacts of man's industrial activities, artifacts that we thought could influence global plant production. We were not able to finish such a model because of many gaps in our knowledge of atmospheric and biologic science. However, trying to build the model was extremely interesting and eventually led us to another procedure for evaluating man's effects on the biosphere, a procedure that gave us some of the answers we were seeking. This chapter summarizes some of the things we learned about the earth and about modeling during this process.

POTENTIAL EFFECTS OF GLOBAL INDUSTRIALIZATION ON BIOSPHERIC PRODUCTIVITY

In order to understand how regional or global primary production might change it is necessary to consider the most important factors that determine primary production. One such factor is soil fertility, that is, the mineral nutrients in the soil, the water-holding capacity, and so on. All other important factors are related to the atmosphere: the carbon dioxide in the air is the basic structural material of new plant tissue, climate determines growing seasons, weather determines the growth on a given day, and rain and sunlight intensity by their abundance and distribution are probably the most important factors controlling the productivity of terrestrial vegetation. There are other factors, but these are the most important.

The use of industrial energy in the world is doubling every 20 years or so, and there are many industrial artifacts that impinge upon and change the productivity of the biota of the Earth. We were curious as to whether these industrial artifacts could change the productivity of the biosphere. This first section reviews some of the ways that industrial fuel burning is changing, and could change, some important atmospheric parameters, and the ways that these changes could influence biospheric productivity.

One extremely well-documented industrial artifact is that the combustion of fossil fuels is contributing to an increase in the concentration of carbon dioxide in the atmosphere (Revelle and Suess, 1957; Bolin and Bischof, 1970; Machta, 1973; Ekdahl and Keeling, 1973; Woodwell et al., 1973). This increase is about 0.2% per year over the past few decades, and 0.3 to 0.4% per year for the past few years. If this trend continues atmospheric carbon dioxide will double by the year 2040 (Bacastow and Keeling, 1973).

Increases in atmospheric carbon dioxide content influence the biosphere in at least two ways. First, carbon dioxide increases the surface temperature of the Earth due to the now well-known "greenhouse effect" (Manabe and Wetherald, 1967), and small to moderate increases in the Earth's temperature would increase the rate of photosynthesis by lengthening the growing season and because many plants photosynthesize better when the temperature is warmer (Lieth, 1973).

Second, an increase in carbon dioxide concentration has been shown experimentally to increase the rate of terrestrial photosynthesis since CO_2 is a plant nutrient. This relation is approximately linear for values in the vicinity of present and projected atmospheric concentrations (Hasketh, 1963; Lieth, 1973; Zelitch, 1967; Wilson and Cooper, 1969). Thus increased carbon dioxide production by the burning of fossil fuels would be a "good thing" for the biosphere. Both the heating and the direct enhancement of available carbon dioxide, at least within certain limits, would make the biosphere more productive.

However, many other effects of industrialization would act to decrease primary production. One such factor results from the increasing acidity of rain, caused by increases in the concentration of combustion-derived oxides of nitrogen and sulfur in the atmosphere. The acid rain problem is widespread and already of serious proportions. For example, the pH of rain over large areas of northern Europe and the entire northeastern United States is now generally between 3.0 and 4.5, versus a normal 5.6 to 6.0, and the pH over large regions of Europe declined by about 1 unit from 1956 to 1965. There are some indications that the pH in some areas has stabilized at 3.0 to 4.5 (Likens and Bormann, 1972, 1973). Localized acid rain has damaged buildings and crops, but the long-range ecological effects are poorly known. However, one laboratory study on the leaching rate of calcium indicates that below a pH of 4.0 rain is much more effective at washing nutrients from the soil (Overrein, 1972). Any decrease in soil fertility would be important in determining regional rates of photosynthesis.

Another complex interaction between industrial fuel burning and the biosphere results from the injection of dust and aerosols into the upper atmosphere. The effects of these materials can be divided into two classes: upper atmospheric effects and lower atmospheric effects. Upper atmospheric dust loading was due primarily to volcanoes in the past, but more recently man has contributed as well via nuclear bomb tests, high-altitude aircraft emissions, and fossil-fuel burning. Lower atmospheric dust loading has been a function of natural factors such as sea spray and volcanoes as well as slash and burn agriculture, overgrazing, mechanized agriculture, and fossil-fuel burning (Kellogg and Schneider, 1974). Benjamin Franklin (1784) was perhaps the first to notice the importance of large volcanoes, for he found during the cold summer of that year (which had been preceeded by several large volcanoes) that the sun's rays "when collected in the focus of a burning-glass they would scarcely kindle brown paper."

Dust loading at either level will decrease solar insolation at the earth's surface due to the absorption and reflection of incident light. One effect of less sunshine reaching the world's ecosystems would be decreased photosynthesis, since plant productivity is a function of sunlight intensity (Hesketh, 1963; Aruga, 1965). However, this relation is complicated by two factors: light inhibition at high intensities and the effect of light quality. Individual plants in the laboratory show a leveling off and decline in photosynthesis at high light intensities, due to saturation and inhibition of enzyme systems. However, light inhibition rarely occurs for natural communities within the range of naturally occurring sunlight. Different species become important at different light levels, and leaves lower in the canopy or water column receive higher levels of light as upper leaves become light saturated.

In addition, although dust loading decreases the direct radiation reaching plants, it increases the diffuse, or scattered rays. While this again might not favor an individual plant, an entire plant community could benefit from such an effect due to increased canopy penetration (Lister, 1974).

Dust loading also affects the temperature of the Earth's surface and atmosphere. Upper atmospheric dust loading both decreases the temperature at the Earth's surface and increases the temperature in the upper atmosphere as a consequence of absorption and reflection. While most of the evidence for this relation is circumstantial, we do have one nice "experiment" performed gratis, as it were, by the eruption of Mount Agung, a medium-sized volcano, in 1963, which injected large quantities of dust into the upper atmosphere. During the year or so that followed, upper atmospheric temperatures in the latitude of Agung increased by some 6 to 7°C due to the increased absorption of solar radiation by the dust particles. There was a simultaneous dramatic decrease in sunlight reaching the Earth's surface (Ellis and Pueschel, 1971). Further validation of the relation between atmospheric dust loading and ground temperature is presented by Budyko (1969), who correlates decade-by-decade global temperature changes and the quantity of solar radiation reaching the Earth's surface. Times of relatively low insolation tend to follow major volcanic eruptions. For example, there was a substantial decline in both insolation and temperature following the eruption of Krakatoa in 1882 and Santa Maria in 1912. Historically the years following major volcanic eruptions have been noted for spectacular sunsets, cold weather, and high food prices.

The effects of lower atmospheric dust loading are less clear. Whether this low-level dust loading has a heating or cooling effect on the Earth's atmospheric system is apparently a function of the physical characteristics of the particles (i.e., size, absorptivity, reflectivity, etc.), its location in the atmosphere, and various properties of the cloud cover and surface reflectivity (Weare et al., 1974; Mitchell, 1970; 1971). The majority consensus, however, seems to be that the effect of man's particulate loading of the atmosphere will be to cool the Earth's surface temperature (Kellogg and Schnieder, 1974). The atmospheric dust loading effect appears to have offset and could continue to offset any potential warming trends due to carbon dioxide production. There exists considerable controversy now between the "dust" and the "greenhouse" advocates.

There are other factors that might be important in determining the effect of industrial activities on the temperature of the earth including changes in the moisture content of the atmosphere or the distribution of rain patterns, of winds and cloud formation, and so forth. It is obvious that the problem is extremely complex and the interaction of the atmospheric processes with the biosphere even more complex. Nevertheless, we have diagrammed what we thought were the most important of these processes and interactions in Figure 1. A further consideration of these factors is given in Hall (1975).

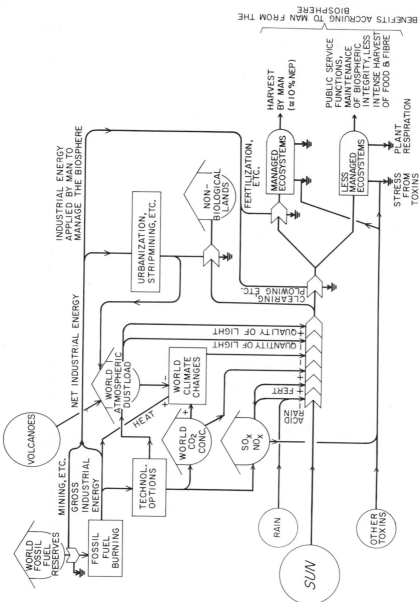

Figure 1 The biosphere, the industriosphere, and their interactions. A conceptual model of the flow of energy through both the biotic ecosystems of the earth and through man's fossil fuel powered industrial systems. In general, the energy from the sun runs the biotic world represented on the bottom half of the diagram. Man takes his own energy, or energy from the industrial sector, and diverts biotic energies from natural pathways to high yield-to-man pathways. However, the burning of industrial fuels inter-acts in many complicated ways with the biosphere. These are considered in the text.

DEVELOPMENT OF THE MODEL

We next constructed a preliminary computer model of the potential effects of industrialization on the biosphere using Figure 1 as a guide. [A more technical description of this model is given in Wartenberg (1974)]. No prior attempts had been made to examine this relation on a global scale to the best of our knowledge. It was necessary to make many assumptions and simplifications to construct a model of this type. We consider the effects of these assumptions on the output of the model and then examine the value of this type of study.

It soon became obvious to us that there was too little known about a number of the interactions given in Figure 1, such as the relation of acid rain to plant growth, to include these effects in our model. However, there was information available on the interactions that we believed were of greatest importance in affecting the productivity of the biosphere, such as the effects of increases in dust and CO_2 (Figure 2). This was our first and most important assumption and simplification, a subjective judgement limiting the scope of

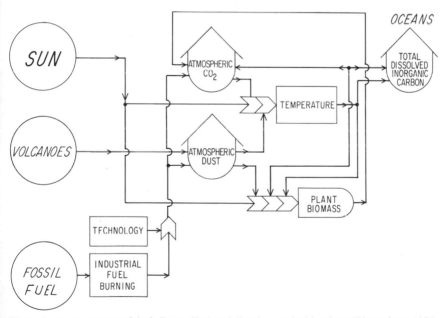

Figure 2 A computer model of effects of industrialization on the biosphere (Wartenberg, 1974). This conceptual framework was derived from Figure 1, but has been greatly simplified to allow for construction of the initial computer model. It incorporates those interactions considered most important by the authors in affecting the primary productivity of the biosphere, but does not attempt to approach the complexity of the real system.

the model by defining its components to those we thought most important and for which we had data.

This simpler model had three basic forcing functions: sunlight, volcanoes, and fossil-fuel burning. We reviewed much of the pertinent information and selected what we thought were the most reliable formulations of these three functions. Part of this information comes from correlations in the record of past events, but some must be considered the speculations of other researchers. Unfortunately, there is often disagreement in the literature on the derivation and formulation of these functions, underscoring some of the uncertainties involved in global modeling.

THE FORCING FUNCTIONS

We assumed that the solar input to the Earth was a constant, since we were interested in annual, not seasonal or daily variations in productivity. In addition, no one has been able to show that the measured year-to-year variations in solar radiation such as those vaguely correlated with sunspot cycles exert significant influence on any of the state variables in which we were interested.

Volcanic eruptions are natural events that are poorly understood, but we think they are an important controlling factor of global photosynthesis. While

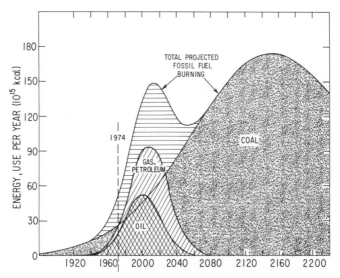

Figure 3 Projections of global industrial fuel burning based on estimates from Hubbert (1969) and a continuation of present trends.

we are beginning to gain a slight understanding of why they occur and where, we are still far from being able to predict the frequency or magnitude of the eruptions.* Therefore, we incorporated a random or "stochastic volcano generator" into our model. The generator can be used to approximate the frequency and intensity of volcanic activity from the historical record or some other hypothetical frequency.

The annual burning rates and global stores of fossil fuels were well studied and understood in contrast to the unpredictability of volcanoes. As a result, fairly good predictions of man's fossil fuel consumption in the future were available, barring any major change in current trends, technologies, or resources (Figure 3).

COEFFICIENTS, STATE VARIABLES, AND RESULTS

Only limited data were available for some important coefficients used in this model. However, we studied the sensitivity of the model to variations in each of these individual functions to determine their relative importance and to see if using values from within the range of possible numbers changed the trend of the model results (e.g., Table 1). The model could also be viewed from a conditional perspective: if the inputs do vary in the prescribed manner what will be the effect on the state variables? In this way, various scenarios or sets of inputs in agreement with the speculations of different researchers may be used as inputs, and the resulting outputs can be examined for differences, similarities, and trends.

The examination of the response and sensitivity of the state variables is the primary goal of this modeling effort. The state variables were broken down into three submodels: atmospheric, physical oceanic, and biotic.

The atmospheric submodel is concerned with three basic parameters: dust loading, carbon dioxide changes, and global mean surface-temperature (as a function of the other two). At the current stage of development, this model utilizes greatly simplified formulations of these complex functions, including a simple one-layer atmosphere. As the understanding of atmospheric phenomena increases, modifications could be made in our model to give a more realistic view of the Earth's atmosphere. Our coefficients were derived from preliminary studies at the University of Wisconsin (Bryson, personal communication).

The oceanic submodel accounts only for the oceanic absorption of carbon dioxide. This was made a function of the difference in partial pressure between

---

* Since we developed this model we have come across an extremely provocative theory that if validated might help to predict volcanoes (Winkless and Browning, 1975). Such information could easily be incorporated into our model.

Table 1   Sensitivity Analysis of the Variation of the Coefficients of the Temperature (TEMP) Function on the Primary Production (PP)[a] of the Earth as Predicted by a Simulation Model.[b]

$PP = BIOMASS \cdot 0.3 \cdot \ln(CO2/100) \cdot (1/1 + e^{-1.315 - 0.119\,TEMP}) \cdot V$

$TEMP = TEMP_0 + 2.7 \cdot ((\ln(CO21/320)) - (\ln(CO22/320))) - DUST1/(45 \cdot 10^9) + 0.02$

Year	Coefficient Multiplication Factor						
	1/10	1/5	1/2	1	2	5	10
Coefficient (1) = 2.7							
1880	66.70	66.70	66.71	66.73	66.76	66.85	67.00
1910	66.28	66.29	66.32	66.38	66.50	66.85	67.42
1940	69.41	69.44	69.53	69.69	70.00	70.94	72.46
1970	74.47	74.55	74.81	75.22	76.05	78.47	82.25
Coefficient (2) = 1/(45 × 10⁹)							
1880	67.58	67.48	67.20	66.73	65.76	62.78	57.62
1910	69.69	69.33	68.24	66.38	62.51	50.09	30.08
1940	73.20	72.82	71.67	69.69	65.55	52.08	30.27
1970	79.75	79.27	77.79	75.22	69.78	51.82	24.77
Coefficient (3) = 0.02							
1880	65.32	65.48	65.95	66.73	63.25	72.56	78.82
1910	63.47	63.80	64.78	66.38	69.48	77.70	87.60
1940	65.26	65.77	67.26	69.69	74.27	85.40	95.88
1970	68.98	69.70	71.82	75.22	81.48	95.28	103.29

[a] Output as net primary productivity, in 10^9 metric tons C yr^{-1}.

[b] This is an example of the type of sensitivity analysis that has been performed on this model. A more extensive treatment is given in Wartenberg (1974).

atmosphere and the seas. Absorption coefficients were derived from Broecker et al. (1971).

Our preliminary treatment of the biosphere submodel used a constant total biomass for the world. The output was primary production as a function of volcanic and industrial activity, including the second-order effects of atmospheric carbon dioxide content, dust loading, and global mean temperature. Coefficients were derived from Lieth (1973) and Whittaker and Likens (1973). The final and most important output of the model was global primary production as a function of industrial fuel burning.

The results of the preliminary simulation indicated that primary productivity was most sensitive to carbon dioxide levels in the atmosphere

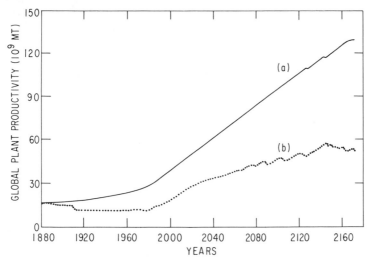

Figure 4   Predicted global primary productivity (after Wartenberg, 1974). Net global primary productivity predicted by our model from 1880 to 2180 is shown for increasing industrial fuel burning and two different frequencies of randomly occurring volcanoes. Curve (*a*) is the prediction for a period of relatively light volcanic activity, and curve (*b*) represents productivity during a period of relatively high volcanic activity as compared to the volcanic record of the past 100 years.

and predicted an increase in primary production over the next 200 years (Figure 4*a*). However, our model was also very sensitive to volcanic dust loading, and substantial volcanic activity changed the results considerably (Figure 4*b*). It is possible that industrial dust loading could have a similar effect.

WHAT WAS GAINED FROM THE MODEL?

What one sees emerging from this rather simplistic modeling attempt is a model insecure in its major assumptions, but one that gives not unexpected predictions. Since we felt that so little was known about the major components of the model, it did not make any sense to refine the model further at that time and we abandoned its development.

What then was the value of such a model, since so little faith can be placed in its predictions? One useful function was to point out the significance of major gaps in both the currently available data base and our knowledge of the most important interactions that affect global primary productivity. This sort of study also is useful in planning the direction of future research and data collection and in determining which interactions are most worthy of additional study in terms of their effects on the biosphere. For example, more comprehensive studies of the limiting factors of plant growth at the biome level are clearly warranted. We also need to understand more clearly the ways in which primary productivity varies through naturally occurring changes in temperature and precipitation and how plant distribution patterns are affected by changes in climate. Additionally, has man changed the biomass of the Earth, and has this had a significant effect on global primary productivity?

This modeling attempt was useful in other ways, ways in which we initially were neither interested nor aware of. Once our interest in the subject had been whetted and after it became apparent that the desired information was unobtainable from a model, we began to look to empirical means for determining what possible changes in global photosynthesis might have occurred so far during man's industrial tenure. Some tree ring analyses (Jonsson and Sonberg, 1972; Whittaker et al., 1974) had indicated a decline in forest growth over much of the northern hemisphere, a decline some attribute principally to the effects of acid rain (a parameter we had not included in the model since we had no information to derive a mathematical function). However, other analyses indicated that the decrease in tree ring annual increments observed in some locations may not be a general feature.

Still frustrated in our attempt to measure trends in global photosynthesis, we devised a procedure based on the annual changes in atmospheric carbon dioxide, a method we believe sufficiently reliable and sensitive for our purposes (Hall et al., 1975). The results of this analysis indicated that there had been no change in primary production of the northern hemisphere during at least the past 15 years (Figure 5). In view of our model predictions that photosynthesis probably should be increasing (there have been no major volcanoes since 1912), this was somewhat surprising and indicates to us that there must be some compensatory mechanism balancing the carbon dioxide effects. This could be the photosynthesis-inhibiting mechanisms resulting from acid rain, or industrial dust loading may be more important than we thought. Another possibility would be the plant community compensating mechanisms elucidated by Botkin (1973) and Lemon (in preparation).

The final way in which this model was useful was as a learning device. It helped us to conceptualize the interactions between the biosphere and industrial activity and led us to a great deal of interesting literature outside of our own field. One approach we are currently developing is the use of multiple correlations of regional weather records and crop production data to gain more information about the major meteorological determinants of crop production. Other people also are beginning to look at this problem, and there will be much more information available in a few years. It is clear that models will be of great help in this process (Lemon, in preparation). However, we think that it will be a long, long time before any manifestation

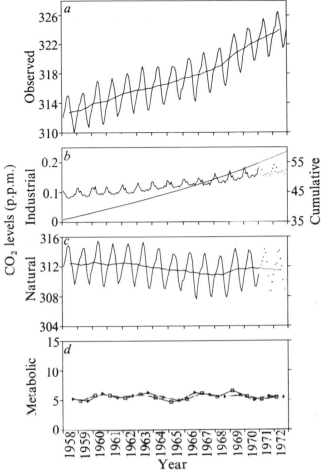

Figure 5 A 15-year record of biotic metabolism in the northern hemisphere. (*a*) Month-by-month variation in the CO_2 concentration of tradewind air at Mauna Loa, Hawaii. This is a good representation of average CO_2 concentration in the atmosphere of the Northern Hemisphere and is the only complete record available. The upwardly sloping smoother curve is the 12-month moving average. (*b*) The month-by-month generation of industrial CO_2 in the Northern Hemisphere, along with the cumulative production above the preindustrial level of 293 ppm (both corrected for oceanic uptake of CO_2). The last 2 years are extrapolated values from the least square fit of data to a third-order polynomial. (*c*) The observed CO_2 variation corrected for net industrial additions. (*d*) Semiannual net ecosystem production and respiration, determined as the difference between the crests and troughs in Figure 2c. This analysis indicates that there has been no statistically significant change in Northern Hemispheric biotic metabolism over this 15-year period (see Hall et al., 1975, for details).

of our model would give believable predictions for primary productivity. No matter how elegant the model is or could be made to be, an appreciation of its limitations is an essential aspect of modeling.

ACKNOWLEDGMENTS

We thank Jeffrey Richey for critical comments on this manuscript. This paper is based in part on Wartenberg (1974), Hall (1975), and Hall et al. (1975).

REFERENCES

Aruga, Y. 1965. Ecological studies of photosynthesis and matter production of phytoplankton. I. Seasonal changes in photosynthesis of natural phytoplankton. *Bot. Mag. (Tokyo)* **78**: 280–288.

Bacastow, R., and C. D. Keeling. 1973. Atmospheric carbon dioxide and radiocarbon in the natural carbon cycle: II. Changes from A.D. 1700 to 2070 as deduced from a geochemical model. *Brookhaven Symp. Biol.* **24**: 86–135.

Baker, J. D. 1970. Models of oceanic circulation. *Sci Amer.* **222** (1): 114–121.

Bolin, B. and W. Bischof. 1970. Variations of the carbon dioxide content of the atmosphere in the Northern Hemisphere. *Tellus* **22**: 431–442.

Borgstrom, G. 1969. *Too many: A study of the earth's biological limitations.* MacMillan, London. 368 pp.

Botkin, D., J. Janak and J. R. Wallis. 1973. Estimating the effects of carbon fertilization on forest composition by ecosystem simulation. *Brookhaven Symp. Biol.* **24**: 328–344.

Broecker, W. S., Y. H. Li, and T. H. Peng. 1971. Carbon Dioxide—Man's unseen artifact. In D. W. Hood (Ed.) Impingement of man on the Oceans. Wiley-Interscience, New York. Pp. 287–327.

Bryson, R. 1974. A perspective on climatic change. *Science* **184**:.753.

Budyko, M. I. 1969. The effect of solar radiation variations on the climate of the earth. *Tellus* **21**(5): 612–619.

Ekdahl, C. A., and C. D. Keeling. 1973. Atmospheric carbon dioxide and radiocarbon in the natural carbon cycle: I. Quantitative deductions from the records at Mauna Loa Observatory and at the South Pole. *Brookhaven Symp. Biol.* **24**: 51–85.

Ellis, H. T., and R. F. Pueschel. 1971. Solar radiation: Absence of air pollution trends at Mauna Loa. *Science* **172**: 845–846.

Franklin, B. 1784. Meteorological imaginations and conjectures. In J. Bigelow (Ed.), *The complete works of Benjamin Franklin*, Putnam's, New York, Pp. 486–488.

Hall, C. A. S. 1974. The biosphere, the industriosphere and their interaction. *Bull. Atom. Sci.* **31**: 11–21.

Hall, C. A. S., C. A. Ekdahl, and D. E. Wartenberg. 1975. A fifteen-year record of biotic metabolism in the Northern Hemisphere. *Nature.* **255**(5504): 136–138.

Hasketh, J. D. 1963. Limitations to photosynthesis responsible for differences among species. *Crop Sci.* **3**: 493–496.

Hubbert, M. K. 1969. Energy Resources. In *Resources and man*. W. H. Freeman, San Francisco. Pp. 157–242.

Hutchinson, G. E. 1954. The biochemistry of the terrestrial atmosphere. In G. P. Kuiper (Ed.), *The earth as a planet*. University of Chicago Press, Chicago.

Jonsson, G., and R. Sonberg. 1972. Has acidification by atmospheric pollution caused a growth reduction in Swedish Forests? *Res. Notes # 20*, Inst. Skogsproduktion, Stockholm. 48 pp.

Kellogg, W. W., and S. H. Schneider. 1974. Climate stabilization: for better or worse? *Science* **186** (4170): 1163–1172.

Lemon, E. (in preparation). The land's response to more carbon dioxide and diffuse light.

Le Roy Ladurie, E. 1971. *Times of feast, times of famine: A history of climate since the year 1000*. Doubleday, New York. 426 pp.

Lieth, H. 1973. Primary production: terrestrial ecosystems. *J. Hum. Ecol.* **1**: 303–332.

Likens, G., and H. Bormann. 1972. Acid rain. *Environment* **14**(2): 33–40.

Likens, G., and H. Bormann. 1973. Acid rain: A serious regional environmental problem. *Science* **184**: 1176–1179.

Lister, R. 1974. Effects of atmospheric carbon dioxide and particulates on plant photosynthesis. PhD. dissertation, Cornell University.

Machta, L. 1973. Prediction of CO_2 in the atmosphere. *Brookhaven Symp. Biol.* **24**: 21–31.

Manabe, S., and R. T. Wetherald. 1967. Thermal equilibrium of the atmosphere with a given distribution of relative humidity. *J. Atmos. Sci.* **24**: 241–259.

Mitchell, J. M., Jr. 1970. A preliminary evaluation of atmospheric pollution as a cause of the global temperature fluctuation of the past century. In S. F. Singer (Ed.), *Global effects of environment pollution*. Reidel, Dordrecht. Pp. 139–155.

Mitchell, J. M., Jr. 1971. The effect of atmospheric aerosols on climate with special reference to temperature near the earth's surface. *J. Appl. Meteorol:* **10**: 703–714.

Overrein, L. N. 1972. Sulphur pollution patterns observed; leaching of calcium in forest soil determined. *Ambio* **1**(4): 145–147.

Revelle, R. and H. Suess. 1957. Carbon dioxide exchange between atmosphere and ocean and the question of an increase of atmospheric CO_2 during the past decades. *Tellus* **9**(1): 18.

Wartenberg, D. E. 1974. A preliminary computer model of effects of industrialization on the ecosphere. Undergraduate honor's thesis, Cornell University. Unpublished. 89 pp.

Weare, B. C., R. L. Temkin, and F. M. Snell. 1974. Aerosol and climate: some further consideration. *Science* **186**: 827–828.

Whittaker, R. H., and G. Likens. 1973. Carbon in the Biota. In G. M. Woodwell and E. Pecan (Eds.), *Carbon and the Biosphere. Brookhaven Symp. in Biology* 24. U.S.A.E.C. Springfield, Va. Pp. 281–300.

Whittaker, R. H., F. H. Bormann, G. E. Likens, and T. G. Siccama. 1974. The Hubbard Brook ecosystem study: Forest biomass and productivity. *Ecol. Mono.* **44**: 233–252.

Wilson, D. and J. P. Cooper. 1969. Effect of light intensity and CO_2 on apparent photosynthesis and its relationship with leaf anatomy in genotypes of *Lolium perenne* L. *New Phytol.* **68**: 627–644.

Winkless, N., and I. Browning. 1975. *Climate and the affairs of man*. Harpers Magazine Press, New York. 228 pp.

Woodwell, G. M., R. A. Houghton, and N. R. Tempel. 1973. Atmospheric CO_2 at Brookhaven, Long Island, New York. Patterns of variation up to 125 meters. *J. Geophys. Res.* **18**: 932–940.

Zelitch. I., 1967. Water and CO_2 transport in the photosynthetic process. In A. San Pietro, F. A. Green, and T. J. Army (Eds.), *Photosynthesis in plant life*. Academic, New York. Pp. 231–248.

16
Modeling Man and Nature in Southern Louisiana

JOHN W. DAY, JR.
CHARLES S. HOPKINSON, JR.
HAROLD C. LOESCH

This is the second chapter in this volume dealing with coastal Louisiana. In Chapter 10 we describe the present state and ecological history of the natural system and present a model of the present biological system. We now make a logical extension and utilize two models to consider the effects of man's activities on the system. The first model deals with long-term impacts of the present flood control system in the lower Mississippi Valley, and the second is based on plans advocated by some to convert marshland to grazing land for cattle.

THE RIVER MODEL

Over the past several thousand years the Mississippi River has formed a broad deltaic plain consisting of several million hectares (see frontpiece). A system of numerous distributaries allowed spring flood water and its sediments to be distributed widely over this area. An important point is that the major flow of the river was in different distributaries at different times, so that at a given time land was built rapidly in one part of the delta while being eroded elsewhere.

These natural processes have been altered considerably. The river is now leveed (diked) and for the most part held in a single course. Of the total Mississippi River water entering Louisiana, about 70% flows in the main channel and discharges at the present bird-foot delta. The other 30% flows through the Atchafalaya River, a distributary of the Mississippi (see Figure 1 in Chapter 10). The Atchafalaya, with its steeper gradient, was in the process of capturing the major flow of the Mississippi when the Corps of Engineers built structures that presently limit the amount of water that can now flow down the Atchafalaya to 30% of the total.

The Model

We constructed a mathematical model of four estuaries based on the existing estuarine system (Figure 1) and had the river flow successively into each. In doing this we are simulating mathematically the natural flow of the river in different channels at different times. In each estuary or hydrologic unit land is built when the river is introducing sediment. In the natural state a major distributary receives about 80% of the total discharge with the remainder divided more or less evenly between the minor distributaries (Gagliano et el., 1970; 1973). Land is eroded at a rate dependent on the total area, the magnitude of erosive forces (wave action and currents), which we calculated from data on the present Mississippi River, and whether there is

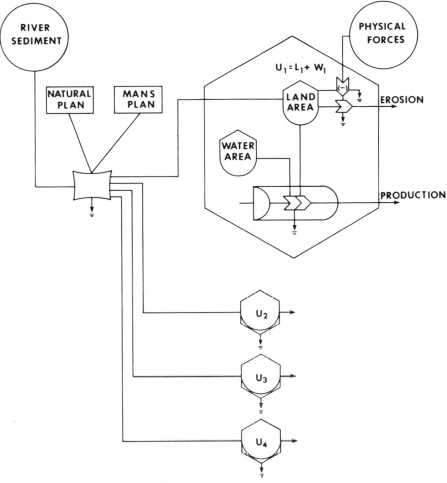

$$U_1 = L_1 + W_1$$

Figure 1 Conceptual diagram of the river simulation model. U_2, U_3, and U_4 have same form as U_1. See text for further explanation.

a vegetation cover (Table 1). A maximum total area (U) was set for each hydrologic unit, and this area divided into land (L) and water (W) portions representing an average estuary. Thus as land is built water area decreases, and as land is eroded water area increases. Productivity of land and water areas were calculated by multiplying the respective areas by measured gross production rates (8.1 kg m^{-2} yr^{-1} and 1.29 kg m^{-2} yr^{-1}, respectively). These were summed to obtain total production for an estuary, and the productivity values of all estuaries were summed for total production of the delta.

Table 1 Parameters of the River Model.

Symbol	Value	Description	Reference
R	47.9 km^2 yr^{-1}	Potential rate of land building	Shlemon, 1973; Gagliano et al., 1970; 1973
k_1	0–1.00	% of basin sediment going into a coastal unit	
P_f	0.01 cal sec^{-1}	Physical wave energy along Louisiana coast	Wright and Coleman, 1973
$k_2 LP_f$	23.5 km^2 yr^{-1}[a]	Rate of erosion of unvegetated mud flat	Gagliano et al., 1973
k_2	0.2897 sec (cal)$^{-1}$ yr^{-1}	Constant	
$k_3 LB$	18.8 km^2 yr^{-1}[a]	80% reduction in erosion rate by vegetation cover	Gagliano et al., 1973
k_3	0.0429 m^2 kg^{-1} yr^{-1}	Constant	
$k_2 LP - k_3 LB$	4.7 km^2 yr^{-1}	Net rate of erosion	
B	0.5 kg m^{-2}	Standing biomass of marsh grass	Day et al., 1973
U	5400 km^2	Total area of each estuarine unit	
L	878 km^2	Initial land area in each unit	

[a] Calculated on basis of land area of 878 km^2.

A series of four differential equations describing changes in land area of each estuary formed the core of the model. Using the terms defined in Table 1, each equation had the form:

$$\frac{dL}{dt} = k_1 R - (k_2 LP_f - k_3 LB)$$

An iteration time of 1 year was used. The program was written in the IBM CSMP language and run on an IBM 360 computer.

The model was simulated for three different conditions:

1 The natural case. Without the influence of man the river would change channels about every 500 years (Gagliano et al., 1970). Thus in the model

any single estuary received a major part of the river flow (80%) for 500 years and a minor flow (5%) for 1500 years.

2 A case where one channel receives 30%, one receives 5%, and the others receive no flow. We believe this simulates the present condition. Thirty percent of the present flow is in the Atchafalaya River, where a new delta is presently being built (Shlemon, 1973). The remaining 70% is in the main channel, but because the river mouth is at the edge of the continental shelf and because the river is leveed to the mouth most sediment is deposited in deep water. The only effective land building in this case is in small subdeltas. We estimate (after Gagliano et al., 1973) that only about 5% of the land-building capacity is being used for this major portion of river discharge. The contrast here is between the Atchafalaya building a delta in shallow water where most sediment is effectively used and the Mississippi building a delta in deep water where most sediment is lost.

3 A case where there is no vegetational erosion control. This is an important question today because large areas of vegetation are being destroyed by dredging activities associated primarily with petroleum extraction and navigation.

The steady-state results for each of the three cases are presented in Table 2 and Figure 2. These results indicate that productivity is increased by channel switching and that the long-term effects of present management practices in the lower Mississippi River system would be to decrease substantially gross primary productivity. The present condition indicates a future decrease of about 40% in total productivity over the next 100 to 300 years, in large part due to the loss of land building sediments to the deep Gulf.

Table 2 Steady-State Total Productivity and Land Area of Four Hypothetical Estuaries under Different Conditions.

Condition[a]	Total Production (mt/yr)	Total Land (km^2)
1 Normal: 80, 5, 5, 5%	79.0×10^6	7071
2 Present condition: 30, 5, 0, 0%	48.8×10^6	2600
3 No erosion control	43.8×10^6	3431

[a] Numbers in this column are percentages of river flow into each estuarine unit.

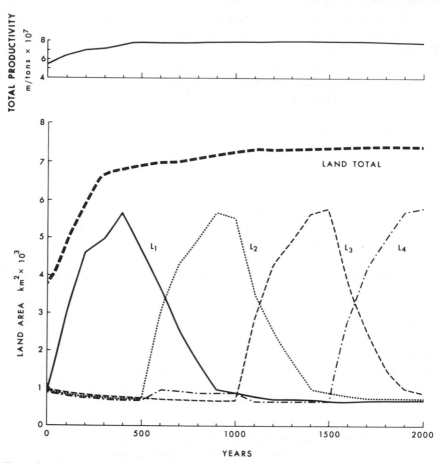

Figure 2   Simulation results of case 1 of the river model showing changes in land area (L_i) of individual hydrologic units, total land area, and total productivity of the coastal zone. First 300 years is start-up period.

A MODEL OF TWO ALTERNATIVES FOR FOOD PRODUCTION FROM COASTAL MARSHES

There is presently a proposal to drain approximately 45,500 hectares of brackish marsh in southwest Louisiana for cattle grazing. This project would likely result in a lowered productivity of marsh grasses due to restricted tidal flow and dewatering; this would in turn lead to lowered fishery production due to lessened detritus production and habitat loss. In this model we consider the effects of the two different strategies for the use of wetlands; natural marsh used to support fisheries versus drained marsh used to support cattle production. No simulation was done for this model, since the construction of a flow diagram and simple calculations were sufficient to show clearly the differences between the two management strategies (Figure 3). In this discussion we consider the two extremes, use of the marsh for either fisheries or beef production. Calculated costs, benefits, and cost:benefit ratios for the two strategies depict the economic value of each (Table 3).

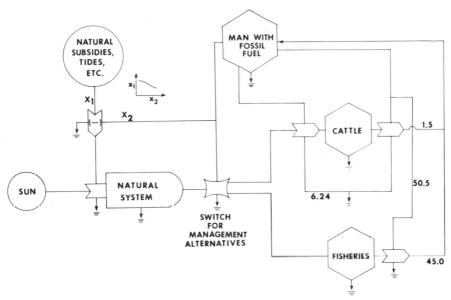

Figure 3 Diagrammatic model showing two alternative management schemes for food production. Flows are in kcal m^{-2} yr^{-1}.

These two alternatives are presented diagrammatically in Figure 3. The flow of biotic energy is channeled by a switch controlled by man into two possible circuits, development or nondevelopment. In the natural state, natural energy subsidies such as tidal flow support a system that yields the nation's most productive fishery. When man switches this flow to support cattle (by the construction of levees) these natural energy subsidies are diminished. This is shown as a negative workgate: when flow x_2 increases, flow x_1 decreases. There is only one fossil-fuel based workgate involved in fisheries; this is the energy used for harvest. Cattle production involves two workgates. The first involves the construction of the project and supplementary feeding, activities that are industrial-energy intensive. The second is energy used for harvest as in the fisheries.

Table 3 Benefit: Cost Analysis of the Use of Coastal Marshland Either for Fisheries or Cattle Production.

	Yield (kcal/m²/yr)	Cost					
			Cost for Harvest (kcal/m²/yr)	Benefit to Cost	Plus Lost Benefit of Alternative (kcal/m²/yr)	Benefit to Cost	Harvest (cost in kcal fuel to produce 1 kcal food)
Fisheries (present yield)	Menhaden 38.5		7.3				0.19
	Shrimp 2.7		39.9				14.8
	Other 3.6		1.8				
	Total 45.0[a]	Total	49.0[c]	1:1.08	49.0 + 1.5 = 50.5	1:1.12	
	(358 lb/acre)						
Cattle (proposed yield)	1.5[b]	Project cost	0.24				
		Feeding	6.0				
	(11.8 lb/acre)	Total	6.24[d]	1:4.2	6.24 + 45.0 = 51.24	1:34.2	4.6

[a] Fisheries yields: Lindall et al. (1972) reported that this area yields 358.1 lb/acre/year of fresh landed fishery products (40 g wet wt/m²). Assuming 75% water = 10 g dry wt/m²/yr, 10 g dry wt/m²/yr × 4.5 kcal/g = 45 kcal/m²/yr.

[b] Cattle yield: Stocking rate of cattle = 4 to 12 acres/cow (U.S. Soil Conservation Service, 1973). Assuming 8 acres/cow (0.125 cows/acre) and one 600-lb calf/cow/year = a production rate of 75 lbs cattle/acre/yr. (0.84 g wet wt/m²/yr). Assuming 60% water = 0.33 g dry wt/m²/yr, 0.33 g dry wt × 4.5 kcal/g = 1.51 kcal/m²/yr.

[c] Fisheries cost: 1973 menhaden catch = 64 metric tons/boat/day. Boats operating 5 to 7 hr/day (average 6) use 5.5 to 6.0 gal diesel fuel/100 hp-hr. Catch rate is 64 metric tons/day at 6 hr/day = 10.7 metric tons/hr. Boat engines average 1260 hp and, therefore, burn (5.5 gals/100 hp-hr × 1260 hp) = 69 gal/hr rounded to 70. Diesel fuel contains about 34,800 kcal/gal: therefore, 243,600 kcal fuel to catch 10.7 m tons fish; assuming 75% water and 4.5 kcal/g = 243,600 kcal fuel for 12,375,000 kcal fish or 0.19 cal fuel for 1 cal menhaden. Menhaden represent 85.4% of the 358-lb/acre catch (Lindall et al., 1972) or 38.5 kcal/m²/yr. 38.5 kcal/m²/yr × 0.19 indicates that a normal offshore shrimp boat will burn 240 gal fuel and catch 1100 lb of shrimp in 24 hr, or 14.8 kcal fuel for 1 kcal shrimp. Shrimp represents 6% of the 358 lb/acre catch or 2.7 kcal/m²/yr × 14.8 = 39.9 kcal/m²/yr (fuel used). Other fishery species represent 8% of the catch. These are primarily fin fish, and we assume a value of 0.5 kcal fuel/kcal fish. Rate of catch = 0.08 × 45 = 3.6 kcal/m²/yr. 3.6 kcal/m²/yr⁻¹ × 0.5 = 1.8 kcal/m²/yr (fuel used). Data are personal communication from NMFS Statistical office in New Orleans and menhaden industry personnel.

[d] Cattle cost: Cost of project estimated at $20 × 10⁶ (from Corps of Engineers and Soil Conservation Service) at 17000 kcal/$ (Odum, 1971 and personal communication) = 340 × 10⁹ kcal over 30-year life = 11.3 × 10⁹ kcal/year. For an area of 45,500 hectare (45.5 × 10⁹ m²) = 0.24 kcal/m²/yr. We neglect maintenance costs here. Cost of feeding: we assume 40% supplemental feeding. 40% of 1.5 = 0.5 kcal/m²/yr. Assuming that 10% of food eaten produces weight gain and 90% is used for maintenance, it would take 6 kcal of food to produce 0.6 kcal of cattle. From Steinhart and Steinhart (1974) we estimate it takes 1 kcal of fossil-fuel energy to produce 1 kcal of food for cattle. Hence the cost of supplemental feeding is 6 kcal/m²/yr.

389

Data on costs and yields of the two alternatives are given in Table 3. Direct costs and yields provide benefit:cost ratios of 1:1.08 and 1:4.2 for fisheries and cattle production, respectively. However, we must include indirect costs also. One indirect cost is the lost benefit of the alternative. For example, if the marsh is drained for cattle, part of the cost is the loss of 45 kcal m^{-2} yr^{-1} of fisheries harvest (Table 3) that would have resulted had the marsh not been drained. When these costs are considered the benefit:cost ratios are strikingly different. About 1.1 cal of fuel energy are required by man to harvest 1 cal of fisheries products, whereas 34.2 cal are required to produce 1 cal of beef.

The costs per unit energy are greatly different for menhaden and shrimp (Table 3). About 0.19 cal of fuel are used to harvest 1 cal of menhaden, while 14.8 cal are consumed to harvest 1 cal of shrimp. This, of course, is reflected in the much higher market price of shrimp. At present menhaden is used primarily as animal feed. We predict that with rising prices of energy and protein, menhaden products will be used directly as human food.

These data support the notion that coastal wetlands can provide inexpensive protein in their natural state. It suggests that cattle production is not well suited for these areas and that in the future alternative, less-energy-intensive methods of shrimp fishing may have to be developed.

Conclusions

What then have we learned from these two models? Both models demonstrate the importance of natural energy subsidies, especially for coastal systems. The natural subsidies emphasized here are the work of the river and the tide. The models also demonstrate what can happen when man diminishes these subsidies. Construction of levees prevented land building by the river and tidal flushing of the marsh, resulting in land loss, lower biotic productivity, and diminished fish harvests. We have attempted to quantify the subsidies and costs mentioned above and believe that this is an important step both in understanding the natural system and in making intelligent natural resource decisions.

These two management models represent our earliest "man-nature" efforts and are in some senses simplistic. More recently we have developed more sophisticated management models based in part on the experience we gained with the earlier models. These later models address such diverse topics as the effects of cultural eutrophication on the oyster fisheries of Barataria Bay, the energy cost of offshore oil production, and an energy analysis of the city of New Orleans. We have found our modeling efforts important in deciding court cases involving the disposition of hundreds of millions of dollars and in protecting and managing the vast wetlands of Southern Louisiana that is our home and that we love.

REFERENCES

Brannon, J. 1973. Seasonal variation of nutrients and physiochemical properties in the salt marsh soils of Barataria Bay, Louisiana. M. S. thesis. Louisiana State University, Baton Rouge, Louisiana.

Day, J. W., Jr., W. G. Smith, P. R. Wagner, and W. C. Stowe. 1973. *Community structure and carbon budget of a salt marsh and shallow bay estuarine system in Louisiana.* Center for Wetland Resources, Louisiana State University, Baton Rouge, Louisiana.

Frazier, D. E. 1967. Recent deltaic deposits of the Mississippi: their development and chronology. *Trans. Gulf Coast Assoc. Geol. Soc.*, **17**: 287–315.

Gagliano, S. M., R. Muller, P. Light, and M. Al-Awady. 1970. Water balance in Louisiana estuaries. *Hydrologic and Geologic Studies of Coastal Louisiana.* Coastal Studies Institute and Department of Marine Science, Louisiana State University, Baton Rouge. Part I, Vol. III, 98 pp.

Gagliano, S., and J. Van Beek. 1970. *Geologic and geomorphic aspects of deltaic processes, Mississippi delta system.* Center for Wetland Resources, Louisiana State University. 140 pp.

Gagliano, S., P. Light, and R. Becker. 1973. Controlled diversions in the Mississippi delta system. Rept. 8. *Hydrologic and Geologic Studies of Coastal Louisiana.* Center for Wetland Resources, Louisiana State University. 146 pp.

Golley, F., H. T. Odum, and R. F. Wilson. 1962. The structure and metabolism of a Puerto Rican red mangrove forest in May. *Ecology*, **43** (1):9–19.

Henderson, G. S. 1974. An ecosystem approach to characterization of the nitrogen cycle in a deciduous forest watershed in *Proceedings of Fourth North American Forest Soils Conference.* Quebec, August 1974. In press.

Ho, C. 1971. Seasonal changes in sediment and water chemistry in Barataria Bay. *Coastal Studies Bull.* **6**: 67–85.

Kirby, C. J. 1972. The annual net primary production and decomposition of the salt marsh grass *Spartina alterniflora* Loisel in the Barataria Bay estuary of Louisiana. Unpublished. Ph.D. dissertation, Louisiana State Univ., Baton Rouge, Louisiana.

Lindall, W. N., J. Hall, J. E. Sykes, and E. L. Arnold. 1972. Louisiana Coastal Zone: Analysis of resources and resource development needs in connection with estuarine ecology. Sects. 10 & 13, Fishery Resources and Their Needs, *Rept. Commer. Fish. Work Unit, Nat'l Marine Fisheries Serv. Biol. Lab.*, St. Petersburg Beach, Fla.

Nixon, S. and C. Oviatt. 1974. Ecology of a New England salt marsh. *Ecol. Monogr.* **43**(4): 463–498.

Odum, E. P., and A. A. de la Cruz. 1967. Particulate organic detritus in a Georgia salt marsh-estuarine ecosystem. In G. H. Lauff (Ed.), Estuaries, AAAS Publ. No. 83. 383–388.

Odum, H. T. 1971. *Environment, power, and society.* Wiley-Interscience, New York.

Patrick, W. H. 1974. Personal communication. Dept. of Agronomy. Louisiana State University, Baton Rouge.

Patten, B. C. (editor), 1971. *Systems analysis and simulation in ecology.* Vol. I. Academic, New York. 607 pp.

Pope, R. M., Jr. 1973. *Selected aspects on recreational use and unique environs of Coastal Louisiana.* LSU–SG–73–03, Report 3, Louisiana Superport Studies (with J. H. Stone and J. M. Robbins).

Shlemon, R. 1973. Development of the Atchafalaya Delta. Rept. 13. *Hydrologic and Geologic Studies of Coastal Louisiana.* Center for Wetland Resources, Louisiana State University, Baton Rouge. 51 pp.

Smalley, A. E. 1958. The role of two invertebrate populations, *Littorina irrorata* and *Orchelium fifficinium* in the energy flow of a salt marsh ecosystem. Ph.D. dissertation, University of Georgia. 126 pp.

Steinhart, J., and C. Steinhart. 1974. Energy use in the U.S. food system. *Science* **184**: 307–316.

Teal, J. M. 1962. Energy flow in the salt marsh ecosystem of Georgia. *Ecology* **43**(4): 614–624.

Thomas, J. P. 1971. Release of dissolved organic matter from natural populations of marine phytoplankton. *Mar. Biol.* **11**: 311–323.

Turner, E. 1974. Community plankton respiration in salt marsh, tidal creek, estuary, and coastal waters. Ph.D. dissertation, University of Georgia, Athens.

Tusneem, M. E., and W. H. Patrick, Jr. 1971. *Nitrogen transformation in waterlogged soil.* Louisiana State University, Dept. of Agronomy, Agricultural Exper. Sta. Bulletin No. 651. 75 pp.

U.S. Soil and Conservation Service. 1973. Environmental Impact Statement for the Cameron Creole Watershed project.

Vinogradov, A. P. 1953. *The elementary composition of marine organisms.* Yale University Press, New Haven. 647 pp.

Wagner, P. R. 1973. Seasonal biomass abundance, and distribution of estuarine dependent fishes in the Caminada Bay system of Louisiana. Ph.D. dissertation, Louisiana State University, Baton Rouge.

Wright, L., and J. Coleman. 1973. Variations in morphology of major river deltas as functions of ocean wave and river discharge regimes. AAPG 57(2): 370–398.

17
War, Peace, and the Computer: Simulation of Disordering and Ordering Energies in South Vietnam

MARK BROWN

reface. This work was undertaken and finished during the height of military conflict in South Vietnam, before withdrawal of U.S. military personnel. The consequences of the United States intervention and subsequent withdrawal are now quite apparent and agree in general with our computer projections. Our original contract did not include provisions for a change in government.

The ending of the "Vietnam War" marked the beginning of the end for "democracy" in South Vietnam. For as the simulation results in this investigation show, the 6-year war from 1965 to 1970 left the country of South Vietnam with little infrastructure to continue prolonged conflict with their neighbors to the north. Without massive aid programs of both military and civilian aid (to increase the internal productivity of the country) there was little doubt that the country would fall. Of course, the external subsidy to the "winning" side, not simulated here, was a very important factor in the ultimate disposition of the war.

The simulation results indicated that even with a twofold increase in aid over those amounts in 1971, it would require as long as 10 years for South Vietnam to recover to the levels of productivity experienced before the "6-year war." As a result the country was particularly vulnerable to continued conflict. With these things in mind it was of little surprise that the government of South Vietnam surrendered to that of North Vietnam on May 7 of this year. Perhaps this simulation will be of some use to the new political structure responsible for the welfare of the people of South Vietnam.

December 10, 1975   MARK BROWN

Any system that maintains continuous life must develop a balance between ordering and disordering processes. For example, ecosystems are dependent upon photosynthesis to build structure and to provide a base for the food chain. But ecosystems also require the catabolic processes of death and bacterial decay to make room for new organisms and to release nutrients needed for new plant growth from dead organisms. If one or the other of these two processes were to dominate for a long period of time, the system eventually would no longer be able to function properly. An important component of this anabolic-catabolic relationship is the ability of a system to rebuild structure after some disruption. For example, when a large tree topples in a forest because of a windstorm the successional communities are able to use the nutrients from the decomposition of that and other trees. New biotic

Studies stimulated by a contract between the National Academy of Sciences and the Department of Environmental Engineering Sciences, University of Florida, Gainesville, for "Models of Herbicide, Mangroves, and War in Vietnam."

structures soon fill the clearing aided by the increased sunlight energy that reaches the forest floor and eventually produce another large tree to replace the one that toppled.

Social systems also are dependent upon both the proper balance between constructive and destructive forces, and the ability to repair damaged structure. The crowded, complex systems of today's civilization continuously require large flows of energy from industrial and biotic sources to repair and replace structures damaged by entropic degradation. If these ordering processes are disrupted by war or other large disorders, the physical structure of society will suffer. On the other hand, if ordering energies are available to rebuild social structure, as was the case with the Marshall plan following World War II, the social effects of war may be mitigated to some extent.

This chapter investigates with the aid of an analog computer the country of South Vietnam, during the period 1960 through 1970, from the point of view of the ordering energies of biotic and industrial processes and the disordering energies of war. It includes calculations of the disordering effects of the extensive United States' bombing, herbicide spraying, and Rome plowing on ecosystems and cities. Quantitative calculations are made on the changes in land quality following these disruptions and the ways in which these changes effect the economy and population distribution patterns of the people of South Vietnam. Finally, we investigate what some of the ramifications of different levels of U.S. foreign aid, and hence new ordering energies, might mean to the reconstruction of Vietnam.

A MODEL OF VIETNAM

Figure 1 is a diagram of some major energy processes of the country of South Vietnam at war. It shows the major compartments and flows of energy and materials throughout the country. The circles to the left represent forcing functions, in this case the *ordering energies* available to South Vietnam. These are the materials and energies that are used in the normal processes of the country for construction and maintenance of structure (e.g., biomass, buildings, agricultural systems, etc.). S_1 is all incoming goods, both United States' material aid and imported goods from the World Market. S_7 is all money sent to South Vietnam in the form of aid, and S_6 is all the natural energies (sun, tides, wind, rain, etc.) that help maintain the natural structure of the country. The circles to the right represent the *disordering energies* that are forcing functions for the country at war, that is, the materials and energies that are used for war and thus increase the processes of disruption and entropy. S_2 is all the energies available to the Viet Cong and North Vietnamese military structure to make war. S_3, S_4, and S_5 are the three major disordering energies of the U.S. and South Vietnam military structure.

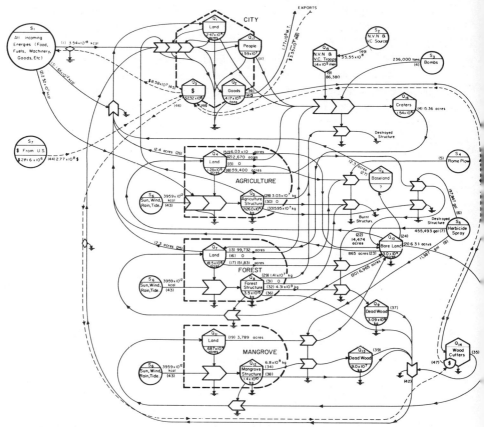

Figure 1 Major energy processes of South Vietnam while at war. Numbers in parentheses (30) indicate reference to a calculation of a rate constant explained in the Appendix. Subscripted letters (Q_4, S_2) indicate reference to calculation of storages and sources in Appendix. Dashed lines (----) are flows of money. Solid lines (——) indicate flows of materials and energies. Details are available from author, or see "The effects of herbicides in South Vietnam," Part B: working papers.

The land categories (city, agriculture, forest, and mangrove) each are separated into their respective components (or storages). The city system has within it land, people, goods, and a storage of money. The natural systems have two components each: land area and biotic structure (i.e., standing crop of biotic structure). The land storage and the forcing functions interact to produce social structure.

The components to the right of the natural systems, those of craters, base land, bare land, and dead wood, are components that are considered storages of disordered land and structure as it is transferred from one land use category to another by the impact of the disordering energies of war.

In Figure 1 the rates of material and energy flows, and the quantities stored in each of the components of the country have been calculated for the year 1965, the beginning of the escalation of the United States involvement in the Indochina War. (Calculations and sources for the calculations are summarized in Appendix I, and detailed calculations may be obtained from the author; see also "The Effects of Herbicides in South Vietnam," Part B: working papers, National Academy of Sciences, February 1974.) Evaluation of the flows of the different disordering energies gives perspective to their effects on the many processes of the country; in all cases herbicides account for the greatest disruption.

Another way to understand the effects that the disordering energies have had on the processes of the country is to calculate the percentage of the ordering energy budget of South Vietnam disrupted by these war activities. Column 1 in Table 1 shows the cumulative energy budget (i.e., the ordering energies) for the 6-year period, 1965 to 1970. Columns four, five, and six show our estimates of the total amounts of disordered energy introduced to the country by each of the disordering operations of bombing, herbicides, and Rome plowing. Column seven shows the percentage of the total ordering energy budget that was disrupted for each of the various subsystems of the country.

During the 6-year period of major United States' involvement in the Indochina War (1965 to 1970), 10.7% of the total natural and man-derived energy budget of South Vietnam was disrupted directly by war. This represents a relatively small fraction of the country's man-derived and natural energy budget, although it does not account for secondary interactions, feedback operations, and time delays that may cause further disruption of normal processes and may increase the overall effects of the war.

ORDER AND DISORDER IN VIETNAM

It is well known that many ecological systems are stress adapted and to a certain extent even depend upon some form of stress to accelerate regenerative processes and recycle nutrients. Just as fire-climax forests and prairies are stress adapted and depend on pulses of fire to recycle nutrients (Chapter 23), the country of South Vietnam possibly could be considered a stress-adapted system. Conflict and large-scale warfare has been the rule rather than the exception since the Geneva Agreements of 1954 and even before with a long and bitter conflict between the Communist-led Vietminh and the French armed forces. In some systems frequent chronic stress, such as grazing in grasslands, results in greater overall production by reducing diversity and releasing disordered materials for reconstruction and repair. Could the same apply to a chronic 30-year war and the country of South Vietnam? Or is the

Table 1 Comparison of Disordered Energy.

	Cumulative Energy Budget 1965–1970[g]	Fossil Fuel Equivalent[h]	Fossil Fuel Work Equivalent	Disordered Energy in F.F.W.E. Bombs[i]	Disordered Energy in F.F.W.E. Herbicides[j]	Disordered Energy in F.F.W.E. Rome Plow[j]	% of Energy Budget in F.F.W.E. Disordered
Human settlements[a]	116.4	1	116.4	9.10[k]	4.9[k]	—	12.1%
Agriculture systems[b]	672.0	0.005	3.4	0.12	0.04	0.01	5.0%
Forest systems[c]	6660.0	0.005	33.3	1.52	0.89	0.20	7.8%
Mangrove systems[d]	366.0	0.005	1.8	0.06	0.68	—	40.4%
Estuarine systems[e]	174.0	0.005	0.9	—	0.05	—	5.7%
Other natural[f] energies (wind, tide, thermal gradient)	15882.0	0.0005	7.9	?	?	—	?
Total	23870.4		163.7	10.80	6.56	0.21	10.7%

[a] The sum of purchased foods, foreign aid, and fuel. Purchased goods and foreign aid not including fuel were \$2.6 × 10[8] and \$5.1 × 10[8], respectively, multiplied by 1.4 × 10[4] kcal/dollar to convert to equivalent fossil fuel energies required to generate the same work (10.7 × 10[12] kcal). Add to this, fuel (8.66 × 10[5] metric tons)(10[6] g ton[−1])(10 kcal g[−1]) = 8.66 × 10[12]. 10.7 × 10[12] + 8.66 × 10[12] = 19.4 × 10[12] kcal.

[b] The chemical potential energy entering the system as agriculture production was estimated by multiplying the land area in agriculture (7.31 × 10[6] acres) by the estimated gross photosynthesis (1.6 × 10[5] kcal acre[−1] day[−1]) and then by 100 days. (Estimated time crops are in leaf each year, assumed two growing seasons of 50 days each because of monsoon climate.) (1 acre = 0.405 ha.)

[c] The gross photosynthesis of inland forests was estimated by multiplying the land area in forests (1.9 × 10[7] acres) by the estimated gross photosynthesis (1.6 × 10[5] kcal acre[−1] day[−1]) (Rodin, 1967) and then by 365 days.

[d] The gross photosynthesis of mangrove systems was estimated by multiplying the area (0.69 × 10[6] acres) by the estimated gross photosynthesis (2.4 × 10[5] kcal acre[−1] day[−1]) (Golley, 1962) and then by 365 days.

^{e} The gross photosynthesis of estuarine systems was estimated by multiplying the area (1.0×10^6 acres) by the estimated gross photosynthesis (8.0×10^4 kcal acre^{-1} day^{-1}) (Odum, 1973) and then by 365 days.

^{f} The sum of the natural potential energies: (rivers (644×10^{12} kcal/yr), tides (152×10^{12} kcal/yr), and rain as runoff (119×10^{12} kcal/yr) (Odum, 1973). See Chapter 21.

^{g} A cumulative energy budget was calculated as prewar values before disordering energies were introduced. It was calculated by multiplying the annual energy budget of each subsystem times the time span (6 years).

^{h} See Chapter 7 for a consideration of energy-quality factors used here. F.F.W.E. here = F.F.E. in Chapter 7.

^{i} Natural and man-derived energies disordered by bombs was calculated by multiplying the land area disordered each year from 1965 to 1970 by the estimated gross production and then by the number of years remaining in the 6-year period. Based on statements by Ewel (1970) it was assumed that recovery of tropical forest systems in 6 years was negligible. Gross production lost was then converted to fossil-fuel equivalents as in Chapter 7.

^{j} Natural energies disordered by herbicides was calculated by multiplying the land area disordered each year from 1965 to 1970 by the estimated gross production and then by the number of years remaining in the 6-year period.* For agriculture sprayed, I assumed only a 1-year loss. I assumed no recovery of forest systems, based on statements by Tschirley (1969), and, again based on statements by Tschirley (1969), no recovery of mangroves. Estuarine system disruption was estimated as that part of the estuary within herbicided area (33%). Gross production lost was then converted to Fossil Fuel Equivalents as in Chapter 7. The Rome plowing operation started in 1968. It involved five companies of 30 Catapillar D-8 bulldozers, each fitted with a special $2\frac{1}{2}$ ton blade made in Rome, Georgia (thus the name). Its function was to scrape clean areas of known or suspected enemy activity by felling and piling all trees and undergrowth.

^{k} City land area disordered by bombs and herbicides was 8.7 and 4.2%, respectively. It was assumed that an equal percentage of the total energy budget was disrupted.

* This area was measured on official Department of Defense herbicide spray-run maps, scale of 1:1,000,000; year by year totals for Mangrove land sprayed are as follows: 1965, 4700 acres; 1966, 96,330 acres; 1967, 197,600 acres; 1968, 68,666 acres; 1969, 90,155 acres; 1970, 11,609 acres.

effect of chronic disordering a seriously degraded economic and cultural system?

The 6-year involvement of the United States in the Indochina War resulted in an escalation of disordering energies that could be represented by an additional stress pulse of 10.7% over and above the disordering resulting from previous military engagements. Far more important than the magnitude of this overall stress are the percentages of the energy budgets that were disrupted from each of the subsystems of South Vietnam. Column 7 in Table 1 lists the percentages of each subsystem disordered. A comparison of these indicates that while the country had a 6-year stress pulse of 10.7%, individual stresses account for major local effects, and interactions of these components may magnify the overall stress.

For example, the mangrove systems were stressed nearly 41% (i.e., productivity was reduced by 41%): a stress that appears to have reduced the ability of the system to recover in a short period of time. Data gathered by NAS personnel while surveying ecological effects of herbicides in Vietnam indicate a lack of recovery, possibly due to loss of seed source. It has been suggested recently that mangrove systems are an important part of the food chain in estuarine systems. A stress of this magnitude could have severe effects on estuarine systems that, in turn, will affect the human settlements by a loss of food source. This may be particularly important since fish (which are in large part dependent upon mangrove-based food chains) are an important protein source in this protein-deficient land, where animal protein makes only 4 to 5% of the per capita calorie intake (Fuller, 1963).

The productivity of the forest systems of the country were reduced nearly 8%; a stress that probably will have little effect on the individual system, but if such factors as quantity of wood destroyed and regeneration time are taken into account, the human settlements depending on the forest systems as an auxiliary energy source may feel the stress far more than the forest itself. It has been estimated by Westing (1972) that 6.2 billion board feet of marketable lumber was destroyed by aerial bombardment and herbicide application from 1965 to 1970 in South Vietnam.

The human settlements of South Vietnam had 13% of their land area and an equal percentage of their fossil fuel-derived energy budget disordered during the 6-year period (1965 to 1970). These effects are increased by normal population growth, an increased movement to urban regions due to relocation of refugees, reduced agriculture production (5.2%), reduced estuarine production (9%), and reduced forest production (8%). Again the direct stress is magnified and will require many years to recover.

The systems diagram is another way of showing these same effects. Figure 2 is a diagram of the gross energy budget for the country of South Vietnam showing all the main energy flows, constructive and destructive, including those of nature and man-derived energies of the cities. Energy of low quality,

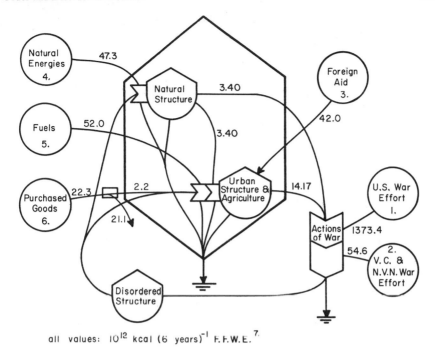

all values: 10^{12} kcal (6 years)$^{-1}$ F.F.W.E.[7]

Figure 2 Simplified energy budget of South Vietnam, 1965 to 1970. Numbered footnotes given below:

* 1   U.S. war effort was calculated by adding the incremental costs of the war in Vietnam for the years 1965 to 1970. The incremental costs are the costs that represent the "net difference between wartime and peacetime needs" of the U.S. Military (U.S. Congress. House Committee on Appropriations, Dept. of Defense Appropriations for 1970, pt. VII. Hearings, Washington, U.S. Government Printing Office, 1970. p. 395). The total war effort (981×10^8 dollars) was then multiplied by 1.4×10^4 kcal dollar^{-1} (the mean industrial energy use per dollar spent for the United States) to obtain the energy expenditure. This figure includes the U.S. industrial energy used to manufacture as well as ship war materials to South Vietnam.

2   Viet Cong and North Vietnamese war effort was estimated by assuming steady increase from $555 million (Thayer, 1969) to $765 million in 1970 (A. P., Gainesville Sun, Gainesville, Florida. April 2, 1972). The total war effort (3.9×10^9 dollars) was then multiplied by 1.4×10^4 kcal dollar^{-1} to obtain the energy expenditure.

3   U.S. Aid was calculated by adding the official aid for the years 1965 to 1970. (Annual Statistical Bulletin, No. 14.) The official aid (3.03×10^9 dollars) then was multiplied by 1.4×10^4 kcal dollar^{-1} to convert to potential energy entering the system.

4   Natural energies are all those energies entering the country from the chemical potential energies of gross photosynthesis of ecosystems and the chemical potential energies of rivers, tides, thermal heating, winds, and rains as runoff.

5   Fuel inputs are from "Vietnam Statistical Yearbook." For the 6-year period (5.2×10^6 metric tons), this was multiplied by 10^6 g metric ton and by 10 kcal g^{-1} to convert to calories of work.

6   Purchased goods were calculated by adding the import arrivals from Annual Statistical Bulletin No. 14 for the years 1965 to 1970. The import arrivals (1.59×10^9 U.S. dollars) were then multiplied by 1.4×10^4 kcal dollar^{-1} to convert dollars to the potential energy equivalent.

such as sunlight, is calculated as the chemical potential energy after trans-formation by photosynthesis and expressed in equivalent fossil-fuel work potential. High-quality energies such as urban technological materials and machines are expressed in equivalent fossil-fuel energy required to generate work necessary for their production (Chapter 7).

When many of the details of small component flows of energy through the country such as in Figure 1 are eliminated by a larger-scale view, certain patterns and some consequences of the war become more obvious. The magnitudes of ordering and disordering energies, the difference in the levels of the U.S. war effort and the Viet Cong/North Vietnam war effort, and the differences in aid as compared to imported energies begin to show the magni-tude of the disordering effect of war on both the biotic and economic struc-tures of South Vietnam.

During the 6-year period from 1965 to 1970 the ratio of destructive energies of war to the constructive energies of nature and the city was approximately 1:10. The impact of a pulse disorder-to-order ratio of 10% will release materials and increase repair mechanisms required for rejuvenation, which in the short run may increase productivity. However, a disordering pulse that alters the landscape by removing 4,722,800 acres from productive lands and reduces productivities in an additional 5,600,000 acres of lands had reduced drastically the stability of the landscape (Appendix I). Increased gross pro-duction in both the natural and urban systems may not produce a stronger or more stable system, but on the contrary the result could be a highly variable, subsidized economy with decreased stability. Certainly, the decreases in rural hamlet and village populations as the inhabitants left the war zones for the urban areas, the decreases in industrial and agricultural production, and increases in imports of consumer goods would reflect an economy having less occupational and industrial diversity—an economy that is highly unstable because of its need for increased flows of high-quality fossil fuels in a world of decreasing availability of these fuels.

While the total disordering energies were significant in themselves, it is interesting to note the differences in magnitude of the United States' war effort compared to the Viet Cong and North Vietnamese. The United States, fighting a war on unfamiliar terrain and using very sophisticated mechanisms, spent 25 times the energy in its war effort as the Viet Cong/North Vietnamese. The Viet Cong and North Vietnamese using the techniques of jungle warfare kept concentrations of troops and supplies at a minimum, thus making it nearly impossible for the concentrated war effort of the U.S. to overpower it. The end result was more a disordering of the natural environment than troops, with less damage to the Viet Cong and North Vietnamese war effort than some people expected.

But more important, the diagram shows where the most stress was inflicted by the actions of war. For example, the natural system, while having a large absolute loss in comparison to the urban systems, was stressed 7.2%. The urban system (i.e., the manbuilt-technological economy) had 12.2% of its energy budget disrupted. Other consequences of the war (secondary inter- actions) such as the shift in population from rural areas to the urban centers, the loss of population as casualties of war, and shifts in land use have magni- fied this stress.

SIMULATION MODEL OF VIETNAM

The war obviously has caused many human casualties and disrupted cities and valuable ecosystems. But the disordering of lands and the accompanying transfer from one land use category to another might or might not be a "bad consequence" of the war in the long run, depending on future aid patterns and one's personal point of view. For example, those who advocate increased agricultural production for the country view the change from forest to bare land as accomplishing the first step in the process of bringing more land into agricultural production. This seems to be the case as long as there are the necessary energies (such as U.S. Aid) available that can be added to the country's own reserves to complete this transfer.

The actions of war have disordering energies on not only the physical, natural, and man-made systems, but on the social and economic systems as well. As a consequence of the war there has been a massive switch from a rural to an urban population. This shift in population has caused the expansion of the urban centers and at the same time a decrease in the nation's industrial and especially agricultural productivity. This in turn has caused increased dependence on U.S. aid and purchased imports.

The effects that disordering energies have had on the country of South Vietnam can be assessed completely only through time. One aid toward this assessment is the use of the systems diagram translated into the language of the analog computer which can make projections into the future. Such a simulation can give new insight into the cumulative effects of all disordering energies. Figure 3 is a simplified version (for the purposes of simulation) of the changing land use model in Figure 1. This diagram shows the action of war accelerating the recycling of minerals and the potential reuse of dis- ordered lands and materials. These are all components that, when fed back with an accompanying energy input, are available for and stimulate re- construction. Evaluation of the rates gives a perspective of those changes that are important, and computer simulations show the cumulative effects on South Vietnam's energy budget as well as the costs of reordering.

Calculations and sources of information for the Rates of Flows and the Storages in Figure 3 are summarized in Appendix I. War Effort (W) and United States Aid (A) are forcing functions generated to depict the escalation,

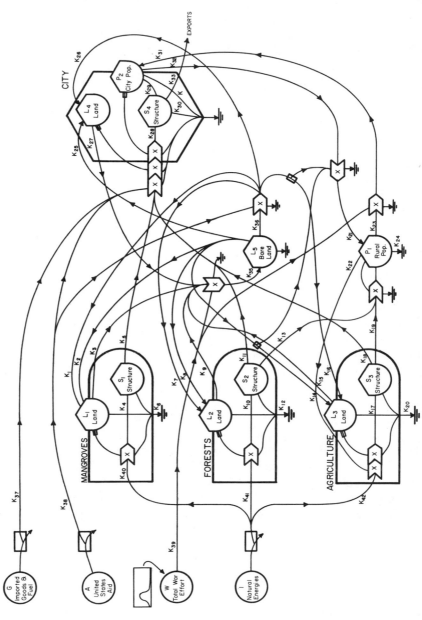

Figure 3 War-stress model, country of South Vietnam. See legend for Figure 1. Calculations are found in Appendix I

404

and later deescalation of U.S. involvement in Indochina. The initial conditions of the War Effort (W) were set at the conditions of 1950, the approximate level of conflict prior to U.S. involvement. In 1965 the level increases
reaching a maximum in 1967, then decreases to an estimated level of conflict
equal to that prior to 1965. The equations used in the model are given in
Appendix C.

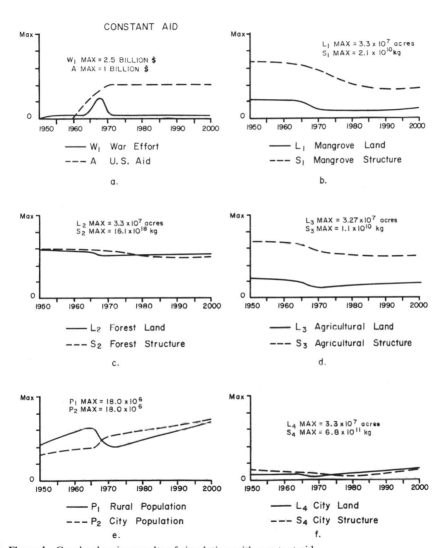

Figure 4   Graphs showing results of simulation with constant aid.

United States aid (A) is set at zero until 1960, then increases at a rate consistent with reported U.S. Aid. In the first simulation (Figure 4a to f) aid was held constant after 1970 at the reported level of aid in 1970. In the second simulation (Figure 5a to f) aid was increased after 1970 at the same rate to a maximum in 1985, then terminated. And in the third simulation (Figure 6a to f) aid was decreased steadily after 1970 terminating in 1980.

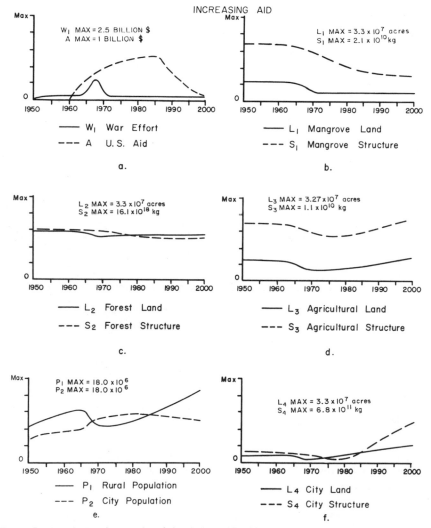

Figure 5   Graph showing results of simulation with aid increased.

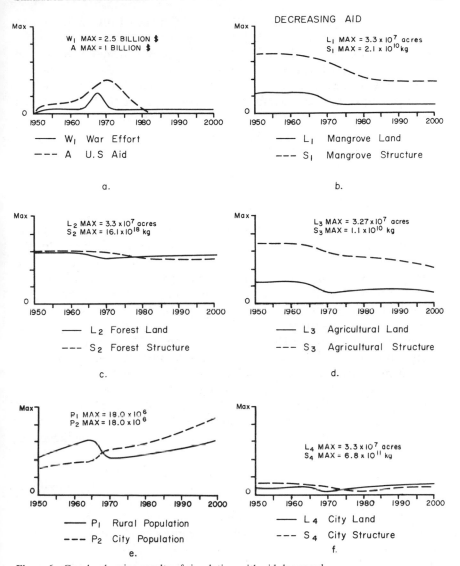

Figure 6 Graphs showing results of simulation with aid decreased.

Figures 4, 5, and 6 are computer-generated graphs from the first, second, and third simulations of the model. The graphs indicate a steady-state system from 1950 to 1965 (with slight decreases in certain storages, L_2, S_2, L_3, S_3) until escalation of the war in 1965, causing decreases in all compartments of the system except the human settlements.

The mangrove systems (Figure 4b) show the expected disruption anticipated by the calculations in Table 1. However, the structure component (S_1)
continues to decrease after the pulse disordering of war, due to increased
demand for wood as charcoal. There is no recovery in the simulation in
agreement with the report of NAS site visit personnel that recovery of de-
forested mangrove systems exhibited little regrowth due to the loss of seed
source. Provisions for an artificial seeding program are not included in this
model. In the second simulation the mangrove structure decreases due to
increased demands of the human settlements. In the third there is little change
as foreign aid supplies much of the nation's fuel requirements.

The forest systems (Figure 4c) of the country exhibit a slight decrease in
biomass in the first simulation due in part to the increased demands of the
growing populations in the human settlements. Recovery, as expected, does
occur, but the system does not reach the level of productivity exhibited at the
beginning of the simulation run. In the model the transfer coefficient for the
flow returning to forest land from bare land as a result of U.S. aid is extremely
small, since there are no active reforestation programs underway in Vietnam
to date. Normal recovery by succession is a long process with bare areas
invaded first by grasses and bamboo, retarding the overall recovery to
initial levels of productivity.

The agricultural systems (Figure 4d) show the greatest change due to the
disruption by war, the decreases in the rural population, and the increased
urban demand. In the first simulation, the structure or biomass does not
recover to the initial level after disruption. Increased aid in the second simulation (Figure 5d) provides the industrial energy and other resources necessary to increase agricultural production to a level slightly higher than the
initial, by transferring more land back into production, increasing the movement of refugees back to the rural areas, and by increasing fossil-fuel subsidy
to crops (i.e., fuel-intensive fertilizers, irrigation, pesticides, tractors, etc.). In
the third simulation (Figure 6d), by cutting off aid production remains lower
than the level obtained in simulation 1, as might be expected; since land
transfers are decreased, the rural populations remain in the cities and increased agriculture production is difficult because of a lack of farmers and a
sufficient fuel subsidy.

Consequences of the population shifts (Figure 4e) are felt throughout the
system and are very important in the overall economic well-being of the
country. In the first simulation the movement of the rural population to
the cities puts additional burdens on the entire system because of a decreased
"producer" population and an increased "consumer" population. Manipulation of this flow during sensitivity analysis caused extreme changes in most
compartments. If the rate of movement to the urban areas was decreased all

components exhibited faster recovery rates, and the loss of agricultural production was not as great. If the rate of movement was increased the opposite occurred. This "system sensitivity" to the movement of the populations begins to indicate the extent to which the country's economic stability is aggravated by the secondary effects of the refugee problem. In the second simulation (Figure 5e) movement back to the rural areas is facilitated by an increased flow of agricultural land back into production. In the third simulation (Figure 6e) diminishing aid retards the return of the rural population, although both components increase due to normal population growth.

The human settlements (Figure 4f) show smaller visible changes, but are interesting in view of the time delay involved in the buildup of city structure (S_4) following the increased urban expansion (L_4). By the year 1970, an increase in city land area is apparent. However, due to shortages of materials, energy, and industrial capacity, and pressure from an increased population, a time lag of 10 years is required before the materials and energies are available for growth of city structure. Recovery to the initial levels of urban structure does occur by the year 2000, but this does not meet the demands of the larger population. Increasing aid in the second simulation (Figure 5f) meets these demands easily. The increase in city structure in this case begins to show a runaway growth, possibly an undesirable consequence. In the third simulation (Figure 6f) decreased aid prolongs the time delay and reduces the final structure value.

OVERALL PERSPECTIVE

Consequences of the war in Vietnam were difficult to evaluate directly. The immediate effects of the disordering energies have caused an 11% disruption of the 6-year cumulative natural and industrial energy budget (1965 to 1970) of the country. The herbicide program accounted for nearly half this disruption: an interesting consequence, indicating the high-amplifier disordering value of the relatively low energy cost herbicides. Delays in ending the present level of conflict may cause delays and reductions in the recovery rate of order in the country. As long as needed materials and energies are diverted from the job of reconstruction to that of fighting brush wars with the Viet Cong and North Vietnamese, reordering of the country may be delayed. Economic stability in many systems is related to the diversity of normal occupations and components; and as suggested earlier, continued conflict may result in maintaining a low diversity and a potentially unstable economical position. The economy of South Vietnam has been seriously distorted by the burden of military spending, inflation, physical destruction, and population dislocation.

These conditions also have distorted South Vietnam's import/export balance. In order to curb the magnitude of runaway inflation caused by both the increased purchasing power of a population that is money subsidized and a shortage of available goods, the Saigon government has encouraged vast increases in imports, particularly consumer goods (Table 2).

Table 2   Balance of Trade[a] (in Millions of U.S. Dollars).

	1965	1967	1968	1969	1970
Exports (based on customs data)	25	16	12	12	12
Imports (based on licensing data)	660.4	581.5	628.8	740.1	641.0

[a] "Impact of the Vietnam War." 1971.

The economy can be brought back into balance with production more nearly equaling consumption only if energies are diverted from war and used as a stimulator (or pump) on reconstruction and reordering. These energies combined with energies and materials released by disordering could stimulate recovery of the disrupted zones of the overall system. As recovery takes place structure is built to obtain and use additional energies, and a diversity of structural components is integrated to more effectively "trap" and utilize (or maximize) these new energies.

On the other hand, with continued conflict the system may be developing special adaptations to war-stress energies so that they are not so destructive to the overall processes. Consider, for example, increases in black-market activities, prostitution, and the abandonment of the village hamlet in favor of individual families scattered throughout the countryside, supporting themselves by subsistence farming. The former two examples are activities that maximize the use of new energies as a result of the war (war-time goods and G.I. dollars); the latter is an adaptation for survival, that is, concentrated villages, when bombed or burned tend to lose more structure per unit of disordering energy than a dispersed population of individual houses scattered throughout a countryside.

In the first sections of this paper it was shown that the war resulted in a measurable disordering of structure and processes by 11% over the 6-year period of U.S. escalation of war activities; and that disordering energies amounted to approximately ten times the ordering natural- and fossil-fuel based energies. To the intuitive mind, figures of this magnitude should have little effect on the overall processes of the country over long time periods.

However, when the war stress model was simulated, additional stresses and sensitivities (loss of food and material sources, and radical changes from producing to consuming populations as a result of population dislocation) were revealed that give new insight into the magnitude of disordering and expected recovery time.

Other models simulated for the country of South Vietnam but not reported here suggest that if aid is provided there will be a regrowth and in some instances increased productivity in a relatively short period of time. This parallels the case of Germany and Japan after the second World War. However, the reordering and recovery of Germany and Japan under the Marshall Plan were accomplished with a great expenditure of outside "ordering energy" during a time of increasing fossil fuel availability and utilization. In addition, Germany and Japan had extensive industrial-based economies with a wide diversity of components and occupations and had fought internally subsidized wars that provided the base for new postwar industrialization. South Vietnam, to the contrary, had little industrialization and was fighting an "externally subsidized" war. It did not reap the benefits of war energies adding new industrial structure.

The war-stress-model simulation indicates that the same rapid reordering experienced by Germany and Japan may be unrealistic in Vietnam; that repair and reconstruction to levels of order equal to those before 1965 can be accomplished only with increasing aid.

U.S. economic assistance to South Vietnam amounted to nearly $3 billion as of 1970 at a level of approximately $400 million per year. The model indicates that the overall system of the country has been distorted to such a degree that reordering aid must increase for the next 15 years to a level of approximately $500 million or a total of nearly $8 billion if the country is to recover rapidly. With constant aid at 1970 levels it will take nearly 30 years to achieve levels of production and structure equal to those of 1970, but even then they will not meet the demands of a growing population. If aid decreases (the most likely circumstance in light of increased demand for and expense of fossil-fuel energies and the surrender of South Vietnam), recovery and reordering within the next 30 years is most unlikely.

The model presented here is a preliminary attempt to simulate a very complex problem. Results of its simulation give insight to overall sensitivities not apparent otherwise, as well as the role of U.S. aid as the stimulator for regrowth. While the model is dependent on data and relationships that are not known with precision, general trends can be drawn that indicate the length of disruption to South Vietnam's economic system under different aid conditions. With more complete data and a greater understanding of the war process and its relationship to order and disorder, this model could be updated and run again.

One basic assumption in our model was that future patterns of, for example, distribution of aid between urban and rural areas, would be similar to existing patterns. This, of course, might change dramatically with changes in the political structure of South Vietnam. While our model was not specifically designed to look at such political parameters they could be included in other runs.

Wars have long been a characteristic of human society and may increase as competition for increasingly limited resources grows stronger. By attempting to understand wars within a broader, more objective context, by studying their long-term effects, and by understanding their relationship to available energies, it might be possible to avoid wars that may be especially counter-productive to various national interests during periods of decreasing energies. In past times of abundant energy supplies war was often a stimulator of domestic economy. However, foreign wars, in particular, require the diversion of large quantities of energy to be successful. Might it not be wise in the future to give any projected war a dry run on the computer? Would this have saved us the agony of the unpopular war in Vietnam with its accompanied destruction of natural and agricultural systems and untold human misery? Could such a model have told us in 1965 that if we were to help the South Vietnamese people we would need to supply extensive ordering energies of aid for many years as well as the disordering energies of military aid? Or, carried to the extreme, might this concept evolve, like the armies of old who chose their Davids and Goliaths to save the slaughter of thousands, into a future state where the wars are fought only on computers? To the simulation victors go the digital spoils!

LITERATURE CITED

Annual Statistical Bulletins, Office of Joint Economic Affairs, U.S. AID, Vietnam. Vols. 11, 12, 13, 14.

A. P. Wire Service. *Gainesville Sun*. Gainesville, Florida. April 2, 1972.

Background Material Relevant to Presentations at the 1970 Annual Meeting of the AAAS, Herbicide Assessment Commission of the American Association for the Advancement of Science. Meselson, Westing, Constable, Revised. 1/4/71.

Congressional Record. Senate. January 28, 1972.

Ewel, J. J. 1970. Experiments in arresting succession with cutting and herbicides in five tropical environments. Ph.D. dissertation. University of North Carolina, Chapel Hill. Unpublished.

Fuller, R. B. 1963. *Inventory of World Resources, Human Trends and Needs*. Phase 1, Document 1 of World Resources Inventory. Southern Illinois University, Carbondale.

Golley, F., Odum, H. T., and R. Wilson. 1962. "The Structure and Metabolism of a Puerto Rican Red Mangrove Forest in May." *Ecology*, **43**: 9–19.

Haitung, M. and G. H. Raets. 1968. Capacidad fisicu y rendiminiento de obreros foretales en diferentes condiciomed climatics del tropico. *Instituto forestal latono America De Incestigacion y Capacitacion. Boletin* No. **26:** 3–31.

Impact of the Vietnam War. 1971. Committee print by Foreign Affairs Division, Congressional Research Service, Library of Congress, U.S. Government Printing Office, Washington.

Kellman, M. C. 1970. *Secondary plant succession in tropical montane Mundanao*. Australia National University, Canberra. 174 p.

Odum, E. P. 1963. *Ecology*. Holt, Rinehart and Winston, New York.

Odum, H. T. 1971. *Environment Power and Society*. Wiley, New York.

Odum, H. T. 1974. *The effects of herbicides in South Vietnam*, Part B: Working Papers. "Models of Herbicide, Mangroves and War in Vietnam." National Academy of Sciences, National Research Council, Washington D.C.

Orians, G. H., and E. W. Pfeiffer. 1970. Ecological effects of the war in Vietnam. *Science*, **168**(8931): 544–554.

Pfeiffer, E. W., and A. Westing. 1971. Land war. Three reports. *Environment*, **13**(9): 2–15.

Report on the War in Vietnam, Commander in Chief Pacific (as of 30 June, 1968). Superintendent of Documents, U.S. Government Printing Office, Wash., D.C.

Richey, J. E. 1970. The role of disordering energy in microcosms. Master of science thesis. University of North Carolina, Chapel Hill. Unpublished.

Rodin, L. E. and N. I. Bazilevic. 1967. *Production and Mineral Cycling in Terrestrial Vegetation*. Oliver and Boyd, Edinburg. 288 pp.

Tschirley, F. H. 1969. Defoliation in Vietnam. *Science* **21**: 779–786.

Snedaker, S. C. 1970. Ecological studies on tropical moist forest succession in eastern lowland Guatemala. Ph.D. thesis, University of Florida. 131 pp.

Tergas, L. E. 1965. Correlation of nutrient availability in soil and uptake by native vegetation in humid tropics. M.S. thesis, University of Florida. 64 pp.

Thayer, G. 1969. *The War Business*. Simon and Schuster, New York.

U.S. Congress. House. Committee on Appropriations. Dept. of Defense Appropriations for 1970 pt. VII. Hearings, Washington, D.C. U.S. Government Printing Office. 1970.

Vietnam subject index maps. 1970. Research files of the Engineer Agency for Resources, Inventories, and Vietnam Research and Evaluation Information Center, Bureau for Vietnam.

Vietnam Statistical Yearbook. 1970. National Institute of Statistics, Republic of Vietnam, Presidency of the Republic, Directorate General of Planning.

Westing, A. H., and E. W. Pfeiffer. 1972. The cratering of Indochina. *Sci. Amer.* **226**(5): 21.

APPENDIX I   CALCULATION OF DISORDERED LANDS BY YEAR

Statistics were not available on the categories and extent of land disordered. In the absence of pertinent data the following assumptions were necessary.

1. *Bombing.* In the 6 years from 1965 to 1970, 5,556,100 tons of air munitions and 128,500 tons of sea munitions were expended (Impact of the Vietnam War, 1971). For the purposes of the model and lack of sufficient data it was assumed that 75% of the air munitions alone were capable of producing

craters equivalent to that of a 500-lb bomb (30 ft in diameter) with a blow down (cleared area) of 100 ft in diameter (Pfeiffer, 1971). This is probably an overestimate of the number of 500-lb bombs dropped. However, when compared with all crater-producing munitions, the calculated cratered area is conservative at best. From these calculations, 16,668,000 five-hundred-pound bombs were dropped on South Vietnam in 6 years. This compares to 21 million estimated by Westing and Pfeiffer (1972).

It was assumed that the bombing was spread evenly throughout the country, since the principal bombing method used was "carpet bombing." From the relative amount of the total area of the country in each of the four major land use categories (city land, 0.06; agricultural land, 17.2%; forest land, 58.8%; mangrove land, 1.6%), the percentages of bombs dropped in each of these was calculated. The number of crater-producing bombs dropped in each land use category was then multiplied by the cleared area produced by one bomb (0.18 acres).

Year	Total Bombs	City Land (cleared acres) $\times 10^2$	Agriculture Land (cleared acres) $\times 10^4$	Forest Land (cleared acres) $\times 10^4$	Mangrove Land (acres) $\times 10^4$
1965	0.96×10^6	1.2	2.9	10.0	0.28
1966	1.5×10^6	2.4	4.8	16.5	0.45
1967	2.8×10^6	3.2	8.8	30.0	0.83
1968	4.3×10^6	4.7	13.4	46.0	1.26
1969	4.2×10^6	4.5	13.0	44.0	1.22
1970	2.9×10^6	3.3	9.1	31.0	0.86

2. *Herbicides.* In the 6 years from 1965 through 1970 a total of 5,092,228 acres of forest and 1,035,882 acres of cropland have been sprayed. Of the 5.1 million acres of forest sprayed 486,140 acres were mangroves.* The acres of defoliated land for each land use category were calculated in the following manner:

First, agriculture land sprayed was assumed to be 90% defoliated for that year (assuming that some land might be replanted in that year) and the following year the land was replanted. Thus only one year's yield is lost. Second, forest land area sprayed was assumed to be 20% defoliated based on statements from Meselson et al. (1970) that "some estimates indicate that one out of every eight or ten trees is killed by a single spraying and that 50 to

* This area was measured on official Department of Defense herbicide spray-run maps, scale of 1:1,000,000; year by year totals for mangrove land sprayed are as follows: 1965, 4700 acres; 1966, 96,330 acres; 1967, 197,600 acres; 1968, 68,666 acres; 1969, 90,155 acres; 1970, 11,609 acres.

80 percent are killed in areas where more than one spraying has occurred."
Therefore, a conservative estimate of 20% killed was assumed. Third, 90% of
mangrove land sprayed was defoliated based on statements by Tschurley
(1969) that mangroves are particularly susceptible to defoliants and that one
application at the normal rate employed in Vietnam is sufficient to kill most
of the trees, and possibly because of loss of seed source there has been little
or no reestablishment of the forest. The following table shows the year-by-
year total land sprayed with herbicides and consequently removed from
production from each of the land-use categories.

Year	Total Area Herbicided (10^4 acres)	City Land Affected (10^2 acres)	Agricultural Land Out of Production (10^4 acres)	Forest Land Out of Production (10^4 acres)	Mangrove Land Out of Production (10^4 acres)
1965	22.2	0.4	6.5	3.0	0.42
1966	84.3	1.5	10.2	13.0	9.6
1967	170.8	2.8	22.1	26.0	19.8
1968	133.1	2.6	6.4	24.0	6.9
1969	128.7	2.4	6.6	22.6	9.0
1970	25.3	0.6	3.3	4.2	1.2

3. *Rome plowing operation.* According to available information, 350,000
acres of forest land were estimated as being cleared from 1968 to 1970 by the
Rome Plow Operation. Westing and Pfeiffer (1971 and 1972) have calculated
the area of plowed land at approximately 750,000 acres. They further estimate
that of this "126,000 acres (were) of prime timberlands accessible to lumber
operations, and 2,500 acres of producing rubber trees." These figures do not
include prime timberland that is not accessible to lumber mills but contributes to the total energy budget of South Vietnam. It was assumed that
350,000 acres of forests and 90,000 acres of agricultural land were cleared by
the Rome Plow Operation. The land areas cleared each year were assumed
to be equal, because it was assumed the operation proceeded at a constant
annual rate.

Year	Agriculture Land (10^4 acres)	Forest Land (10^4 acres)
1968	3.00	11.66
1969	3.00	11.66
1970	3.00	11.66

Appendix B

Imports (thousands of metric tons)

1965	2,159
1966	2,423
1967	2,269
1968	2,387
1969	3,469
1970	3,358 *

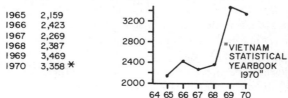

"VIETNAM STATISTICAL YEARBOOK 1970"

Fuel to South Vietnam (thous. metric tons) (partial data)

1966	1.005 [2.]
1967	0.862 [2.]
1968	1.062 [1.]
1969	1.128 [1.]
1970	1.210 [1.] *

* Only six month figures were given
for these years, so figure was
doubled for year total.

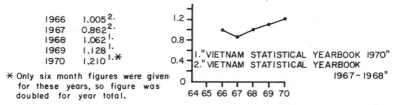

1. "VIETNAM STATISTICAL YEARBOOK 1970"
2. "VIETNAM STATISTICAL YEARBOOK
1967–1968"

Money Spent for Goods & Fuel by Vietnam
(inport arrivals) (millions U.S. $)

1965	387.7
1966	607.2
1967	744.0
1968	707.5
1969	837.7
1970	715.1

"ANNUAL STATISTICAL
BULLETIN NO. 14"

Structure Destroyed by War (thousands of acres)

1965	2.12
1966	7.34
1967	12.12
1968	12.87
1969	8.89
1970	4.54

U.S. Aid (US $ millions)

1965	202.3	265
1966	478.9	406
1967	271.6	356
1968	295.5	370
1969	206.7	477
1970	369.4	504

●――――● ●――●――●
         Official Aid

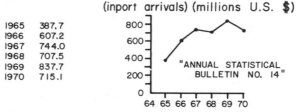

"ANNUAL STATISTICAL
BULLETIN NO. 14"

416

Appendix B cont'd.

U.S. War Appropriations   (US $ millions)

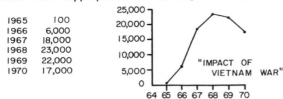

1965	100
1966	6,000
1967	18,000
1968	23,000
1969	22,000
1970	17,000

"IMPACT OF
VIETNAM WAR"

Fuel to War Effort (millions of barrels) (partial figures)

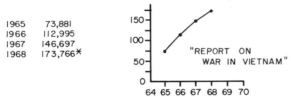

1965	73,881
1966	112,995
1967	146,697
1968	173,766*

"REPORT ON
WAR IN VIETNAM"

Military Goods (thousands of tons)(partial figures)

1965	3,300*
1966	7,280
1967	11,800*
1968	14,840

"REPORT ON
WAR IN VIETNAM"

* Only six month figures were given
for these years, so figure was
doubled for year total.

Appendix C

MANGROVE: $\dot{L}_1 = + K_1 L_5 A + K_2 L_5 - K_3 [L_1 + L_2 + L_3 + L_4] \ W$

$\dot{S}_1 = + K_4 L_1 I - K_5[(s_1 + s_2 + s_3)(G+A)(L_4)(P_2)] - K_6 S_1$

FORESTS: $\dot{L}_2 = + K_7 L_5 A + K_8 L_5 - K_9[(L_1 + L_2 + L_3 + L_4) \ W]$

$\dot{S}_2 = K_{10} L_2 I - K_{11}[(s_1 + s_2 + s_3)(G+A)(L_4)(P_2)] - K_{12} S_2 - K_{13}[(s_2 + s_3)(P_1)]$

FARMS: $\dot{L}_3 = + K_{14} L_5 + K_{16} L_5 A - K_{15}[(L_1 + L_2 + L_3 + L_4) \ W]$

$\dot{S}_3 = K_{17} L_3 I - K_{18}[(s_1 + s_2 + s_3)(G+A)(L_4)(P_2)] - K_{19}[(s_2 + s_3)(P_1)]$

$\dot{P}_1 = K_{21}[(K_{14} L_5 + K_{16} L_5 A) K_{32} P_2] - K_{23} P_1 \ W - K_{24} P_1$

CITIES: $\dot{L}_4 = + K_{25} L_5 + K_{26} L_5 A. - K_{27}[(L_1 + L_2 + L_3) \ W]$

$\dot{S}_4 = K_{28}[(s_1 + s_2 + s_3)(G+A)(L_4)(P_2)] - K_{29} S_4 - K_{30} S_4 - K_{33} S_4$

$\dot{P}_2 = + K_{31} P_1 \ W. + K_{29} S_4 - K_{32} P_2 [K_{14} L_5 + K_{16} L_5 A]$

$\dot{L}_5 = K_{35} - L_1 - L_2 - L_3 - L_4$

417

18
Application of an Ecosystem Model to Water Quality Management: The Delaware Estuary

ROBERT A. KELLY
WALTER O. SPOFFORD, JR.

The wise management of our natural resources is no easy task. It depends not only on individual preferences regarding a balance between our material wants and aesthetically pleasant surroundings, but also on our ability to be able to predict the impacts of various courses of action, both on the economy and on the natural environment. We do not explore how decisions regarding the natural environment are or should be made in this chapter. Instead, we concentrate on providing those responsible for making these decisions with information on the costs and impacts on the environment of various courses of action.

The purpose of this chapter is to outline an ecosystem model of the Delaware Estuary and show how it can be linked with an economic model to form a management model. Such an integrated model then could be used to obtain a set of politically and socially feasible regional waste management strategies.

An Historical Perspective

Many different mathematical models have been used in the management of water quality. The first was that of Streeter and Phelps (1925), which examined the impact of organic waste discharges from industrial and municipal sources on the oxygen balance of various water bodies. Over the years, many changes have been made to the Streeter-Phelps model, mostly to account for various processes that were known to occur in natural systems but were intentionally omitted from the earlier models (O'Connor, 1967).

Aquatic models of the Streeter-Phelps type are mathematically attractive since they consist of linear functions relating ambient concentrations to waste discharges. These models have been incorporated into more encompassing management (optimization) models, which have been used since the early 1960s to find least-cost alternatives for achieving ambient water quality standards.

Increased sophistication in both optimization techniques and aquatic models have led us to examine the possibility of incorporating a nonlinear ecosystem model within an optimization framework so that (1) tradeoffs among discharges other than organic material could be examined, (2) ecologically realistic relationships describing the behavior of the aquatic system could be used, and (3) ambient standards could be set on primary (fish populations, algal densities) rather than secondary (DO level) indicators of "environmental quality."

At the time this chapter was prepared, both authors were Research Associates, Quality of the Environment Program, Resources for the Future, Inc., Washington, D. C. These remarks reflect the views of the authors and not necessarily those of Resources for the Future.

In order to develop these considerations more fully, a brief description of the ecosystem model is presented, followed by a discussion of linear and nonlinear programming management models and the implications of a mathematical device called the "environmental response matrix."

A more complete discussion of the regional residuals (waste) management model with particular emphasis on including nonlinear environmental models within an optimization framework is given by Spofford (1973). The ecosystem model summarized here is presented in considerably more detail by Kelly (1976).

ECOSYSTEM MODEL

The ecosystem model was developed as an integral part of a research program at Resources for the Future (RfF) on regional-environmental quality management. This situation constrained the model in two ways: only existing data on the Delaware Estuary could be used, and the resulting model had to be adapted for use within an optimization framework (Spofford et. al., 1976).

The Delaware Estuary extends from the head of tide at Trenton, N.J., to (for our purposes) Artificial Island (Figure 1), a distance of about 130 km (80 miles) (DRBC, 1970). It is characterized by a mean annual flow at Trenton of 310 m^3 sec^{-1} (11,000 CFS) and low salinity throughout its length [at mean flow, about 10 mg liter^{-1} chloride at Marcus Hook, 30 km (20 miles) above Artificial Island (Keighton, 1966)].

The Delaware River Basin Commission (DRBC), a regulatory agency formed by agreement among the states bordering the river and the federal government, has placed major emphasis on the dissolved oxygen distribution in the estuary. Such an emphasis excluded consideration of other important factors which can be of considerable interest to recreational and other users of the Delaware (e.g., fishermen, boaters, municipal water users, etc.). Such other uses require a consideration of resident fish population, presence of algal blooms, general aesthetic appearance, and the presence or absence of toxic materials.

Since ecological research on the estuary has been done piecemeal, with only a few studies made concerning its entire length (as defined here), much of the model is based on data aggregated from different locations and even on inference. It is exploratory, and in some respects rather simplistic, and hence it should be considered at this time as only a very crude prediction tool.

Our model falls within the general class of simulation models, with time differential equations describing rates of change in state variables (compartments) for several physically related sections of the estuary. Each section

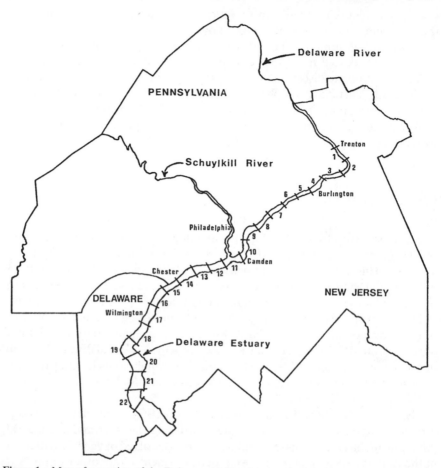

Figure 1 Map of a portion of the Delaware Valley showing the locations of the reaches used for the Estuary Model. The area drawn represents the geographical boundary for an integrated waste management model. Source: Kelly (1976).

(reach) is treated as a homogeneously mixed body of water, with net advective water flow serving as the only link between reaches. Longitudinal dispersion is not incorporated as a transport mechanism in this version of the estuary model. The general overall model is represented schematically as follows:

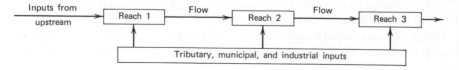

Because the model is to be used in a static economic framework, only steady-state conditions were of interest.

A serious data limitation prevented more than a simple representation of the aquatic ecosystem. The choice of compartments was determined by a trophic (i.e., food) level approach, because this is the pathway the pollutants would affect. One compartment in each reach represents primary producers (algae), one compartment herbivores (zooplankton), one compartment omnivores (fish), and one compartment decomposers (bacteria). Nutrients (nitrogen and phosphorus), organic matter (BOD), and oxygen compartments are included for their role in regulation of the rates of transfer of matter or energy among the living components. These eight compartments are the true state variables, for they affect and are affected by other compartments. Three other compartment—turbidity, toxics (phenolic equivalent), and temperature (heat)—are included as state variables in the mathematical sense for ease in computing the environmental response matrix but can be considered exogenous to the biotic system since the biota do not significantly affect their behavior.

The equations describing the time rate of change behavior of each of the consumer compartments (zooplankton, bacteria, and fish) can be developed from Figure 2. The only differences between zooplankton (illustrated) and the other consumers are the number and identification of food sources and the number and identification of predators.

There are two inputs of material to each living compartment, one from upstream sources [exogenously determined (e.g., the downstream drift of algae)] and one by feeding (ingestion). The latter's rate is determined by a product term that attempts to account for (1) the availability of food, (2) the size of the population, and (3) the effects on the feeding rate of temperature, toxic materials, and oxygen concentration (Figure 3). All of these relationships are themselves nonlinear with the exception of the effect of the feeder's population size. The general shapes of the nonlinear expressions are shown within each multiplier (Figure 2). The expressions describing these functions are given in Figure 3b.

There are five losses of material from a living compartment due to (1) respiration, (2) death, (3) excretion (egestion), (4) predation, and (5) loss downstream. Respiration and death are mathematically similar in that their rates can be expressed as products of three nonlinear expressions and the population size. However, as shown within the multipliers (Figure 2), the effects of the different factors on the rates are not the same, and the ultimate fate of materials generated in these two processes is different. Excretion (egestion) is assumed to be proportional to the square of the population size, and loss downstream, a linear function of the population size. Predation is codetermined by the predator and predatee population sizes as indicated in the

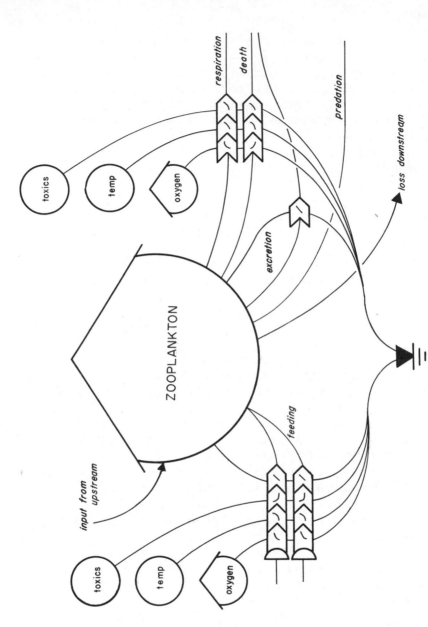

Figure 2 Representation of the zooplankton compartment in energy circuit language. The configuration shown is the same for fish and bacteria with these exceptions: fish, no input from upstream, no loss downstream, three food sources, no predation loss; bacteria, one food source, two predation losses. This diagram is represented by ⬡ in Figure 5.

Figure 3a General mathematical representation of the model for one reach:

$$\dot{A} = f_A - r_A - d_A - e_A - p_A + y_A - \omega A$$

$$\dot{Z} = f_Z - r_Z - d_Z - e_Z - p_Z - y_Z - \omega Z$$

$$\dot{B} = f_B - r_B - d_B - e_B - p_B + y_B - \omega B$$

$$\dot{F} = f_F - r_F - d_F - e_F$$

$$\dot{O} = 2.319\{f_A - r_A - r_Z - r_B - r_F\} + \psi(O^* - O) + y_O - \omega O$$

$$\dot{L} = d_A + e_A + d_Z + e_Z + d_B + e_B + e_F + d_F - f_B + y_L - \omega L$$

$$\dot{N} = 0.1087\{r_A + r_Z + r_B + r_F - f_A\} - 0.1N\{1 - e^{-0.60}\{1 - e^{-0.9(10-\varphi)}\}\{0.17146e^{0.05877\theta}\} + y_N - \omega N$$

$$\dot{P} = 0.02174\{r_A + r_Z + r_B + r_z - f_A\} - \frac{0.02P}{\psi} + y_P - \omega P$$

$$\dot{\sigma} = y_\sigma - \frac{0.1\sigma}{\psi} - \omega\sigma$$

$$\dot{\varphi} = y_\varphi - 0.4\varphi - \omega\varphi$$

$$\dot{\theta} = y_\theta - \frac{0.683}{z}(\theta - \theta^*) - \omega\theta$$

Figure 3b Coefficients and functional forms of equations used for various biological processes:

$$f_A = \{6.0\}\{A\}\{1 - e^{-3.00}\}\{0.17146e^{0.05877\theta}\}\{1 - e^{-0.4(10-\varphi)}\}\left\{\frac{1}{(0.151+N)(0.0151+P)}\right\}\left\{\frac{2.66}{Z(0.05 + 0.17A + 0.10B + 0.10\sigma)}\right\}$$

$$f_Z = \{1.5\}\{Z\}\{1 - e^{-0.80}\}\{0.17146e^{0.05877\theta}\}\{1 - e^{-0.3(10-\varphi)}\}\left\{\frac{1}{2}\left(\frac{1}{4.0 + A} + \frac{1}{1.25 + B}\right)\right\}$$

$$f_B = \{3.3\}\{B\}\{1 - e^{-1.00}\}\{0.17146e^{0.05877\theta}\}\{1 - e^{-0.9(10-\varphi)}\}\left\{\frac{1}{7.0 + L}\right\}$$

$$f_F = \{1.2\}\{F\}\{1 - e^{-0.60}\}\{0.17146e^{0.05877\theta}\}\{1 - e^{-0.25(10-\varphi)}\}\left\{\frac{1}{3}\left(\frac{1}{3.0 + A} + \frac{1}{2.0 + Z} + \frac{1}{4.5 + B}\right)\right\}$$

Figure 3 (*Continued*)

Figure 3b (*Continued*)

$$p_A = \cfrac{f_Z}{1 + \cfrac{4.0 + A}{1.25 + B}} + \cfrac{f_F}{1 + \cfrac{3.0 + A}{2.0 + Z} + \cfrac{3.0 + A}{4.5 + B}}$$

$$p_Z = \cfrac{f_F}{\cfrac{2.0 + Z}{3.0 + A} + 1 + \cfrac{2.0 + Z}{4.5 + B}}$$

$$p_B = \cfrac{f_Z}{\cfrac{1.25 + B}{4.0 + A} + 1} + \cfrac{f_F}{\cfrac{4.5 + B}{3.0 + A} + \cfrac{4.5 + B}{2.0 + Z} + 1}$$

$$p_F = 0$$

$$r_A = \{0.8\}\{A\}\{1 - e^{-0.8O}\}\{0.125e^{0.069315\theta}\}\{0.1 + e^{0.5(\varphi - 10.2107)}\}$$

$$r_Z = \{0.6\}\{Z\}\{1 - e^{-0.6O}\}\{0.125e^{0.069315\theta}\}\{0.1 + e^{0.4(\varphi - 10.2634)}\}$$

$$r_B = \{1.5\}\{B\}\{1 - e^{-1.5O}\}\{0.125e^{0.069315\theta}\}\{0.1 + e^{0.8(\varphi - 10.1317)}\}$$

$$r_F = \{0.5\}\{F\}\{1 - e^{-0.5O}\}\{0.125e^{0.069315\theta}\}\{0.1 + e^{0.3(\varphi - 10.3512)}\}$$

$$d_{A'} = \{1.0\}\{A\}\{0.1 + e^{4.0(O + 0.0263)}\}\{0.09391e^{0.078846\theta}\}\{0.1 + e^{1.0(\varphi - 10.1054)}\}$$

$$d_Z = \{2.0\}\{Z\}\{0.1 + e^{0.9(O + 0.1171)}\}\{0.09391e^{0.078846\theta}\}\{0.1 + e^{0.6(\varphi - 10.1749)}\}$$

$$d_B = \{3.0\}\{B\}\{0.1 + e^{5.0(O + 0.0211)}\}\{0.09391e^{0.078846\theta}\}\{0.1 + e^{3.0(\varphi - 10.0351)}\}$$

$$d_F = \{0.5\}\{F\}\{0.1 + e^{0.7(O + 0.1505)}\}\{0.09391e^{0.078846\theta}\}\{0.1 + e^{0.5(\varphi - 10.2634)}\}$$

$$e_A = \{0.04\}\{A\}\{A\}$$

$$e_Z = \{0.05\}\{Z\}\{Z\}$$

$$e_B = \{0.09\}\{B\}\{B\}$$

$$e_F = \{0.10\}\{F\}\{F\}$$

Notation:

A = algal concentration (mg liter^{-1})
Z = zooplankton concentration (mg liter^{-1})
B = bacteria concentration (mg liter^{-1})
F = fish concentration (mg liter^{-1})
O = oxygen concentration (mg liter^{-1})
L = organic matter concentration (mg liter^{-1})
N = nitrogen (TKN) concentration (mg liter^{-1})
P = phosphorus concentration (mg liter^{-1})
σ = turbidity (Hellige units)
φ = toxic materials (phenol concentration equivalents)
θ = temperature (°C)
ψ = reaeration rate constant (day^{-1}), specific for each reach
z = depth (m), specific for each reach
O^* = oxygen saturation value ($= 14.652 - 0.41022\theta + 0.0079910\theta^2 - 0.00007777740\theta^3$)
θ^* = equilibrium temperature of water $= 23.3°C$
O = upper case of alphabetic o
f = feeding
r = respiration
d = death
e = excretion
p = predation
y = input from outside reach
ω = water turnover rate

Figure 3 Model for specific reach.

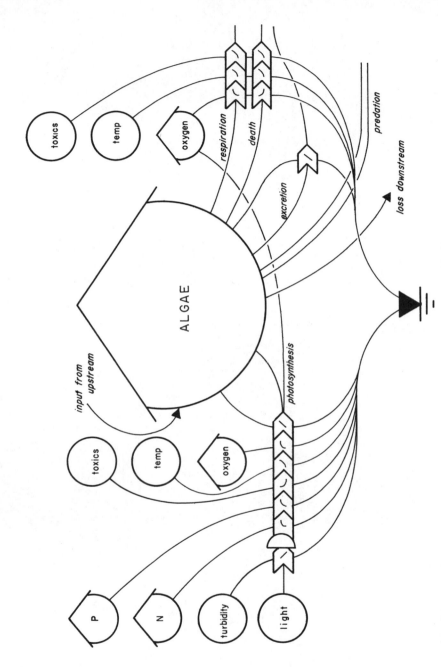

Figure 4 Representation of the algae compartment in energy circuit language. This diagram is represented by ⌓ in Figure 5.
P = phosphorus

428

feeding term. For further clarification and justification of coefficients, see Figure 3a and 3b, and Kelly (1976).

The equation describing the time rate of change of algae is similar to those of zooplankton, bacteria, and fish for all five loss terms and for the input of material from upstream. However, the nutrient uptake term for algae is of necessity quite different from the ingestion term of the consumers (Figure 4). The major differences are two: (1) light determines in part the rate of photosynthesis and thus the need for nutrients, and (2) unlike the consumers where all food sources are assumed to be completely substitutable, plants need all nutrients concurrently in order to grow. For these reasons, the rate of nutrient uptake is expressed as a product of nonlinear expressions relating the availability of light (as a function of surface illumination, turbidity, and depth), the availability of nutrients (a product of two Michaelis-Menten expressions—one for nitrogen and one for phosphorus), and other controlling variables as in the ingestion terms for the consumers. In this we are following the simplest generally accepted method of modeling algal photosynthesis.

Figures 3a and 5 show the overall model for a typical section of the estuary. The living components in Figure 5 are simplified for clarity, while the other state and exogenous variables are shown in their full complexity. The rate of oxygen loss or gain from the water column through diffusion to the atmosphere is made proportional to the reaeration coefficient and the oxygen concentration. Rates of sedimentation of phosphorus and suspended materials are assumed to be inversely proportional to the amount of mixing. Loss of nitrogen (measured as total Kjeldahl nitrogen) due to nitrification (conversion to NO_2^- and NO_3^-) is assumed to be related to temperature, oxygen concentration, and toxics concentration, as well as to the concentration of nitrogen (TKN). The rate of heat loss by the processes of radiation and evaporation is made a function of the surface area of the section of the estuary. Toxic materials are converted to nontoxic materials at a rate proportional to their concentration. The rationalization for these relationships is discussed in Kelly (1976).

The output of the model compared with measured data is given in Figure 6 for the month of September 1970. It should be pointed out that the relatively good fit of predicted values to these data, although enhancing the believability of the model, in no way confirms its validity. For another time period, using input data obtained in the same way as for the September runs, the fit of observed data to model output is considerably less encouraging. Until these two sets of data give equally good fits, the results of the model for management purposes can be viewed as only preliminary. Had we the time and money to gather the needed data ourselves consistent with the requirements of our model, we suspect that our overall results would have been more reliable.

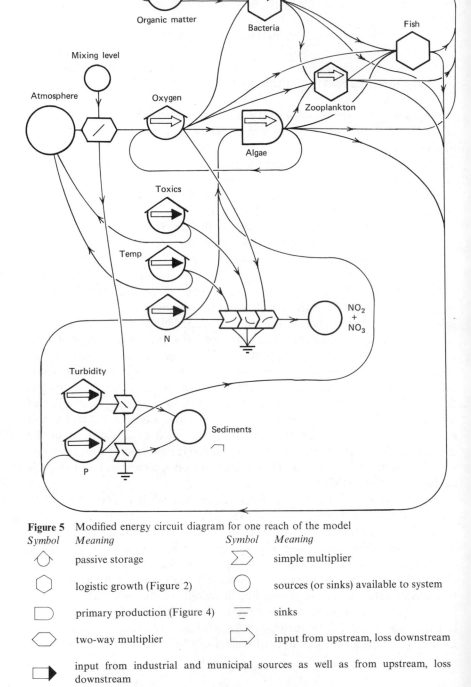

Figure 5 Modified energy circuit diagram for one reach of the model

Symbol	Meaning	Symbol	Meaning
△	passive storage	⇒	simple multiplier
⬡	logistic growth (Figure 2)	○	sources (or sinks) available to system
▢	primary production (Figure 4)	≡	sinks
⬦	two-way multiplier	⇨	input from upstream, loss downstream
▬▶	input from industrial and municipal sources as well as from upstream, loss downstream		

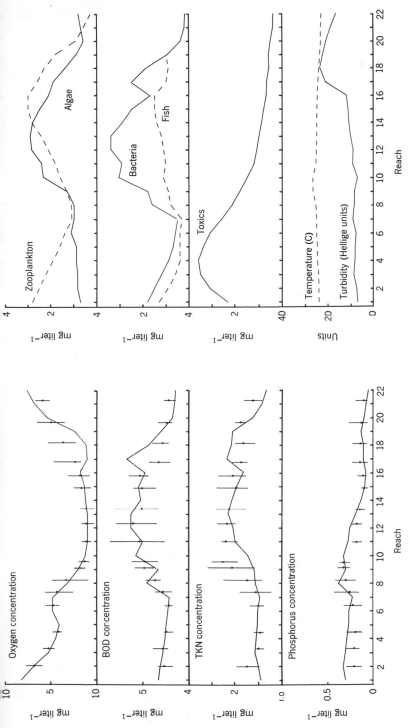

Figure 6 Output of the model plotted along the length of the estuary. Vertical lines on the left side of the figure are observed ranges of the endogenous variables; dots on those lines are the means of the observations.

REGIONAL WASTE MANAGEMENT MODELS

Regional waste management models are useful for analyzing the simultaneous impacts on costs and on the environment of alternative waste management strategies. These models are used to rank sets of management strategies according to a given economic criterion, such as least cost to the region. Programming or optimization models are particularly useful for this purpose since they are specifically designed to locate the optimal, or in some sense "best," strategy (see Chapter 4 for an introduction to optimization models). There are two basic types of programming models: (1) linear programming (LP) models and (2) nonlinear programming (NLP) models. These models are classified according to their mathematical characteristics.

There are two major parts of all programming models, whether they be of the linear or of the nonlinear type: (1) the objective or criterion function and (2) the constraint set. If the objective function and the constraint set are both comprised of linear functions, a linear programming model results. If either the objective function or the constraint set is comprised of nonlinear functions, even a single nonlinear function, the result is a nonlinear programming model. Nonlinearities in the objective function are generally easier to deal with than nonlinearities in the constraint set. Often nonlinear programming problems can be converted to the simpler linear programming problems by careful redefinition of some of the variables and piece-wise linearization of the nonlinear functions.

The purpose of the objective function is to provide a basis for ranking feasible combinations of values for the decision variables. In the management models discussed here, the objective function is based on costs, although it need not be for other purposes.

The constraint set provides the necessary structural relationships among the decision variables, as well as the upper and lower allowable bounds on these variables. Environmental models form an important part of the constraint set of regional waste management models. They provide the necessary relationship between waste discharges and ambient concentrations.

To provide the reader with a little better understanding of programming models and how they can be used in regional waste management analyses we discuss in more detail both the linear and nonlinear programming models.

The Linear Management Model

Assume for the moment that in a given region the relevant portion of man's activities and the natural environment can be represented by a management model which consists entirely of linear functions. A linear programming model that minimizes the cost of achieving a set of ambient environmental standards within some geographic area can be expressed, in matrix notation, as

$$\text{minimize}\{C = cW\} \tag{1}$$

$$A_1 W + A_2 Z \geq B \tag{2}$$

$$HZ = X \tag{3}$$

$$X \gtrless S \tag{4}$$

$$Z \geq 0, \quad W \geq 0, \quad X \geq 0 \tag{5}$$

where cW, the objective function (Chapter 4), is an expression that sums the total costs for all firms and municipalities within the region of internal production process changes and wastewater treatment associated with reducing wastewater discharges to level Z; $A_1 W + A_2 Z \geq B$ represents a set of linear relationships that ensure mass balance of materials for industries and municipalities, as well as constraints on mix of product outputs, resource availability, production levels—in short, models of man's waste generation and discharge activities within the geographic bounds; B is a vector of constants, many of which are equal to zero; $HZ = X$ are steady-state linear models of the natural world, with X representing the steady-state spatial distribution of concentrations of various materials in the environment (constrained to be greater than or equal to zero) that must be greater than or less than ambient standards S depending on the environmental quality indicator selected for the analysis; W are activity levels of processes used to reduce wastewater discharges; and Z (which are constrained to be greater than or equal to zero) are discharges into the environment of specific materials from spatially located industries and municipalities in the region.

The objective of the analysis is to determine the wastewater discharge levels, Z, that minimize the total regional costs of wastewater management, C, and at the same time meet the ambient standards, S. This example represents a steady-state analysis. Note that the linear environmental models are included in the constraint set. Given that the objective function and the constraint set are both linear, this problem can be solved using linear programming (LP) techniques (e.g., Gass, 1958; Dorfman et al., 1958). A number of linear programming computer codes are available. Examples of the linear programming formulation for wastewater management purposes can be found in Sobel (1965), Revelle et al. (1968), and Thomann (1972).

The Nonlinear Management Model

If on the other hand either the objective function or the constraint set consists of nonlinear relationships and if these nonlinear relationships cannot easily be linearized, linear programming techniques cannot be employed for the analysis. The ecosystem model we have described consists of nonlinear relationships. It is not possible using algebraic techniques, even for the steady-state case, to convert this model to a linear form where linear programming techniques would be applicable. Because of this, nonlinear programming techniques must be used. The purpose of this section is to provide such an example.

For the discussion that follows, it is assumed that only the environmental model (Eq. 3) consists of nonlinear functions and that the rest of the constraints are linear. Since for most NLP algorithms, a nonlinear objective function is easier to handle than a nonlinear constraint set, the nonlinear environmental model is removed from the constraint set and incorporated in the objective function. This requires the use of penalty functions (frequently used under a variety of names; see Zangwill, 1969; Fiacco and McCormick, 1968) for exceeding ambient standards. These penalties, which are strictly a mathematical devise for ensuring that the standards are met, are appended to the objective function. The modified nonlinear formulation of the regional management model may be expressed as

$$\text{minimize}\{F = cW + P(Z)\} \tag{6}$$

$$\text{subject to } A_1 W + A_2 Z \geq B \tag{2}$$

$$Z \geq 0, \quad W \geq 0 \tag{5}$$

where

$$P(Z) = \sum_{i=1}^{n} p_i[s_i, x_i = h_i(Z)] \tag{7}$$

which is the penalty function associated with violating ambient standards, S. The total regional penalty, $P(Z)$, represents the sum of all the individual penalties, p_i, associated with exceeding any, or all, of the individual standards, s_i. Instead of requiring that the management problem be bounded by the set of ambient standards, in a mathematical sense the standards are allowed to be violated but only at a high (monetary) penalty to the objective function. If the penalty is high enough, the management model will respond by cutting back wastewater discharges, thus reducing violations of the standards. As the violations are reduced so is the total penalty. When an optimum solution is reached, the environmental standards may still be violated in a mathematical sense, but only by small percentages of their values. This violation, however, can be made as small as desired by manipulation of the penalty function.

It is important to emphasize that the penalty function is merely a mathematical device used here to ensure that the ambient standards are met (within a given tolerance). No economic interpretation should be attached to it. The penalty function is not used to select socially desirable levels of ambient quality. For the analysis here, the ambient standards are assumed to be given.

All nonlinear programming algorithms start from a trial feasible solution and using an iterative search process, select increasing better solutions until the best possible or optimal solution is found. In this discussion, total region wastewater management costs are used to judge whether one solution is better than another. Several NLP algorithms use the gradient, ∇F, of the objective function (the gradient being the direction of the maximum increase, or decrease, in the objective function value) to determine the most efficient direction of travel along the response or cost surface toward an optimum solution. At each iteration in the search process a step is taken in the direction of the gradient, and this is continued until no further improvement in the objective function value can be obtained. Once the direction of travel is determined the size of the step to be taken during that iteration can be computed by a variety of methods (e.g., Saaty and Bram, 1964), but a discussion of this is beyond the scope of this simple exposition.

The gradient of the nonlinear objective function, Eq. 6, is defined as

$$\nabla F \equiv \left\{ \frac{\partial F(W,Z)}{\partial W} + \frac{\partial F(W,Z)}{\partial Z} \right\} \equiv c + \frac{\partial P(Z)}{\partial Z} \tag{8}$$

where c is a vector of changes in costs (marginal costs) for a unit change in the activity levels, W, and $\partial P(Z)/\partial Z$ is a vector of changes in penalties (marginal penalties) associated with violation of the standards for a unit change in the discharge levels, Z.

To obtain an expression for the marginal penalty with respect to a given discharge, z_j, we see from Eq. 7 that

$$\frac{\partial P(Z)}{\partial z_j} = \sum_{i=1}^{n} \frac{dp_i}{dx_i} \frac{\partial x_i}{\partial z_j}, \quad j = 1, m \tag{9}$$

where $\partial P(Z)/\partial z_j$ is the change in total penalty for a change in discharge, z_j; dp_i/dx_j is the change in penalty for a change in concentration, x_i; $\partial x_i/\partial z_j$ is the change in concentration of material, x_i, for a change in discharge, z_j; m is the total number of discharges in the region; and n is the total number of water quality indicators and critical locations where ambient standards are set. The latter differential is an element of the "environmental response matrix."

Environmental Response Matrix

In order to calculate the marginal penalties (Eq. 9) for use in the regional management model (see previous discussion) or to examine the relative impact on water quality of changing individual waste discharges into the estuary, it is necessary to compute an environmental response matrix whose elements are

$$\frac{\partial x_{i,j}}{\partial z_{k,l}}, \quad \begin{array}{l} i = 1, 11 \\ j = 1, 22 \\ k = 1, 11 \\ l = 1, 22 \end{array} \tag{10}$$

where $\partial x / \partial z$ signifies the change in ambient concentration of a given material or biota (i.e., the state variables) with a change in discharge of a given waste, i specifies the state variable, k specifies the waste, j specifies the reach location of the state variable, and l the reach location of the waste discharge.

The ecosystem model at steady state can be expressed functionally as 22 sets (one for each reach) of the following eleven equations:

$$\begin{aligned} f_1 &= g_1(x_1, x_2, \dots, x_{11}) + y_1 = 0 \\ f_2 &= g_2(x_1, x_2, \dots, x_{11}) + y_2 = 0 \\ &\vdots \\ f_{11} &= g_{11}(x_1, x_2, \dots, x_{11}) + y_{11} = 0 \end{aligned} \tag{11}$$

or in matrix notation,

$$F = G(X) + Y = 0 \tag{12}$$

where x_i for a given reach are concentrations of compartments, and y_i for the same reach are inputs to compartments.

For a given reach we may calculate a matrix whose elements are

$$\frac{\partial x_i}{\partial y_k}, \quad \begin{array}{l} i = 1, 11 \\ k = 1, 11 \end{array}$$

which represents the change in concentration of compartment i due to a change in input of material k, implicitly by the relation

$$\left[\frac{\partial F}{\partial Y} \right] + \left[\frac{\partial F}{\partial X} \right] \left[\frac{\partial X}{\partial Y} \right] = 0 \tag{13}$$

(Sokolnikoff and Redheffer, 1958), where the change in each function value with a change in a compartment concentration is defined as

$$\left[\frac{\partial F}{\partial X} \right] \equiv \begin{bmatrix} \dfrac{\partial f_1}{\partial x_1} & \cdots & \dfrac{\partial f_1}{\partial x_{11}} \\ \vdots & & \\ \dfrac{\partial f_{11}}{\partial x_1} & & \end{bmatrix} \tag{14}$$

where the change in each function value with a change in the input of a material is defined as

$$
\left[\frac{\partial F}{\partial Y}\right] \equiv \begin{bmatrix} \dfrac{\partial f_1}{\partial y_1} & \cdots & \dfrac{\partial f_1}{\partial y_{11}} \\ \vdots & & \\ \dfrac{\partial f_{11}}{\partial y_1} & & \end{bmatrix} \tag{15}
$$

and where the desired expression is

$$
\left[\frac{\partial X}{\partial Y}\right] \equiv \begin{bmatrix} \dfrac{\partial x_1}{\partial y_1} & \cdots & \dfrac{\partial x_1}{\partial y_{11}} \\ \vdots & & \\ \dfrac{\partial x_{11}}{\partial y_1} & & \end{bmatrix} \tag{16}
$$

Since $[\partial F/\partial X]$ and $[\partial F/\partial Y]$ can both be computed directly from Eq. 11, $[\partial X/\partial Y]$ can be calculated as

$$
\left[\frac{\partial X}{\partial Y}\right] = -\left[\frac{\partial F}{\partial X}\right]^{-1}\left[\frac{\partial F}{\partial Y}\right] \tag{17}
$$

where $[\partial F/\partial X]^{-1}$ is the inverse of the matrix defined as Eq. 14.

This computation yields for a given reach the change in concentration of each compartment with a change in input to each compartment when Y is in units of mg liter^{-1} day^{-1}. To convert the $[\partial X/\partial Y]$ matrix to units of discharges (lb day^{-1}), the following relation may be used:

$$
Y_j = \frac{aZ_j}{v_j} + \frac{\omega_{j-1}v_{j-1}X_{j-1}}{v_j}, \quad j = 1, 22 \tag{18}
$$

where Y_j is the input to each compartment of reach j (mg liter^{-1} day^{-1}), a is the conversion factor for pounds to mg, Z_j is the discharge of materials to reach j (lb day^{-1}), v_j is the volume of reach j in liters, and the second term accounts for inputs from the upstream reach. (Recall that longitudinal dispersion is not considered in this example.) Using Eq. 18, the change in total inputs to reach j with a change in waste discharge to reach j may be expressed as

$$
\frac{\partial y_{k,j}}{\partial z_{k,j}} = \frac{a}{v_j}, \quad k = 1, 11 \tag{19}
$$

and the change in the state variables in reach j with a change in the waste discharges to reach j may be expressed as

$$
\frac{\partial x_{i,j}}{\partial z_{k,j}} = \frac{\partial x_{i,j}}{\partial y_{k,j}}\frac{\partial y_{k,j}}{\partial z_{k,j}}, \quad \begin{array}{l} i = 1, 11 \\ k = 1, 11 \end{array} \tag{20}
$$

Similarly, from Eq. 18 it follows that the change in inputs to reach j with a change in the concentration of materials and biota in the upstream reach $j - 1$ may be expressed as

$$\frac{\partial y_{k,j}}{\partial x_{i, j-1}} = \frac{\omega_{j-1} v_{j-1}}{v_j}, \quad i = k$$

$$\frac{\partial y_{k,j}}{\partial x_{i, j-1}} = 0, \quad\quad\quad i \neq k$$

(21)

where ω_j is the turnover rate of water in reach j, and the vs represent volumes of the respective reaches. Thus the environmental response matrix stated in Eq. 10 for the entire 22-reach estuary system can be calculated as

$$\left[\frac{\partial X}{\partial Z}\right]_{j,l} = \begin{cases} \left[\frac{\partial X}{\partial Y}\right]_{j,j}\left[\frac{\partial Y}{\partial X}\right]_{j, j-1}\left[\frac{\partial X}{\partial Y}\right]_{j-1, j-1} \cdots \\ \left[\frac{\partial X}{\partial Y}\right]_{l+1, l+1}\left[\frac{\partial Y}{\partial X}\right]_{l+1, l}\left[\frac{\partial X}{\partial Z}\right]_{l,l} \quad j > l \\ \\ \left[\frac{\partial X}{\partial Z}\right]_{j,j} \quad\quad\quad\quad\quad\quad\quad\quad j = l \\ \\ \begin{bmatrix} 0 \cdots 0 \\ \vdots \\ 0 \end{bmatrix} \quad\quad\quad\quad\quad\quad\quad j < l \end{cases}$$

where the first subscript of the matrices refers to the reach of the numerator and the second subscript refers to the reach of the denominator.

A graphical representation of parts of the environmental response matrix is presented in Figure 7. From these graphs, one can see the relative change in the concentration of fish in any downstream reach resulting from an increased discharge of one unit of organic matter (Figure 7a), nitrogen (Figure 7b), or phosphorus (Figure 7c) into any other reach, all other inputs constant. Each horizontal axis represents the effect on all downstream reaches of a change in discharge into the first reach appearing on that axis.

DISCUSSION

We have shown how the environmental response matrix enters into the calculation of marginal penalties for a nonlinear programming (optimization) algorithm. The marginal penalties at the optimal solution can be considered analogous to effluent charges, which have the intent of reducing discharges

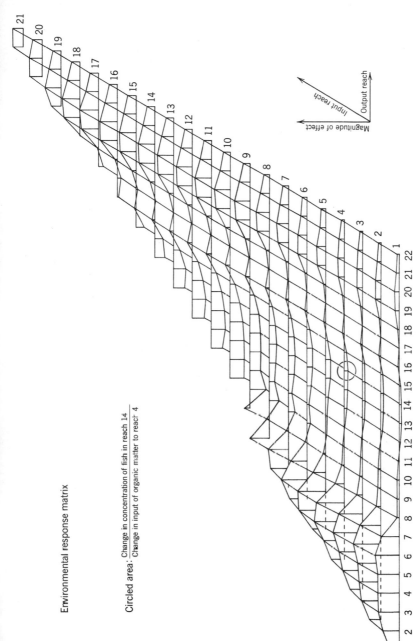

Environmental response matrix

Circled area: Change in concentration of fish in reach 14 / Change in input of organic matter to reach 4

Magnitude of effect

Input reach

Output reach

Figure 7a Three-dimensional plot of the relative effect on fish concentrations due to a unit increase in organic matter discharge into any given reach.

439

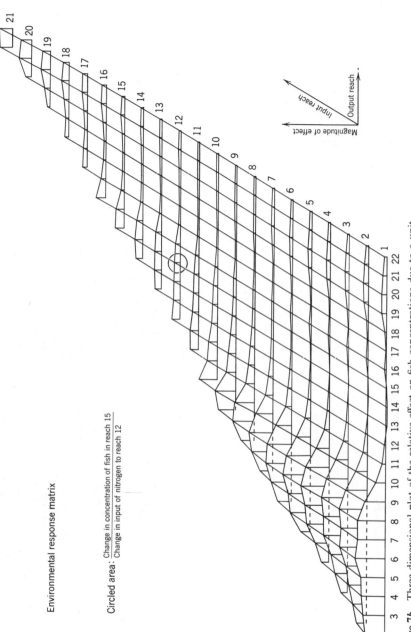

Environmental response matrix

Circled area: $\dfrac{\text{Change in concentration of fish in reach 15}}{\text{Change in input of nitrogen to reach 12}}$

Figure 7b Three-dimensional plot of the relative effect on fish concentrations due to a unit increase in nitrogen discharge into any given reach.

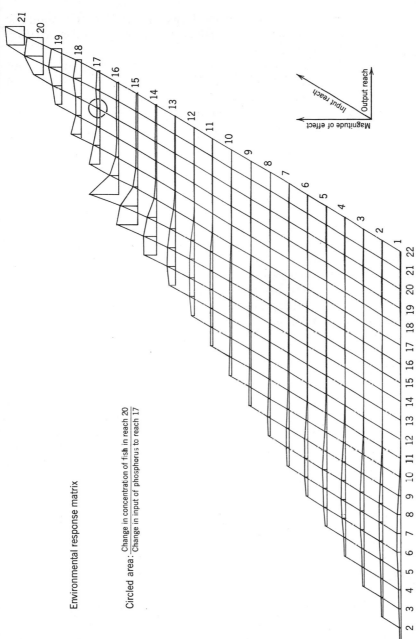

Environmental response matrix

Circled area: $\dfrac{\text{Change in concentration of fish in reach 20}}{\text{Change in input of phosphorus to reach 17}}$

Magnitude of effect

Input reach

Output reach

Figure 7c Three-dimensional plot of the relative effect on fish concentrations due to a unit increase in phosphorus discharge into any given reach.

441

in the most efficient manner (Kneese and Bower, 1968). For more complete discussions of the significance of marginal penalties, see Spofford et al. (1975) and Russell and Spofford (1972).

The use of the environmental response matrix is not limited to mathematical manipulation. Plots of different portions of the matrix can be used to determine what discharges are having the greatest impacts on the system and thus where to first examine discharge reduction possibilities. For example, Figure 7a indicates that to increase fish populations in the middle reaches of the estuary (reaches 10 to 15) one could decrease discharges of organic matter to reaches 5 to 15. Reaches 9 to 15 are apparently where organic matter discharges are having the greatest deleterious effect, due in large part to the low oxygen concentrations already existing there. Discharges upstream as far as reaches 5 or 6 could be reduced, but the change in fish population in the middle reaches with a change in discharge of organic material upstream becomes increasingly less significant the further upstream one goes. Also, it would seem unnecessary to decrease discharges downstream from reach 16, since organic matter increases the production of fish below this point. For the predicted state of the ecosystem used in these computations, the plots in Figures 7b and 7c show that nitrogen and phosphorus have no deleterious effects on the fish population.

SUMMARY

The increasing sophistication of ecological models provides potentially useful tools for making decisions on water resources. Although ecological models have traditionally been used for predicting transient behavior through simulation, we have demonstrated how these types of models can also be incorporated into wider-scope management (optimization) models. This is made possible by means of an environmental response matrix, which also serves a useful function as an independent qualitative tool. However, lack of appropriate data remains the major limiting factor in constructing and validating ecological models to be used for management purposes.

REFERENCES

Delaware River Basin Commission. 1970. *Delaware estuary and bay water quality sampling and mathematical modeling project.* Final Progress Report to the FWPCA. Trenton, N. J.

Dorfman, R., P. A. Samuelson, and R. M. Solow. 1958. *Linear programming and economic analysis.* McGraw-Hill, New York. 523 pp.

Fiacco, A. V., and G. P. McCormick. 1968. *Nonlinear programming: Sequential unconstrained minimization techniques.* Wiley, New York, 210 pp.

Gass, S. I. 1958. *Linear Programming: Methods and applications*. McGraw-Hill, New York. 223 pp.

Keighton, W. B. 1966. *Fresh-water discharge: Salinity relations in the tidal Delaware River*. Geological Survey Water-Supply Paper 1586-G. U.S. Government Printing Office, Washington, D.C.

Kelly, R. A. 1976. Conceptual ecological model of the Delaware Estuary. In B. C. Patten (Ed.), *Systems analysis and simulation in ecology*, Vol. IV. Academic, New York. Pp. 3–46.

Kneese, A. V., and B. T. Bower 1968. *Managing water quality: Economics, technology, institutions*. Published for Resources for the Future, Inc. by The Johns Hopkins Press, Baltimore. 328 pp.

O'Connor, D. J. 1967. The temporal and spatial distribution of dissolved oxygen in streams. *Water Resour. Res.* 3(1): 65–79.

Revelle, C. S., D. P. Loucks, and W. R. Lynn. 1968. Linear programming applied to water quality management. *Water Resour. Res.* 4(1): 1–9.

Russell, C. S., and W. O. Spofford, Jr. 1972. A Quantitative framework for residuals management decisions. In A. V. Kneese and B. T. Bower (Eds.), *Environmental quality analysis: Theory and method in the social sciences*, pp. 115–179. The Johns Hopkins University Press for Resources for the Future, Inc., Baltimore, London. 408 pp.

Saaty, T. L., and J. Bram. 1964. *Nonlinear mathematics*. McGraw-Hill, New York. 381 pp.

Sobel, M. J. 1965. Water quality improvement programming problems. *Water Resour. Res.* 1(4): 477–487.

Sokolnikoff, I. S., and R. M. Redheffer. 1958. *Mathematics of physics and modern engineering*. McGraw-Hill, New York. 812 pp.

Spofford, W. O., Jr. 1973. Total Environmental Quality Management Models. In R. A. Deininger (Ed.), *Models for environmental pollution control*. Ann Arbor Science Publishers, Ann Arbor. Pp. 403–436.

Spofford, W. O., Jr., C. S. Russell, and R. A. Kelly. 1975. Operational Problems in Large Scale Residuals Management Models. In E. S. Mills (Ed.), *Economic analysis of environmental problems*. National Bureau of Economic Research, New York.

Spofford, W. O., Jr., C. S. Russell, and R. A. Kelly. 1976. *Environmental Quality Management: An application to the Lower Delaware Valley*. Resources for the future, Washington, D.C.

Streeter, H. W., and E. B. Phelps. 1925. *A study of the pollution and natural purification of the Ohio River*. Public Health Bulletin No. 146. U.S. Public Health Service, Washington, D.C.

Thomann, R. V. 1972. *Systems analysis and water quality management*. Environmental Science Services Division of Environmental Research and Applications, Inc., New York. 286 pp.

Zangwill, W. I. 1969. *Nonlinear programming: A unified approach*. Prentice-Hall, Englewood Cliffs, New Jersey. 356 pp.

PROLOGUE TO
SECTION IV

We are all familiar with various pronouncements of deterioration and doom made by environmentalists, and we, for example, often have been asked "Well, why don't you suggest something better?" This section provides examples of how scientists have attempted to provide something better by suggesting ways in which we can understand or predict better ways of managing a region or even a country. Florida is a good place to develop resource management models because the natural resources of the state are the basis of the tourist-, agriculture-, and fishing-based economy, and because water and fuel availability have become critical factors in further development of the state. Several of the papers here are an integration of ecology, modeling, and regional planning. For example, Littlejohn investigates the ways in which the maintenance of natural areas actually helps maintain the economic vitality and livability of a region. He makes a point that we think is much more general than is generally recognized: "In contrast to widely publicized conflicts between environmental and energy policies, natural water management provides for the conservation of both energy and the environment." It is very encouraging that the County Commissioners of Collier County have taken the results of this model as a basis for planning in the county.

In some cases, even when large parts of a region are still in their natural state there may be only limited development that can be supported without adversely affecting the whole economy of that region. Boynton, Gray, and Hawkins in Chapter 20 demonstrate how important it is to carefully consider the total consequences of economic development. Although the model presented here uses economic criteria to measure the potential effects of development, the same authors have developed more recently a similar model based on energy as a criteria of value. They have found that the two approaches give similar results. The third example from the state of Florida concerns a very ambitious attempt by Kemp and other ecologists and engineers to quantify the total impact of a series of nuclear power plants located on the Crystal River estuary of west-central Florida. This study is important for at least two reasons. The first is that it is one of the first attempts to incorporate the use of energy-related measurements of value into a total assessment of a large region. The second is that the study attempts to measure all the effects of a power plant and from this assess the net worth of the power plant to society. As is pointed out in Chapter 14 on the Hudson River, this is important but rarely done. Often the decision-making process gets sidetracked by whatever aspects are the easiest to measure or that have the greatest emotional impact.

Recent studies have found that American agriculture is a very energy-intensive process and that, in addition, certain agricultural materials, such as pesticides, have deleterious environmental consequences. Computer models coupled with detailed biological knowledge may reduce the use of both energy and pesticides, as shown by Shoemaker in Chapters 4 and 22.

Kessell's fire model represents to us an almost ideal combination of modern concepts of resource management and modeling theory, and, in addition, it is actually being applied on a large scale to Glacier National Park. To a certain extent Kessell is very wise. Neither the exact processes of his model nor the criteria for value are prescribed in advance. Thus the operation and even the information desired from the model can be changed at any time, allowing an integration of the vast-data processing powers of a computer with the flexibility of human judgement.

Professor Singer uses economic criteria in looking at the future of the quality of life in the United States. He comes to some interesting conclusions about the material quality of our life in the future and how this would be affected by the rate at which the United States population grows. While we are not quite as optimistic about some aspects of resource availability as Singer, we think that the model is an excellent device to explore the consequences of national decisions about such things as energy policies. Of course, like all models it is flexible and can be run again with new information or assumptions if that appears desirable.

Chapter 25, by Nixon and Kremer, integrates many of the economic and ecological concepts contained in the rest of this book. We were particularly interested to see that their detailed, physiologically based model was basically successful in simulating ecosystem properties. This, and other chapters, have convinced us that with good fieldwork combined with good modeling it really is possible to understand and simulate even quite complicated ecosystems.

IV
CASE HISTORIES
Designing Optimal Interactions of
Man and Nature

19
An Analysis of the Role of Natural Wetlands in Regional Water Management

CHARLES LITTLEJOHN

Channelization and drainage of wetlands for agricultural and urban development have raised many questions concerning the best use of wetlands, their role in water management, and their contribution to Florida's economy. Recent fuel shortages in Florida and the nation as a whole raise still another question: What are the initial and long-term fuel commitments associated with floodplain channelization, dredge and fill, flood control, and land reclamation projects? As fossil fuels become increasingly depleted and/or expensive, state leaders soon may be searching for low-energy alternatives to traditional energy-intensive water-management practices.

One such alternative is more effective utilization of nature's programs of water management. Rather than replacing natural ecological systems with fuel-driven environmental technologies (canals, dams, weirs, pumps, dredges, tertiary treatment, etc.), Florida may be wiser during times of industrial energy scarcity to utilize natural meandering channels rather than man-made canals for flood discharge and water distribution, seasonal floodplain fluctuations rather than tertiary treatment plants for nutrient and sediment removal, and swamps and floodplains rather than reservoir systems for water storage and retention.

In contrast to widely publicized conflicts between environmental and energy policies, natural water management provides for the conservation of both energy and the environment.

THE GORDON RIVER

An opportunity for quantitative assessment of the role of natural ecological systems in regional water management arose from controversies concerning channelization of the Gordon River near Naples, Florida. The region of consideration as it existed before modern human settlement is shown in Figure 1, which shows areas of upland scrub, pine-palmetto, swamp wetlands, mangroves, and estuarine waters.

Much of the area has been converted first to agriculture and then to housing in the last three decades with the accelerated growth of Naples as a retirement and tourist resort center. The water needed by these developments has come from well fields in a sandy ridge that runs north and south between the Gordon River and the beach. Developments to date have already involved some channelization within the Gordon basin, and to the east an extensive network of canals discharges through the "Golden Gate" canal, which connects with the Gordon River near its estuarine junction. The pattern of land use that existed in 1972 is given in Figure 2 showing remnants of the earlier zones. Main water courses are given in Figure 3,

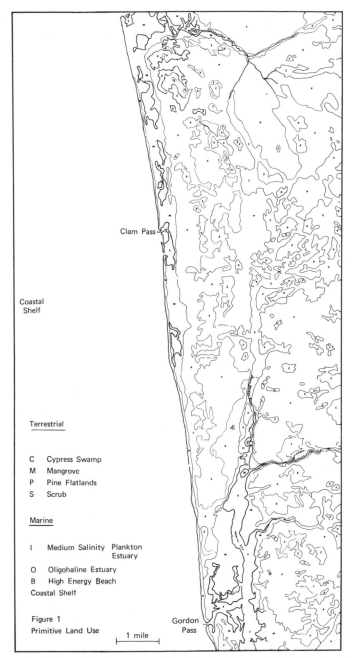

Coastal
Shelf

Terrestrial

C   Cypress Swamp
M   Mangrove
P   Pine Flatlands
S   Scrub

Marine

I   Medium Salinity Plankton
                Estuary

O   Oligohaline Estuary
B   High Energy Beach
Coastal Shelf

Figure 1

Primitive Land Use

Clam Pass

Gordon
Pass

1 mile

Figure 1   Reduced view of land use map of the Gordon River region in its primitive state as estimated from 1937 soil maps. Original has a scale of 1 in. = 2000 ft.

453

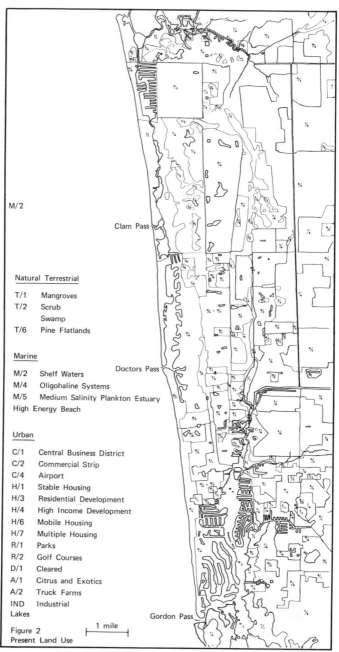

M/2

Clam Pass

Natural Terrestrial

T/1   Mangroves
T/2   Scrub
      Swamp
T/6   Pine Flatlands

Marine

M/2   Shelf Waters
M/4   Oligohaline Systems
M/5   Medium Salinity Plankton Estuary
High Energy Beach

Doctors Pass

Urban

C/1   Central Business District
C/2   Commercial Strip
C/4   Airport
H/1   Stable Housing
H/3   Residential Development
H/4   High Income Development
H/6   Mobile Housing
H/7   Multiple Housing
R/1   Parks
R/2   Golf Courses
D/1   Cleared
A/1   Citrus and Exotics
A/2   Truck Farms
IND   Industrial
Lakes

Gordon Pass

Figure 2
Present Land Use

|⟵ 1 mile ⟶|

Figure 2   Reduced view of present land use map of the greater Naples area as determined from 1971 aerial photographs. Original has a scale of 1 in. = 2000 ft (1 cm = 240 m).

454

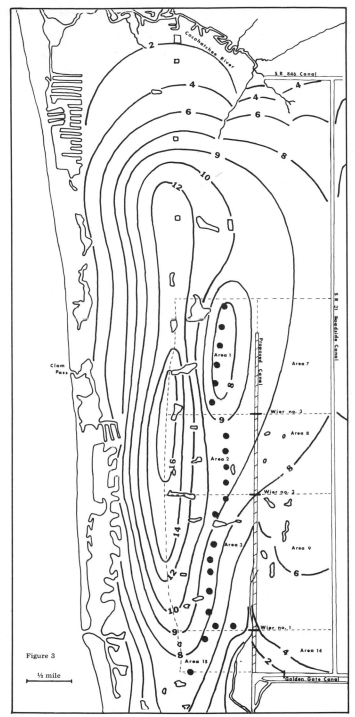

Figure 3 Base map of Naples region showing shallow aquifer (1971 USGS contours), Naples municipal well field (solid circles), proposed channelization including locations of water control structures, and outline of areas (dashed lines) considered during analog and digital computer simulations. Original has a scale of 1 in. = 2000 ft.

including the Gordon River, Golden Gate Canal, the proposed channelization of the basin, contours of water in the city aquifer, and municipal wells.

Aquifer recharge is entirely from local rainfall, which is abundant during summer but relatively scarce from November to May. Seasonal rainfall results in corresponding seasonal fluctuations in aquifer storage.

Consumptive uses of water include municipal pumpage from the 21 city wells shown in Figure 3 and spray irrigation of agricultural crops, residential grasses, and golf courses from private wells scattered throughout the region. The remaining water is either transpired by natural terrestrial communities or discharged to sea via surface and ground water flow.

Pumpage rates provided by the Naples Public Works Director indicate both seasonal fluctuations due to the winter influx of tourists and a steadily increasing yearly pumpage corresponding to rapid overall population growth. During a long dry season occurring in the winter of 1971, pumpage resulted in drawdowns of up to 2 ft below mean sea level in southern portions of the well field, posing the threat of salt-water contamination of municipal water. Proponents of the proposed channelization argued that a canal with appropriate control structures would have the effect of shifting large amounts of water from the northern to southern portions of the well field, providing some well recharge and protection against salt-water intrusion. Opponents pointed out that with any increased channelization, more water would be discharged from the total drainage area, which would in turn result in reduced aquifer storage.

Under the pressures of differences in public opinion as to the desirability of channelization and development of the remaining swamp area, the Collier County Board of Commissioners requested an environmental evaluation that would consider alternatives for this region that might maximize the quality of life, economic stability, and best use of water resources. The final contract report to the Collier County Commissioners by Odum et al. (1972) (and a master's thesis by Littlejohn, 1973) has recently been adopted in principle by the Commissioners as a development plan for the region. This paper is a summary of that work with an emphasis on the hydrological questions and models.

NATURAL WATER MANAGEMENT

During primitive times (i.e., before the intensive development of the past century), natural vegetation in the Gordon River region was completely adapted to local water regimes. The coastal ridge, sparsely populated with dry-adapted coastal scrub vegetation that intercepted relatively little rainfall, was the primary aquifer recharge area. The rain that fell on the ridge rapidly entered aquifer storage through dry permeable soils to become part of ground-water flow systems that discharged slowly westward to the sea and eastward into the Gordon River basin.

Within the basin, large communities of cypress grew that transpired heavily during summer when water was in excess, but dropped their leaves and became dormant in winter as rainfall sharply decreased. Large bowl-shaped storage areas and swamp vegetation retarded overland flow, filtered the "noise" of random daily rainfall, and provided a more stable discharge of water into Naples Bay. Retention of water also allowed more time for nutrient uptake by cypress and associated plant communities which stabilized the quality of water discharged. The Gordon River and flood plain provided for rapid removal of water during peak flooding caused by hurricanes and severe storms. The channel was protected by mangroves that held the shoreline and provided a source of organic nutrients to Naples Bay. A meandering channel design provided maximum bank area on which mangroves could develop subject to the constraint of minimum discharge capability.

The natural design of Naples Bay provided for adequate tidal flushing that supplied inorganic nutrients to mangroves and prevented an excessive concentration of organics which would lead to less efficient biochemical reactions. Since tidal energy was low, an extremely shallow, wide bay evolved.

Nature's design also allowed for protection against hurricane winds and tides. Extensive growth of mangroves in the Gordon Pass area protected the bay by absorbing wind and wave energy, while mangroves and dune vegetation guarded the coastal ridge from the west. Fresh-water storage in the shallow aquifer provided protection against inland salt water surges, thereby preventing damage to cypress populations within the basin. The water level in the aquifer tended to be highest during the hurricane season, thus providing maximum protection only when needed. The aquifer also maintained fresh-water supplies during periods of low rainfall by continuously providing a slow discharge of fresh water from the coastal ridge into the Gordon Basin. In this way slower discharge rates from the aquifer system prevented salt water intrusion and provided long-term stability while surface interactions took place at a much more rapid rate.

TRANSITION TO PRESENT

Present patterns of development have drastically altered natural programs of water management. Coastal-ridge scrub vegetation has been largely displaced by urban systems that both reduce aquifer recharge (by impervious areas and storm drainage systems) and require additional water from aquifer storage for municipal consumption and spray irrigation of ornamental plants.

Stabilizing cypress communities have been displaced in many areas by golf courses that have highest water demand during the winter dry season when management procedures induce the heaviest growth of Bahia grass to meet the increased tourist pressure. Because of development patterns like this, municipal pumpage has become greatest during the winter tourist season when Naples' population nearly triples (producing the opposite effect of the adapted vegetation). Golf-course construction, in addition, has required some channelization that accelerates fresh-water discharge into Naples Bay.

In short, the process of modern development has been the displacement of systems "in phase" with seasonal rainfall with urban systems often "out of phase" with local climatic conditions. In addition, regional nutrient cycles are now altered due to increased concentrations of urban wastes, reduced natural areas for nutrient uptake, and fertilizer loads required for nonadapted ornamental vegetation.

REGIONAL WATER BUDGETS

In order to assess quantitatively the relationships between changing patterns of land use and regional water regimes, "unit" water budgets were constructed as shown in Figure 4 for each category of land use. Overall water budgets were developed for the coastal ridge and the Gordon River basin (areas 1, 2, 3, 15 and areas 7, 8, 9, 14, respectively, as shown in Figure 3) by multiplying the various unit budgets by the amount of land in each category of use (Tables 1 and 2). Results of this exercise for present land use patterns are summarized in Figure 5.

Similar analyses were performed for three alternate land-use contingencies. These included primitive land use (extrapolated from regional soil maps), full development (according to present zoning regulations), and partial development that did not disrupt the important ecological functions of the region (Figures 1, 6, and 7, respectively). The latter includes preservation of the coastal ridge recharge areas, construction of a cypress swamp water conservation area within the Gordon River basin, and retention of buffer zones of natural vegetation between contrasting areas of development.

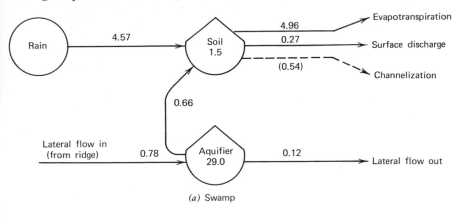

(a) Swamp

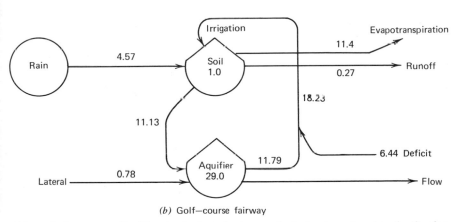

(b) Golf—course fairway

Figure 4 Per acre or "unit" water budgets for two ecological systems found in the Gordon River Basin. Flows are in acre-ft/year; storages in acre-ft (1234 cu. meters = 1 acre foot).

ANALOG COMPUTER SIMULATION OF REGIONAL WATER FLOW

Seasonal variations in water supply and demand were assessed by constructing mathematical models of ridge and basin water budgets (as shown in Figure 8) and programming these on an analog computer. By approximating seasonal rainfall and urban demand fluctuations with sine-wave

Table 1 Ridge Unit Water Budgets for Present Land Use.[a]

Community	R^b	GW_b^c	SR^d	P_m^e	P_{si}^f	ET	GW_i^g	I^h	Area[i] (acres)	% Total Area
Natural terrestrial[j]	4.57	−0.71	0	−1.69	0	−1.03[v]	−1.21	3.52	1780	63.2
Residential[k]	4.57	−0.71	−1.14[p]	−1.69	−6.5[r]	−1.95[w]	0.92	7.97	730	25.9
Golf-course fairway[l]	4.57	−0.71	0	−1.69	−15.2[s]	−7.60[x]	5.43	12.17	230	8.2
Commercial and industrial[m]	4.57	−0.71	−3.65[q]	−1.69	−2.6[t]	−0.78[y]	2.26	2.74	21	0.7
Intensive agriculture[n]	4.57	−0.71	0	−1.69	−5.67[u]	−5.67[z]	3.50	4.57	55	2.0
Combined ridge[o]	4.57	−0.71	−0.34	−1.69	−3.02	−1.83	0	5.42	2816	100.0

R = Rainfall

GW_b = Ground water discharge into Gordon River basin

SR = Surface runoff due to paving and development

P_m = Pumpage from municipal wells

P_{si} = Pumpage from private wells for spray irrigation

ET = Evapotranspiration

GW_i = Ground-water flow induced by local gradients within ridge

I = Infiltration including rain and spray irrigation

[a] Ridge includes areas 1, 2, 3, and 15 as shown in Figure 3. Flows are in ft/year-unit area. 1 ft = 0.305 M.

[b] Based on 50 years Naples weather records.

[c] Based on average water table difference of 2 ft between ridge and basin as calculated from USGS water contours; estimated 100-ft depth of ground water flow; estimated permeability of 7.4×10^3 ft/month based on USGS tests; assumed yearly flow distributed evenly over ridge as calculated using Darcy's Law; $(7.4 \times 10^3$ ft/month) (12 month/year) (100 ft) (2 ft)/(5.0 $\times 10^4$ ft) (2.46 $\times 10^4$ ft) $(2$ ft)/(5.0 $\times 10^3$ ft) (1.23 $\times 10^8$ ft^2); assumed southerly ground water flows not significant in area of consideration.

[d] Surface runoff of excess water into Naples Bay through storm drainage sewers.

e Municipal pumpage of 2.97×10^8 ft^3/year divided by ridge area of 1.23×10^8 ft^2 and adjusted by assuming 20% of total pumpage drawn from undeveloped areas north and west; 10% pumpage drawn from basin based on flow net analysis of 1971 contours.

f Spray irrigation estimates based on personal communication from local agricultural extension agents; represents total pumpage from private wells to spray irrigate crops, ornamental vegetation, and golf-course fairways.

g Back calculated from continuity equation; positive value represents water deficit that must be made up by ground water recharge from surrounding natural areas.

h $I = P_m + GW_b + P_{si}$.

i As calculated from land use maps at scale of 1 in. = 2000 ft.

j Includes pine flatlands, cypress, scrub, mangroves, and old field successional systems.

k Estimated 50% of residential lot covered based on 1970 Collier County zoning regulations and aerial photographs.

l Includes 70% of golf course estimated from aerial photographs.

m Estimated 80% of commercial zone paved based on aerial photographs and Collier County zoning regulations.

n Includes 2/7 of total farmland since land usually intensively farmed for 2 years, left dormant 5 years (per personal communication with local agricultural extension agents).

o Unit ridge budget as calculated from unit community budgets considering percent of total area occupied by each community.

p Assumed 50% of rain falling on covered areas leaves flow system via storm drainage and evaporation.

q Assumed 100% of rain falling on covered areas leaves flow system via storm drainage and evaporation.

r Estimated 3 in./week spray irrigation of Bahia grass based on personal communication of local agricultural extension agents.

s Estimated $\frac{1}{2}$ in./day spray irrigation of Bermuda grass based on personal communication of local agricultural extension agents.

t Same as q except only 20% Bahia Grass, rest covered.

u Estimated 40 in./150 day overlapping seasons from August 15 to May 1 based on personal communication of local agricultural extension agents.

v $ET = P_{si} + R - SR$.

w Estimated 30% of spray irrigation based on personal communication of local agricultural extension agents.

x Estimated 50% of spray irrigation based on personal communication of local agricultural extension agents.

y Same as v.

z Estimated 100% of irrigation assimilated or evaporated.

461

Table 2 Basin Unit Water Budgets for Present Land Use.[a]

Community	R[b]	GW_b[c]	GW_sea[d]	SD_sea[e]	SR[f]	P_m[g]	P_si[h]	ET	GW_i[i]	I[j]	Area[k] (acres)	% Total Area
Natural terrestrial[e]	4.57	0.78	−0.12	−0.27	0	−0.26	0	−4.24[x]	−0.48	0.33	2100	78.6
Residential[m]	4.57	0.78	−0.12	−0.27	−2.77[r]	−0.26	−6.5[t]	−1.95[y]	0	6.35	82	3.1
Golf-course fairway[n]	4.57	0.78	−0.12	−0.27	0	−0.26	−15.2[u]	−7.60[z]	2.90	12.17	302	11.3
Commercial and industrial[o]	4.57	0.78	−0.12	−0.27	−3.94[s]	−0.26	−2.6[v]	−0.78[aa]	0	2.45	84	3.1
Intensive agriculture[p]	4.57	0.78	−0.12	−0.27	0	−0.26	−5.67[w]	−5.67[bb]	0.97	4.57	104	3.9
Combined basin[q]	4.57	0.78	−0.12	−0.27	−0.20	−0.26	−2.15	−4.50	0	2.02	2672	100.0

R = Rainfall
GW_b = Ground water received from ridge
GW_{sea} = Ground water discharge to Naples Bay
SD_{sea} = Surface discharge through Gordon River
SR = Surface runoff due to paving and development
P_m = Pumpage from municipal wells
P_{si} = Pumpage from private wells for spray irrigation
ET = Evapotranspiration
GW_i = Ground water flow induced by local gradients within ridge
I = Infiltration including rain and spray irrigation

[a] Basin includes area 7, 8, 9, and 15 as shown in Figure 3. Flows in ft/year-unit area.
[b] See Table 1.
[c] Discharge from ridge corrected for ridge/basin area differences $= (0.71 \text{ ft/year})(1.23 \times 10^8 \text{ ft}^2/1.12 \times 10^8 \text{ ft}^2)$.
[d] Assume only southerly flow significant; $GW_{sea} = (7.4 \times 10^3 \text{ ft/month})(12 \text{ month/year})(100 \text{ ft})(4.7 \times 10^3 \text{ ft/month})(5.45 \text{ ft} - 1.35 \text{ ft})/(1.23 \times 10^4 \text{ ft})$
$(1.12 \times 10^8 \text{ ft}^2)$.

[e] Estimated 10 cfs discharge through existing drainage slough based on USGS control weir discharge data.; (10 cfs) (60 sec/min) (60 min/hr) (24 hr/day) (365 days/year)/(1.12 × 10⁸ ft²).

[f] Runoff due to paving plus excess standing water lost to storm drainage and evaporation.

[g] Estimated 10% of total pumpage corrected for ridge/basin area difference = (0.24 ft/year) (1.23 × 10⁸ ft²/1.12 × 10⁸ ft²).

[h] Spray irrigation estimates based on personal communication of local agricultural extension agents; represents total pumpage from private wells to spray irrigate crops, ornamental vegetation, and golf-course fairways.

[i] Back calculated from continuity equation; positive value represents water deficit that must be made up by ground water recharge from surrounding natural areas.

[j] $I = SD_{sea} + P_m + GW_{sea} + P_{si} - GW_b$.

[k] As calculated from land use maps at scale of 1 in. = 2000 ft.

[l] Includes pine flatlands, cypress, scrub, mangroves, and old field successional systems.

[m] Estimated 50% of residential lot covered based on 1970 Collier County zoning regulations and aerial photographs.

[n] Includes 70% of golf course estimated from aerial photographs.

[o] Estimated 80% of commercial zone paved based on aerial photographs and Collier County zoning regulations.

[p] Includes 2/7 of total farmland since land usually intensively farmed for 2 years, left dormant 5 years (per personal communication with local agricultural extension agents).

[q] Unit basin budget as calculated from unit community budgets considering percent of total area occupied by each community.

[r] 1.14 ft/yr plus 1.63 ft/yr excess calculated from continuity equation.

[s] 3.65 ft/yr plus 0.29 ft/yr excess calculated from continuity equation.

[t] Estimated 3 in./week spray irrigation of Bahia grass based on personal communication of local agricultural extension agents.

[u] Estimated ½ in./day spray irrigation of Bermuda grass based on personal communication of local agricultural extension agents.

[v] Same as q except only 20% Bahia grass, rest paved.

[w] Estimated 40 in./150 day overlapping seasons from August 15 to May 1 based on personal communication of local agricultural extension agents.

[x] $ET = P_{si} + R - SR$.

[y] Estimated 30% of spray irrigation based on personal communication of local agricultural extension agents.

[z] Estimated 50% of spray irrigation based on personal communication of local agricultural extension agents.

[aa] Same as v.

[bb] Estimated 100% of irrigation assimilated or evaporated.

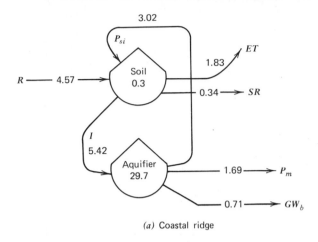

(a) Coastal ridge

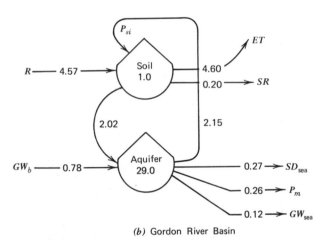

(b) Gordon River Basin

Figure 5   Unit water budgets for "typical" acre of land located on the coastal ridge (a) and Gordon River Basin (b). Flows in acre-ft/year; storages in acre-ft.

forcing functions, analog computer simulations were performed for each contingency of land use. This process is shown in detail for one land type in the Appendix. The results of simulated primitive coastal ridge conditions in which consumption is limited to evapotranspiration are shown in Figure 9. Water use (even at a high level) in phase with rainfall results in extremely stable aquifer storage.

The results of simulation of the coastal ridge region, using parameters corresponding to present development, are shown in Figure 10. Less stable

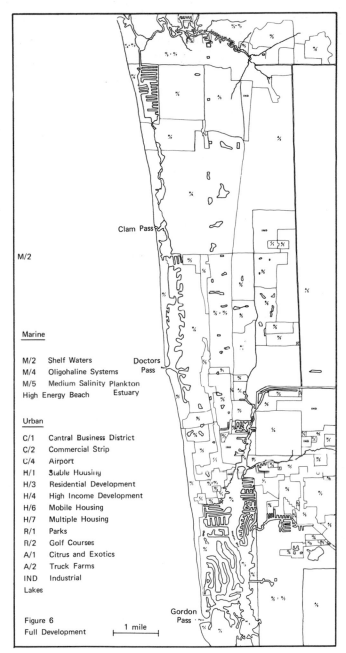

M/2

Clam Pass

Marine

M/2	Shelf Waters	Doctors
M/4	Oligohaline Systems	Pass
M/5	Medium Salinity Plankton	
High Energy Beach	Estuary	

Urban

C/1	Central Business District
C/2	Commercial Strip
C/4	Airport
H/1	Stable Housing
H/3	Residential Development
H/4	High Income Development
H/6	Mobile Housing
H/7	Multiple Housing
R/1	Parks
R/2	Golf Courses
A/1	Citrus and Exotics
A/2	Truck Farms
IND	Industrial
Lakes	

Gordon
Pass

Figure 6
Full Development

1 mile

Figure 6 Reduced view of land use map showing full development of Naples region according to present Collier County zoning regulations. Original has a scale of 1 in. = 2000 ft.

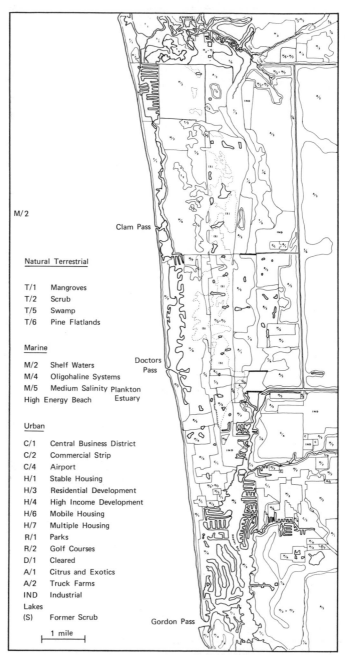

M/2

Clam Pass

Natural Terrestrial

T/1   Mangroves
T/2   Scrub
T/5   Swamp
T/6   Pine Flatlands

Marine

M/2   Shelf Waters
M/4   Oligohaline Systems
M/5   Medium Salinity Plankton
High Energy Beach   Estuary

Doctors Pass

Urban

C/1   Central Business District
C/2   Commercial Strip
C/4   Airport
H/1   Stable Housing
H/3   Residential Development
H/4   High Income Development
H/6   Mobile Housing
H/7   Multiple Housing
R/1   Parks
R/2   Golf Courses
D/1   Cleared
A/1   Citrus and Exotics
A/2   Truck Farms
IND   Industrial
Lakes
(S)   Former Scrub

1 mile

Gordon Pass

Figure 7 Reduced view of land use map of an alternative plan of development retaining recharge and swamp retention areas for natural water management.

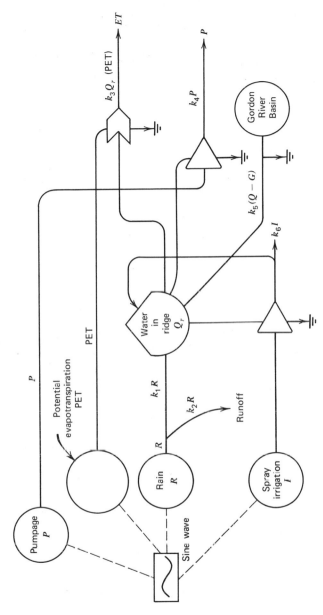

Figure 8 Coastal ridge unit water model in circuit language with water flows expressed algebraically.

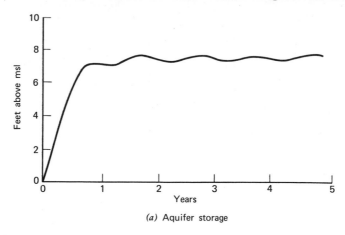

(a) Aquifer storage

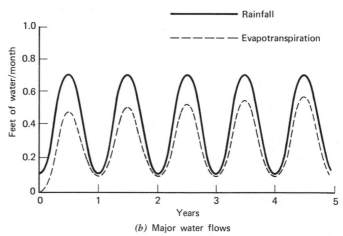

(b) Major water flows

Figure 9 Coastal ridge aquifer storage (a) and corresponding flows (b) simulating primitive conditions. Steady-state conditions are reached from a zero-storage initial condition in approximately 2 years.

aquifer behavior is observed with large quantities of water stored during the summer followed by greatly reduced aquifer storage during the relatively dry winters.

Results of analog computer simulations of the coastal ridge ecosystem under four different conditions are summarized in Figure 11. Included are contingencies of full development and partial development making maximum

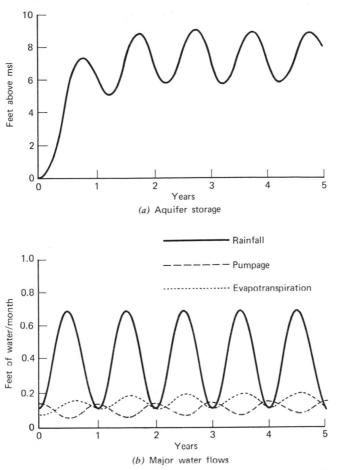

Figure 10 Coastal ridge aquifer storage (*a*) and corresponding flows (*b*) for a simulation of present conditions. Note how natural water usage is in phase with water supply and present use patterns are out of phase.

use of natural programs of water management. Note that the swamp-conservation alternative provides aquifer storage similar to present conditions even though pumpage is increased by approximately 50%.

Gordon River basin simulations corresponding to each assumed land use are summarized in Figure 12. Decreasing aquifer stability with increased development is again observed. Full development simulations show "negative" aquifer storage during dry periods that if allowed to occur would

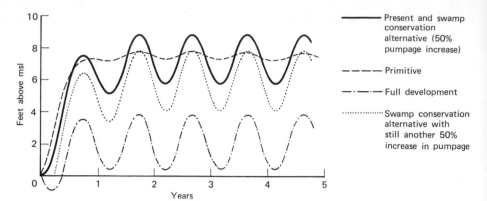

Figure 11 Coastal-ridge aquifer storage for each condition shown simultaneously for comparison. The swamp-conservation alternative results in aquifer behavior almost identical to that presently observed, although municipal pumpage is increased by 50%.

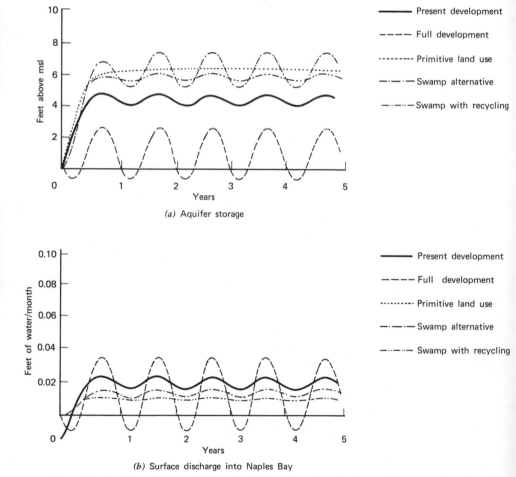

(a) Aquifer storage

(b) Surface discharge into Naples Bay

Figure 12 Basin aquifer storage (a) and variations in surface discharge (b) for each alternative of land use.

result in depleted wells, salt water intrusion, and municipal well contamination. Partial development retaining natural vegetation for water management results in aquifer behavior similar to present conditions, and when secondarily treated sewage is recycled through retention swamps (rather than discharged into the estuary as is the case now) extremely stable aquifer behavior is established. The most stable Naples Bay salinity regimes occur in simulations that include primitive and sewage recycling alternatives (Figure 12b).

DISCUSSION

It is apparent that the evolutionary adaptation of undeveloped ecological systems, or at least this one, results in the internal stabilization of potentially unstable water regimes. Thus it is no surprise that the highest model stability of water flow and storage occurs when parameters calculated for undeveloped conditions are simulated.

Results of simulations using present land-use parameters indicate a reduction in aquifer storage and a decrease in water table stability caused by present development patterns. Water deficits caused by man's activities are reflected in lower natural terrestrial metabolisms and, therefore, lower water consumption by natural systems. In this sense, remaining natural systems serve as a "buffer" that allows man to obtain net water supplies necessary for urban activities but at the expense of biotic productivity.

Full development according to present zoning regulations represents a reasonable extrapolation of present patterns of development. Since natural terrestrial buffers would be eliminated, water deficits could not be absorbed even partially. Severe aquifer drawdowns would result as predicted by both unit calculations and simulations. Unless present development patterns are drastically altered, full development would, therefore, not be possible without water subsidies obtained from outside this region. Unfortunately, intensive development in southeast Florida has already created enormous water deficits that require supplements from natural recharge areas extending as far north as the upper Kissimmee in central Florida. Given continued rapid development of this and other regions throughout the state, a region such as that containing the Gordon River cannot depend on obtaining water from remote recharge areas because other regions of even denser development are rapidly creating demands for the same water. Ultimately each area of development may be entitled only to that water falling within development boundaries.

Options available under conditions of full development and water supply include desalinization, recycling with tertiary treatment, or regulated water consumption during dry periods. Each option represents a regional fuel or administrative cost that must be financed by increased local taxes or federal subsidies (even if adequate fuel supplies are available).

Swamp conservation land use as shown in Figure 7 represents an alternative approach to water management that makes maximum use of work performed by natural systems rather than fossil-fuel-driven environmental technology. Retention of cypress swamps, reduction of channelization, wastewater recycling through natural systems, and use of ecological concepts for development that preserve recharge characteristics of the coastal ridge all contribute to greater stability of water regimes and increased aquifer storage.

REFERENCES

Black, Crow, and Eidsness. 1966. *Water system improvements for the city of Naples*. Engineering report.

Black, Crow, and Eidsness. 1969. *Collier County, Florida, Comprehensive Area Plan for water and sewer development*. Engineering report.

Black, Crow, and Eidsness. 1970. *Future water supply studies for the city of Naples, Florida*. Engineering report.

Black, Crow, and Eidsness. 1971. Preliminary plans for Gordon River Canal, Water Management District No. 7, Collier County, Florida. Engineering report.

Black, Crow, and Eidsness. 1971. *Water and wastewater studies for the city of Naples, Florida*. Engineering report.

Chow, Ven Te. 1964. *Handbook of applied hydrology*. McGraw-Hill, New York.

Coastal engineering study of the proposed clam pass improvement. 1970. Department of Coastal and Oceanographic Engineering, University of Florida.

Craighead, Frank C. 1971. *The trees of south Florida*. University of Miami Press, Miami.

Feiss, Carl T. 1972. *A plan for preservation of Rookery Bay in view of urban development of surrounding areas*. A report to the Collier County Conservancy, Urban Studies Department, University of Florida, Gainesville.

Florida Almanac, 1972. Dukane Press, Hollywood, Florida.

Florida Coastal Coordinating Council, 1972. *Coastal zone management in Florida—1971*, A report to the governor of Florida, Tallahassee.

Florida Statistical Abstract. 1970. Bureau of Economic and Business Research, College of Business Administration, University of Florida.

Florida Tourist Report. 1971. Florida Department of Commerce, Tallahassee.

Golley, Frank B. 1965. Structure and function of an old field broomsedge community. *Ecol. Monogr.* **35:** 113–137.

Golley, F. B., H. T. Odum, and R. F. Wilson. 1962. The structure and metabolism of a Puerto Rican red mangrove forest in May. *Ecology* **43:** 9–19.

Klein, H. 1954, *Ground water resources of the Naples area, Collier County, Florida*. Florida Geological Survey Report of Investigations No. 11.

LaRoe, E. T. 1971. *Position statement on the proposed Gordon River Canal*. Report to the Collier County Board of Commissioners, Collier County, Florida.

LaRoe, E. T. 1972. *Alternatives to the Gordon River Canal*, Report to the Collier County Board of Commissioners, Collier County, Florida.

Leighty, R. G. 1954. *Soil survey, Collier County, Florida*. U.S. Department of Agriculture and Florida Agricultural Extension Service.

Little, J. A., R. F. Schneider, and B. J. Carroll. 1970. *A synoptic survey of limnological characteristics of the Big Cypress Swamp, Florida*, Environmental Protection Agency, Southeast Water Laboratory.

Littlejohn, C. B. 1973. *An Energetics Approach to Land Use Planning in a Coastal Community of Southwest Florida*, Master thesis, Department of Environmental Engineering, University of Florida, Gainesville. Unpublished.

Lugo, Ariel E., S. E. Snedaker, S. Bayley, and H. T. Odum. 1971. *Models for planning and research for the South Florida environmental study*. Center for Aquatic Sciences, University of Florida, Gainesville.

McCoy, H. J. 1962. *Ground water investigation of Collier County, Florida*. Florida Geological Survey Report of Investigations No. 31.

McCoy, H. J. 1972. *Hydrologic data and studies, Collier County, Florida*. U.S. Geological Survey, Miami, Florida.

Morrow, B. W., and J. A. Stevens. 1971. *Hydrological effects of the proposed Gordon River Canal*, Report to the Collier County Board of Commissioners, Collier County, Florida.

Naples-on-the-Gulf and Collier County, Florida Statistical Fact Report 1971. Naples Area Chamber of Commerce, Naples, Florida.

Odum, E. P. 1971. *Fundamentals of Ecology*. Sanders, Philadelphia. P. 51.

Odum, E. P., and H. T. Odum. 1972. Natural areas as necessary components of man's total environment. *Transactions of the Thirty-seventh North American Wildlife and Natural Resources Conference, March 12–15*. Published by the Wildlife Management Institute, Washington, D.C.

Odum, H. T. 1971. *Environment, power, and society*. Wiley, New York.

Odum, H. T. 1972a. Use of energy diagrams for environmental impact statements. *Tools for Coastal Management*, Proceedings of the Conference February 14–19, 1972, Washington, D.C. Marine Technology Society, Washington, D.C.

Odum, H. T. 1972b. An energy circuit language for ecological and social systems: Its physical basis. In Bernard C. Patten (Ed.), *Systems analysis and simulation in ecology, Vol. II*. Academic, New York.

Odum, H. T., B. J. Copeland, and E. A. McMahan. 1969. *Coastal ecological systems of the United States*. A Report to the Federal Water Pollution Control Administration, Washington, D.C.

Odum, H. T., and C. F. Jordan. 1970. Metabolism and evaporation of the lower forest in a giant plastic cylinder. In H. T. Odum, (Ed.), *A tropical rain forest* Division of Technical Information, U.S. Atomic Energy Commission, Washington, D.C.

Odum, H. T., C. Littlejohn, and W. C. Huber. 1972. *An environment evaluation of the Gordon River area of Naples, Florida and the impact of development plans*. Department of Environmental Engineering Sciences, University of Florida, Gainesville.

Odum, H. T., and L. L. Peterson, 1972. "Relationship of energy and complexity in planning," *Archit. Des.* October 1972.

Odum, H. T., and R. F. Wilson. 1962. Further studies on reaeration and metabolism of Texas bays, 1958–60. *Publications of the Institute of Marine Science*, Vol. 8 pp. 23–55.

Parker, G. G., 1971. *A tentative water budget analysis of southwest Florida water management district*. Southwest Florida Water Management District, Brooksville.

Peterson, L., A. Lugo, S. Snedaker, H. T. Odum, et al. 1970. Towards a general theory of planning design. *SWEPT*, Dept. of Architecture, University of Florida.

Pyne, R., David A., and J. I. Garcia-Gengochea. 1971. *Hydrological study of the proposed Gordon River Canal*, Report to the Collier County Board of Commissioners, Collier County, Florida.

Sherwood, C. B., and H. Klein. 1961. *Ground water resources of northwestern Collier County Florida*. Florida Geological Survey Information Circular No. 29.

Todd, D. K. 1971. *Ground water hydrology*. Wiley, New York.

Wetterquist, Peterson, Odum, Christenson, Snedaker, et al. 1972. *Identification and evaluation of coastal resource patterns in Florida*. Department of Architecture, University of Florida, Gainesville.

Zoning, City of Naples, Florida, 1971. Municipal Code Corporation, Tallahassee, Florida.

Zoning Regulations, Collier County, Florida, 1970. Adley Associates, Inc., Tampa, Florida.

APPENDIX

Equations, calculations, and the analog computer diagram (Figure 13) for one simulation (coastal ridge, present land use) are as follows.
Equation:

$$\eta \dot{Q}_r = k_1 R - k_3(\text{PET})(Q_r) - k_4(P) - k_5(Q_r - G) - k_6 I$$

where R = rainfall
η = porosity*
PET = potential evapotranspiration[†]
Q_r = average ridge water table
G = average basin water table*
P = municipal pumpage
I = spray irrigation water loss[†]

Time constants:

$$k_1 R = \frac{4.23}{12} \rightarrow k = \frac{4.23/12}{4.57/12} = 0.926$$

$$k_3(\text{PET})(Q_r) = \frac{(1.03)(0.632)}{12} \rightarrow k_3 = \frac{(1.03)(0.632^{\ddagger})}{(4.57)(7)} = 0.0204$$

$$k_4 P = \frac{1.69}{12} \rightarrow k_4 = \frac{1.69/12}{2.41/12} = \frac{1.69}{2.41} = 0.70$$

* Assumed 30% based on USGS studies.
[†] Assumed approximately equal to and following the same distribution of rainfall.
* Assumed constant to simplify simulation.
[†] Assumed constant over year (even distribution) as conservative estimate.
[‡] Represents evapotranspiration of natural vegetation only; therefore obtained from Table 1 by multiplying natural ET (1.03) by fraction of area occupied (0.632).

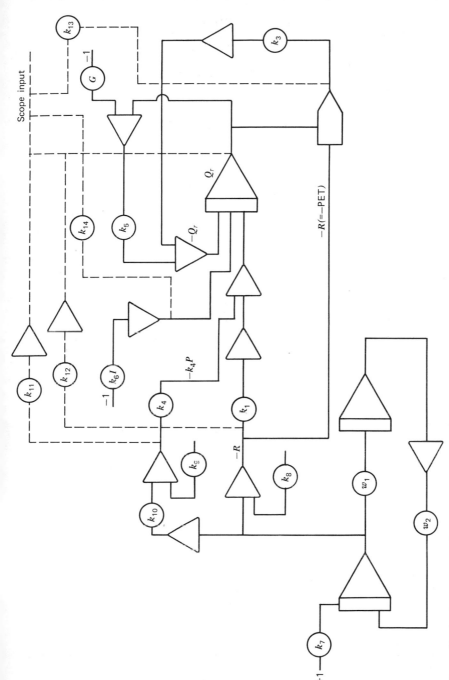

Figure 13 Analog computer diagram for simulation of coastal ridge water regimes.

475

$$k_5(Q_r - G) = \frac{0.71}{12} \rightarrow k_5 = \frac{0.0592}{7.00 - 5.45} = 0.0382$$

$$k_6I = \frac{1.21**}{12} \rightarrow k_6I = 0.101 \quad \text{(constant)}$$

Scaled variables:

Variable	Average Value (ft)	Max Expected Value (ft)	Scaled Variable
Q_r	7.00	10	$\{Q_r/10\}$
R	0.380	1	$\{R\}$
P	0.202	1	$\{P\}$
G	5.45	10	$\{G/10\}$
k_6I	0.101	1	$\{k_6I\}$
PET	0.380	1	$\{PET\}$

Scaled* equation:

$$\eta\dot{Q}_r = k_1R - K_3(PET)(Q_r) - k_4(P) - k_5(Q_r - G) - k_6I$$

$$0.3\dot{Q}_r = 0.926\{R\} - 0.0204\{PET\}\left\{\frac{Q_r}{10}\right\}10 - 0.70\{P\}$$

$$\qquad -0.0382\left\{\frac{Q_r - G}{10}\right\}10 - 0.101$$

$$\dot{Q}_r = 3.09\{R\} - 0.68\{PET\}\left\{\frac{Q_r}{10}\right\} - 2.33\{P\} - 1.27\left\{\frac{Q_r - G}{10}\right\} - 0.337$$

$$\left\{\frac{\dot{Q}_r}{10}\right\} = 0.309\{R\} - 0.068\{PET\}\left\{\frac{Q_r}{10}\right\} - 0.233\{P\}$$

$$\qquad -0.127\left\{\frac{Q_r - G}{10}\right\} - 0.034$$

Pot settings:

Equation	Forcing Functions	Output
$k_1 = 0.309$	$k_7 = 0.34$	$k_{11} = 1.00$
$k_3 = 0.068$	$w_1 = 0.524$	$k_{12} = 1.00$
$k_4 = 0.190$	$w_2 = 0.524$	$k_{13} = 0.204$
$k_5 = 0.127$	$k_8 = 0.460$	–
$k_6I = 0.034$	$k_9 = 0.220$	–
$G = 0.545$	$k_{10} = 0.240$	–

** 1.21 obtained by multiplying ET losses (due to spray irrigation) from man-managed systems by the fraction of total area occupied.
* Scaled for purposes of analog patching. Output readjusted to true values.

20
A Modeling Approach to Regional Planning in Franklin County and Apalachicola Bay, Florida

WALTER BOYNTON
DAVID E. HAWKINS
CHARLES GRAY

W hat role do coastal resources have in the ecological balance and economic vitality of coastal regions? What are the best ways to use resources? What degree of development creates a stable economy, continued viable resources, and a high quality of life in the long run?

Some recent experience has shown that increases in new development begin to stress natural estuarine and coastal resources and in many cases may even eliminate some of the resources that once were a vital part of the original economy, or that were a part of the original attraction for development (Cronin, 1967; Ryther, 1954). Additionally, when natural areas come into short supply, the taken-for-granted contributions they make to the overall economy decrease, requiring substitute expenditures of special work by man to replace their functions. The loss of natural system work may be reflected in higher taxes necessary to provide clean water, waste disposal and recreation, as well as the loss of resource-based commercial activities. The increasing cost of industrial fuels now may force us to focus attention on retaining self-renewing natural systems. Some specific controversies that have arisen recently in coastal regions include such issues as channelization, nutrient loading, and the development of wetlands and coastal areas for industrial, commercial, and residential uses.

We present here the preliminary results of a project aimed at determining the best mix of developed areas and self-renewing fisheries for a coastal county in Florida. Our work to date has focused on evaluating the sensitivity of the local oyster fishery to additional developments of the tourist, retirement, shipping, and cattle industries.

Study Area—Apalachicola Bay Region

Franklin County is located on the northwest Gulf Coast of Florida (Figure 1). The county is rural with most of the 7000 residents living along the coast. The physical characteristics of the county are dominated by flat, generally swampy forests that cover over 90% of the land area, and Apalachicola River and Bay. The bay is a medium-salinity estuary extending nearly the entire length of the county and is separated from the Gulf of Mexico by several offshore islands (Figure 1). The Apalachicola River has an average yearly discharge of 716 m^3 sec^{-1} (25,300 cfs) draining a watershed of 46,620 km^2 (18,000 miles^2) including parts of Georgia and Alabama (Hawkins, 1973). About 87% of the land area is owned by commercial

Work was supported on Project R/EM-3 of Florida Sea Grant Program, "Models for Coastal Management." H. T. Odum, principal investigator. Parts of this paper are based on graduate theses and dissertations in the Department of Environmental Engineering Sciences (Gray, 1972; Hawkins, 1973; Boynton, 1975).

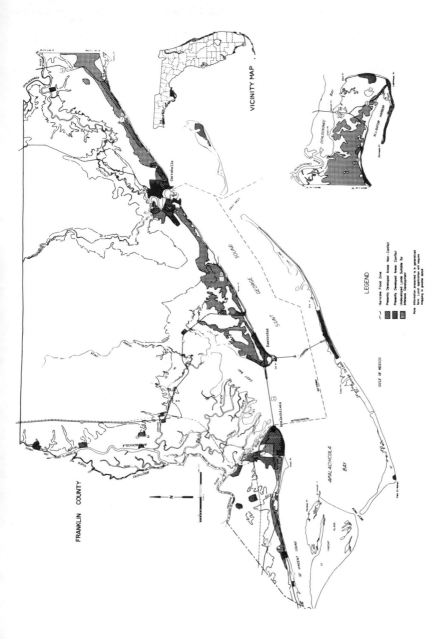

Figure 1 Franklin County and Apalachicola Bay including land use suggestions made by the Florida Coastal Coordinating Council.

479

pulp and paper companies or is part of the Apalachicola National Forest and is thought to be unavailable for industrial or residential development in the foreseeable future. Paper-company activities account for a very small part of the economic base of the county (Colberg and Windham, 1965; Rockwood, 1973).

The local economy is predominantly dependent on water-resource-based activities centered on Apalachicola Bay. The oyster fishery accounts for about 50% of total county income. The bay produces 90% of the commercial oyster harvest in the state and also supports shrimp, crab, and fin fish industries with a total annual dockside value for all species of about five million dollars (Rockwood, 1973). The largely undeveloped forest lands, inland and coastal waters, and barrier islands in Franklin County support a tourist business and provide recreational opportunities for the residents (Florida Tourist Study, 1970). Agriculture and manufacturing activities are of negligible importance at present, although a large cattle ranch now in the building stages may change this. Most basic goods, fuels, and services are purchased from outside the county. Per capita income is low compared to other areas of Florida and the nation, and population level is generally stagnant due in part to emigration of residents searching for more productive or stable jobs (Colber and Windham, 1965).

Fairly recently a variety of developments have occurred or have been proposed for the county. A cut was made in an off-shore island to facilitate access to the open Gulf, and this modification may have changed the salinity patterns and adversely affected oyster production. Future proposals include expansion of the tourist trade, addition of retirement homes, dredging or diversion of the Apalachicola River to expand shipping operations to a terminal in Jackson County, Florida, and the addition of a 40,000-acre cattle ranch just north of the town of Apalachicola. Local leaders in Franklin County view at least some of the above changes as further threats to the oyster industry, which is already threatened by pollution from sewage. They fear that if they allow or stimulate additional development they may lose the way of life they know and trust, namely, the fisheries. However, the question remains as to whether some additions can be made to the economy that will not conflict with or eliminate the fishery economy.

MODELING METHODOLOGY

A description of the symbols used in this section is found in Chapter 2 and some of the underlying philosophy in Chapter 7.

Lotka's maximum power principle states that the system that tends to prevail over alternative systems is the one that maximizes the use of all energy flows available to it, and it is the one which develops useful feedback roles for all participants. These feedbacks serve the purpose of assuring continued energy flows and capturing any additional useful flows that become available. In a system such as Franklin County containing both man and nature (towns, estuaries, fisheries, tourism, and forests), this means using the variety of components wisely in partnerships that will obtain maximum power flows through the total system, survive periods of stress, and build means for a vital economy of both man and nature in the long run. The welding of Lotka's principle to the type of system modeling reported here is best summarized in a recent paper by Odum (1974). In summary here, we represent corollaries of Lotka's principle in our models by showing all energy sources available to an area, especially those that are dominant and/or expected to change. Interaction processes that degrade part of the inflowing energy to build higher quality flows and structure are also shown. The diagrams emphasize pathways that recycle materials suspected of limiting major processes. Both energy and economic models have been developed from these diagrams.

The specific steps followed in constructing an energy model are listed in Table 1. In general, this involves constructing a model of the area of interest including all important input energies (both natural and manre-lated), materials and money, interactions (additive, multiplicative, switching, etc.), and sinks and exports. The modeling combines all component processess in the system of interest. The final diagram is both an easily understood synthesis of processes in the study area and a visual picture of a system of nonlinear differential equations describing the behavior of each state variable. Terms for each equation can be taken directly from the diagram as specified by the configuration of each pathway. The full set of equations are shown in Appendix Table 1. Note that each subscripted constant (k_i) in the equation corresponds to a subscripted flow (J_i) in Figure 3.

In this model, evaluated equations were simulated on an Applied Dynamics AD-80 analog computer. Data needed to evaluate flow rates, initial conditions, transfer coefficients, and some validation graphs were obtained from ongoing field studies, literature sources, and personal interviews. The basic data are given in Appendix Tables 2 to 4 and are treated more fully in our referenced work.

The model was tested for responses to expected changes in the study area. In each case, changes expected in the region were translated into new computer settings and run. Each computer run was similar to a controlled experiment in which one or more factors were changed. The model was tested against field observations to determine if the factors chosen for in-clusion in the model were the important ones and whether the coefficients were approximately correct. When this process has been completed the

Table 1   Modeling Methods, the General Procedure for Model Development and Simulation.[a]

---

1. Define the main system components (state variables) and forcing functions (outside energy sources).

2. Using discussions and existing understanding of processes, diagram the pathways of interaction and energy flow including switching and threshold actions. These pathways constitute algebraic terms when translated into equations (Appendix Table 1).

3. For a particular or average situation, include the numbers for stocks and flows. Solve term for each pathway for its coefficient after substituting the flow and stock values existing at that time. The relative importance of pathways becomes evident. If data is available over time develop time-series coefficients.

4. Usually, the first draft of a model having much of what we know of details and minor as well as major flows may have 50 components and many more pathways. Next, we try to group, simplify, and eliminate by setting many effects constant for the study consideration. This makes a complex model into a simplistic macromodel that focuses on the main issue in terms of its main driving functions and causative internal pathways.

5. For each storage unit (integrator) write a differential equation as the sum of the input and outflow pathway terms. Each term has a numbered coefficient.

6. Translate differential equations into analog computer diagram (wiring diagram). Scale the analog diagram and patch wiring according to the wiring diagram.

7. Simulate the model, testing for steady states and transient patterns, for sensitivity of response to varied forcing functions and coefficients, and for similarities to observed time graphs (validation). Relate to public issues at question.

8. Revise models to incorporate improvement to fit observations, and do it in a way that is dynamically explainable in terms of known mechanisms of the systems parts rather than as empirical curve fitting.

9. Draw conclusions about the consequences of the models and actions on them. Try to find test examples that will validate in real world tests the experiments done on the models. Often, this requires looking at historical trends.

---

[a] Adapted from Odum (1972).

resulting graphs of system behavior are predictions which are probably more accurate and contribute more understanding than the present intuitive methods of decision making. Although it is impossible to validate prefectly such predictions, these modeling efforts can at least explore what some intuitive choices might mean in the future.

The economic model presented here is aimed at evaluating the effects of changing levels of natural inputs and urban development on the oyster-based economy of Franklin County. To be sure that all important interactions and storages were included within the model, system boundaries were defined one level larger than the fishery, namely the county level.

Other, smaller scale models have been developed in this study for purposes of investigating the detailed interactions contained in the regional model and include a daily metabolism model and a model of oyster-salinity-predator relationships.

These simulation models are the first part of a three-part methodology. The second and third phases include energy-value optimization calculations for studying total system work under various alternatives (Odum et al., 1973) and aerial mapping to show spatial relationships of system components. Reports on these extensive applications of energy-value theories to Franklin County are available from the author.

THE MODEL

We developed a regional model of Apalachicola Bay, its principle economic factors, and what we perceive to be the most important interactions of these factors (Figure 2). This model is divided into three principle sections: Apalachicola Bay, including the principal biotic and physical parameters; the oyster industry (lower right); and the developed portion of the county (upper right). River flow transports inorganic nutrients, organic matter, toxins, and coliform bacteria into the bay in "baseline" concentrations independent of man's activities. Contributions to each of the above are also made from local sources that change as a function of development (J_{48}, J_{52}). Tidal exchange flushes the system with an average turnover time of several days. It should be noted that in this model the bay is considered to be a homogeneous body of water with no local gradients. Phytoplankton production is shown as pathway J_{39}, which is influenced by sunlight and nutrient concentration. The multiplicative workgate was chosen for this process because the response gives a limiting factor curve that is a well established photosynthetic response to limiting nutrients or sunlight (Rabinowitch and Govindjee, 1969). We concluded that nitrogen was the factor most likely to be limiting photosynthesis based on a review by Pomeroy et al. (1972) and on data from Apalachicola Bay provided by Estabrook (1973). Other decisions on establishing realistic configurations were made in a similar fashion, based on literature reviews and field work conducted in the bay.

The organic matter in the bay (Q_{12}) supports a population of oysters (Q_{11}), other benthic organisms (J_{41}), and predators (Q_{16}) that also feed on the oysters. Most oyster predators in this area are stressed or eliminated by salinity fluctuations. This is shown as flow J_{54}, which relates predator losses to salinity changes. Toxins stress the oyster population directly (J_{37}), while coliforms (Q_{15}) indicate sewage pollution and operate a switch in the model regulating oyster harvest (J_{35}).

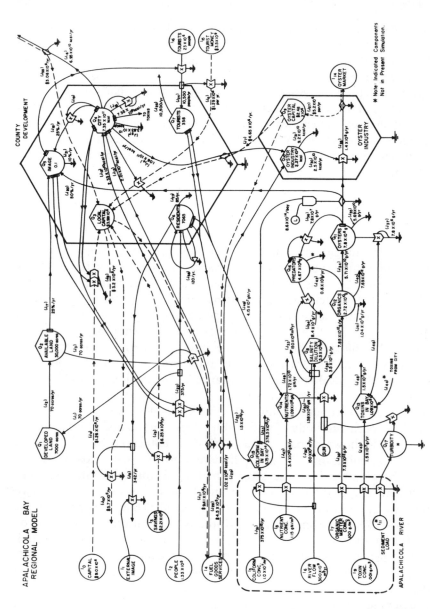

Figure 2 Model of Franklin County and Apalachicola Bay, Florida, with data determined for 1970.

Oysters are harvested in proportion to the standing crop of oysters and the size of the oyster industry. The oysters are sold in an external market, and the money received is used to buy needed goods and materials from both within the county (J_7) and from outside sources (J_{28}). Prices regulate the ratio of dollars to goods in each exchange and have an important effect in determining the economic position of each sector in the model. All prices are held constant in this simulation.

The developed sector in the model receives money from the oyster industry (J_7) in exchange for goods and services (J_{14}), outside investment (J_4), tourists (J_6), new residents (J_5), and from the sale of exported goods and services (J_{55}). The money received by this sector is exchanged for goods, fuels, and services (J_{11}) needed to maintain and add new structure. Several of the money flows are influenced by interactions with the regional image (Q_5) and city structure (Q_4) as shown by the multiplier symbols. Available land (Q_2) is shown interacting with fuels (I_4) to produce more developed land (Q_1) and a flow into city structure (J_{12}). The number of people living in the county (Q_6) is increased or decreased by immigration (J_{22}), which is dependent on the regional image and city structure, by losses due to mortality (J_{23}) and emigration from the county (J_9). The birth rate is approximately constant. Dollar flows are associated with the movement of people.

As an example of the model development process used here, the image (Q_5) "storage" is examined in more detail. This is a very important, but difficult to quantify, portion of our model. An early question that emerged in our study concerned the factors that attracted money and people to the Franklin County region. It was hypothesized that there is in every region a factor that represents the ability of an area to attract outside investments and thus amplify that region's ability to do work. We chose to call this factor image and to make it proportional to both the appropriate storages and flows. In interviews, some suggested that image was principally a function of the natural resources of the area. Others felt it was related to the existing community structure in the area. Several combinations of rates and storages were suggested. Thus image was a state variable that should integrate the diverse factors that may attract people and capital to the county. In our model image had inputs from estuarine and terrestrial metabolism and urban activities as represented by the level of city structure (Q_4). Each input to image was estimated from conversations with local leaders. Of the total input, 25% was estimated to result from the amount of natural land available, 25% from the level of city structure, and 50% from environmental quality as indicated by estuarine water quality and continued oyster harvesting. As a state variable image also had a decay time and could influence other pathways for a period of time after inputs stopped (J_{21}). In the model image is shown as affecting decisions to reside in the county, visit, or invest. It was assumed that losses from image were negligible in attracting outside materials to the county because promotion and advertising were not large county activities. No special feedback inter-actions are used between image and image inputs for this same reason. In addition, overdevelopment of the area can cause a decline in image input as shown by pathway J_{18}. Further details concerning the image concept, justification, and uncertainties may be found in Boynton (1975).

RESULTS OF SIMULATIONS

We present here representative simulation results for the purpose of showing general trends, sensitive portions identified to date, and the kind of output obtained from our models. These results have already been used in planning decisions.

Figures 3 to 7 summarize the behavior of selected model variables over a specific time period for the conditions and assumptions described in the respective figure caption and text. The vertical scale on each figure indicates percent of full scale, while actual values for each variable at full scale are given in the captions.

Model Simulation for 1970 Levels of Forcing Functions, Rates, and Storages

Figure 3 shows model responses under conditions corresponding to 1970. In this simulation and in all simulations to date, forcing function inflows were held constant, and inflationary trends were not considered. This simulation shows very gradual increases in residents, developed land, and oyster harvest rate over a 30-year period. Sharper increases were shown for local capital, and oyster industry capital and structure. Image and city structure declined slightly, while populations of coliforms, oysters, and tourists remained approximately the same. When we decreased capital investment inflow in the model by 50% all growth trends leveled. In a simulation run going approximately 150 years into the future (Figure 4), the system under the influence of constant inputs reached a steady state with the oyster industry, oyster capital, and harvest rate up by factors of about four, and local capital and developed land up by factors of about two. Oysters, residents, image, and city structure remained essentially constant. More recent simulations using a detailed model of the oyster industry indicate that fourfold increases in harvest rate may not be possible. A two-fold increase now seems more realistic, and hence the growth shown in Figure 4 may be too large.

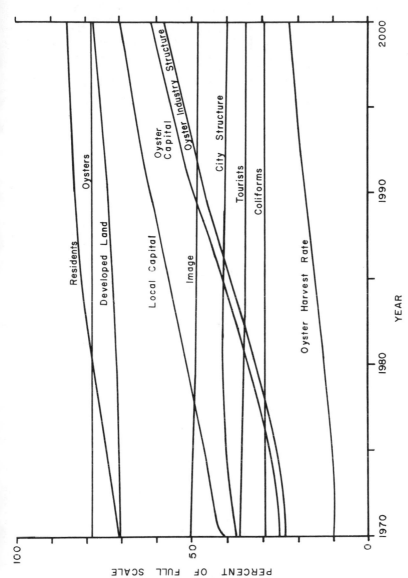

Figure 3 Simulation response of the model under constant (1970) levels of external forcing functions. (Note: Full scale corresponds to residents, 10,000 people; developed land, 10,000 acres; local capital, 10×10^6; city structure, 2×10^{12} kcal; tourists, 1000 people; image, 0.2; oysters, 1×10^9 g dry organic matter yr^{-1}; oyster-industry capital, 10×10^6; oyster-industry structure, 10×10^{12} kcal.)

487

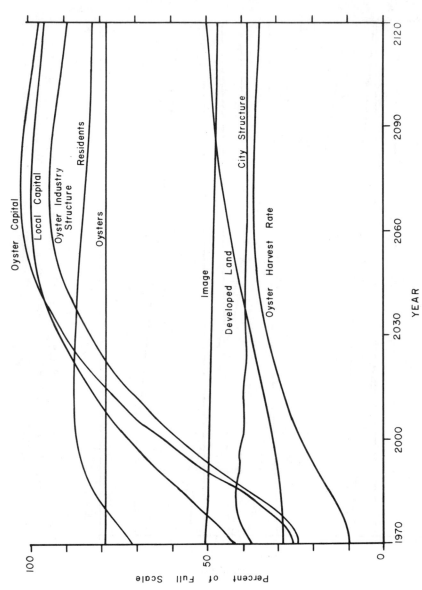

Figure 4 Extended simulation of the model under constant (1970) levels of external forcing functions (Note: See Figure 3 legend for scale interpretations).

488

Effects of Repeated 1-Year Shutdowns of Commercial Oyster Harvesting in Apalachicola Bay

Federal and State laws require that harvesting of oysters be discontinued whenever coliform bacteria counts in the bay exceed 79 MPN (most probable number) per 100 ml. In this run we increased coliform inputs from the river such that the legal limit was exceeded for a period of 1 year in every 4 years. This simulation experiment was conducted because there is a possibility that a large tourist development and cattle ranch may be established on St. George Island and along the river, respectively. Preliminary calculations indicated that a doubling of coliform input would be sufficient to exceed legal limits. Further calculations and simulations (Boynton, 1975) suggested that either of these development possibilities would be large enough to potentially cause such input changes. And since portions of the bay near the river mouth frequently have coliform counts above 79 MPN, the doubling needed for the experimental closure was considered realistic. During the 1-year closure no oysters were harvested. The results are shown in Figure 5. Oyster capital, after one year of closure, is 20% of its former level, while oyster industry structure and local capital each drop by 20%. In an earlier version of this simulation, a single 1-year closure indicated a complete recovery time of 7 to 10 years. In this simulation each closure is initiated at a lower level. This simulation assumes no alternative sources of income for those involved in the fishery and perhaps indicates a larger loss to the local economy than what might be expected to actually occur.

Effects or Reduced River Flow

This simulation explores possible consequences of decreased river flow on the local economy of Franklin County. It was run in response to a river diversion project on the Apalachicola River proposed by the U.S. Army Corps of Engineers. As shown in Figure 6, when river flow is decreased by 50% similar declines are seen in organic matter and oysters. The oyster decline is caused mainly by decreases in available organic material from the river. As shown in Figure 2, most of the food for oysters comes from the river organic material rather than from the phytoplankton production. Both oyster capital and the oyster industry show rapid declines to very low levels within 10 years. If the predation effect had been stronger, as recent data indicates that it should be, the decline of the industry would have been even more rapid. Thus we consider these results a minimum estimate. Local capital also exhibits a sharp decline to about 30% of its former level reflecting the dependence of the local economy on the oyster industry.

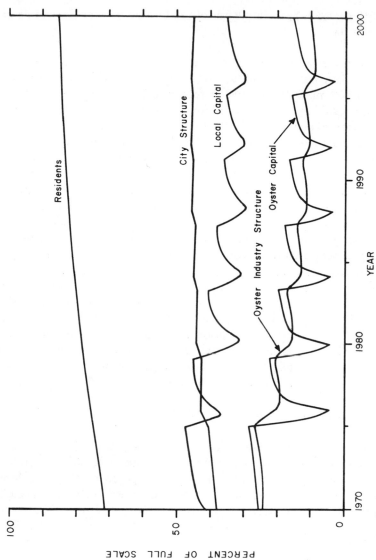

Figure 5 Effect of repeated 1-year shutdowns of commercial oyster harvesting in Apalachicola Bay on future growth in Franklin County (Note: See Figure 3 legend for scale interpretation).

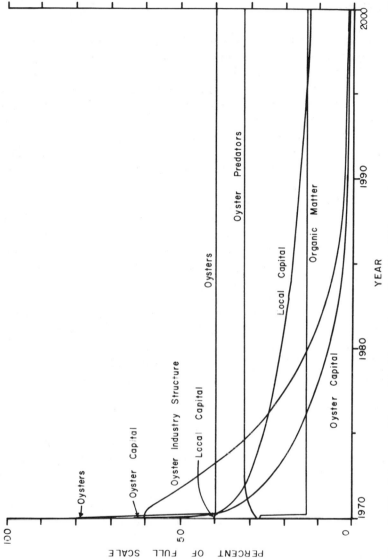

Figure 6 Stimulation run of the model with river flow reduced by 50% as would be the case if the river diversion plan were activated (Note: See Figure 3 legend for scale interpretation).

491

DISCUSSION

Our effort has sought to increase our understanding of a large, open system of man and nature. The modeling methodology has served the multiple purpose of synthesizing many diverse pieces of information, showing the relative importance of the major inputs, drains, exports, and interactions, and during simulations testing our knowledge of the real system as represented by the model. In addition, the simulations have helped us explore what might be the consequences of different management decisions. If in the future no radical changes occur in the study area that were not considered in the model, the following considerations would seem to apply to Franklin County.

As shown in Figure 3, all man-related sectors of the economy are either growing very slowly (1% per year) or are presently leveled and are likely to remain so in the future. The changes in numbers of residents, local capital, developed land, and city structure are all fairly accurate representations of recent trends in Franklin County. This figure is provided to show the near-steady-state nature of most of the county. Figure 4 shows the extent of growth that might be expected if inputs remain constant. Under these conditions, at a point about 100 years into the future, losses (repair, replacement, etc.), balance inputs, and a slowly oscillating steady state results.

If investments from outside the county are cut in half or if the buying power of money invested is halved (such as by inflation), no real growth is observed in the model. With capital inflow doubled a mild boom is observed, assuming again no inflation. While growth may be adjusted by outside capital inputs, the present economy is basically supported by the oyster fishery, and growth is closely related to increases in oyster harvest rate. If any new development interferes in a negative way with oyster production (via coliform increases, salinity modifications, etc.), the model output suggests a substantial decline in the economic well-being of the region.

A critical problem, yet to be resolved, concerns the point where current inflationary trends force a leveling of growth in the county. Building only to the sustainable level avoids the drop that would occur if cheap fuels and goods are no longer available. In this regard, the slow growth tradition and use of abundant natural resources in Franklin County may be a viable management guideline in times such as these when fuel supplies are uncertain. Many are now advocating a return to a sustained resource use and leveled growth in South Florida where environmental and fiscal problems are severe (Marshall, 1972).

Unfortunately, there is no long history by which to judge the predictive value of large open-system simulation models such as the one presented here. This being the case, we have used historical data where available to validate our simulation models. Our modeling approach assumes that if all important time-varying energy sources are included in the model, then general trends in the past and on into the future will be representative of the real situation.

We did run one such "historical" simulation as a partial validation of our model using 1970 values for forcing functions and coefficients, but starting from conditions present in 1820. Although the simulated population growth of the region was similar to the real population growth, the development of land and the growth of the oyster fishery were both considerably less than what actually has occurred (Figure 7). Since completing this simulation, however, we have found that Apalachicola was a busy port in the early 1800s serving cotton producers in parts of Alabama and Georgia. Thus a good deal of the early growth in the county was supported by this function rather than oystering. Also, hurricanes and yellow fever destroyed and slowed growth during the early times in Franklin County (Martin, 1944). If these features were to be added to the validation runs, growth would be in closer agreement to the observed values. At this point trends seem correct, indicating that major interactions and forcing functions have been included. Improved data, historically varying forcing functions, and the addition of some detail would improve the timing of events in the model.

Figures 5 and 6 show the response of the county economy to coliform increases beyond the legal limits and a 50% reduction in river flow, respectively. The simulations illustrate the importance of the oyster fishery to the overall economy and the industry's sensitivity to legal and physical modifications. In the case of closures due to excessive coliform levels the industry and local economies gradually decline, with sharp declines during closures, but do rebound whenever oystering is permitted. However, when river flow is decreased, the overall decline is sharper and permanent as the standing crop of oysters is reduced to lower levels. The industry is sensitive to both types of modifications but is more permanently affected by changes in the river flow. Reduced river flow is likely should a proposed river diversion canal be built or consumptive uses increase.

An additional question to be resolved involves the value of the fishery relative to other types of industry. If the value to man of all county resources is dependent on the total work done by industry and nature, how much new work would be needed to replace the work associated with the oyster industry or the fisheries in general if they were lost? In the case of the oyster industry, much of the work done in producing the product is done by the natural estuarine system without dollar costs. Preliminary estimates indicate this free subsidy to man's economy to be approximately 50% of the total work or about 3×10^{10} kcal yr^{-1}, based on energy quality concepts, the details of which can be found in Odum et al. (1973), in Chapter 7, and in

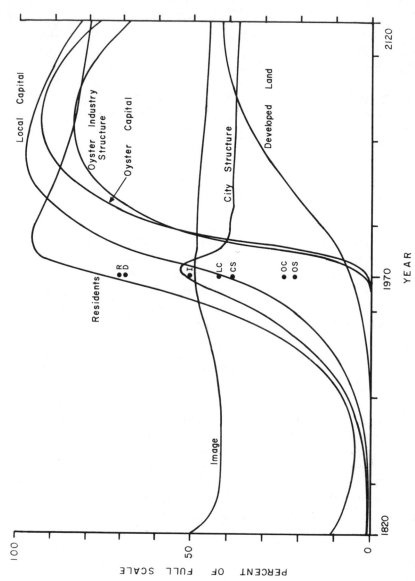

Figure 7 Historical simulation of Franklin County.

494

our forthcoming work. It seems clear that in a period when net power from conventional fuels may be declining and the addition of more fossil-fuel intensive work is less desirable or perhaps even not possible (Odum, 1973), a policy of retaining the mutualistic man-nature relationship inherent in the fishery is highly desirable. Future simulations, energy-value calculations, and mapping will be aimed at testing this concept and other management alternatives. In the meantime our preliminary simulations have indicated that the proposed additions to the county's economy have a large potential for adversely affecting the total economic well-being of the county and if instituted should be undertaken only very cautiously.

ACKNOWLEDGMENTS

We express our thanks to Dr. H. T. Odum and The Systems Ecology Group in Gainesville for many helpful discussions and suggestions. Mr. James Estes, Franklin County Agricultural Extension Agent, and Mr. Lanse Anderson, former City Manager of Apalachicola, provided financial information about the county. We also acknowledge the cooperation extended to us by Florida State University in the use of their lab and in the exchange of information about Apalachicola Bay.

REFERENCES

Boynton, W. 1975. Energy basis of a coastal region: Franklin County and Apalachicola Bay, Florida. Ph. D. dissertation, University of Florida.

Colberg, M. R. and D. M. Windham. 1965. *The oyster based economy of Franklin County, Florida*. Florida State University, Tallahassee.

Cronin, I. F. 1967. The role of man in estuarine processes. In G. H. Lauff (Ed.), *Estuaries*. American Association for the Advancement of Sciences. Publ. No. 83. Washington, D.C. Pp. 667–689.

Estabrook, R. H. 1973. Phytoplankton ecology and hydrography of Apalachicola Bay. M. S. thesis, Florida State University, Tallahassee.

Florida Department of Commerce. 1971. *The 1970 Florida Tourist Study*. Tallahassee.

Gray, C. 1972. Simplified model of Apalachicola Bay oyster industry and retirement develop ment plans. Research report submitted as part of the requirements for the degree of master of engineering, University of Florida.

Hawkins, D. E. 1973. Simulation and evaluation of a regional model of Franklin County, Florida and the Apalachicola Bay oyster industry. Research report submitted as part of the requirements for the degree of master of engineering, University of Florida.

Lotka, A. J. 1922. A contribution to the energetics of evolution. *Proc. Natl. Acad. Sci.* **8:** 147–155.

Marshall, A. R. 1972. South Florida—A case study in carrying capacity. AAAS section on geology and geography, human impact on the Atlantic Coastal Zone. AAAS annual meeting 1972, Washington, D.C.

Martin, S. W. 1944. *Florida during the territorial days.* The University of Georgia Press, Athens.

Odum, H. T. 1971. *Environment, power, and society.* Wiley, New York. 331 pp.

Odum, H. T. 1972. Use of energy diagrams for environmental impact statements, In *Tools for coastal management. Proceeding of the conference.* Marine Technology Society, Washington, D.C. Pp. 197–213.

Odum, H. T., C. Littlejohn, and W. C. Huber. 1972. *An environmental evaluation of the Gordon River area of Naples, Florida and the impact of developmental plans.* Department of Environmental Engineering Sciences, University of Florida.

Odum, H. T. 1973. Energy, ecology and economics. *Ambio* **2**(6): 220–227.

Odum, H. T., W. Smith, H. McKellar, D. Young, M. Lehman, and M. Kemp. 1973. Preliminary presentation of models to show interactions of power plants and estuary at Crystal River, Florida and energy costs and benefits for alternatives of management of cooling waters. *Progress report to Florida Power Corporation and licensing agencies concerned with planning at Crystal River, Florida.* Environmental Engineering Sciences, University of Florida, Gainesville.

Odum, H. T. 1974. Combining energy laws and corollaries of the maximum power principle with visual systems mathematics. Siam Institute for Mathematics and Society Research Application Conference on Ecosystems, July 1–4, 1974, Alta, Utah.

Pomeroy, L. R., L. R. Shenton, R. D. H. Jones, and R. J. Reimold. 1972. Nutrient flux in estuaries. In G. E. Likens (Ed.). *Nutrients and Eutrophication: The Limiting Nutrient Controversy.* Special Symposia, Vol I. *Limnol. Ocean* Lawrence, Kansas.

Rabinowitch, E. and Govindjee. 1969. *Photosynthesis.* Wiley, New York. 273 pp.

Rockwood, C. E. 1973. *A management program for the oyster resource in Apalachicola Bay, Florida.* The Florida State University, Tallahassee.

Ryther, J. H. 1954. The ecology of phytoplankton blooms in Moriches Bay and Great South Bay, Long Island, New York. *Biol. Bull*; **106**: 198–209.

Appendix Table 1 System Equations Corresponding to Storages and Flows in Figure 2 and Tables 2, 3, and 4.

1. Q_1 = Developed land

$$\dot{Q}_1 = k_1 Q_2 Q_3 - k_2 Q_1$$

2. Q_2 = Available land

$$\dot{Q}_2 = k_2 Q_1 - k_3 Q_2 Q_3 \qquad (k_3 = k_1)$$

3. Q_3 = Local capital

$$\dot{Q}_3 = k_4 Q_5 Q_4 I_0 + k_5 I_2 I_3 + k_6 Q_4 Q_5 I_{15} I_{16} + k_{55} Q_4 + k_7 Q_9 - k_8 Q_3 - k_9 k_{10} Q_6 - k_{11} Q_3$$

4. Q_4 = City structure

$$\dot{Q}_4 = k_{12} Q_2 Q_3 + k_{13} Q_3 + k_{14} Q_4 Q_9 - k_{15} Q_4^2 - k_{16} Q_4 - k_{57} Q_4$$

5. Q_5 = Image

$$\dot{Q}_5 = k_{17} Q_2 + k_{18}(Q_4 - k_{19} Q_4^2) + k_{20} Q_{11} - k_{21} Q_5$$

6. Q_6 = Residents

$$\dot{Q}_6 = k_{22} I_2 Q_4 Q_5 - k_9 Q_6 - k_{23} Q_6 + k_{56} Q_6$$

7. Q_7 = Tourists

$$\dot{Q}_7 = k_{24} I_{16} Q_5 - k_{25} Q_7$$

8. Q_8 = Freshwater in bay

$$\dot{Q}_8 = k_{32} I_6 - k_{33} Q_8$$

9. Q_9 = Oyster industry capital

$$\dot{Q}_9 = k_{27} Q_{10} Q_{11} - k_7 Q_9 - k_{28} Q_9$$

10. Q_{10} = Oyster industry structure

$$\dot{Q}_{10} = k_{29} Q_9 + k_{14} Q_4 Q_9 - k_{30} Q_{10} Q_{11} - k_{31} Q_{10}$$

11. Q_{11} = Oysters

$$\dot{Q}_{11} = k_{34} Q_{12} - k_{35} Q_{11} - k_{36} Q_{11} Q_{16} - k_{26} Q_{10} Q_{11} - k_{37} Q_{11} Q_{13}$$

12. Q_{12} = Organic matter in bay

$$\dot{Q}_{12} = k_{38} I_7 I_6 + k_{39} I_{10} Q_{14} - k_{40} Q_{12} - k_{41} Q_{12} - k_{42} Q_{12} - k_{34} Q_{12}$$

13. Q_{13} = Toxins in bay

$$\dot{Q}_{13} = k_{43} I_6 I_9 + k_{44} Q_4 - k_{45} Q_{13}$$

14. Q_{14} = Nutrients in bay

$$\dot{Q}_{14} = k_{47} I_6 I_8 + k_{48}(Q_6 + Q_7) - k_{49} Q_{14} - k_{50} I_{10} Q_{14}$$

15. Q_{15} = Coliforms in bay

$$\dot{Q}_{15} = k_{51} I_6 I_5 + k_{52}(Q_6 + Q_7) - k_{53} Q_{15}$$

16. Q_{16} = Oyster predators

$$\dot{Q}_{16} = k_{42} Q_{12} + k_{36} Q_{11} Q_{16} - k_{54} I_6 Q_{16}$$

Appendix Table 2 System Component Storages

Storage	Name	Description	1970 Storage Level, Remarks	Ref.[a]
Q_1	Developed land	All land currently in use for the purpose of structural siting. Includes buildings, airport, roads, and other significant modifications to the natural lands in Franklin County	7000 acres Area of land presently developed determined by planimeter measurement	6
Q_2	Available land	All land in an undeveloped condition in the county, but not including those lands within the National Forest or held by pulp and paper companies	30,000 acres Area determined by planimeter measurement from Coastal Coordinating Council Maps. This area excludes land that according to the CCC should not be developed	5
Q_3	Local capital	Considered to be a common deposit of all money, both private and public which enters the county by way of export sales, tourist business, resident bank accounts, capital investment from outside the county and outside subsidy. A measure of the county wealth in terms of immediately liquidable money assets, but excluding the money associated with oyster industry	$\$3.91 \times 10^6$ Total bank deposits = $\$6.39 \times 10^6$. Approximately 38.8% of total county personal income is associated with the oyster industry; remainder with other exports and services. For proportional savings, $Q_3 = (61.2\%)(\$6.39 \times 10^6) = \$3.91 \times 10^6$	2, 7, 8
Q_4	City structure	A measure of the nonliquid assets of the county. Includes buildings, services, capital goods, and marketable goods, but not those associated with the oyster industry	7.35×10^{11} kcal Assessed county value = $\$42.4 \times 10^6$. Value of oyster equipment = $\$10 \times 10^6$ (see Q_{10}). Energy value of money in county 2.27×10^4 kcal dollar^{-1}. Then, $Q_4 = [\$(42.4 - 10)(10^6)][2.27 \times 10^4$ kcal dollar$^{-1}] = 7.35 \times 10^{11}$ kcal	8, see Table 4 (flow J_{14}) and Table 2 (storage Q_{10})

Q_5	Image	A measure of the attractiveness of Franklin County relative to other areas of the country with regards to capital investment, personal residence, and recreational opportunities. Image is a function of available land, degree of city structure, and the viability of the oyster industry as an indicator of water quality. The above are taken in the ratio of 25, 25, and 50% respectively	0.1 units (1.0 is full scale) This setting is somewhat arbitrary at present. The setting of 0.1 is based on the observation that the county has remained relatively undeveloped, while other areas of the state have been rapidly developed. As other areas fill up and so forth, the image of Franklin County may improve	
Q_6	Residents	The total permanent population of the county	7065 people	9
Q_7	Tourists	The average number of tourists in Franklin County at any one time	358 people; 10,500 yr^{-1} at 12.4 days $visit^{-1}$; no seasonal variation considered	3, 10
Q_8	Salinity dilution	Volume of fresh water in Apalachicola Bay. Variations in this variable are taken as a measure of salinity changes	5.50×10^8 m^3 at $S = 15$ ppt	1
Q_9	Oyster industry capital	Total liquid dollar assets of the oyster industry. Includes fishermen, house operators, transporters, others directly employed	$\$2.48 \times 10^6 = (38.8\%)(6.39 \times 10^6)$	2, 7, 8 Table 2 (Q_3)
Q_{10}	Oyster industry structure	Capital investment value in structure, facilities, and material goods associated with the oyster industry. Includes boats, motors, shucking houses, cars, and homes of those associated with the industry	2.27×10^{11} kcal = $(\$10 \times 10^6)(2.27 \times 10^4$ kcal $dollar^{-1})$	2, 11, 12, 13

Appendix Table 2 (*Continued*)

Storage	Name	Description	1970 Storage Level, Remarks	Ref.[a]
Q_{11}	Oysters	Standing crop of oysters (excluding shells) in the bay	7.8×10^8 g dry organic matter bay^{-1} = (715 bushels acre^{-1})(6000 acres)(4 lb meat bushel^{-1})(454 g lb^{-1})(0.1 g dry organics/g biomass (net))	14, 18
Q_{12}	Organic matter in bay	Total amount of organic matter in the bay available to oysters	2.72×10^{10} g dry organic matter. (25 g dry organic/m^3)(2.33 m)(467 × 10^6m^2) = 2.72×10^{10}	15, 22
Q_{13}	Toxins in bay	Total toxins in the bay including pesticides and herbicides	1.09×10^5 g = $(1086 \times 10^6 \quad$ m^3) $(0.0001$ g m$^{-3})$	16, 23, 28, 30
Q_{14}	Nutrients	Total amount of NO$_3$–N in the bay	$(0.1$ g NO$_3$–N m$^{-3})(1085.8 \times 10^6$ m$^3)$ = 1.086×10^8 g NO$_3$–N bay^{-1}	17, 22
Q_{15}	Coliforms in bay	Average number of coliform bacteria in the bay. Taken at period of average river flow (S = 15 ppt in bay)	53 MPN at 15‰ salinity. $(530 \times 10^3$ m$^{-3})(1086 \times 10^6$ m$^3)$ = 5.75×10^{14} coliforms bay^{-1}	17
Q_{16}	Predators	Biomass of organisms which prey on oysters. Includes conch, drills, and stone crabs. The biomass is calculated at 15‰ S, when few predators are present	4.67×10^6 g dry organic matter bay^{-1} $(0.01$ g m$^{-2})(467 \times 10^6$ m$^2)$ = 4.67×10^6 g bay^{-1} (Predator density taken as 0.01 g m^{-2} over the area of the total bay)	16, 18

[a] References cited are reported in Table 5.

500

Appendix Table 3   External Energy Sources

Energy Source	Name	Description	1970 Source Level, Remarks	Ref.[a]
I_0	Outside capital	Pool of outside capital potentially available to Franklin County	9.0×10^6 yr^{-1}	13
I_1	External image	Measure of the attractiveness of areas outside of Franklin County as viewed by people living in the county	Not currently employed in model Effect is accounted for through J_{10} in Table 4	–
I_2	People	Number of people who potentially could be attracted to Franklin County	0.133×10^6 people yr^{-1} (This is roughly the state-wide immigration rate yr^{-1} adjusted for area)	19
I_3	Savings	Liquid assets associated with people moving into the county	$5000 household^{-1}. Household of 3 persons Total of 2.21×10^8 yr^{-1}	20
I_4	Goods, services, and fuel	Material goods, fuel, and services obtained from sources outside of the county in exchange for money	I_4 does not appear explicitly in the equations. See J_{13} (Table 4)	–
I_5	Coliform concentration in river	Average coliform concentration in the Apalachicola River	1.0×10^7 m^3. 250 MPN = average value. 250 MPN = 2500 liter $\times$ 1000 liter m^{-3} = 2.5×10^6 m^{-3}. I_5 was set at 1.0×10^7 m^{-3} as the highest expected value of I_5	17, 21
I_6	River flow	Average yearly input of fresh water to Apalachicola Bay from the river	300×10^9 m^3 yr^{-1} maximum expected Average yearly river flow is 150×10^9 m^3 yr^{-1}	11
L	Permissible level of coliform bacteria	Level above which Department of Public Health closes waters to commercial oyster harvest	8.6×10^{14} coliforms bay^{-1} (Based on 79 MPN 100 ml^{-1}) 790 liter^{-1} $\times$ 1000 liter m^{-3} $\times$ 1085.8 $\times$ 10^6 m^3 = 8.6×10^{14}	18
I_7	Organic matter concentration in river	Average concentration of organic matter (particulate and dissolved) in the river	100 g dry organic matter m^{-3} Average of 50 g dry organic matter m^{-3}. The value of I_7 is set at 100 g dry organic m^{-3} as the highest expected value	17, 23
I_8	Nutrient concentration in river	Concentration of NO$_3$–N m^{-3} in the river	0.23 g NO$_3$–N m^{-3} average I_8 is set at 1.15 g NO$_3$–N m^{-3} as the highest value expected	17, 22
I_9	Toxin concentration in river	Concentration of pesticides, herbicides, and industrial toxins in river water	0.0001 g m^{-3} I_9 is set at 0.001 g m^{-3} as highest expected value	23, 30
I_{10}	Sunlight	Total solar radiation incident on surface area of Bay	660×10^{12} kcal yr^{-1} (467×10^6 m^2 bay^{-1}) (3900 kcal m^{-2} day^{-1})(365 days)	24
I_{15}	Tourist capital	Money potentially available to the county via tourist trade	3.5×10^9 yr^{-1} annually Total spent by all Florida tourists = (23 $\times$ 10^6 tourists) ($150 person^{-1}). Current expenditures in Franklin County are 1.29×10^6 yr^{-1}	10
I_{16}	Tourists	Tourists potentially available to Franklin County	23×10^6 yr^{-1} Total number of tourists visiting Florida. The current flow is about 10,500 yr^{-1} to Franklin County	10

[a] References cited are reported in Table 5.

Appendix Table 4 System Flow Rates

Flow Path	Flow Name	Description	1970 Flow Rate, Remarks	Ref.[a]
J_1	Land being developed	Rate of conversion of available land to developed land in county	70 acres yr^{-1}	12
J_2	Land depreciation	Rate of degradation of developed land structures to undeveloped or natural state without maintenance	70 acres yr^{-1} Based on 1% of developed land degrading per year: $(0.01)(7000 \text{ acres}) = 70 \text{ yr}^{-1}$	20
J_3	Land being developed	Same as J_1	70 acres yr^{-1}	12
J_4	Capital inflow	Flow of capital from outside county to local capital stock for investment or use within county	$1,260,000 yr^{-1} $(70 \text{ acres yr}^{-1})(\$22,500 \text{ acre}^{-1})(80\% \text{ from outside county})$	12, 20
J_5	Savings inflow	Flow of savings associated with people moving into county	$625,000 ($5000 household^{-1})(125 households moving into county per year)	20
J_6	Tourist expenditures	Total money spent by tourists while in Franklin County	1.29×10^6 yr^{-1} (10,500 tourists yr^{-1})(12.4 days visit^{-1})($10.00 person-day^{-1})	3, 10, 13
J_7	Local expenditures by oyster industry employees	Money spent for goods, fuels, and services by the oyster industry within Franklin County	4.65×10^6 yr^{-1} $(0.90)(\$5.2 \times 10^6 \text{ yr}^{-1})$	2, 20
J_8	Investment return	Return on county investments	$320,000 yr^{-1}. 5% of total liquid assets $= (0.05)(\$6.39 \times 10^6)$	20
J_9	Emigration	Rate of residents leaving Franklin County	342 yr^{-1} 375 yr^{-1} moving into county (see J_{22}). $+35$ yr^{-1} as net birth (see J_{56} and J_{23}). -68 yr^{-1} as reported yearly —population increase in 1972 $= 342 \text{ yr}^{-1}$	20, 25
J_{10}	Savings outflow	Money associated with people leaving the county	$570,000 yr^{-1} $= (\$5000 \text{ household}^{-1})(1 \text{ household}/3 \text{ residents})(342 \text{ residents/year})$	20
J_{11}	Purchases from outside county	County expenditures, excluding direct purchases by oyster industry, for goods, fuel, and services produced outside the county	9.41×10^6 yr^{-1} (Total income $-$ oyster industry costs $-$ savings $-$ investment return) $= [(\$10 + 0.45) \times 10^6 - \$0.45 \times 10^6 - 0.32 \times 10^6 = \$9.41 \times 10^6]$	2, 7, 8, 20
J_{12}	Developed structure	Contribution to county structure from land development and construction	2.33×10^{10} kcal yr^{-1} $= (\$1.015 \times 10^6 \text{ yr}^{-1}$ spent on land development)(at 0.27×10^4 kcal dollar^{-1}). (see J_{14} remarks for explanation)	12, 20
J_{13}	Goods, fuel services	Contribution to county structure from goods, fuels, and services purchased outside the county and used to maintain or improve current facilities	2.01×10^{11} kcal yr^{-1} Total energy value of goods purchased outside county $- J_{12} = (\$9.41 \times 10^6 \text{ yr}^{-1})(2.27 \times 10^4 \text{ kcal yr}^{-1}) - 2.33 \times 10^{10} \text{ kcal yr}^{-1}$	20, 31–34
J_{14}	Goods, fuel services	Goods, fuels, and services purchased within the county by the oyster industry	1.06×10^{11} kcal yr^{-1} J_7 (local energy value of money). Local energy value of money = total fuel energy/total outside purchases $= 2.24 \times 10^{11}$ kcal/9.41×10^6 dollars $= 2.27 \times 10^4$ kcal dollar^{-1}	18, 29–32
J_{15}	Structure degradation	Rate of depreciation of city structure	This is density dependent and in view of the sparse population in Franklin County we set this flow at 7.35×10^9 kcal yr^{-1} or 1% yr^{-1}	13, 26
J_{16}	Export flow of energy	Flow of exports (excluding oysters) out of county	6.95×10^{10} kcal yr^{-1} $= (\$3.06 \times 10^6 \text{ yr}^{-1})(2.27 \times 10^4 \text{ kcal dollar}^{-1})$	2, 20

Flow Name	Description	1970 Flow Rate, Remarks	Ref.[a]
Image	See Table 2 (Q_5)	25% of total image contribution at present due to land availability	11–13
Image	See Table 2 (Q_5)	25% of total image contribution at present level of structure	11–13
Image	See Table 2 (Q_5)	Established to reduce the image contribution of city structure to a level of zero at three times the current structure level	13
Image	See Table 2 (Q_5)	50% of total image contribution at present water resources quality level	11–13
Image	The linear decay of image over time	10% yr^{-1} degradation	26
Immigration	Total inflow of new residents to the county	Assume 125 households yr^{-1} × 3/household = 375/yr^{-1}	20, 25
County death rate		85 yr^{-1}	8
Tourist influx	Total inflow of tourists to the county each year	10,500 yr^{-1}	10
Tourist efflux	Required to balance J_{24}	10,500 yr^{-1}	10
Oyster harvest	Total harvest of oysters on a yearly basis	1.4 × 10^8 g dry organic matter yr^{-1} (or 1.4 × 10^9 g meat yr^{-1})	27
Oyster industry income	Total income to oyster industry computed as dockside value × 4. The 4 is the multiplier factor for oyster sales.	\$5.2 × 10^6 yr^{-1} (4 times dockside value)	2, 27
Oyster industry expenditures	Money flow to outside of county for fuels and so forth spent directly by the oyster industry	\$450,000 yr^{-1} \$500 household^{-1} or 8.65% of total oyster industry income	12, 20
Goods, fuels, services	The fuels, goods, and services entering the oyster industry directly from outside the county	1.02 × 10^{10} kcal yr^{-1} (J_{28})(2.27 × 10^4 kcal dollar^{-1})	12, 20
Fuels	The fuels (gasoline) consumed in the operation of the boats in harvesting oysters	2.5 × 10^{10} kcal yr^{-1}	14, 16, 18
Depreciation of oyster industry	Yearly depreciation rate of all equipment and personal property associated with the oyster industry	9.2 × 10^{10} kcal yr^{-1} This figure includes depreciation of boats, motors, shucking houses, homes, cars, and so forth	3, 11, 12, 16
River flow	Average yearly input of fresh water in the Bay from the Apalachicola River. Rainfall is not included as a separate source of fresh water	150 × 10^9 m^3 yr^{-1} = (4750 m^3 sec^{-1})(3600 sec hr^{-1})(24 hr day^{-1})(365 day yr^{-1})	1
Fresh water flushing rate	This figure assumes no great evaporative loss. It balances inflow rates on a yearly basis	150 × 10^9 m^3 yr^{-1}	1
Organic matter uptake by oysters	Calculated by summing the oyster biomass (Q_{11}) utilization of organics: respiration + production + mortality + feces + pseudofeces + harvest	5.71 × 10^{10} g yr^{-1}	14, 16, 18
Oyster metabolism	Calculated by subtracting toxin stress, predation, and harvest rate from J_{34}	5.69 × 10^{10} g yr^{-1}	14, 16, 18
Loss of oyster biomass to predators	This flow is regulated by both the level of predators and oysters. Data indicates this flow is small at 15‰ (due to there being few predators present); we chose 10%/yr of oyster biomass as the flow at present	7.8 × 10^7 g yr^{-1} (0.10)(7.8 × 10^8) = 7.8 × 10^7 g yr^{-1} 10% of oyster stock yr^{-1} at a salinity of 15‰	14, 16
Oyster biomass loss due to toxin stress	Calculations indicate that for present free water and tissue concentrations of DDT, a decrease (or loss) of 1% of oyster biomass a year is reasonable	7.8 × 10^6 g yr^{-1} 1% of oyster stock yr^{-1} at present toxin concentrations	16, 23
Organic matter inflows	Total organic matter inflow to the bay from the river	7.50 × 10^{12} g yr^{-1} (50 g organic matter m^{-3})(150 × 10^9 m^3 yr^{-1}) = 7.5 × 10^{12} g yr^{-1}	16, 22

Flow Path	Flow Name	Description	1970 Flow Rate, Remarks	R
J_{39}	Production of organic matter	Phytoplankton and marsh production of organics in the bay per year	3.5×10^{11} g yr^{-1} (0.8 g organic matter m^{-2} day)(365 days) $(1085.8 \times 10^6$ m^3 bay$^{-1}) = 3.5 \times 10^{11}$ g yr^{-1}	22
J_{40}	Organic matter outflow	This was calculated as total inflows − total metabolism = total outflow	7.69×10^{12} g yr^{-1} $7.85 \times 10^{12} - 0.16 \times 10^{12} = 7.69 \times 10^{12}$ g yr^{-1}	16,
J_{41}	Organic matter uptake by all organisms except oysters and oyster predators	Calculated as total uptake − oyster uptake	1.04×10^{11} g organic bay^{-1} yr^{-1}. $1.611 \times 10^{11} - 0.571 \times 10^{11} = 1.04 \times 10^{11}$	16,
J_{42}	Organic matter uptake by predators	Calculated as that organic matter necessary to fill metabolic needs of predators in addition to oyster input	0.6×10^7 g yr^{-1}	14,
J_{43}	Toxin inflow	Inflow of toxins to the bay from the river. At present this includes pesticides and pesticide residues	1.5×10^7 g day$^{-1} = (0.0001$ ppm) $\times (150 \times 10^9$ m^3 yr$^{-1}) = 1.5 \times 10^7$ g yr^{-1}	16,
J_{44}	Toxin inflow	Runoff of toxins from Apalachicola and Carrabelle	Negligible at present	
J_{45}	Toxin outflow	Calculated to balance toxin inflows	1.5×10^7 g yr^{-1}	16,
J_{46}	Toxin uptake		Negligible. Treated as sensor stress on oysters	16,
J_{47}	Nutrient inflow	Nutrient inflow (NO$_3$–N) to the bay from the river	3.4×10^{10} g NO$_3$–N yr$^{-1} = (0.23$ g NO$_3$–N m$^{-3})(1085.6 \times 10^6$ m^3 yr$^{-1}) = 3.4 \times 10^{10}$ g yr^{-1}	17,
J_{48}	Nutrient inflow	Nutrient input to the bay from the town of Apalachicola and Carrabelle	4.15×10^7 g NO$_3$–N yr$^{-1} = (30$ mg N liter$^{-1})(3785 \times 10^3$ liter day$^{-1})(365$ days yr$^{-1}) = 4.15 \times 10^7$ g NO$_3$–N yr^{-1}	22,
J_{49}	Nutrient outflow	Calculated as difference between inputs and portion utilized by phytoplankton	1.72×10^{10} g NO$_3$–N yr^{-1} See J_{50}	1, 2
J_{50}	Nutrient uptake	Uptake by phytoplankton calculated by assuming the organic matter fixed by phytoplankton was 5% nitrogen	1.58×10^{10} g NO$_3$–N yr^{-1}	16,
J_{51}	Coliform inflow	Coliform inputs from the Apalachicola River	375×10^{15} yr^{-1} $(2.5 \times 10^6$ m$^{-3})(150 \times 10^9$ m^3 yr$^{-1}) = 375 \times 10^{15}$ yr^{-1}	1, 1
J_{52}	Coliform inflow	Coliform inputs from local sewage	1.5×10^{15} yr$^{-1} = (10^6$ gal day$^{-1})(4 \times 10^6$ coliforms/gal^{-1}) (365 days)	17,
J_{53}	Coliform outflow	Calculated to balance the total inflows of coliforms to the bay	376.5×10^{15} yr^{-1}	1, 1
J_{54}	Stress on oyster predators	Calculated to balance the total inputs to predators per year	8.4×10^7 g yr^{-1}	14,
J_{55}	Income from exports	Income from all exports including labor but excluding oysters	\$3.06 $\times 10^6$ yr^{-1}. \$10 $\times 10^6$ yr$^{-1} - ($5.2 \times 10^6$ yr$^{-1} + 1.29×10^6 yr$^{-1} + 0.45×10^6 yr$^{-1}) = 3.06×10^6 yr^{-1} Total income − (oysters + tourists + investment)	20,
J_{56}	County birth rate	Rate of births in Franklin County	120 yr^{-1}	8
J_{57}	Linear degradation of county structure	The accounts for the continual decay of structure over time. We assumed a value of 5% yr^{-1}	3.83×10^{10} kcal yr^{-1} (5% yr^{-1})	13

[a] References cited are reported in Table 5.

Appendix Table 5   Sources of Data Reported in Appendix Tables 2–4.

1. Gorsline, D. S. 1963. Oceanography of Apalachicola Bay, Florida, *in* T. Clements, Ed. Essays in Marine Geology in Honor of K. O. Emery, University of Southern California, Los Angeles, California.

2. Colberg, M. R., and D. M. Windham. 1965. The Oyster Based Economy of Franklin County, Florida. Florida State University, Tallahassee, Florida.

3. Colberg, M. R., T. S. Dietrich, and D. M. Windham. 1968. The Social and Economic Values of Apalachicola Bay, Florida. Florida State University, Tallahassee, Florida.

4. Odum, H. T. 1971. Environment, Power and Society. Wiley-Interscience, New York.

5. Northwest Florida Development Council and Economic Development District. 1972. Commercial Tourism Land Absorption Study. Tallahassee, Florida.

6. Florida Coastal Coordinating Council. 1970. Florida Coastal Zone Land Use and Ownership. Tallahassee, Florida.

7. Carrabelle and Apalachicola State Bank records. 1970–1972.

8. Florida Statistical Abstract. 1971. University of Florida Press, Gainesville, Florida.

9. Florida Department of Commerce, Bureau of Industrial Development. 1970. Tallahassee, Florida.

10. Florida Department of Commerce. 1971. 1970 Florida Tourist Study. Tallahassee, Florida.

11. Estes, J., Franklin County Agricultural Extension Agent. Personal communication.

12. Anderson, L., Apalachicola City Manager. Personal communication.

13. Boynton, W., and D. Hawkins. Estimates. University of Florida, Gainesville, Florida.

14. Menzel, R. W., N. C. Hulings, and R. R. Hathaway. 1966. Oyster Abundance in Apalachicola Bay, Florida in Relation to Biotic Associations Influenced by Salinity and Other Factors. *Gulf Res. Rep.* 2.

15. Boynton, W. 1974. Phytoplankton Productivity in Chincoteague Bay, Maryland-Virginia. M. S. Thesis. University of North Carolina, Chapel Hill, North Carolina.

16. Boynton, W. Estimates. University of Florida, Gainesville, Florida.

17. Keyes, V. Bureau of Sanitary Engineering, Division of Health, Department of Health and Rehabilitative Services, Jacksonville, Florida. Personal communication.

18. Day, J., W. G. Smith, P. R. Wagner, and W. C. Stowe. 1972. Community Structure and Energy Flow in a Salt Marsh and Shallow Bay Estuarine System in Louisiana. Office of Sea Grant Development. Louisiana State University, Baton Rouge, Louisiana.

19. Littlejohn, C. Division of Planning, State Department of Natural Resources, Tallahassee, Florida. Personal communication.

20. Hawkins, D. Estimate. University of Florida, Gainesville, Florida.

21. Taylor, J. Department of Health and Rehabilitative Services, Apalachicola, Florida. Personal communication.

22. Iverson, R. Department of Oceanography, Florida State University, Tallahassee, Florida. Personal communication.

23. The Effects of Pesticides in Fish and Wildlife. 1964. U.S. Department of Interior, Circular 226.

24. Odum, E. P. 1971. Fundamentals of Ecology, Saunders, Philadelphia, Pa.

25. Division of Population, University of Florida, Gainesville, Florida.

Appendix Table 5 *(Continued)*

26. Odum, H. T. Estimate. University of Florida, Gainesville, Florida.
27. Summary of Florida Commercial Marine Landings. 1970. Florida Department of Natural Resources, Division of Marine Resources, Bureau of Science and Technology, Tallahassee, Florida.
28. Livingston, R. Biology Department, Florida State University, Tallahassee, Florida. Personal communication.
29. Fox, J. L., H. D. Putnam, and M. Keith. Estimates. University of Florida, Gainesville, Florida.
30. Thompson, N. Pesticide Research Lab, University of Florida, Gainesville, Florida.
31. Florida Power Corporation, March 13, 1973, St. Petersburg, Florida.
32. Thomas, L. Department of Revenue, Tallahassee, Florida.
33. Stapleton, S. Insurance Commission, Liquified Petroleum Gas Division, Tallahassee, Florida.
34. Rutledge, R. Revenue Officer, Finance and Accounting, Department of Agriculture, Tallahassee, Florida.
35. Brezonik, P. Department of Environmental Engineering, University of Florida, Gainesville, Florida.

21
Energy Cost-Benefit Analysis Applied to Power Plants Near Crystal River, Florida

W. MICHAEL KEMP
WADE H. B. SMITH
HENRY N. MCKELLAR
MELVIN E. LEHMAN
MARK HOMER
DONALD L. YOUNG
HOWARD T. ODUM

The need to evaluate natural systems quantitatively in designing human systems is becoming increasingly apparent. This intuitive proposition has been discussed previously in certain scientific circles (see, e.g., Oglesby et al., 1972; Darling and Milton, 1966), and in 1970 it was officially incorporated into the structure of legal processes in this country when Congress passed the National Environmental Policy Act (NEPA, 1969). This law calls for environmental impact assessment to "insure that presently unquantified environmental amenities and values may be given appropriate consideration in decision making along with economic and technical considerations." This would require a quantitative understanding of the coupled systems of man and nature. Recently a theory has been developed toward that end and formulated as an energy evaluation methodology (Odum 1971, 1976; Chapters 2 and 7). This general theoretical framework is applied here to questions of fitting power plants into coastal zone systems.

Prior to NEPA the Army Corps of Engineers and various resource economists had widely employed a dollar cost-benefit procedure whereby the positive and negative effects of a given project within the human sector are quantified with economic tools. Details of this technique and its application are reviewed elsewhere (James and Lee, 1971; Prest and Turvey, 1966; Dorfman, 1963; Chapter 6). This method allows planners and decision makers to choose between project alternatives for the one that apparently maximizes dollar efficiency. However, it measures the value of nature only for processes that involve some work service done by man as with fisheries products or recreational expenditures. The method does not measure the total work contribution of nature to man's economy.

In response to the NEPA mandate several formal techniques for environmental impact assessment have been suggested. Leopold et al. (1971), for instance, proposed the use of a large evaluation matrix with natural system components as matrix rows and categories of impact as column headings. Intersection points are used to tabulate numbers, which subjectively indicate the relative magnitude and importance of each impact on a system component, thus providing an indication as to the value of nature. However, the intersections of the Leopold matrix are assumed linear, so that chains of interactions and feedback properties cannot be considered (Odum, 1971). Only environmental effects are considered, and no format is offered for comparing various economic and environmental factors.

This work was supported by the Florida Power Corporation under contract No. GEC–159, 914–200–188.19 with the Dept. of Environmental Engineering Sciences, University of Florida, H. T. Odum, principal investigator.

Modifications to this basic matrix evaluation technique have been suggested, as in the use of the "Delphi decision process" proposed by Dee et al. (1973), which employs an interdisciplinary "panel of experts" to decide on the functional relationships and relative importance at matrix intersections. The Atomic Energy Commission (1972) has developed still another cost-benefit method to be used in environmental impact statements for nuclear power plants and other AEC regulated projects. This method calls for the tabulation of effects both to man and to nature for various project alternatives. Evaluations are done in terms of different units of measurement ranging from pounds $year^{-1}$, to acres, to rems $year^{-1}$, to hours $year^{-1}$, to dollars $year^{-1}$, to qualitatively descriptive words, thus making a decision between alternatives difficult because different kinds of effects are not directly comparable.

Any useful evaluation technique must have at its foundation some measurable objective criterion by which alternative project plans can be compared. Moreover, the analysis itself must be done in terms of a general measurement parameter that is common to all system functions and external influences being considered. In the process of human systems planning, which includes assessing the impact of planned activities on the regional system of man and nature, the ultimate objective must be to design vital, prosperous systems that will survive in competition with alternative systems, be they neighboring economic systems or other options for the system in question. The "Lotka maximum power principle" described in Chapter 7 states that this goal can be achieved if the plan allows for the maximum development of the system's energy resources into useful functions. This is a logical objective criterion for planning and impact analysis. Since all "real-world" processes can be described and compared in terms of their energy involvement, we propose a new cost-benefit analysis technique that employs energy-measurement units.

This paper describes a procedure based on these criteria, using as a case study the evaluation of the role of a coastal power plant in its environment. The evolution of this "energy cost-benefit" methodology can be traced in a series of recent papers by Odum (1972, 1974a, 1974b, 1974c, Chapter 7), Odum et al. (1973, 1974), and Kylstra (1974).

CONCEPTS OF ENVIRONMENTAL IMPACT ANALYSIS
AND REGIONAL PLANNING

The following section explains some basic principles underlying the energy cost-benefit method. A more complete discussion is given in Odum (1971, 1976; Chapter 7).

Lotka Principle of Maximum Power for Regional Design

According to a general principle of system evolution and survival first suggested by A. J. Lotka (1922) and later extended and developed in Odum (1971, 1974a, 1976), the system that will survive in the long run against the forces of natural selection is that one that maximizes the use of available energies in work processes. This requires that a system build adequate structure from which to channel energies into feedback pathways that further increase its total power budgets. Furthermore, a system of man and nature, such as a region containing estuaries, industries, power plants, and human settlements makes the most competitive use of its total available energies (both natural and fossil fuel) by developing patterns that emphasize partnerships, symbioses, and diverse functions, and by avoiding unnecessary waste within the energy constraints prevailing at any moment.

Energy as a Common Denominator

Previously, most quantitative planning and project evaluation has been done in terms of dollars, and within the sphere of man's activities money usually serves as an adequate index of energy value. When extended, however, to analyze the larger coupled system of man and nature it is inadequate because it does not consider the large energy flows of nature in support of man's activities for which man makes no direct money payment. By definition all work processes involve flows of energy (see any physics text, e.g., Richards et al., 1960), and since matter and information have energy equivalents, energy units are most appropriate to serve as common denominator in analysis of complex systems involving man and nature. In human economies money and energy (goods, services, fuels, etc.) flow countercurrent to one another, and the ratio of the total U.S. fuel consumption to its GNP gives a relationship of energy to dollars within the general U.S. economy. The various sectors of our economy are so intertwined that this ratio can be used adequately to convert most dollar flows into energy flows. The diverse energy flows in both natural ecosystems and human economies can thus be put on an equal footing for integrated analysis.

Converting Energy Values to Work Values Using Energy Quality

In any large and diverse region there is a broad spectrum of many kinds of energies driving system processes. But since Lotka's principle deals with the useful work accomplishments resulting from energy flowing in a system and not just with the heat equivalent value of that energy, the general ability of a given flow of a given kind of energy to do a unit of work needs to be estimated quantitatively. Previous workers have recognized that different types of energy do not possess equal ability to perform work (Evans, 1969; Tribus and McIrvine, 1971), implying that energy flows have a "quality" factor associated with them.

Within the network of both natural systems and the economic systems of man, as well as within the combined systems of both, many components of varied energy qualities occur working together. The work of lower energy quality units and functions develops higher quality ones, which in turn feed back special services of management and recycling for the effective functioning of the whole system. For example, in biotic systems one observes food chains converging in the high quality functions of larger complex animals that have management roles in the ecosystem (Figure 1). This concept of "energy quality" and methods for calculating energy quality conversion factors are further explained by Odum in Chapter 7. Derivations of conversion factors used in this analysis are outlined in footnotes to Table 1, and details of calculations are available elsewhere (Odum et al., 1974).

Transition and Steady State

When a modification is made in a regional pattern, there may be a transition period in which subsystem patterns change as part of the adapting and selecting processes, ultimately leading to a new kind of steady state that maximizes useful power under the new conditions. These temporary losses of system work must be charged as a cost to the modification made, but must not become the primary issue determining the desirability of one planning alternative over another. A new or changed energy regime, while displacing some previous values, is also the basis for additional work value to the region. The main issue should be the long range nature of the new steady state after adaptation has occurred as compared to the one displaced.

Ultimate Evaluation on Large Scale

Because all real systems are open systems with energy flowing across their boundaries, modifications in a given system will affect the larger system in which it is embedded. Therefore, any planning or project impact analysis must consider not only the responses of the system to local, direct effects, but also the larger system that includes the smaller as a functioning subcomponent. This hierarchical arrangement of systems and subsystems for a power plant and its region of interaction is illustrated in Figure 2. In an energy evaluation of management options using the Lotka principle, the ultimate choice among alternatives must be based on the maximization of

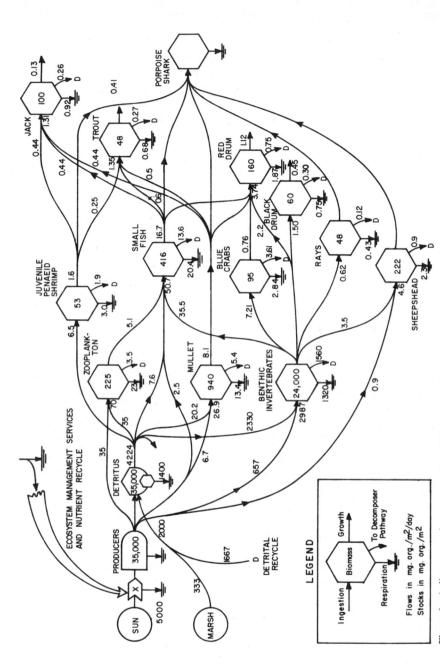

Figure 1 A diagrammatic representation of the food web for the Crystal River estuary illustrating the convergence of different trophic pathways to larger complex animals that have management roles in the ecosystem. This diagram based on extensive quantitative field sampling was used to calculate energy-quality ratios for organisms lost from entrainment and entrapment.

Table 1 Annual Energy Budget for Region Affected by Power Plants (3 units) at Crystal River, Florida.

Energy Source (Figure 4)[a]	Flow of Heat Content (kcal $\times 10^{12}$)	Energy Quality Ratio[b]	Work Value in Fossil-Fuel Equivalents[c] ($\times 10^{12}$)
Natural energies:			
Wind[d]	2.3	0.13	0.3
Tides[e]	2.5	0.4	1.0
Waves[f]	0.3	0.2	0.1
Fresh water head from rain[g]	0.6	1.7	1.0
Fresh water (as dilutant)[h]	0.3	1.0	0.3
Sun-based productivity			
Land[i]	343.0	0.07	24.0
Estuary[j]	15.8	0.07	1.1
Subtotal free	364.4		27.8
Purchased energies:			
Power plant fuels[k]	+23.0[p]	1.0	23.0
Other fuels[l]	+22.3[p]	1.0	22.3
Goods and services imported[m]	+56.3[p]	1.0	56.3
Exported[n]	−9.9[p]	1.0	−9.9
Tourists and capital investment[o]	+56.0[p]	1.0	56.0
	144.1		147.7
Total	506.5		175.5

[a] See text for calculation of size of affected area.

[b] Energy quality ratios (EQR) are tentative. Details of how they were calculated are available elsewhere (Odum et al., 1974), but the basis of each is as follows:

Wind. Used data for the energy balance of the earth's atmosphere from Hess (1959) after Phillips (1956). EQR based on the amount of solar energy required to generate the resulting wind energy.

Tides. Based on the amount of tidal energy to generate electric power at La Rance, France and proposed power plant on the Bay of Fundy (Anonymous, 1966; Lawton, 1972a, b).

Footnotes of Table 1 (*Continued*)

Waves. Considered sediment transport work done by ocean waves and the equivalent fossil fuel work (dredging) required to transport the same quantity of sediment over the same distance (Bascom, 1964; Weigel, 1964; Ingle, 1966).

Fresh water head from rain. Based on generation of electricity from proposed dam on Apalachicola River, Florida (Boynton, 1975).

Fresh water as dilutant. From fossil fuel and dollar cost of desalinating sea water (Odum, 1970; United Nations, 1967, 1969, 1970).

Productivity. Calculated from conversion of photosynthetically reduced sugars to wood biomass to electricity with a back calculation to coal (Odum, 1974a).

[c] Raw heat content value (column 1) times EQR (column 2).

[d] *Wind.* Yearly work done by winds was based on the kinetic energy of wind. An eddy diffusion coefficient of 10,000 cm^2 sec^{-1} was used. Average wind velocity was 8.7 mph (Bureau Economic Business Res., 1973). This wind speed was assumed to occur 10^4 cm above ground. Cross-sectional area assumed to be 3.7×10^{13} cm^2.

$$\text{Power, } P_w = 1.2 \times 10^{-3} \text{ g/cm}^3)(371 \text{ cm/sec})^2(1 \times 10^4 \text{ cm}^2/\text{sec})$$

$$\times \frac{(2.39 \times 10^{-11} \text{ kcal/erg})(3.15 \times 10^7 \text{ sec/yr})(3.7 \times 10^{13} \text{ cm}^2)}{(2)(1 \times 10^4 \text{ cm})}$$

$$= 2.3 \times 10^{12} \text{ kcal/yr}$$

[e] *Tides.* Power, $P_r = \rho g A h^2 / 2$

$$h = 91.4 \text{ cm}; A = 1.77 \times 10^{13} \text{ cm}^2; g = 980 \text{ cm sec}^{-2}; \rho = 1.025 \text{ g cm}^{-3}$$

$$P_r = (1.025 \text{ g cm}^{-3})(980 \text{ cm sec}^{-2})(1.77 \times 10^{13} \text{ cm}^2)\frac{(91.4 \text{ cm})^2}{2}$$

$$(2.38 \times 10^{-11} \text{ kcal erg}^{-1})(705 \text{ tides yr}^{-1})(2 \text{ cycles tide}^{-1})$$

$$P_r = 2.49 \times 10^{12} \text{ kcal/yr}$$

[f] *Waves.* Power, $P_{wv} = 1/8 \rho g^{3/2} H^{5/2} L$ (for shallow waves, $H/L < 1/20$)

$$\rho = 1.025 \text{ g cm}^2; g = 980 \text{ cm sec}^2; H = 30 \text{ cm}; L = 1.88 \times 10^7 \text{ cm}$$

$$(3.15 \times 10^7 \text{ sec yr}^{-1})(2.38 \times 10^{-11} \text{ kcal erg}^{-1})$$

$$P_{wv} = 0.27 \times 10^{12} \text{ kcal yr}^{-1} = 3.8 \times 10^6 \text{ ft-lb day}^{-1} \text{ ft}^{-1}$$

[g] *Fresh water head.* Power, $P_{FH} = \rho \cdot g \cdot V \cdot h$

$$P = 1.0 \text{ g cm}^{-3}; g = 980 \text{ cm sec}^{-2}$$

V = flow of fresh water running off and infiltrating in region

$$= (\text{rainfall})(\text{area})(\text{fraction runoff}) = (50 \text{ in yr}^{-1})(0.083 \text{ ft in}^{-1})(2.53 \times 10^{11} \text{ ft}^2)0.28$$

$$= 0.29 \times 10^{12} \text{ ft}^3 \text{ yr}^{-1} = 0.82 \times 10^{16} \text{ cm}^3 \text{ yr}^{-1}$$

Footnotes of Table 1 (*Continued*)

h = mean height of water = mean elevation of region = 100 ft = 30.48×10^3 cm

$$P_{FH} = (1 \text{ g cm}^{-3})(980 \text{ cm sec}^{-2})(30.48 \times 10^2 \text{ cm})(0.82 \times 10^{16} \text{ cm}^3 \text{ yr}^{-1})$$

$$(2.38 \times 10^{-11} \text{ kcal erg}^{-1})$$

$$= 0.58 \times 10^{12} \text{ kcal yr}^{-1}$$

[h] *Fresh water dilutant.* Power, $P_{FD} = (\Delta F)(V)(m_s) = \left(nRT \ln \dfrac{C_1}{C_2}\right)(V)(m_s)$

n = 1 mole/35 g; R = gas constant = 1.99 cal/mole °K

T = annual mean water temp = 20°C = 293°K

C_1 = delta freshwater concentration of dissolved solute = 120 ppm

C_2 = solute concentration of seawater as sink = 35,000 ppm

V = total freshwater in region = 2.95×10^{10} m^3 yr^{-1} (rain)

$$P_{FD} = \left(\frac{1 \text{ M}}{35 \text{ g}}\right)\left(1.99 \times 10^{-3} \frac{\text{kcal}}{\text{M}^\circ\text{K}}\right)(293^\circ\text{K}) \ln \left(\frac{120}{35,000}\right)$$

$$(120 \text{ g m}^{-3})(2.95 \times 10^{10} \text{ m}^3 \text{ yr}^{-1})$$

$$= 0.34 \times 10^{12} \text{ kcal yr}^{-1}$$

[i] *Land productivity.* Power, $P_{LP} = (A_L)(M)$

A_L = area of affected land region = 2.35×10^{10} m^2

M = typical mean metabolism for pine flatwoods = 40 kcal m^{-2} day^{-1}

P_{LP} = $(2.35 \times 10^{10} \text{ m}^2)(40 \text{ kcal m}^{-2} \text{ day}^{-1})(365 \text{ day yr}^{-1})$

$$= 343 \times 10^{12} \text{ kcal yr}^{-1}$$

[j] *Marine productivity.* Power, $P_{MP} = (A_m)(M)$

A_M = area of affected marine region = 1.74×10^9 m^2

M = typical outer bay metabolism = 25 kcal m^{-2} day^{-1}

P_{MP} = $(1.74 \times 10^9 \text{ m}^2)(25 \text{ kcal m}^2 \text{ day}^{-1})(365 \text{ day yr}^{-1})$

$$= 15.8 \times 10^{12} \text{ kcal yr}^{-1}$$

[k] *Power plant fuels.* Power, $P_{PC} = (C)(p)(f)(K_{FF})$

C = per capita electric consumption = 8.5×10^3 kW-hr cap^{-1} yr^{-1} (Zucchetto, 1975)

p = population of region = 2.3×10^6 (Bureau Economic Business Res., 1973)

f = fraction of population served by Crystal River = 0.38

K_{FF} = fossil fuel equivalent of electric power = 3.6

$P_{PC} = (8.5 \times 10^3 \text{ kW-hr cap}^{-1} \text{ yr}^{-1})(860 \text{ kcal kW-hr})(2.3 \times 10^6 \text{ cap})(0.38)(3.6)$

$\qquad = 23 \times 10^{12} \text{ kcal yr}^{-1}$ (FFE)

[l] *Other fuels.* (gasoline, natural gas, liquid fuels):

$P_{OF} = (C_G + C_{NG} + C_{LF})(P)$

$\quad P = (p)(f) = $ population in Crystal River power plant region.

$\quad C_G = $ consumption of gasoline $= 15.9 \times 10^6 \text{ kcal person}^{-1} \text{ yr}^{-1}$

$\quad C_{NG} = $ consumption of natural gas $= 1.98 \times 10^6 \text{ kcal person}^{-1} \text{ yr}^{-1}$

$\quad C_{LF} = $ consumption of liquid fuels $= 7.66 \times 10^6 \text{ kcal person}^{-1} \text{ yr}^{-1}$

$P_{OF} = [(15.9 + 1.98 + 7.66) \times 10^6](0.88 \times 10^6 \text{ cap})$ (data from Zucchetto, 1975)

$\qquad = 22.3 \times 10^{12} \text{ kcal yr}^{-1}$

(Fuels burned in electric power generation are subtracted.)

[m] *Imported goods and services.* Power, $P_I = I(f)(1 - e)(S)$

$\quad I = $ total dollars paid for imports in Florida $= \$16.88 \times 10^9 \text{ yr}^{-1}$ (Littlejohn, 1974)

$\quad f = $ fraction of Florida population in region $= 0.13$

$\quad e = $ fraction of budget spent on fuels $= 0.10$ (Littlejohn, 1974)

$\quad S = $ conversion of dollars to kcal $= 30,000 \text{ kcal dollar}^{-1}$

$\qquad = $ national average of (natural $+$ industrial energy use) GNP in dollars

$P_I = \$16.06 \times 10^9 \text{ yr}^{-1} (0.13)(0.90)(30,000 \text{ kcal dollar}^{-1}) = \$1.88 \times 10^9 \text{ yr}^{-1}$

$\quad = 5.63 \times 10^{13} \text{ kcal yr}^{-1}$

[n] *Exported goods and services.* Power, $P_E = E(f)(S)$

$\quad E = $ total dollar value of exports from Florida $= \$2.54 \times 10^9 \text{ yr}^{-1}$

$\quad f = 0.13; S = 30,000 \text{ kcal dollar}^{-1}$ (Littlejohn, 1974)

$P_E = (\$2.54 \times 10^9 \text{ yr}^{-1})(0.13)(30,000 \text{ kcal dollar}^{-1})$

$\quad = (\$3.31 \times 10^8 \text{ yr}^{-1})(30,000) = 9.9 \times 10^{12} \text{ kcal yr}^{-1}$

[o] *Tourists and capital investments.* Power, $P_{TC} = (D_I)(f)(S) - P_E$

$\quad D_I = $ total dollars coming into Florida (Littlejohn, 1974)

$\quad f = $ fraction of Florida population in region

$P_{TC} = (\$16.88 \times 10^9 \text{ yr}^{-1})(0.13)(30,000 \text{ kcal dollar}^{-1}) - 9.9 \times 10^{12} \text{ kcal yr}^{-1}$

$\quad = 56 \times 10^{12} \text{ kcal yr}^{-1}$

[p] *Fossil-fuel equivalents.* See Chapter 7.

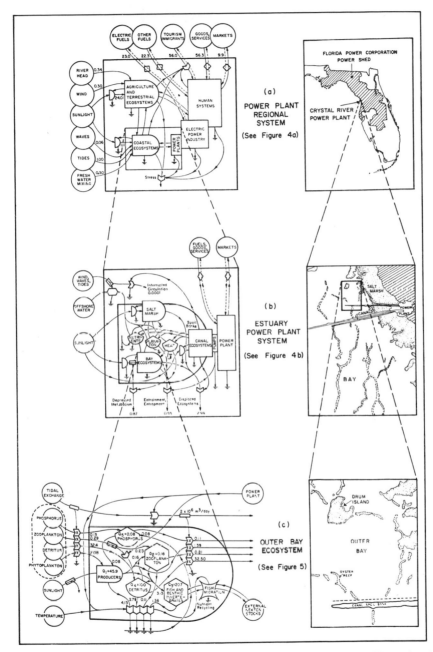

Figure 2 Energy diagrams and maps for the Crystal River power-plant region illustrating the hierarchical nature of systems and subsystems.

517

work in the larger-scale system in which the natural and human economies are considered together.

CALCULATION OF ENERGY COST-BENEFIT FOR
THE POWER PLANTS AT CRYSTAL RIVER, FLORIDA

An important issue at the Crystal River power-plant site has been the suggested need for closed-cycle cooling to reduce the environmental effects on the estuary of the once-through system now in use. In this chapter we develop an energy cost-benefit evaluation of alternative cooling water systems, combining the energy quality concept with earlier studies and extensive new data. The basic question addressed in this analysis is whether the total energy in the region is greater with cooling towers or without.

Description of Study Area

The electric power generating plants considered in this study are located on the west coast of Florida near the town of Crystal River (Figure 3). These plants are a major generating facility for the Florida Power Corporation and are connected into the company's electric power grid serving a 32-county region in the northwestern part of the state (inset of Figure 3). This "powershed" region is a relatively rural area of Florida dominated by lake, river, and estuarine systems, sandy soil forests, marshes and swamps, agricultural fields, and pine plantations, as well as several major urban, industrial, and academic centers.

The power plants are sited in Citrus County on the low-wave-energy portion of the Florida Gulf Coast (Tanner, 1960). The shallow sloping bottom along the coast (46 km to the 5-fathom contour) is part of the drowned karst topography typical of this portion of west central Florida. The immediate coastal region is comprised of a series of shallow basins separated by oyster reefs. Major sources of freshwater to the area are the Crystal River 4.8 km to the south and the Withlacoochee River 6.4 km to the north.

The estuary adjacent to the power plant is characterized by five basic ecological subsystems: (1) an inner bay, (2) an outer bay, (3) the salt marsh, (4) the oyster reefs, and (5) the power plant intake and discharge canals. After passing through the 4-km-long discharge canal, the heated plume flows over the shallow inner bay, which is about 1 m in depth, with mixed benthic communities of sea grass, algae, oyster, and mud associations. Rising tides push the warmed water over the adjacent tidal salt marsh, which is dominated by *Juncus roemerianus*. Seaward of the inner bay are deeper outer basins (about two meters mean depth) in which plankton and reef ecosystems

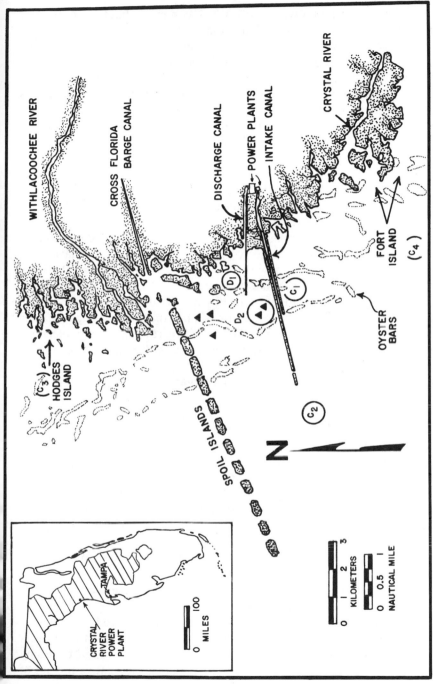

Figure 3 Map of study area showing power plants and their intake and discharge canals. Service area of the Florida Power Corporation is given in the inset.

519

become more important. Oyster reefs, which comprise about 5% of the bay area are scattered throughout the immediate offshore area.

Two oil-burning power generating units have been in operation during this study, unit 1 since July 1966, and unit 2 since November 1969, giving a combined total output of 987 MW electrical. A nuclear powered unit of 855 MW electrical output is under construction. The two operating units circulate water for once-through cooling at a combined flow of 640,000 gpm ($2422 \text{ m}^3 \text{ min}^{-1}$) through canals dredged across the saltmarsh and shallow inshore area. Maximum condenser temperature rise is $6.1°C$ ($11°F$).

The power plant has had various effects on the adjacent coastal environment. The intake canal and its dredge-spoil banks, which extend some 12.5 km into the Gulf of Mexico, have displaced both salt marsh and bay ecosystems, and may have altered the estuarine hydrography (Kemp, 1976). The pumping of large volumes of cooling water has created problems of plankton entrainment and fish entrapment at the plant intake (Snedaker, 1974; Maturo et al., 1974). Local turbidity and salinity patterns also have been changed, and these combined with the thermal effluent appear to have reduced metabolic activities in the discharge area (McKellar, 1975; Smith, 1976). The specific issue addressed in this case study is whether these environmental costs resulting from the use of once-through cooling in power plant operation would justify the investment of energy capital into a technological means of mitigating these losses (such as closed-cycle cooling towers).

Model Formulation for Understanding and Organizing

At the outset of any rational procedure for studying such complex systems as the region in which a power plant interacts, effort must be made to organize available information into a concise and presentable form that assists in understanding the system studied. This can be done by developing diagrammatic representations of the system in question including all important system components and interactions. These initial diagramming attempts should include as much of the detailed understanding as possible, but subsequent drawings should group related components and pathways into aggregates. The process of organizing and condensing the initial highly detailed models tends to produce a hierarchical order of systems and subsystems. In this way the smallest scale models may focus on the most fundamental detail without losing the holistic point of view, since each subsystem model is component to a larger system model (see Chapter 3).

In our study of the Crystal River power plant system, three basic scales of analysis were identified, and these are illustrated in Figure 2. At the largest scale (Figure 4a) the power plant interacts with the general regional economy of human and natural systems. Here we investigated the relative impact of

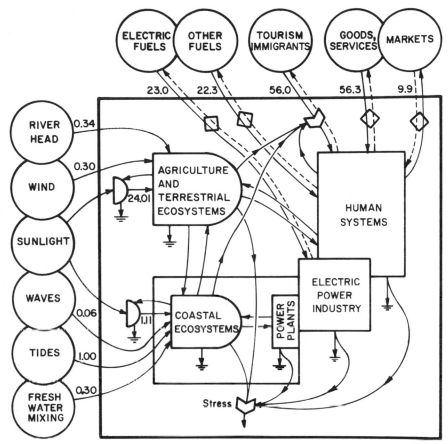

Figure 4 Energy diagrams. (*a*) For region served by Florida Power Corporation with major energy inflows and outflow given in 10^{12} kcal FFE/year; (*b*) for estuary affected by power plant with major energy impacts given in 10^9 kcal FFE/year (See page 522).

power plant or cooling tower construction and operation on the regional balance between natural energies and fossil-fuel-based energies. At the intermediate scale (Figure 4*b*) we identified and analyzed specific positive and negative energy effects of the power plant on the estuary as a whole, under present operating conditions and under proposed alternative situations. To do this, however, a more detailed analysis of the estuary's subsystems was required, and in Figure 5 a model of the "outer bay" subsystem is given as an example of this third level of examination.

Mathematical configurations for these subsystem models, which were used for computer simulation, were developed on the basis of ecological

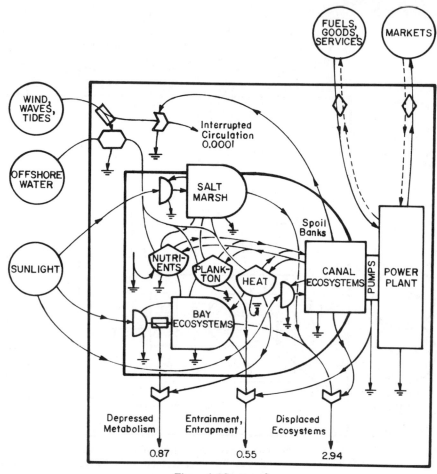

Figure 4 (*Continued*)

and general systems concepts pertaining to the organization and interactions of ecosystem components. For example, the primary productivity flow (J_{GP} in Figure 5) is shown as a multiplicative function of temperature, sunlight, phosphorus (Q_5), and plant biomass (Q_1). A large literature corroborates the fact that each of these parameters plays an important role in photosynthesis (see, e.g., Hill, 1963; Hutchinson, 1967). McKellar (1975) shows that this configuration generates a function of the form,

$$J_{GP} = \frac{K_1 J_0 Q_5 Q_1 T}{1 + K_2 Q_5 Q_1 T}$$

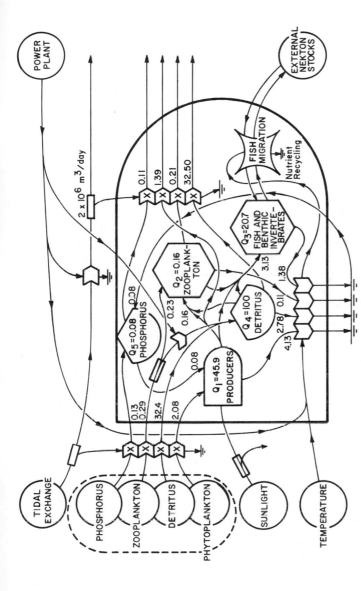

Figure 5 Diagrammatic representation of the simulation model for the outer discharge bay. Values given for forcing functions, stocks, and flows are the initial set of conditions representing the outer control bay during the summer. Modeled import of phosphorus (J_{05}), zoo-plankton (J_{02}), and phytoplankton (J_{01}) and the forcing functions of temperature (I_2) and sunlight (I_1) were programmed as sinusoidal functions representing seasonal changes. Total water exchange represents the combined effects of tides, currents, and power-plant pumping. Temperature-preference emigrations are controlled by temperature-logic gates. When the simulated temperature exceeds 30°C or drops below 15°C then fish migrate out of the bay. Stocks = g/m²; Flows = g/m²/day. Values for stocks and flows for phosphorus (Q_5) are in units of *total* phosphorus. All other stocks and internal flows are in units of *organic matter* (from McKellar, 1975).

523

This is a multiple-parameter extention of the Michaelis-Menten or Monod function derived from enzyme kinetics theory. It is an expression commonly used in ecological modeling (Wiegert, 1975) that generates a hyperbolic response curve. The model system has numerous other realistic features including recycling of nutrients, temperature-controlled metabolism, and temperature-keyed fish migration. However, further discussion of these is beyond the scope of this chapter. Such discussions are available in the appropriate reports and dissertations given at the end of this chapter.

Data Collection and Field Program

Numerical values for energy and dollar flows into and out of the large region were compiled primarily from various economic and fuel statistics, from vegetation maps, and from oceanographic, hydrological, and climatological data records. Data to quantify the estuary models was gathered in an extensive field program undertaken by a large group of researchers. As Steele (1974) suggests, these models served as a format for organizing and synthesizing field data into a coherent view of the estuarine system. This synthesis effort was our primary task; however, in order to obtain all the important numbers for the modeling phase and address questions not included in other projects our synthesis program was also involved in field measurements.

The field studies were also aimed at developing an understanding of the character of the estuarine systems presently influenced by the operation of two electric-power-generating units compared to unaffected control areas, and to document any additional effects once the third unit commences operation. The seasonal patterns of community metabolism was one of the basic measurements that we made for each of the estuarine subsystems. The remarkable difference in community gross primary productivity (daytime net productivity plus night respiration) between the inner bay ecosystem receiving power plant discharge and an unaffected control system is provided in Figure 6 as an example of the data from our field program. Details of this data are available elsewhere (Lehman, 1974; Smith et al., 1974; Young, 1974; Homer, 1976; Kemp, 1976; McKellar, 1975; Smith, 1976). Data from other investigators is available in Florida Power Corporation reports (1974, 1975).

Simulations for Understanding and Prediction

Values were placed on the storages and pathways of the model diagrams using data gathered in the field at Crystal River and from the literature. The differential equations inherent in the diagrams were written, scaled, and programmed on an analog computer. Model simulation experiments furnished at least three essential types of information for the overall study.

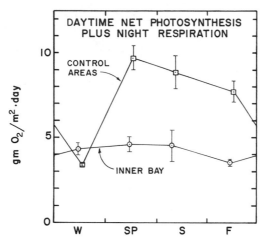

Figure 6 Seasonal graph of community primary productivity for shallow inner bay ecosystem affected by thermal discharge (circles) and unaffected control bay (squares). Graph represents gross primary productivity as defined as daytime net productivity plus night respiration (from Smith, 1976).

First, simulations were performed using pathway coefficients calculated for conditions in one particular season without power plant and with just units 1 and 2, and these provided important insight into the character of system structure and behavior. The models were validated for data from the other seasons and from other years when available. Next, observations on the response of model components to changes in external forcing functions and internal pathway coefficients enabled the identification of sensitive ecosystem parameters, helping to guide the field research effort. Third, adjusting model input functions to reflect conditions expected for operation with unit 3 on line allowed some prediction of system response for the new conditions. An example of this procedure is given for the outer bay model illustrated in Figure 5. Simulation curves are given in Figure 7 for control conditions with no power plant (solid lines), and for conditions approximating units 1 and 2 operating (dotted lines). The solid dots represent field measurements for validating the control condition simulation. Further details on the results of model simulations for this project are provided elsewhere (Odum et al., 1973; Lehman, 1974; McKellar, 1975; Smith 1976; Kemp, 1976).

Energy Budget for the Coastal Zone Region Influenced by the Crystal River Power Plant

Figure 4*a* gives a general model of the main energy flows in the region influenced by the operation of the Crystal River power plant with flow given

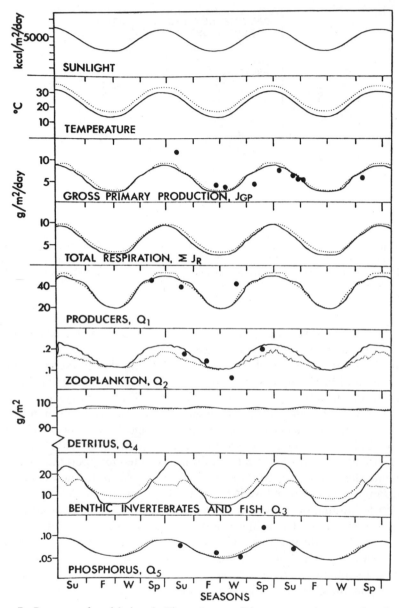

Figure 7 Response of model given in Figure 5 to conditions representing control study area (solid lines) and increased temperature conditions of Figure 5*c* (dotted line). Solid dots represent measured field data for control stations.

in units of fossil fuel equivalents (FFE), that is, with energy quantities corrected for their relative value. The main components of the regional energy budget corresponding to this diagram are listed in Table 1 with the details of their calculation given in footnotes to the table.

The region considered to be influenced by the Crystal River power plant operation (i.e., the "powershed") is calculated as the 32-county service area of the Florida Power Corporation times the fraction of the total generating capacity represented by Crystal River units 1, 2, and 3. The marine area considered in regional analysis is that area delineated by a smoothed coastline of 73 km extending seaward 9.2 km. This area encloses a volume of water equal to the maximum that could possibly be affected by the power plant pumping during its design life.

Figure 4a and Table 1 provide some perspective on the value and importance of the power plant system to the overall system of man and nature, in terms of both calories of heat and calories fossil-fuel equivalents value. The diagram includes the work of the sun in photosynthesis and stirring air masses, the input of potential energy of rain both as a flowing mass and as a chemical dilutant, the energy transferred to the earth by wind friction, coastal contributions of waves and tides, the input of various fossil fuels including coal and oil to run the power plants, the import and export of goods and services, and the flux of tourists and investment dollars.

Fuel consumption of the Crystal River power plants is calculated to be about 5% of the region's total energy budget in terms of heat calories, but accounts for over 13% of the work done when converted to FFE. Total natural energies represent about 72% of the regional energy budget in terms of heat content, but only 15% of the regional work value. Conversely purchased energy accounts for 28% considering heat content but 85% in terms of work equivalents. The total work value of the region is equivalent to some 171.2×10^{12} FFE kcal yr^{-1} or in its monetary equivalents, 5.7 billion 1973 U.S. dollars.

Energy Cost-Benefit Calculation of Cooling Alternatives

We have assessed the various alternative schemes for managing the power plant cooling water flow at Crystal River and have calculated losses to the regional work economy associated with each. Government regulatory agencies suggested that cooling towers be considered as a possible means for mitigating losses to natural systems from using the coastal water for cooling. Table 2 provides a comparison of the total work value lost to the system resulting from the three-unit power plant operation at Crystal River under three management options. Numbers were calculated from field data and

Table 2 Changes in Annual Regional Energy Budget Associated with Management Alternatives for Power Plants at Crystal River, Florida.

Energy Flow	Estuary Cooling			Cooling Tower, Unit 3			Cooling Towers, All Units		
	Kcal of Heat ($\times 10^9$)	EQR[a]	FFE[b] ($\times 10^9$)	Kcal of Heat ($\times 10^9$)	EQR[a]	FFE[b] ($\times 10^9$)	Kcal of Heat ($\times 10^9$)	EQR[a]	FFE[b] ($\times 10^9$)
Natural energies									
Physical									
Wind, waves, tides, sun	no change			no change			no change		
Potential energy in residual heat of discharge water[e]	(560)[c]	0.0001	(±0.056)[c]	(160)[c]	0.0001	(±0.016)[c]	0.0	0.001	0.0
Plant stirring[f]	+0.13	0.014	+0.0004	+0.004	0.014	+0.00006	0.0	0.014	0.0
Interrupted circulation[g]	−0.01	0.014	−0.0001	−0.01	0.014	−0.0001	−0.1	0.014	−0.0001
Terrestrial									
Productivity of land lost to construction[h]	−22.6	0.07	−1.58	−23.2	0.07	−1.62	−23.8	0.07	−1.66
Productivity effects of salt drift[i]	0.0	0.07	0.0	−4.33	0.07	−0.30	−8.65	0.07	−0.61

	$110,000 yr⁻¹ scenario			$9,300,000 yr⁻¹ scenario			$18,070,000 yr⁻¹ scenario		
Estuarine									
Metabolism of ecosystems displaced by canals[j]	−19.4	0.07	−1.36	−19.4	0.07	−1.36	−19.4	0.07	−1.36
Canal metabolism[k]	+13.2	0.07	+0.92	+11.0	0.07	+0.77	+6.6	0.07	+0.46
Depression of inner bay metabolism[l]	−7.8	0.07	−0.55	−3.9	0.07	−0.27	0.0	0.07	0.0
Screen wash mortality[m]	−1.0	0.07	−0.07	−0.5	0.07	−0.035	0.0	0.07	0.0
Entrainment mortality[n]	−11.45	0.07	−0.8	−5.61	0.07	−0.39	−1.15	0.07	−0.1
Subtotal of changes in ecosystem			−3.44			−3.21			−3.27
Purchased energies									
Fuels, imports, exports, income, tourists		no change			no change			no change	
Construction and maintainance of power plants[o,d]		(230,000)[c]			(230,000)[c]			(230,000)[c]	
Construction and maintainance of cooling towers[p,d]	0.0	1.0	0.0	−276.0	1.0	−276.0	−539.0	1.0	−539.0
Total change in fossil fuel work value			−3.44			−279.21			−542.27
			($110,000 yr⁻¹)[d]			($9,300,000 yr⁻¹)[d]			($18,070,000 yr⁻¹)[d]

[a] EQR, energy quality ratio for conversion to fossil-fuel equivalents (FFE). For discussion of energy quality see Chapter 7. For documentation of factors used see individual footnotes.

[b] FFE, fossil-fuel equivalents.

[c] Numbers not included in totals.

[d] Converted to/from 1973 U.S. dollars by 30,000 kcal dollar⁻¹ (Odum, 1974c; Kylstra, 1974).

Footnotes of Table 2 (*Continued*)

[e] *Potential energy in residual heat.* $P_t = \Delta T \times C \times Q \times \rho \times \Delta T/T$

ΔT = change in temperature across condensers, °C

C = specific heat of water = 1 cal g^{-1} deg C^{-1}

Q = flow rate of water = 7.1×10^6 m^3 day^{-1}

ρ = density of water = 1.02 g cm^{-3}

T = absolute temperature = 293 °K

Units 1 to 3.

$$P_t = (8°C)\left(1\,\frac{\text{cal}}{\text{g}°C}\right)(3.5 \times 10^6 \text{ m}^3 \text{ day}^{-1})(1.02 \text{ g cm}^{-3})(10^6 \text{ cm}^3 \text{ m}^{-3})(10^{-3} \text{ kcal cal}^{-1})\left(\frac{8°}{293°}\right)365 \text{ day yr}^{-1}$$

$$= 5.6 \times 10^{11} \text{ kcal yr}^{-1}$$

Units 1 to 2.

$$P_t = (6°C)\left(1\,\frac{\text{cal}}{\text{g}°C}\right)(3.5 \times 10^6 \text{ m}^3 \text{ day}^{-1})(1.02 \text{ g cm}^{-3})(10^6 \text{ cm}^3 \text{ m}^{-3})(10^{-3} \text{ kcal cal}^{-1})\left(\frac{6°}{293°}\right)365 \text{ day yr}^{-1}$$

$$= 1.6 \times 10^{11} \text{ kcal yr}^{-1}$$

[f] *Plant stirring.* $P_{ke} = \frac{1}{2}\rho Q v^2$

ρ = density of water = 1.02 g cm^{-3}

v = velocity of plume = 30 cm sec^{-1}

Q = flow rate of water = 7.1×10^6 m^{-3} day^{-1}

Units 1 to 3.

$Pke = \frac{1}{2}(1.02 \text{ g cm}^{-3})(7.1 \times 10^{12} \text{ cm}^3 \text{ day}^{-1})(30 \text{ cm sec}^{-1})^2(365 \text{ day yr}^{-1})(2.38 \times 10^{-11} \text{ kcal erg}^{-1})$

$$= 0.3 \times 10^8 \frac{\text{kcal}}{\text{yr}}$$

Units 1 to 2.

$Pke = \frac{1}{2}(1.02)(3.5 \times 10^{12})(15)^2(365)(2.38 \times 10^{-11})$

$= 0.04 \times 10^8 \text{ kcal yr}^{-1}$

Circulation interrupted by spoil banks. This is the calculated loss from estuarine water of kinetic energy diverted into frictional heat loss due to drag of canal spoil banks.

$Pke = \frac{1}{2}\rho RdV^3$ = power of kinetic energy

ρ = density of water g cm^3

R = east-west horizontal length of affected zone

d = depth (cm)

V = velocity (cm sec^{-1})

A. Water movement not interrupted by spoil banks. Assume a horizontal velocity profile of constant mean velocity from end of spoil banks to 0.5 m from coast (boundary layer).

1 Area outside boundary layer.

$V = 5 \text{ cm sec}^{-1}$

$Pke_1 = \frac{1}{2}(1.020 \text{ g cm}^{-2})(5 \text{ cm sec}^{-1})^3(4.5 \times 10^5 \text{ cm})(200 \text{ cm})$

$= (5.74 \times 10^9 \text{ erg sec}^{-1})(3.15 \times 10^7 \text{ sec yr}^{-1})(2.38 \times 10^{-11} \text{ kcal erg}^{-1})$

$= 4.30 \times 10^6 \text{ kcal yr}^{-1}$

2 Area inside boundary layer.

$V = 2.3 \text{ cm sec}^{-1}$

531

Footnotes of Table 2 (Continued)

$Pke_2 = \frac{1}{2}(1.020 \text{ g cm}^{-3})(2.5 \text{ cm sec}^{-1})^3(0.5 \times 10^5 \text{ cm})(100 \text{ cm})$

$= (3.98 \times 10^7 \text{ ergs sec}^{-1})(3.15 \times 10^7 \text{ sec yr}^{-1})(2.38 \times 10^{-11} \text{ kcal erg}^{-1})$

$= 0.30 \times 10^6 \text{ kcal yr}^{-1}$

Total P_A (area inside + area outside) $= Pke_1 + Pke_2 = 4.60 \times 10^6 \text{ kcal yr}^{-1}$

B. Water movement interrupted by spoil banks. Assume that as flow streamlines move west to sweep around spoil banks, they create a back-eddy on the southwest side of the spoil bank/shore intersection. Velocities are reduced due to frictional drag.

1 Zone outside of back-eddy ($R = 3.5 \times 10^5 \text{ cm}$)

$V = 2.5 \text{ cm sec}^{-1}$

$Pke = \frac{1}{2}(1.02)(2.5)^3(3.5 \times 10^5)(200)$

$= (5.6 \times 10^8)(3.15 \times 10^7)(2.38 \times 10^{-11})$

$= 4.2 \times 10^5 \text{ kcal yr}^{-1}$

2 Zone within back-eddy ($R = 1.5 \times 10^5 \text{ cm}$)

$V = 1.0 \text{ cm sec}^{-1}$

$Pke = \frac{1}{2}(1.02)(1)^3(1.5 \times 10^5)(100)$

$= (0.77 \times 10^7)(3.15 \times 10^7)(2.38 \times 10^{-11})$

$= 0.58 \times 10^5 \text{ kcal yr}^{-1}$

Total $P_B = Pke_1 + Pke_2 = 0.468 \times 10^6 \text{ kcal yr}^{-1}$

Difference in Pke with spoil banks $= P_B - P_A$

$Pke_{net} = 0.468 \times 10^6 \text{ kcal yr}^{-1} - 4.60 \times 10^6 \text{ kcal yr}^{-1} = -4.12 \times 10^6 \text{ kcal yr}^{-1}$

Total loss in available power prior to spoil bank emplacement equals two times calculated P_{net}, assuming mirror image effect on other side of spoil banks.

Total $(P_{ke})_{net} = 2(-4.12 \times 10^6 \text{ kcal yr}^{-1}) = -0.0824 \times 10^8 \text{ kcal yr}^{-1}$

h Construction land productivity loss.

Unit 3.   Power, $P_{C_1} = Ac_1 \times M$

Ac_1 = Area covered by construction = $3 \times 10^5 \text{ ft}^2 = 1.58 \times 10^6 \text{ m}^2$

M = Metabolism of ecosystems displaced = $60 \text{ kcal m}^{-2} \text{ day}^{-1}$

$Pc_1 = (1.58 \times 10^6 \text{ m}^2)(60 \text{ kcal m}^{-2} \text{ day}^{-1})(365 \text{ day yr}^{-1}) = 232 \times 10^8 \text{ kcal yr}^{-1}$

Units 1 to 3. Power, $Pc_2 = Ac_2 \times M$

$Ac_2 = $ Area covered by construction $= 1.61 \times 10^6 \text{ m}^2$

$Pc_2 = (1.61 \times 10^6 \text{ m}^2)(60 \text{ kcal/m}^2/\text{day}) \left(365 \dfrac{\text{day}}{\text{yr}} \right) = 238 \times 10^8 \text{ kcal/yr}$

^{i} *Depressed land productivity from salt spray.*

Unit 3. Power, $Ps_1 = (R^2\pi)(\delta)M$

$R = $ radius of area in which salt spray addition equals background rate of salt deposition

$\delta = $ assumed mean rate of productivity inhibition $= 0.25$

$M = $ metabolism of affected terrestrial ecosystems $= 60 \text{ kcal m}^{-2} \text{ day}^{-1}$

Background salt deposition rate $= 0.125 \text{ lb m}^{-2} \text{ yr}^{-1}$ (Yaalon and Lomas, 1970; Gutfreund and Urone, 1972)

Maximum salt deposition rate from towers $= 4.8 \text{ lb acre}^{-1} \text{ mo}^{-1}$ (Dames and Moore, 1974; Gilbert Assoc., 1974)

$= 0.140 \text{ lb m}^{-2} \text{ yr}^{-1}$

R (at maximum rate) $= 0.5 \text{ km}$

$Ps_1 = (7.9 \times 10^5 \text{ m}^2)(0.25)(60 \text{ kcal m}^{-2} \text{ day}^{-1})(365 \text{ days yr}^{-1}) = 4.33 \times 10^9 \text{ kcal yr}^{-1}$

Units 1 to 3.

$Ps_2 = 2 \times Ps_1 = 8.65 \times 10^9 \text{ kcal yr}^{-1}$

^{j} *Ecosystems displaced by canals.* Production $= A \times P \times D$

$A = $ area displaced (m^2)

$P = $ productivity of displaced system

$= (40 \text{ kcal m}^{-2} \text{ day}^{-1} \text{ for land})$

$= (25 \text{ kcal m}^{-2} \text{ day}^{-1} \text{ for marine})$

$D = $ time in days

Footnotes of Table 2 (*Continued*)

A. Terrestrial systems.

Area of plant $= 1.68 \times 10^6 \text{ ft}^2 = 1.55 \times 10^6 \text{ m}^2$

$Pt = (1.55 \times 10^6 \text{ m}^2)(40 \text{ kcal m}^{-2} \text{ day}^{-1})(365 \text{ days yr}^{-1})$

$= 2.26 \times 10^{10} \text{ kcal yr}^{-1}$

B. Marine systems.

1 Area of discharge canal $= 5.3 \times 10^6 \text{ ft}^2 = 4.93 \times 10^5 \text{ m}^2$

2 Area of intake canal $= 17.5 \times 10^6 \text{ ft}^2 = 1.62 \times 10^5 \text{ m}^2$

$Pm = (2.13 \times 10^6 \text{ m}^2)(25 \text{ kcal m}^{-2} \text{ day}^{-1})(365 \text{ days yr}^{-1})$

$= 1.94 \times 10^{10} \text{ kcal yr}^{-1}$

Total production displaced $= Pt + Pm = 4.20 \times 10^{10} \text{ kcal yr}^{-1}$

[k] *Canal metabolism.* Production $= PP \times A \times K \times D$

PP = gross primary production (mean annual)

A = area in canals (m^2)

K = kcal $(gO_2)^{-1} = 4.5$

D = time in days (365)

A. Units 1 and 2 operating.

1 Intake canal. $(A = 5.62 \times 10^5 \text{ m}^2)(PP = 9 \text{ gO}_2 \text{ m}^{-2} \text{ day}^{-1})$

$P_I = (9)(5.62 \times 10^5)(4.5)(365)$

$= 8.32 \times 10^9 \text{ kcal yr}^{-1}$

2 Discharge canal. $(A = 1.47 \times 10^5 \text{ m}^2)$ $PP = 11 \text{ gO}_2 \text{ m}^{-2} \text{ day}^{-1}$

$P_D = (11)(1.47 \times 10^5)(4.5)(365)$

$\quad = 2.64 \times 10^9 \text{ kcal yr}^{-1}$

3 Total metabolism $= P_I + P_D = 1.10 \times 10^{10} \text{ kcal yr}^{-1}$

B. Units 1, 2, and 3 operating. Based on model predictions metabolism increases by 20%

$P = (1.2)(1.10 \times 10^{10} \text{ kcal yr}^{-1}) = 1.32 \times 10^{10} \text{ kcal yr}^{-1}$

C. No circulating water flow. Based on model predictions metabolism decreases by 40%

$P = (0.6)(1.10 \times 10^{10} \text{ kcal yr}^{-1}) = 0.66 \times 10^{10} \text{ kcal yr}^{-1}$

[l] *Depressed inner bay metabolism.* The annual mean total community metabolism for the shallow inner bay ecosystem was measured to be 50% lower for the discharge area than for the control area. Production bay $= (P_C - P_D)A_B$

$P_C = $ metabolism of control area $= 1.22 \times 10^4 \text{ kcal m}^{-2} \text{ yr}^{-1}$

$P_D = $ metabolism of discharge area $= 0.65 \times 10^4 \text{ kcal m}^{-2} \text{ yr}^{-1}$

$A_B = $ area of inner bay system $= 6.9 \times 10^5 \text{ m}^2$

$P = [(1.22 \times 10^4) - (0.65 \times 10^4)](6.9 \times 10^5)$

$\quad = [0.57 \times 10^4](6.9 \times 10^5)$

$\quad = 3.9 \times 10^9 \text{ kcal yr}^{-1}$

[m] *Screen wash mortality.*

$P = B \times EQR$

$B = $ Biomass lost in 1 yr (extrapolated from g wet wt/52 days)

$EQR = $ Energy Quality Ratio

Footnotes of Table 2 (*Continued*)

1 Batfish

$P = (6.61 \times 10^5)(29.8) = 8.9 \times 10^7$ kcal yr^{-1}

2 Burrfish

$P = (1.57 \times 10^5)(25.3) = 1.8 \times 10^7$ kcal yr^{-1}

3 Blue Crab

$P = (0.98 \times 10^5)(29.7) = 1.3 \times 10^7$ kcal yr^{-1}

4 Cowfish

$P = (0.49 \times 10^5)(29.5) = 0.7 \times 10^7$ kcal yr^{-1}

5 Pinfish

$P = (0.41 \times 10^5)(23.7) = 0.4 \times 10^7$ kcal yr^{-1}

6 Tunicate

$P = (0.41 \times 10^5)(11.2) = 0.2 \times 10^7$ kcal yr^{-1}

7 Silver Jenny

$P = (0.38 \times 10^5)(27.8) = 0.5 \times 10^7$ kcal yr^{-1}

8 Squid

$P = (0.32 \times 10^5)(36.4) = 0.5 \times 10^7$ kcal yr^{-1}

9 Silver Perch

$P = (0.29 \times 10^5)(31.3) = 0.4 \times 10^7$ kcal yr^{-1}

10 Scaled Sardine

$P = (0.25 \times 10^5)(24.4) = 0.3 \times 10^7$ kcal yr^{-1}

11 Jack

$P = (0.17 \times 10^5)(33.5) = 0.3 \times 10^7$ kcal yr^{-1}

12 Mullet

$P = (0.11 \times 10^5)(12.7) = 0.1 \times 10^7$ kcal yr^{-1}

13 Atlantic Threadfin

$P = (46.2 \times 10^5)(24.4) = 50.7 \times 10^7$ kcal yr^{-1}

14 Other

$P = (1.43 \times 10^5)(25) = 1.67 \times 10^7$ kcal yr^{-1}

$P_{total} = 0.67 \times 10^9$ kcal yr^{-1} = work equivalent loss yr^{-1}

Biomass total = 59.6×10^5 g yr^{-1} = amt lost yr^{-1}

Total loss of biomass through screen wash = $B_L = (59.6 \times 10^5 \text{ g yr}^{-1})(5 \text{ kcal g}^{-1}) = 29.8 \times 10^6$ kcal yr^{-1}

Value of this mass as detritus $= V_m = (29.8 \times 10^6 \text{ kcal yr}^{-1})(\text{EQR detritus}) = 0.17 \times 10^9 \text{ kcal yr}^{-1}$

Total annual loss of value to region $= P_{\text{total}} - V_m = 0.67 \times 10^9 \text{ kcal yr}^{-1} - 0.17 \times 10^9 \text{ kcal yr}^{-1} = 0.5 \times 10^9 \text{ kcal yr}^{-1}$

" Entrainment mortality for zooplankton. $P_E = N \times m \times Q \times M \times K \times R \times \text{EQR}$

N = numerical density (individuals m^{-3})

m = mass per individual (kg individuals^{-1})

M = metabolism per mass (kcal kg^{-1} day^{-1})

K = entrainment mortality

R = loss of metabolism (day replace^{-1})

Q = daily circulating water flow (m^3 day^{-1})

EQR = energy quality ratio

Units 1 and 2 operating.

Copepods.

$P = (10.769 \text{ ind m}^{-3})(0.68 \times 10^{-8} \text{ kg ind}^{-1})(500 \text{ kcal kg}^{-1} \text{ day}^{-1})(0.3 \text{ kill})(10 \text{ day replace}^{-1})(3.4 \times 10^6 \text{ m}^3 \text{ day}^{-1})(11.1)(365 \text{ days yr}^{-1})$

$= 1.51 \times 10^9 \text{ kcal yr}^{-1}$

Fish eggs and larvae.

$P = (22 \text{ ind m}^{-3})(1.18 \times 10^{-8} \text{ kg ind}^{-1})(250 \text{ kcal kg}^{-1} \text{ day}^{-1})(0.90 \text{ kill})(20 \text{ day replace}^{-1})(3.4 \times 10^6 \text{ m}^3 \text{ day}^{-1})(11.1)(365 \text{ days yr}^{-1})$

$= 1.61 \times 10^7 \text{ kcal yr}^{-1}$

Chaetognaths and medusae.

$P = (171 \text{ ind m}^{-3})(1.98 \times 10^{-8} \text{ kg ind}^{-1})(250 \text{ kcal kg}^{-1} \text{ day}^{-1})(0.30 \text{ kill})(20 \text{ day replace}^{-1})(3.4 \times 10^6 \text{ m}^3 \text{ day}^{-1})(24.0)(365 \text{ days yr}^{-1})$

$= 1.51 \times 10^8 \text{ kcal yr}^{-1}$

Footnotes of Table 2 (*Continued*)

Veligers, trochophores, mysids, and so forth.

$P = (2279 \text{ ind m}^{-3})(0.68 \times 10^{-8} \text{ kg ind}^{-1})(250 \text{ kcal kg}^{-1} \text{ day}^{-1})(0.30 \text{ kill})(20 \text{ day replace}^{-1})(3.4 \times 10^6 \text{ m}^3 \text{ day}^{-1})(20.4) (365 \text{ days yr}^{-1})$

$= 4.13 \times 10^8 \text{ kcal yr}^{-1}$

Juvenile fish.

$P = (0.4 \text{ ind m}^{-3})(10^{-4} \text{ kg ind}^{-1})(250 \text{ kcal kg}^{-1} \text{ day}^{-1})(0.90 \text{ kill})(20 \text{ day replace}^{-1})(3.4 \times 10^6 \text{ m}^3 \text{ day}^{-1})(24)(365 \text{ days yr}^{-1})$

$= 4.56 \times 10^9 \text{ kcal yr}^{-1}$

Total entrainment.

Total 1 and 2 operating $= 6.76 \times 10^9 \text{ kcal yr}^{-1}$

Total 1, 2, and 3 operating $= 1.38 \times 10^{10} \text{ kcal yr}^{-1}$

Percent value of biomass converted to detritus $= 0.17$

Total loss of value in kilocalories (1 and 2 operating) $= 5.61 \times 10^9 \text{ kcal yr}^{-1}$

Total loss of value (1, 2, and 3 operating) $= 11.45 \times 10^9 \text{ kcal yr}^{-1}$

[o] Number quoted by J. R. Hall, Florida Power Corporation, St. Petersberg.
[p] Number calculated by Gilbert Assoc. (1974).

derivations as given in the table footnotes. Energy quality factors for organisms killed in screen wash and entrainment were derived from calculations based on food web energy relationships such as given in Figure 1.

These calculations indicate that there is a substantial loss of ecosystem work (3.44×10^9 FFE kcal yr^{-1}, or \$110,000 yr^{-1}) for the proposed three-unit operation at Crystal River. This is about 0.002% of the total regional work budget. The total loss in work value to the ecosystem is less than 1% of the diversion of fossil fuel capital resulting from the cooling tower alternative (for all three units). Based on the procedure we have described the use of mechanical draft cooling towers—the cheapest cooling tower technology (Gilbert Associates, 1974)—for unit 3 predicted a loss to the region eighty times greater than using the estuary for cooling, while cooling towers for all three units predicted losses about 160 times that incurred without cooling towers.

The money that would be invested in cooling-tower construction, operation, and maintenance represents a diversion of fossil-fuel energy from other possible investments into the economy of man and nature that could return a greater income to the system. By way of perspective, the cost of cooling towers for all units 1 and 2 is about equivalent (in terms of work value) to loss of primary productivity in an estuarine area of about 2.6×10^8 m^2, or 100 mi^2 assigned to the power plant (about 1.8×10^9 m^2 or 600 mi^2). The energy flowing throughout the national economy to supply 17 million dollars worth of goods and services to build and maintain a cooling tower is estimated to be 70% from purchased fossil fuels and 30% from the free services to the economy from work of the environment such as absorbing and recycling wastes. In other words, 30% of the energy cost of the cooling towers is environmental impact elsewhere. This effect of 100×10^9 kcal of fossil-fuel work equivalents per year is about two times greater than the 64×10^9 FFE kcal yr^{-1} projected as the impact on the estuary of three-unit operation at Crystal River. By the objective criteria presented in this analysis, cooling towers are a poor investment for overall national environmental well-being as well as for the total work budget (and thus economic viability) of this region.

Power Needs for a Vital Economy Estimated from the Ratio of Energy Invested to Work Returned

The ratio of the total work output (work of nature, W_n, plus work of man, W_{ff}) in a system to the fossil-fuel investment (W_{ff}) in that system (all converted to fossil fuel equivalents) gives an indication of the overall return for invested fossil fuel capital:

$$\frac{\text{Total work output (natural plus fossil fuel)}}{\text{Fossil fuel work investment}} = \frac{W_n + W_{ff}}{W_{ff}}$$

The ratio is large in primitive societies and declines toward unity as the system becomes more and more dependent on bought energies such as fossil fuel.

Among competing systems of man and nature with approximately equivalent amounts of fossil fuel to invest, the most economically competitive system would be that which invests the fossil fuel energy in such a way as to produce the maximum system work without destroying the balance between man's and nature's work value. The adaptive investment ratio to guide the design of one system is determined by the average investment ratio within the larger system with which there is competition. Referring to numbers given in Table 1 and considering the fossil and nuclear fuels burned versus terrestrial productivity, the investment ratio for the Crystal River power ·plant region is

$$\frac{W_n + W_{ff}}{W_{ff}} = \frac{(24.0 + 45.3) \times 10^{12} \text{ FFWE kcal yr}^{-1}}{45.3 \times 10^{12} \text{ FFWE kcal yr}^{-1}} = 1.53$$

Since this number is higher than the ratio calculated for Florida (1.22) and the United States (1.33)(Kylstra, 1974), there still may be some opportunities for more fossil-fuel investment in the region. If this were to occur until the ratio for the region matched that of the United States the present value of 45.3×10^{12} FFE kcal yr^{-1} would increase to 52.1×10^{12} FFE kcal yr^{-1}. Allowing 43%* of this for electric-power generation and converting to electrical units, the ultimate power needs of the region served by the Crystal River units would be 1927 MW. This is similar to the capacity already on line and under construction, suggesting that these power plants and their environmental effects are approaching saturation values for the region. Additional units could make this region less competitive, particularly if the price of fuels rise.

In summary, an energy evaluation showed that the system of power plants and estuarine cooling at Crystal River after an adaptation period was economically and ecologically more competitive for that site than the proposed alternative of cooling towers. The adapted ecosystems were somewhat different from unaffected ones, but were within the range of the energy budgets, metabolism, diversity, and productivity of other Gulf coast estuaries. This outcome was specific for the energy conditions at Crystal River; another site with different factors might have produced another result. With fuels

---

* Approximate percentage for Florida in 1970 (Federal Power Commission, 1970).

that support so much of our economy becoming scarce, it is mandatory that we plan our fossil-fuel investments in terms of priorities based on those combinations of monetary and natural work value that will maximize the total value of the regional system of man and nature.

REFERENCES

Anon. 1966. Tidal power comes to France. *Eng.* **202**: 17–24.

Atomic Energy Commission. 1972. *Proposed AEC guide to the preparation of benefit-cost analysis to be included in applicant's environmental reports.* U.S. Government Printing Office: Washington, D.C. 483–026/159.

Bascom, W. 1964. *Waves and beaches.* Doubleday, Garden City, N.Y. 267 pp.

Boynton, W. R. 1975. Energy basis of a coastal region, Franklin County and Apalachicola Bay, Florida. Ph. D. thesis. University of Florida, Gainesville.

Bureau of Economic and Business Research. 1973. *Florida statistical abstract.* College of Business Administration. University of Florida Press, Gainesville.

Dames and Moore. 1974. Terrestrial survey, salt drift evaluation, Crystal River, Florida Power Corporation. Unpublished Report, Dames and Moore, Inc. Atlanta.

Darling, F., and J. P. Milton (Eds). 1966. *Future environments of North America.* Natural History Press, Garden City, N.Y. 767 pp.

Dee, N., J. Baker, N. Drobny, K. Duke, I. Whitman, D. Fahringer. 1973. An environmental evaluation system for water resources planning. *Water Resour. Res.* **9**: 523–535.

Dorfman, R. 1963. *Measuring benefits of government investments.* Brookings Institution, Washington, D.C.

Evans, R. B. 1969. A proof that essergy is the only consistent measure of potential work (for work systems). Ph. D. thesis, Dartmouth College.

Federal Power Commission. 1970. *The 1970 national power survey.* Part III. U.S. Gov. Printing Office. Washington, D.C.

Florida Power Corporation. 1974. Crystal River power plant. Environmental considerations. Final report to the Interagency Research Advisory Committee. St. Petersburg, Florida.

Florida Power Corporation. 1975. *Summary analysis and supplemental data report to the interagency research advisory committee.* St. Petersburg, Florida.

Gilbert Associates, Inc. 1974. Condenser cooling system study for Florida Power Corporation Crystal River Units 1, 2, and 3. Unpublished Report, Gilbert Assoc., Reading, Penn.

Gutfreund, P. D. and P. Urone. 1972. Salt deposition from salt water cooling towers. Dept. of Environmental Engineering Sciences, University of Florida, Gainesville. Florida. (Unpublished.)

Hess, S. L. 1959. *Introduction to theoretical meteorology.* Holt, Rinehart, and Winston, New York. 362 pp.

Hill, M. N. 1963. *The sea,* Vol. 2. *The composition of sea water, comparative and descriptive oceanography.* Wiley, New York. 554 pp.

Homer, M. 1976. Seasonal abundance, biomass, diversity and trophic structure of fish in a salt marsh tidal creek affected by a coastal power plant. In G. W. Esch and R. W. McFarland (Eds.), *Thermal ecology, II.* AEC Symposium Series. AEC Publ., Oak Ridge, Tenn.

Hutchinson, G. E. 1967. *A treatise on limnology Vol. II. Introduction to lake biology and limnoplankton.* Wiley, New York. 1115 pp.

Ingle, J. R. 1966. *The movement of beach sands. An analysis using flourescent grains.* Elsevier, New York.

James, L. D. and R. R. Lee. 1971. *Economics of water resource planning.* McGraw-Hill, New York. 615 pp.

Kemp, W. M. 1976. Ecological and energetic evaluation of a coastal power plant Ph. D. thesis, University of Florida, Gainesville.

Kylstra, C. D. 1974. Energy analysis as a common basis for optimally combining man's activities and nature. Paper presented to the National Symposium on Corporate Social Policy, October 5, 1974, Chicago.

Lehman, M. E. 1974. Oyster reefs at Crystal River, Florida and their adaptation to thermal plumes. Masters thesis, University of Florida, Gainesville. 197 pp.

Lawton, F. L. 1972a. Economics of tidal power. In T. J. Gray and O. K. Gashus (Eds.), *Tidal power.* Plenum, New York. Pp. 105–132.

Lawton, F. L. 1972b. Tidal power in Bay of Fundy. In T. J. Gray and O. K. Gashus (Eds.), *Tidal power.* Plenum, New York. Pp. 1–104.

Littlejohn, C. 1974. Bureau of State Planning, Department of Administration. Tallahassee, Florida. Personal communication.

Leopold, L. B., F. E. Clarke, B. B. Hanshaw, and J. R. Balsley. 1971. A *procedure for evaluating environment impact.* U.S. Geol. Survey Circ. 645. Washington, D.C.

Lotka, A. J. 1922. Contribution to the energetics of evolution. *Proc. Nat. Acad. Sci.* **8**: 147–151.

Maturo, F. J., Jr., J. W. Caldwell, and W. I. Ingram, III. 1974. Effect of power plant operation on shallow water coastal zooplankton. In *Crystal River power plant. Environmental considerations. Final report to the interagency research committee.* Florida Power Corporation, St. Petersburg.

McKeller, H. N. 1975. Metabolism and models of estuarine bay ecosystems affected by a coastal power plant. Ph.D. thesis, University of Florida, Gainesville. 270 pp.

National Environment Policy Act. 1969. Public Law 91–190, 91st Congress, S. 1075 Jan. 1, 1970.

Odum, H. T. 1970. Energy values of water resources. In *Proc. 19th Southeastern Water Res. Poll. Contr. Conf.* pp. 56–64.

Odum, H. T. 1971. *Environment, power, and society.* Wiley, New York. 331 pp.

Odum, H. T. 1972. Use of energy diagrams for environmental impact statements. In Marine Technology Society, *Tools for coastal management, proceedings of the conference.* February 14–15, 1972.

Odum, H. T. 1974a. *Energy quality concentration factors for estimating equivalent abilities of energies of various types to support work.* University of Florida, Gainesville. (Unpublished.)

Odum, H. T. 1974b. Energy cost-benefit models for evaluating thermal plumes. In J. W. Gibbons and R. R. Sharitz (Ed.), *Thermal ecology,* AEC Symposium Series (CONF 730505). Pp. 628–649. AEC Publ., Oak Ridge, Tenn.

Odum, H. T. 1974c. Terminating fallacies in national policy on energy, economics, and environment. In A. G. Schmalz (Ed.), Energy: Today's choices, tomorrow's opportunities. World Future Society, Washington, D.C.

Odum, H. T. 1976. *Energy basis for man and nature.* McGraw-Hill, New York.

Odum, H. T., W. Smith, H. McKellar, D. Young, M. Lehman, and W. Kemp. 1973. Preliminary presentation of models to show interactions of power plant and estuary at Crystal River, Florida and energy costs and benefits for alternatives of management of cooling waters. Progress report to Florida Power Corporation and Licensing Agencies concerned with planning at Crystal River, Florida.

Odum, H. T., W. M. Kemp, W. H. B. Smith, H. N. McKellar, D. L. Young, M. E. Lehman, M. L. Homer, L. H. Gunderson, and A. D. Merriam. 1974. An energy evaluation of the system of power plants, estuarine ecology and alternatives for management. In *Crystal River power plant. Environmental considerations. Final report to the interagency research committee.* Florida Power Corporation St. Petersburg.

Oglesby, R. T., C. A. Carlson and J. S. McCann (Eds). 1972. *River ecology and man.* Academic, New York. 465 pp.

Phillips, J. 1956. *Quart. J. Royal Met. Soc.* **82**: 123–64.

Prest, A. R. and R. Turvey. 1966. Cost-benefit analysis: a survey. In *Amer. Econ. Assoc. surveys of economic theory, resource allocation.* Vol. 3. St. Martin's Press, New York. Pp. 155–207.

Richards, J. A., F. W. Sears, M. R. Wehr, M. W. Zemansky. 1960. *Modern university physics.* Addison-Wesley, Reading, Mass. 991 pp.

Smith, W. H. B., 1976. Measurements and simulation models of a shallow estuarine ecosystem receiving a thermal plume at Crystal River, Florida. Ph.D. thesis, University of Florida. Gainesville.

Smith, W. H. B., H. N. McKellar, D. L. Young, and M. E. Lehman. 1974. Total metabolism of thermally affected coastal systems on the west coast of Florida. In J. W. Gibbons and R. R. Sharitz (Eds). AEC Symposium Series (CONF 730505). AEC Publ., Oak Ridge, Tenn. Pp. 475–489.

Snedaker, S. C. 1974. Evaluations of interactions between a power generation facility and a contiguous estuarine ecosystem. In *Crystal River power plant. Environmental considerations. Final report to the interagency research committee.* Florida Power Corporation St. Petersburg.

Steele, J. H. 1974. *The structure of marine ecosystems.* Harvard University Press, Cambridge, Mass.

Tanner, W. F. 1960. Florida coastal classification. *Gulf Coast Assoc. Geol. Soc. Trans.* **10**: 259–266.

Tribus, M., and E. C. McIrvine. 1971. Energy and information. *Sci. Amer.* **224**: 179–190.

United Nations. 1967. *Proc. of the interregional seminar of the economic application of water desalination.* U. N. Publications, New York.

United Nations. 1969. *First United Nations desalination plant operation survey.* U. N. Publications, New York.

United Nations. 1970. *Solar Distillation as a means of meeting small-scale water demands.* U. N. Publications, New York.

Weigel, R. L. 1964. *Oceanographical engineering.* Prentice-Hall, Inc. Englewood Cliffs, N.J.

Wiegert, R. G. 1975. Simulation models of ecosystems. In R. F. Johnson, P. W. Frank, and C. D. Michener (Eds.), *Ann. Rev. Ecol. Systematics.* Vol. 6. Pp. 311–338. Annual Review, Inc., Palo Alto.

Yaalon, D. H. and J. Lomas. 1970. Factors controlling the supply and the chemical composition of acrosols in a near-shore and coastal environment. *Agric. Meteor.* **7**: 443 454.

Young, D. L. 1974. Studies of Florida Gulf coast salt marshes receiving thermal discharges. In J. W. Gibbons and R. R. Sharitz (Eds.), *Thermal Ecology,* AEC Symposium Series (CONF 730505). Pp. 532–550. AEC Publ., Oak Ridge, Tenn.

Zucchetto, J. J. 1975. Energy basis for Miami, Florida, and other urban systems. Ph.D. thesis, University of Florida, Gainesville.

22
Pest Management Models of Crop Ecosystems

CHRISTINE A. SHOEMAKER

Models of crop ecosystems can aid in the development of pest-management programs. Such models are used to predict the effect on yield of pest control measures for a range of population densities and environmental conditions. The examples given in this chapter emphasize the management of an insect pest of alfalfa. However, mathematical models developed for other insect pests and for plant pathogens are also discussed.

The first section is a general description of pest-management methods and models. The second section describes an optimization model that predicts the best pest-management policies to use against the alfalfa weevil, *Hypera postica*. The last section discusses current implementation of pest-management models and their potential for future use.

PEST-MANAGEMENT SYSTEMS

Pest Control—Economic and Environmental Significance

Each year crop yields are significantly reduced by damage from arthropods, pathogens, weeds, and other pests. It has been estimated that in the United States crop losses due to pest damage have been over 30% (USDA, 1965). The monetary value of this loss in potential production in 1974 has been estimated to have been $55 billion (Pimentel, 1976). Losses occur in spite of the extensive use of pesticides and nonchemical means of pest control. On pesticides alone farmers spent almost one billion dollars in 1971 (Andrilenas, 1974).

Pest control is an even more serious problem in many other countries, especially those with a tropical climate, where conditions are more favorable for pest growth. As a result, the use of pesticides is much more intensive in those countries. For example, the average rate of insecticide use for cotton pest control in Central America is over twice the average rate in the United States (ICAITI, 1976). Malnutrition and starvation in many developing countries could be significantly reduced if more effective and economical means of pest control could be developed.

Attempts to control pest infestations have also caused environmental pollution. It is estimated that between 800 million and 1 billion pounds of pesticides (over 4 lb per person) are used annually in the United States (NAS, 1975). About 59% of this amount is for agriculture (Andrilenas, 1974). These materials vary widely in their characteristics and do not all represent a threat to the environment. However, some pesticides have clearly been harmful to nontarget organisms. A number of insecticides are highly toxic to bees, fish, and birds (Metcalf, 1975). Nontarget populations are also adversely affected if their food source is killed by pesticides.

The deleterious effects of a toxic material are minimized if it is converted into a nontoxic metabolite before it comes into contact with nontarget organisms or if its toxic forms are sufficiently diluted as they move through the environment. Persistent chemicals may cause additional problems if they are concentrated in the bodies of animals near the top of a food chain. Persistent pesticides have been found to be thousands of times more concentrated in the bodies of some aquatic carnivores than in the water in which they live (Woodwell et al., 1967). Such concentrations have been shown to increase the mortality or to decrease the reproductive capability of many species (Pimentel, 1971; Burdick et al., 1964). Because of the ability of many organisms to concentrate persistent chemicals, registrations for several persistent insecticides, including DDT, Aldrin, and Dieldrin, have been withdrawn for most purposes. However, they have been replaced in many instances by insecticides such as Parathion, which although they are not persistent are acutely toxic to most organisms including man. It has been estimated that the number of fatal poisonings from these materials is about 200 per year in the United States. In some developing countries the number of deaths caused by insecticide poisonings is much higher.

The problems associated with the cost and effectiveness of pest control methods and the associated risk of environmental contamination have been exacerbated by the ability of many pests to develop resistance to pesticides. As a result, application rates for many pesticides have increased. This has caused corresponding increases in costs and environmental contamination. Even with higher rates of application, the control of resistant populations is frequently less than had been obtained with standard applications against susceptible populations.

In an effort to reduce continuing crop losses due to pest damage, agriculturalists have attempted to manage pest populations by successfully combining a number of control methods. The types of methods available for insect control and their integration into an effective management program are discussed in the following section.

Integrated Pest Management

The use of chemical pesticides is the method of pest control most familiar to the layman, but there are many other methods of control including biological control, cultural practices, and resistant plant varieties. Methods of control either attempt to suppress the size of a pest infestation or they attempt to prevent the pest from inflicting damage. Methods aimed at the suppression of population size can be divided into those that provide a long-term containment of the pest population to low levels and those methods that provide a rapid reduction in pest population size but do not necessarily provide long lasting control.

Biological control uses naturally occurring or imported enemies of pest species to reduce the size of pest populations. Insects are an important class of natural enemies of pests, and as such are frequently called "beneficial insects." Many cases of successful biological control by beneficial insects have been documented (Huffaker, 1972). Perhaps the earliest example of introduced biological control was the release in California in 1888 of the predatory beetle, *Rodolia cardinalis*, from Australia. The beetle successfully controlled the cottony cushion scale, *Icerya purchasi*, which had been causing severe losses in the citrus industry. A more recent example of biological control by a beneficial insect is described in the next section in a discussion of parasitism of the alfalfa weevil, *Hypera postica*, by *Bathyplectes curculionis*.

Biological control is also provided by pathogens that attack insects. Natural microbial control can be augmented by artificial mass culture and dissemination. The most widely used commercially produced microbial control agent is *Bacillus thuringiensis*. It provides quick and short-lived control and is thus applied in a manner similar to insecticide treatments. *B. thuringiensis* is used against more than twenty insect pests including such major pests as the European corn borer, *Ostrina nubilalis*, the cotton bollworm, *Heliothis zea*, and the gypsy moth, *Porthetria dispar*.

Another important means of pest control is the use of plant varieties that are resistant to pest damage. Mechanisms of resistance are quite varied. Some plants contain chemicals that are toxic to insects feeding upon them. Others have a physical means of preventing damage (e.g., a tougher surface). Some varietes can compensate for attack by quickly replacing damaged tissue. A plant may also gain resistance by maturing quickly. For example, cotton growers have reduced damage from the boll weevil by planting varieties that can be harvested before the peak of a boll weevil infestation.

Cultural practices can reduce pest damage by physically manipulating a crop environment. The rotation of crops and the removal of crop stubble (sanitation) are examples of cultural practices that reduce pest populations by interfering with their ability to continuously produce new generations. For example, the European corn borer overwinters in corn stalks left in the field. Thus by destroying the stalks after harvest the number of corn borers that survive the winter is greatly reduced. Insect damage also can be avoided by timing the planting and harvest of the crop so that the stages during which the crop plant is susceptible to damage do not coincide with a pest infestation. Harvesting itself can be a major cause of pest mortality, and choosing the time of harvest to maximize pest mortality is another means of control.

There are also many pest control methods that have been developed more recently. Among these are the use of sex attractants, the release of sterilized males, and the application of materials containing juvenile hormones that interfere with the insect's ability to mature.

The availability of so many pest-control methods complicates their selection in an integrated pest-management program. In addition, the effective integration of several pest-control methods can be difficult because of the complexity of the effects of several pest-control methods on a crop ecosystem. Some methods may even interfere with one another. Chemical insecticide treatments, for example, may kill predators or parasites that were providing biological control.

Insecticide treatments are an important tool in integrated control programs. For some pests and crops, insecticides are the only effective means of control available. A major advantage of many insecticides is that control is achieved quickly following their application. Because pest densities fluctuate, it is an advantage to have pest-control methods like insecticides that can be used only during those periods when pest densities reach damaging levels. Most of the other methods of control discussed such as resistant plant varieties, cultural practices, and control by beneficial insects provide longer-term control but also require long-term planning. Implementation of these methods must occur many months before the farmer knows what the intensity of a pest infestation will be.

To fully utilize the advantages of short-term control, one must determine when a pest infestation is serious enough to justify a pesticide treatment. Efforts by entomologists to determine when an insecticide treatment is necessary have been based upon the *economic injury level*, which is defined as the lowest pest density for which an insecticide treatment is economically justified. By definition, if the pest density exceeds the economic injury level, the cost of an insecticide treatment must be less than the value of the difference between the yields expected with and without treatment. If the pest density is below the economic injury level, an insecticide application costs more than it can be expected to save by preventing crop losses.

Because of the dynamic nature of population growth, it may be necessary to apply an insecticide before a pest population reaches its economic injury level. For example, an economic injury level is frequently defined in terms of peak larval density. Consequently it may be more effective to apply an insecticide before larvae have reached their peak densities. For this reason Stern et al. (1959) introduced another term, the *economic threshold*, which is defined to be "the density at which control measures should be applied to prevent an increasing pest population from reaching the economic injury level."

Thus farmers are advised to use insecticides only when the pest density exceeds the economic threshold. Many pest populations fluctuate in size from year to year. Hence insecticide recommendations based upon an economic threshold can result in a considerable reduction in the number of insecticide treatments compared to the number applied under conventional programs that recommend treatments at a fixed time regardless of the size of the pest density. Such a reduction in insecticide use has several advantages: (1) the farmer saves the cost of insecticide treatments, (2) the risk of environmental contamination is reduced, and (3) the target pest is less likely to develop resistance to an insecticide. As a consequence, insecticide treatments used only when truly necessary will be able to kill a maximum percentage of the population.

The economic threshold was originally defined in terms of pest density (Stern et al., 1959). However, the advisability of an insecticide treatment also depends upon other factors governing both the effectiveness of the insecticide and the population dynamics of the pest and plant populations following insecticide treatment. Such factors include the age distribution of the pest population, the maturity and vigor of the crop plant, the size of beneficial insect populations, and weather.

The age distribution of the pest population is very important because the damage an insect inflicts and its susceptibility to insecticide, predators, and parasites depends upon its stage of development. For example, some parasites attack only eggs, others only larvae of a specific size. Similarly, an insect may be invulnerable to an insecticide during part of its life cycle. This is especially true of periods spent buried in plant material.

The damage inflicted on crops also varies as an insect develops. For example, more than half of an alfalfa weevil's total consumption of alfalfa leaves is eaten while it is a fourth-instar larva. Other insects cause maximum damage as adults.

The rate at which insects mature from one life stage to the next depends upon temperature. Thus changes in the age distribution of the pest population can be forecast with less error if the temperature can be accurately predicted. Temperature forecasts are also useful in estimating the rate of plant maturation.

Besides age distribution of the pest population, another factor that can influence the advisability of implementing a pest-control measure is the maturity and vigor of the crop plant. Vigorous plants can often compensate for moderate damage so that pest infestations have little effect on yield. However, if the plant is under other stresses as well, such as a shortage of water, light, or nutrients, it may not be able to compensate for pest damage as effectively. A plant's susceptibility to damage also varies as it matures. For example, following a harvest, alfalfa is very susceptible to feeding. When there is no foliage, alfalfa weevils feed upon the crown buds, thereby delaying regrowth and reducing yields. Feeding by weevils later in the season is much less damaging because the plant has more foliage and can replace the plant tissue that is consumed (Fick, 1976).

Potential damage to the crop is reduced if the pests are controlled by beneficial insects. However, the size of this reduction depends upon the size of the pest and beneficial insect populations, the effectiveness of the beneficial insects, and the age of the pest when it is killed by the natural enemy. For example, effectiveness of the parasite *Bathyplectes curculionis* (see the next section) in reducing alfalfa weevil damage is diminished by the fact that the weevil is not killed by the parasite until the weevil becomes a pupa. By this time, the weevil has already inflicted a considerable amount of damage.

Pest-Management Simulation Models

It is clear that the decision of when to implement an insect-pest-management strategy can be made more accurately if age distribution of the pest population, densities of beneficial insect populations, the condition of the crop, and temperature are considered as well as pest density. Thus our pest-control decisions can be based upon a *multidimensional economic thresholds* that depend upon some or all of these factors.

However, to determine multidimensional economic thresholds requires a quantitative understanding of the effects on yields of the interactions among many components of a crop ecosystem. In studies such as Koehler and Rosenthal (1975), designed to establish economic thresholds based solely upon results from field tests, it is understandable why thresholds of only one variable were established. It would not be feasible to try to estimate multidimensional economic thresholds simply by empirical studies. Some variables like temperature cannot be controlled in field studies. In addition, testing all combinations of different values for each of the variables would require an inordinate number of field trials.

It is for this reason that mathematical models have been developed to aid in the understanding and prediction of the interactions between pest control measures and the crop ecosystem. These models attempt to synthesize an understanding of the behavior of the crop ecosystem from mathematical descriptions of relations between components of the ecosystem, (Watt, 1964; Waggoner et al., 1972).

A mathematical description of pest insect populations is usually based upon the age distribution of the population. The population may be considered to be continuously distributed along an age spectrum or may be described as a collection of discrete age classes. An example of a model of a population with a continuous age distribution is given in the next section.

Most simulation models, however, are based upon a division of the population into discrete age classes. The more classes included, the more accurately the age range in a population can be modeled. Let $N_i(t)$ be the number of insects in the ith age class at time, t. The number of individuals in this class will change as its members mature and move to the next age class, N_{i+1}, and as individuals from the previous age class, N_{i-1}, mature and move into class N_i. This process is illustrated in Figure 1. The fraction of individuals that mature in one unit of time is dependent upon temperature and therefore is written $\alpha_i(T)$, where T is temperature.

Some of the insects will not survive from time t to time $t + 1$. The rate of survival depends upon both environmental conditions (T) and pest control decisions (v). Let $s_i(v,T)$ be the fraction of insects in age class i that survive one unit of time. Then the number of individuals in age class i at time, $t + 1$, equals the number that do not mature, $(1 - \alpha_i(T))N_i(t)$, times the survival rate plus those that mature from class $i - 1$ into class i. Thus

$$N_i(t + 1) = s_i(v,T)N_i(t)(1 - \alpha_i(T)) + s_{i-1}(v,T)N_{i-1}(t)\alpha_{i-1}(T) \quad (1a)$$

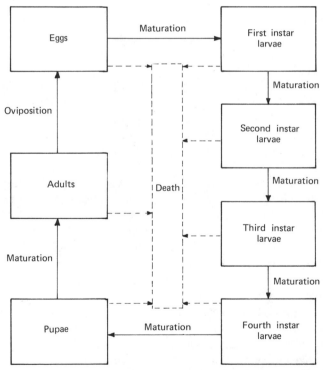

Figure 1   Flow chart description of an insect simulation model. The number of individuals maturing in one time period from the ith age class is $s_i(v,T)\,\alpha_i(T)\,N_i(t)$. The number dying in age class i is $(1 - s_i(v,T))\,N_i(t)$.

The number of eggs, $N_0(t)$, is

$$N_0(t) = s_0(v,T)N_0(t)(1 - \alpha_0(T)) + \sum_{i=m}^{M} N_i(t)r_i(T) \qquad (1b)$$

where $N_m, \ldots, N_M$ are the age classes that oviposite and r_i is their rate of oviposition.

Most pest population simulation models are based upon equations similar to $(1a - b)$. However, each of the functions, s_i, r_i, and α_i, may have a rather complicated form. For example, the survival function may depend upon the effectiveness of predators. In order to determine s_i, it may be necessary to develop a submodel of the predator population. The function, $\alpha_{i-1}(T)$, is the reciprocal of the percent development that would occur if the temperature equalled T for time t to time $t + 1$. For example, if it takes 5 days at 15°C for an insect to pass through the ith age class, then $\alpha_i(15°C) = \frac{1}{5}$, where the units of t are days.

Models used to predict plant growth and yield usually are based upon a division of plant biomass into some of the following classes: photosynthate, leaves, stems, fruit, roots, and carbohydrate reserves. The amount of carbohydrate produced by photosynthesis is calculated from equations based on temperature, light, and leaf area. The rates at which photosynthate is transferred to leaves, stems, and other plant parts are mathematically described as functions of weather conditions and physiological status. Effects of harvesting or defoliating insects are simulated by removing material from the model variables describing the plant parts affected. Figure 2 illustrates the flow of materials in an alfalfa model developed by Fick (1975). The variable *photosynthates* describes the amount of material that is available for top growth and storage. The amount of photosynthetic input used for respiration and parts of the plant not in the model is subtracted from the total photosynthetic input before photosynthates are available for leaves, stems, and TNC (total nonstructural carbohydrates in the taproot). When a harvest is simulated, all leaves and stems are removed. Thus there is no photosynthetic input to support the growth of new leaves and stems. The source of material for regrowth of leaves and stems is from buds which are elongated into leaves and stems. The buds are formed from the TNC. Hence the primary source of material to leaves and stems is photosynthates when the crop has sufficient leaf area; otherwise the primary source of material is TNC. If there are no buds and the taproot carbohydrate supply is exhausted, harvesting the leaves and stems will kill the crop.

Flows between each component of the alfalfa model are determined by light intensity, day length, and the plant's physiological status. Rates of photosynthesis, bud elongation, and leaf and stem growth all depend upon at least one of these variables.

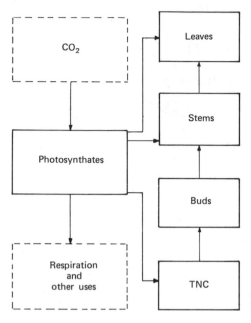

Figure 2 Flow chart description of a model of alfalfa growth (adapted from Fick, 1975).

In order to determine the usefulness of a pest control measure, the change it causes in the pest population and the effect on the crop must be estimated. Changes in the pest population's dynamics can be estimated by simulation of the pest population model with the survivorship rates, $s_i(v,T)$, changed to correspond to the pest control measure v to be implemented.

Usually the effect of the pest on the crop during one unit of time can be determined from

$$D(t) = \sum_{i=1}^{N} f_i N_i(t) \tag{2}$$

where $D(t)$ = total damage done at time t

f_i = amount of damage done by each insect in age class i in one unit of time

$N_i(t)$ = number of insects in age class i at time t

When used to describe weevil damage to alfalfa, f_i represents the grams of alfalfa removed by each age class of alfalfa weevil. The rate of feeding by fourth instar larvae, f_4, is much larger than any of the other feeding rates; and f_1, the first-instar feeding rate, is very small.

To determine the effect of feeding on the crop, the total damage, $D(t)$, then is subtracted from the variables in the plant model that represent the parts subject to damage. For example, in the alfalfa model discussed earlier, the mass of leaves consumed, $D(t)$, is subtracted each day from the alfalfa model variable describing leaf mass. Insect pests may also damage fruiting parts. In this case, $D(t)$ measures fruit damage and is subtracted from the component describing fruiting parts in the corresponding plant model.

By this method, insect damage and plant growth can be integrated. The results of one modeling study (Gutierrez et al., 1974) that describe plant growth and insect damage are listed in Table 1. The predictions ("complete

Table 1   Comparison of Actual Cotton Yields and Those Predicted by a Model of Cotton Growth for Four Different Insecticide-Treatment Policies (adapted from Gutierrez et al., 1975).

Treatment	Observed Yield (bales)	Simulation with[a] BAW and CL Effects (bales)	Bollworm Damage (bales)	Complete Simulation (bales)
I (untreated check)	2.78	2.77	0.004	2.766
II	2.74	2.75	0.019	2.731
III	2.618	2.67	0.014	2.656
IV	2.626	2.74	0.062	2.678

[a] BAW is a beet armyworm and CL is cabbage looper worm.

simulation") and the observed values are very close. This model predicts only plant growth dynamics, not insect population dynamics. Insect damage was calculated from observed numbers of beet army worms and cabbage loopers and from the number of bolls observed to have bollworm damage.

Optimization Methods in Pest-Management Modeling

The models discussed above have been simulation models, based upon the numerical calculation of a set of difference or differential equations. To examine a number of management possibilities, a simulation model must be numerically solved for each possible alternative. If one considers the possibility of applying an insecticide a number of times under different conditions of weather, plant growth, and pest and beneficial insect densities, the number of possible management programs (and thus the number of numerical simulations) becomes very large. For example, to evaluate directly all the possibilities in the alfalfa-weevil-management program discussed in the next section would require over one million simulations.

Because of the expense of testing each management option directly, pest-management models utilizing optimization techniques have been developed (Shoemaker, 1973; Regev et al., 1976). Such techniques determine the management policy which is optimal relative to some criterion of success. The optimal policy is found by examining the mathematical structure of the model rather than by exhaustive simulations of all possible management programs. (Chapter 4 discusses optimization techniques and their advantages and disadvantages relative to simulation methods.) An insect pest-management model that uses an optimization method called *dynamic programming* is described in the next section.

ALFALFA WEEVIL MANAGEMENT: A DYNAMIC-PROGRAMMING MODEL

Damage caused by alfalfa weevils can be reduced by an integrated pest-management program utilizing chemical, biological, and cultural methods of alfalfa weevil control. The objective of the model described in this section is to determine the best method of integrating these control measures in response to changes in population densities and weather.

The Alfalfa Ecosystem

The main components of the alfalfa ecosystem are the alfalfa plants, the alfalfa weevil population, the *B. circulionis* population, and weather. The interactions between these components are illustrated in Figure 3.

In New York alfalfa weevils mate in the spring and lay eggs in the stems of alfalfa plants from March through June. After the eggs hatch, the young larvae move up the stem into the compressed leaves in the growing tips. By the third instar the weevils are feeding on the exposed leaves. After the fourth-instar larval stage, the weevil pupates and 4 to 14 days later emerges as an adult. The adult weevil also feeds on the leaves, or if leaves are not available, the weevil feeds on the buds. In order to combat low humidity the adult weevil diapauses in August through October. Females may lay eggs in November, but very few of the eggs are viable. In December through March the adults are inactive except for occasional warm spells.

Because oviposition occurs over a considerable range of time, there is an overlap between the times of occurance of different larval instars (Figure 3a). The rate of weevil development depends upon temperature, so that

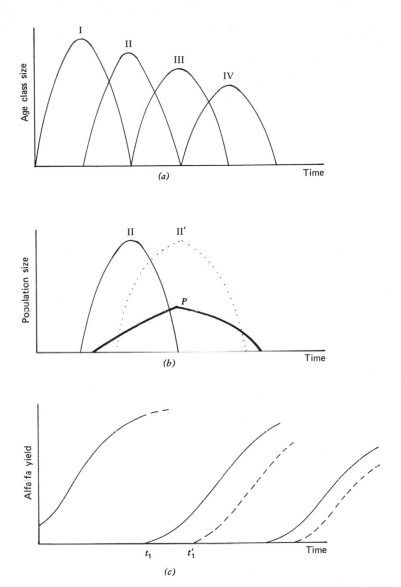

Figure 3 A schematic description of dynamic interactions among populations in an alfalfa ecosystem. (*a*) Represents the size of each larval instar population. The Roman numerals I through IV refer to the first through the fourth larval instar. (*b*) Illustrates the importance of synchrony between the weevil population and its parasite (*P*). In cooler years the second-instar population (as represented by the dotted line) will be subjected to more parasite attacks than will the second-instar population (represented by the solid line) in warm years. (*c*) Represents the amount of harvestable hay in the alfalfa crop if the first harvest occurs at time t_1 (solid line) or t_1' (dotted line).

the calendar dates corresponding to the occurrence of an age class will vary with temperature (Figure 3b).

A means of biological control of the alfalfa weevil is a larval parasite, *Bathyplectes curculionis*. The adult female attacks the second-instar larvae while they are inside the compressed leaves of the growing tip. The effectiveness of the parasite is reduced if some of the weevils pass through the susceptible stage before the parasites emerge (Figure 3b). The parasite eggs mature into larvae within a weevil until the weevil starts to pupate. After the weevil has produced its cocoon, *B. curculionis* kills and eats the weevil. The parasite then overwinters inside its cocoon and emerges the following spring as an adult.

Alfalfa is a perennial crop and is harvested a number of times each growing season. In New York State farmers cut alfalfa an average of three times per year. Many of the weevils are killed or removed when the hay is harvested. Most of the eggs and the first- and second-instar larvae are removed with the hay. The fraction of large larvae, pupae, and adults that survive a harvest is much larger. Since all the weevils have matured past the larval stage by the time the second and third harvests are made, it is the timing of the first harvest that is most important for pest control. Weevil damage can be reduced by harvesting before most of the members of the weevil population become large larvae, which are less susceptible to harvest mortality and cause more damage to the crop.

However, the timing of the first harvest can affect the total yield eventually harvested in three cuttings during the growing season. If hay is cut too early, the period of maximum productivity during the first growth is shortened. The second growth then is given more time to mature, but this may not compensate for the loss in the first growth which is always the highest yielding. As illustrated in Figure 3c, a late cutting (at time t_1' rather than at time t_1) can also result in reduced yields.

Parasite effectiveness also influences pest-control decisions. Clearly, if the parasite will successfully control the weevil, there is less need to implement further controls. In addition, pest-control practices that remove most of the parasitized larvae before they mature into parasite cocoons will reduce parasite populations the following year. Thus one would expect a control decision to be influenced by not only the amount of parasitization but also the age of the parasitized individuals. This depends upon synchrony between the weevil and parasite populations. This relationship is illustrated in Figure 3b.

The optimization technique used in the alfalfa weevil management model is dynamic programming, which is discussed in Chapter 4. Dynamic programming is particularly appropriate because the management decision is implemented at discrete points in time, the predictive equations are nonlinear, and temperature is assumed to be a random variable.

As mentioned in Chapter 4, a shortcoming of dynamic programming is the difficulty of applying it to systems with a large number of state variables. An alfalfa ecosystem model should have state variables describing the weather, the number of individuals in each age class of the weevil and parasite populations, and several variables describing different parts of an alfalfa plant. Such a description would require ten to fifteen state variables. Dynamic programming cannot be used for a system with so many state variables.

The difficulties associated with the large dimension of the alfalfa system can be circumvented by decomposing a management model into two coupled models. The first, which is referred to as the *decision model*, determines the time of harvest (t_h^n) and the insecticide application policy (v^n) that will maximize the long-term net income. The choice of t_h^n and v^n depends upon the values of three state variables, the number of weevils (Q_1^n), the number of parasites (Q_2^n), and the temperature pattern (Q_3^n).

In order to calculate net income we must be able to estimate the yield (Y) and the number of weevils and parasites $(Q_1^{n+1}$ and $Q_2^{n+1})$ that survive to reproduce in the following year. The yield and size of the populations in the following year are determined by the interactions among weevil age classes, parasites, temperature, and plant growth. To predict the effects of these interactions is the purpose of the second model, which is referred to as the *population model*.

Decision Model

The management decisions to be made in year n are the time of harvest and the insecticide treatment. The best choices are assumed to be those that maximize long-term net income. The net income in year n is

$$r_n Y(Q_1^n, Q_2^n, Q_3^n, t_h^n, v^n) - C_n(v^n) \tag{3}$$

where Q_1^n = number of adult weevils mating in the spring of the nth year
Q_2^n = number of adult parasites mating in the spring of the nth year
Q_3^n = the temperature pattern in the nth year
t_h^n = calendar date of the first harvest in year n
v^n = insecticide treatment in year n
r_n = price per ton of alfalfa in year n
C_n = cost of insecticide treatment v^n
$Y(Q_1^n, Q_2^n, Q_3^n, t_h^n, v_n)$ = the total yield of alfalfa obtained in three cuttings in the summer of year n

The costs C_n and prices r_n are expressed in terms of their present value in year 1. Fixed costs that do not affect pest control decisions are omitted from Eq. 3.

The net income over N years is

$$\sum_{n=1}^{N} [r_n Y(Q_1^n, Q_2^n, Q_3^n, t_h^n, v^n) - C_n(v^n)]$$

The objective is to maximize the expected net income by choosing the best combination of decisions t_h^n, v^n, for $n = 1$ to N. Thus the criterion function is

$$\underset{\substack{t_h^n, v^n \\ n=1, N}}{\text{Max}} E\left\{\sum_{n=1}^{N} [r_n Y(Q_1^n, Q_2^n, Q_3^n, t_h^n, v^n) - C_n(v^n)]\right\} \tag{4}$$

where $Q_1^{n+1} = G_1(Q_1^n, Q_2^n, Q_3^n, t_h^n, v^n)$
$Q_2^{n+1} = G_2(Q_1^n, Q_2^n, Q_3^n, t_h^n, v^n)$

The values of the functions, Y, G_1, and G_2, are determined from the population model. The temperature pattern, Q_3^n, is a random variable that assumes a value of η_i with a probability, ρ_i.

As discussed in Chapter 4, a problem of the form of Eq. 4 can be formulated as a dynamic-programming problem

$$H_n(Q_1^n, Q_2^n, Q_3^n) = \underset{t_h^n, v^n}{\text{Max}}[r_n Y(Q_1^n, Q_2^n, Q_3^n, t_h^n, v^n) - C(v^n)$$

$$+ \sum_{n=1}^{N} \rho_i H_{n+1}(Q_1^{n+1}, Q_2^{n+2}, \eta_i)] \tag{5}$$

for

$$Q_1^{n+1} = G_1(Q_1^n, Q_2^n, Q_3^n, t_h^n, v^n) \tag{6}$$

$$Q_2^{n+1} = G_2(Q_1^n, Q_2^n, Q_3^n, t_h^n, v^n) \tag{7}$$

The choice of which t_h^n and v^n are best will depend upon the values of the state variables. Let $w_n = (t_h^n, v^n)$ be the decision vector in stage n and w_n^* be the optimal values of w_n. Then w_n^* is a function of Q_1^n, Q_2^n, Q_3^n and can be written as $w_n^*(Q_1^n, Q_2^n, Q_3^n)$.

The values of the functions, Y, G_1, and G_2, which appear in Eqs. 5, 6, and 7 above, are calculated in the population model. This model is described in the following sections. Except where noted otherwise, the parameter values in the population model are based on field data collected in central New York by Professor Robert Helgeson, Department of Entomology, and Professor Gary Fick, Department of Agronomy, Cornell University.

Alfalfa Weevil Development

The three state variables, Q_1, Q_2, and Q_3, appear in both the decision model and in the population model. As mentioned earlier, the temperature pattern, Q_3, is assumed to be a random variable. Temperature is an important factor because it affects the rate of development of both the alfalfa and the alfalfa weevil. The effect of temperature on the rate of alfalfa weevil development is estimated by Ruesink (1976) to have the form

$$R(t) = \frac{A}{B + Ce^{-DT(t)}} \tag{8}$$

where $R(t)$ = rate of development
$T(t)$ = temperature at time t

and A, B, C, and D are constants.

The length of time a weevil requires to complete its development depends upon the temperature. For example, a weevil oviposited in early April may require twice as many days to reach adulthood as a weevil oviposited in late May when the weather is much warmer. Since the elapse of calendar time cannot be used to determine the physiological age of an individual, it is useful to define a new variable, s, called *physiological time*, which measures the accumulation of heat units that contribute to growth. Since $R(t)$ measures the rate of development, the physiological time, s, corresponding to calendar time, t, is

$$s = s(t) = \int_0^t R(t)\, dt \tag{9}$$

Equation 9 converts from units of calendar time, t, into units of physiological time, s. It will sometimes be necessary to convert in the opposite direction, from units of physiological time, s, into units of calendar time, t. Equation 9 defines a one-to-one correspondence between s and t. Thus the inverse function, which we call h, is well defined and

$$t = h(s) \tag{10}$$

Since the rate of development, $R(t)$, depends upon the temperature, $T(t)$, the functions, $s(t)$ and $h(s)$, depend upon the state variable, Q_3^n, which describes the temperature pattern in year n. The values for Q_3 were obtained by calculating the mean and standard deviation of the physiological time, $s(t)$, from 6 years of weather data for Ithaca, New York. The average temperature pattern ($Q_3 = 2$) is the daily mean of $s(t)$. The cool weather pattern ($Q_3 = 1$) is the average $s(t)$ minus one standard deviation, and the warm weather pattern ($Q_3 = 3$) is the average $s(t)$ plus one standard deviation.

Let $N(s,b)$ be the number of unparasitized weevils that are oviposited at physiological time, b, that are alive at physiological time s. The (physiological) age of these individuals is $s - b$.

As it matures, the alfalfa weevil passes through seven distinct stages. Age class zero represents eggs, age classes one through four represent the first through the fourth instars, age class five represents the pupal stage, and age class six represents the adult stage. Let A_i be the age at which a weevil enters the ith age class. Thus if $A_i \leq s - b < A_{i+1}$ then all of the individuals, $N(s,b)$, are in the ith age class.

Mortality Factors

Alfalfa weevil mortality is due to many factors, which we divide into three classes: parasitism, pest control, and natural causes. Mortality due to the last factor is represented as a function of age and is represented by $D(a)$. It is defined to be the instantaneous death rate due to factors other than parasitism and pest control practices. Included in $D(a)$ are the effects of extreme temperatures and low humidity. In the absence of parasites, harvesting, and insecticide treatments, the change in the population of weevils can be described by

$$\frac{dN(s,b)}{dt} = -D(s-b)N(s,b) \tag{11}$$

In order to solve Eq. 11, the number of weevil eggs oviposited at each physiological time b, $N(b,b)$, must be specified. As mentioned earlier, oviposition occurs over a considerable range of time. Let $\theta_0(b)$ be the fraction of total eggs laid that are oviposited at physiological time b. The total number of eggs laid is the number of weevil females ($Q_1^n/2$) times the average number of eggs laid per female. Then the number of eggs oviposited at time b is

$$N(b,b) = \frac{Q_1^n}{2} m\theta_0(b) = Q_1^n\theta(b) \tag{12}$$

where m is the average number of eggs per female and $\theta(b) = (m\theta_0(b))/2$. Let $\bar{b}$ represent the end of oviposition. Then $\theta(b) > 0$ only for $\theta \le b \le \bar{b}$. The values of the function, $\theta_0(b)$, were estimated from field data for Ithaca, New York (Smith and Shoemaker, 1977).

The solution to Eqs. 11 and 12 gives the expected number of weevils if the only causes of mortality are those that are represented by $D(a)$. This solution is

$$N(s,b) = N(b,b) \exp\left[-\int_0^{s-b} D(a)\, da \right] \tag{13}$$

Substituting Eq. 12 into Eq. 13, we obtain

$$N(s,b) = Q_1^n\theta(b) \exp\left[-\int_0^{s-b} D(a)\, da \right] \tag{14}$$

Parasitism

It is assumed that weevil mortality due to parasitism does not occur until the weevil starts to pupate. At this time the parasite kills the weevil and remains in its cocoon until it emerges the following year. (The fraction of parasites that emerge in the same year was not considered because they do not appear to provide significant biological control in central New York State.)

The adult parasites prefer to attack second-instar larvae. Thus a fraction of the second-, third-, and fourth-instar weevil larvae will have been parasitized but not yet killed. Let $N_p(s,b)$ be the number of parasitized weevils that were oviposited at physiological time b and are alive at physiological time s. It follows that $N_p(s,b)$ is greater than zero only when $A_2 < s - b \le A_5$. As before, $N(s,b)$ is the number of unparasitized weevils.

The changes in the parasitized and unparasitized weevil populations are illustrated in Figure 4. Both populations are being decreased by mortality. In addition, the unparasitized population is decreased as its members become parasitized.

The rate of parasite attack is the product of the parasite's searching rate, the prey density, the predator density, and the fraction of encountered prey that are attacked. (See Chapter 4 for a discussion of this description of predation.) Let $\alpha(a)$ be the product of two terms, the searching rate and the fraction of encountered prey that are attacked. We assume that the parasite attacks only second-instar larvae. Thus $\alpha(a) = 0$ if $a < A_2$ or $a > A_3$ and $\alpha(a)$ equals a constant, α, for all a, $A_2 < a < A_3$.

The rate of parasitism at time t depends upon the number of female parasites, $P(t)$, that are searching for and attacking weevil larvae at time t. Adult parasites do not all emerge at the same time. Thus the number of ovipositing parasites will vary with time as schematically represented in Figure 5. Let $P_0(t)$ be the fraction of the total parasite population, Q_2^n, that are ovipositing females at time t. Then

$$P(t) = \begin{cases} Q_2^n P_0(t) & p \le t \le \bar{p} \\ 0 & \text{otherwise} \end{cases} \tag{15}$$

where $(p,\bar{p})$ is the period of time during which the parasites are attacking alfalfa weevils.

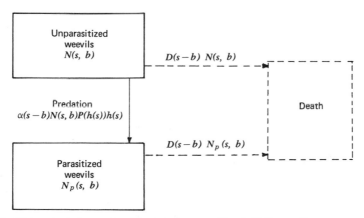

Figure 4   Flow rates between parasitized and unparasitized alfalfa weevils.

Parasite attack is assumed to be a function of calendar time. However, since the change in the weevil population is described in terms of physiological time s, $P(t)$ must also be expressed in terms of physiological time. From Eq. 10,

$$P(t) = P(h(s)) \tag{16}$$

where s is the physiological time that corresponds to calendar time t.

The parasite attack rate, $\alpha(a)$, is also in terms of attacks per unit of calendar time. In order to convert this rate to units of physiological time, we must multiply it by the instantaneous ratio of Δt to Δs that is

$$\lim_{\Delta s \to 0} \frac{\Delta t}{\Delta s} = \lim_{\Delta s \to 0} \frac{h(s + \Delta s) - h(s)}{\Delta s} = \frac{dh}{ds}(s) = \dot{h}(s) \tag{17}$$

The rate of predation can now be computed. It is the product of the attack rate, $\alpha(a)$, a factor converting to physiological time units ($\dot{h}(s)$), the density of ovipositing parasites, and the density of unparasitized weevils. Thus

$$\text{rate of predation} = \alpha(s - b)\dot{h}(s)P(h(s))N(s,b) \tag{18}$$

Differential equations describing the dynamics of the weevil population can be developed from Figure 4. The unparasitized population is decreased both by death and parasitization so

$$\frac{dN(s,b)}{ds} = -D(s-b)N(s,b) - \alpha(s-b)N(s,b)P(h(s))\dot{h}(s) \tag{19}$$

The parasitized population is decreased by death and increased by parasitization. Thus

$$\frac{dN_p(s,b)}{ds} = -D(s-b)N_p(s,b) + \alpha(s-b)N(s,b)P(h(s))\dot{h}(s) \tag{20}$$

The solution to Eq. 12, 19, and 20 is

$$N(s,b) = Q_1^n \theta(b)e^{-(K(s-b) + L(s,b,P))} \tag{21}$$

$$N_p(s,b) = Q_1^n \theta(b)e^{-(K(s-b))}(1 - e^{-L(s,b,P)}) \quad \text{for } s - b < A_5 \tag{22}$$

where

$$K(a) = \int_0^a D(r)\, dr \tag{23}$$

Thus $e^{-K(a)}$ is the probability that a weevil will survive to age a. The function L measures the effect of parasitism, and if $\alpha(a)$ is constant on the integral (A_2, A_3), then

$$L(s,b,P) = \alpha \int_{\underline{t}(s,b)}^{\bar{t}(s,b)} P(t)\, dt \tag{24}$$

The interval $(\underline{t},\overline{t})$ is the period of time before time s when individuals oviposited at time b are susceptible to predation. Thus

$$\underline{t}(s,b) = h(\text{Min}(s,b + A_2)) \tag{25}$$

$$\overline{t}(s,b) = h(\text{Min}(s,b + A_3)) \tag{26}$$

Since $\overline{t}(s,b) = h(b + A_3)$ for all $s > b + A_3$, it follows that $L(s,b,P) = L(b + A_3,b,P)$ for all $s > b + A_3$.

The physical significance of the value of $L(s,b,P)$ is illustrated in Figure 5. From Eq. 24 we see that L is the area under the graph of the parasite function, $P(t)$, over the period of weevil susceptibility. This period of susceptibility depends on the time of oviposition, b, and on the function, h, which is determined by the temperature pattern, Q_3. In the example in Figure 5a, s is less than $b_1 + A_3$. Hence, individuals $N(s,b_1)$ have not yet completed the second-instar larval stage. Thus $\overline{t} = h(s)$. In the second example, Figure 5b, the time of oviposition, b_2, is earlier and $b_2 + A_3$ is less than s. Thus by physiological time s, the individuals, $N(s,b_2)$, have completed the second-instar stage. Their period of susceptibility ended when they matured from the second instar at time, $\overline{t} = h(b^2 + A_3)$.

Note in Figure 5b that for individuals oviposited at time b_2, there is a period of time from $h(b_2 + A_2)$ to p when the alfalfa weevils are second instars but are free from parasitism because no parasites have yet emerged. Weevils oviposited sufficiently early (so that $b + A_3 < p$) will be subject to very little parasitism because they have completed the susceptible stage before the parasites are active. Because of this the fraction of alfalfa weevils that will escape parasitism increases in warm years, in which the weevils complete their second-instar stage earlier than in cool years.

Harvest Mortality

Both parasitized and unparasitized weevils are killed by harvesting and insecticide treatments. This mortality is age specific. Harvesting removes most of the eggs and small larvae. A much larger fraction of the older age classes survive.

Let $\phi(v,a)$ be the fraction of weevils of age a that survive a harvest. If $v = 1$, $\phi(v,a)$ is the probability of surviving both a harvest and an insecticide treatment following the harvest. If $v = 0$, $\phi(v,a)$ is the probability of surviving the harvest when no insecticide is applied.

If the crop is harvested at calendar time, t_h, the physiological time of harvest is $s_h = s(t_h)$. The age at the time of harvest of weevils that were oviposited at physiological time b is $s_h - b$. Thus for $s > s_h$ the number of individuals alive at time s that were born at time b is the product of $\phi(v, s_h - b)$ and the number that would be alive if no harvest mortality took place.

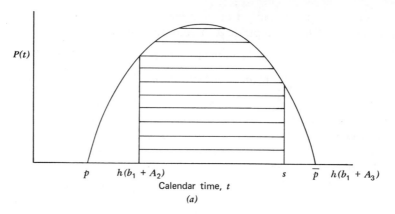

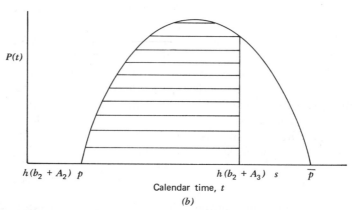

Figure 5 Schematic description of the density of adult *B. curculionis* which attack alfalfa weevil second instar larvae. The hatched area is the value of the function $L(b, s, P)$. (*a*) and (*b*) represent $L(b, s, P)$ for two different values of b.

The latter value is given in Eqs. 21 and 22. Thus after harvest

$$N(s,b) = \phi(v, s_h - b)Q_1^n\theta(b)e^{-[K(s-b)+L'(s,b,P)]} \tag{27}$$

$$N_p(s,b) = \phi(v, s_h - b)Q_1^n\theta(b)e^{-K(s-b)}[1 - e^{-L'(s,b,P)}] \quad \text{for } s > s_h \tag{28}$$

Function L has been replaced with function L', which incorporates the effect of the harvesting and insecticide treatments on the number of parasites.

Thus L' is α multiplied by the area under the parasite curve, $P'(t)$, where

$$P'(t) = \begin{cases} P(t) & t < t_h \\ \phi_p(v)P(t) & t > t_h \end{cases} \qquad (29)$$

and $\phi_p(v)$ is the fraction of parasites that survive the harvest.

Yield

Equations 21, 22, 27, and 28, which describe the size and age structure of the weevil population, can now be used in conjunction with a model of alfalfa growth to estimate the effect of pest control on yield. The alfalfa model used was developed for New York State conditions by Fick (1975) and is described at the beginning of this chapter.

Alfalfa weevil feeding damages the crop in two ways. The first is to remove leaf material and thereby directly reduce yield. The total amount of feeding by all individuals oviposited time b is the integral over $(b, b + A_7)$ of the feeding rate for an individual of age $s - b$ times the number of individuals, $N(s,b)$. To obtain the total amount of feeding by all individuals we must integrate the latter term over all times of oviposition b. Thus

$$\text{total leaf feeding} = \int_0^{\bar{b}} \int_b^{b+A_7} [f(s \quad b)N(s,b) + f_p(s - b)N_p(s,b)] \, ds \, db$$
$$(30)$$

where $f(a)$ and $f_p(a)$ are rates of feeding by unparasitized weevils and parasitized weevils of age a and A_7 is the age at which they enter diapause. Since the parasitized weevils die after reaching age A_5, $f_2(a) = 0$ for $a > A_5$. Functions N and N_p are obtained from Eqs. 21, 22, 27, and 28.

Besides removing leaf material, alfalfa weevil feeding can also delay regrowth of the alfalfa after it has been cut. Immediately after harvest the surviving weevils have no leaves to feed upon so they feed instead upon the buds. Although the total mass of material removed by this feeding may be small, its detrimental effects are significant. Fick (1976) has shown that weevil feeding on buds can delay the start of regrowth by as much as ten days. He has also shown that for regrowth delays of less than seven days, the length of the delay is a linear function of larval density. Therefore, for the population model it is assumed that

$$\text{delay} = k\left[\int_0^{\bar{b}} \{f(s_h - b)N(s_h,b) + f_p(s_h - b)N_p(s_h,b)\} \, db\right] \qquad (31)$$

where s_h is the physiological time of the harvest and $N(s_h,b)$, $N_p(s_h,b)$, $f(a)$, and $f_p(a)$ are defined as above. The value of the constant, k, was obtained from Fick (1976).

The yield of alfalfa depends upon the temperature pattern, Q_3^n, as well as on the number of weevils and parasites and the pest-control measures. Symbols Y or $Y(Q_1,Q_2,Q_3,t_h,v)$ represent the yield as a function of state variables and decision variables.

The calculation of Y is based upon the alfalfa model by Fick (1975). The delay in regrowth is calculated from Eq. 31. The alfalfa model was simulated with this delay in regrowth, with the temperature pattern, Q_3^n, and with cutting time, t_h. The final yield was calculated by subtracting the total leaf feeding (Eq. 30) from the simulated yield.

Overwintering

Oviposition is the result of mating primarily in the early spring by adult weevils who passed the winter as adults. The number of such adults in the spring of the $n + 1$st year, Q_1^{n+1}, is the number of weevils who completed their life cycle in the summer of nth year multiplied by the rate, λ_w, at which they survive the winter. Thus

$$Q_1^{n+1} = \lambda_w \int_0^{\bar{b}} N(b + A_7, b)\, db \tag{32}$$

Where $\bar{b}$ is the end of oviposition. By substituting Eq. 27 for $N(b + A_7, b)$

$$Q_1^{n+1} = \lambda_w Q_1^n e^{-K(A_7)} \int_0^{\bar{b}} \theta(b)\phi(v, s_h - b)e^{-L'(b + A_3, b, P)}\, db$$
$$= G_1(Q_1^n, Q_2^n, Q_3^n, t_h^n, v^n) \tag{33}$$

Thus the number of weevils at the beginning of the growing season in year $n + 1$ is linearly related to the number of weevils at the beginning of the previous season, Q_1^n. Terms λ_w and $e^{-K(A_7)}$ are constants that are independent of the state vectors, Q_i^n, and the decision variables, t_h^n and v^n. The terms inside the integral describe the combined effects of Q_2^n, Q_3^n, t_h^n, and v^n on subsequent densities of weevils.

The number of parasites that oviposite in year $n + 1$ is the product of the number of weevils parasitized in year n and the survivorship of parasites over the winter. Thus the number of parasites in year $n + 1$ is

$$Q_2^{n+1} = \lambda_p \int_0^{\bar{b}} N_p(b + A_5, b)\, db \tag{34}$$

where λ_p is the fraction of parasites which survive the winter. Substituting Eq. 28 into Eq. 34, we obtain

$$Q_2^{n+1} = \lambda_p Q_1^n e^{-K(A_5)} \int_0^{\bar{b}} \theta(b)[1 - e^{-L'(b + A_3, b, P)}]\phi(v, s_h - b)\, db \tag{35}$$

or

$$Q_2^{n+1} = G_2(Q_1^n, Q_2^n, Q_3^n, t_h^n, v^n).$$

Thus the number of parasites in year $n + 1$ is linearly related to Q_1^n, the number of weevils at the beginning of the nth year. The integral term depends upon Q_2^n, Q_3^n, t_h^n, and v^n. The values of the functions, G_1 and G_2, from Eqs. 33 and 35 are substituted into Eqs. 6 and 7 of the decision model to obtain the optimal times for harvesting and insecticide treatments.

Results

Figures 6 and 7 present the optimal management practices, $w_n^*(Q_1^n, Q_2^n, Q_3^n)$, calculated from the alfalfa weevil management model described above in Eqs. 3 through 35. Figure 6a gives w_n^* for cool weather and for all possible combinations of parasite and weevil densities. Figures 6b and c give the results for average and warm weather, respectively. All the results presented in Figure 6 are for a 1-year planning horizon, that is, $N = 1$. Thus the optimal policies presented in Figure 6 do not consider the effect of the current year's pest-control policies on the pest density in the following year. This would be an appropriate model if the area of alfalfa being managed is infested primarily by a migration of adult alfalfa weevils and B. curculionis from outside areas.

In Figure 6a we see that in the absence of weevils the maximum yield is obtained by harvesting the first growth at the last possible date, June 17. However, as the weevil density increases the optimal harvesting time becomes earlier. This is as expected since an early first harvest is a weevil-control measure. The model results are useful because they give quantitative information as well as qualitative. The results indicated how early a harvest should be in response to a specific pest density, parasite density, and temperature pattern.

Notice that early harvesting is preferred over insecticide treatments in most cases. Only when weevil densities are very high and parasite densities are very low are insecticide treatments recommended.

In Figure 6b the temperature, Q_3^n, is higher than in Figure 6a. This increase affects both the rate of insect development and plant growth. For moderate-to-high weevil densities, the optimal time of harvest is earlier than in Figure 6a, where the weather is cooler. This change in policy occurs because the warmer temperatures in Figure 6b cause the weevil as well as the crop to develop more quickly. Therefore, there is more need for early weevil control. Since the crop also matures more quickly, an early harvest causes less of a reduction in total yield.

The optimal cutting times in Figure 6c are even earlier than those in Figure 6b because of the warmer temperature. Insecticide applications are also recommended at lower densities of weevils than recommended for cooler weather.

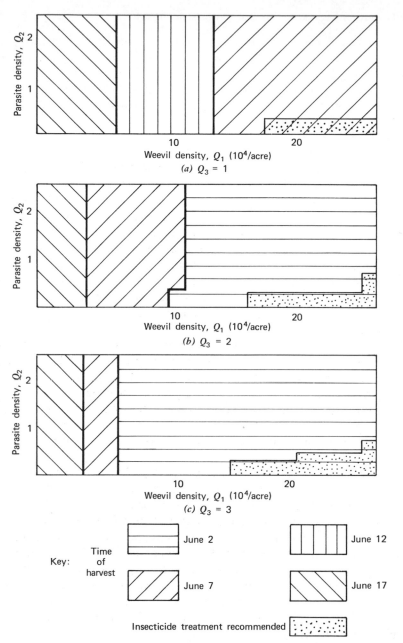

Figure 6 Optimal alfalfa weevil management policies w_1^* (Q_1, Q_2, Q_3) for a 1-year planning horizon for (a) cool weather $(Q_3 = 1)$, (b) average weather, and (c) warm weather. Hatching represents the optimal time of harvest for a given parasite and weevil density. Superimposed dots note where insecticide treatments are recommended.

The pest-management policies calculated to be optimal for a 3-year planning period and warm weather are presented in Figure 7. Because of the longer-term planning horizon, the effect of management decisions on population levels in subsequent years influences the choice of harvest date and insecticide treatment. Because of this effect the optimal results presented in Figure 7 recommend a more intense control of the alfalfa weevil: harvest dates are generally earlier and insecticide treatments more frequent than in Figure 6c.

The results presented in Figures 6 and 7 assume that the value of the alfalfa hay is proportional to the tonnage harvested. However, the value of the alfalfa hay also depends upon its nutritive value, which decreases with late cutting. Research is currently underway to incorporate quality changes in the alfalfa hay into the alfalfa weevil management model. It is likely that when this additional factor is incorporated, the last cutting date, June 17, will be replaced as an optimal policy by an earlier cutting time in warm weather.

The low cost of obtaining the results in Figures 6 and 7 illustrate the advantage of using an optimization method over exhaustive simulation to choose the best management program from a large number of options. To obtain the optimal policies for all values of Q_1^n, Q_2^n, and Q_3^n for 1-, 2-, and 3-year planning horizons from the model described above cost less than \$12 on an IBM 370-168 computer. The economy of these calculations is based upon an algorithm developed in Shoemaker (1977).

To obtain the same results by simulation would involve the calculation of a 3-year model of insect and plant growth for each combination of values of Q_1^1, Q_2^1, Q_3^1, Q_3^2, Q_3^3, t_n^1, $t_n^2 \cdot t_n^3$, v^1, v^2, and v^3. In the above model we consider ten values of Q_1 and Q_2, three values of Q_3^1, Q_3^2, and Q_3^3, four values of t_n^1, t_n^2, and t_n^3, and two values of v^1, v^2, and v^3. Thus the total number of necessary

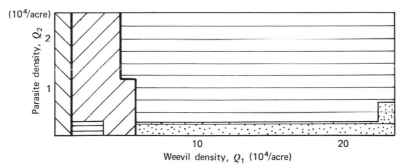

Figure 7 Optimal alfalfa weevil management policies w_1^* (Q_1^3, Q_2^3, Q_3^3) for a 3-year planning horizon and warm weather $(Q_3 = 3)$ in the first year. See Figure 6 for key.

simulations would be

$$10 \times 10 \times 3 \times 3 \times 3 \times 4 \times 4 \times 4 \times 2 \times 2 \times 2 = 1.38 \text{ million}$$

The cost of such a large number of simulations makes simulation an impractical method for calculating the best policies for alfalfa weevil management over a wide range of population densities and weather conditions.

IMPLEMENTATION

Mathematical models can aid in the development and implementation of a pest-management program in several different ways. One way is through pest-control recommendations published by extension agencies. Mathematical models can be an integral part of the research programs upon which the extension agency's recommendations are based.

Mathematical models have had a more direct role in pest-management programs that utilize immediate access to an on-line computer model. Such systems are usually based upon a network of weather monitoring systems, the data from which is stored on computer files. The weather data is used in a simulation model to predict the likelihood of a pest outbreak and the advisability of a pesticide application.

Two established programs that utilize on-line computer models are the potato late blight program in Pennsylvania and the alfalfa weevil program in Indiana. In addition to weather data, the alfalfa weevil program also utilizes current samples of alfalfa weevil densities obtained by insect scouts throughout the region covered by the program. The pest-management recommendations are based in part upon the results of a simulation model that includes an alfalfa weevil and an alfalfa plant model similar to those described in Figures 1 and 2. Extension agents are then able to make recommendations for each region on the basis of the model forecast combined with their own knowledge of each locality. The regional recommendations are entered into a computer file that can be accessed from any one of a number of teletype terminals located throughout the state (Giese et al., 1975).

Unlike the alfalfa weevil management program, the potato late blight program is not based upon models of insect and plant populations. Instead, it uses techniques developed over a decade ago by Hyre and Wallin to estimate the weather conditions under which a blight outbreak is likely. The computer system analyzes weather data from many regions throughout the state and calculates those areas in which fungicide treatments are warranted on the basis of the forecasting model. Growers can obtain the forecasts and recommendations by telephone. The Hyre and Wallin forecasting techniques have been proven to be very reliable, but they had not been widely used probably because local forecasts were not available before. Since the implementation of the computer-based program, the use of forecasts to determine the timing of fungicide applications has increased. This has resulted in a reduction of fungicide use and a consequent savings to the grower (Krause and Massie, 1973).

It is anticipated that the role of mathematical models in the development and implementation of pest-management programs will continue to increase. Models are a valuable tool in understanding interactions among populations and weather in a crop ecosystem. As we have seen in the examples discussed, the understanding gained from mathematical models is very useful in integrating a number of pest-management techniques and in estimating the best timing for pest-control methods.

ACKNOWLEDGMENTS

Research on the application of dynamic programming to insect pest management was supported in part by a grant from the National Science Foundation Engineering Research Initiation Grant GK42147. This grant also supported the work of Albert Valocchi and Leland Baskin who were responsible for computer programming.

Most of the parameters in the alfalfa weevil model were based on field studies by Professors Robert Helgeson and Gary Fick of Cornell University, on a literature review by Dr. William Ruesink of the Illinois Natural History Survey, and on statistical analyses by Gary Smith and the author. All of the above work was supported by a grant to the University of California from the National Science Foundation and the Environmental Protection Agency (NSF GB-34718).

The author is grateful for access to unpublished data provided by her colleagues and for the support of the National Science Foundation and the Environmental Protection Agency. Appreciation is also expressed to the following who provided comments on parts of the manuscript: Keith Porter, Gary Fick, Robert Rovinsky, Gary Smith, and David Pimentel. The findings, opinions, and recommendations expressed herein are those of the author and are not necessarily those of the University of California, the National Science Foundation, or the Environmental Protection Agency.

REFERENCES

Andrilenas, P. A. 1974. *Farmers use of pesticides in 1971—quantities.* Economic Research Service, U.S. Dept. of Agric., Agricultural Economic Report No. 252.

Burdick, G. E., E. J. Harris, H. J. Dean, T. M. Walker, J. Shed, and D. Colby. 1964. The accumulation of DDT in lake trout and the effect on reproduction. *Trans. Amer. Fish. Soc.* **93:** 127–136.

Central American Research Institute for Industry (ICAITI). 1976. *An Environmental and economic study of the consequences of pesticide use in Central American cotton production.* ICAITI Project No. 1412.

Fick, G. W. 1975. ALSIM I (*Level 1*) *User's Manual.* Cornell University, Department of Agronomy, Mimeo 75–20.

Fick, G. W. 1976. Alfalfa weevil effects on regrowth of alfalfa. *Agronomy Journal* **68**: 809–812.

Giese, R. L., R. M. Pest, and R. T. Huber. 1975. Pest management. *Science* **187**: 1054–52.

Fick, G. W. and B. Liu. 1976. Alfalfa weevil effects on root reserves, development rate, and canopy structure of alfalfa. *Agronomy Journal* **68**: 595–599.

Gutierrez, A. P., L. A. Falcon, W. Loew, P. A. Leipzig, and R. van den Bosch. 1975. An analysis of cotton production in California: a model for Acala cotton and the effects of defoliators on its yields. *Environ. Entom.* **4**: 125–36.

Huffaker, C. B. (Ed.). 1971. *Biological control.* Plenum, New York.

Koehler, C. S. and S. S. Rosenthal. 1975. Economic injury levels of the Egyptian alfalfa weevil or the Alfalfa weevil. *J. Econ. Entom.* **68**: 71–75.

Krause, R. A. and L. B. Massie. 1973. Application and implementation of computerized forecasts of potato late blight. *Phytopathology* **63**: 203. (Abstr.)

Metcalf, R. L. 1975. Insecticides in pest management. In R. L. Metcalf and W. H. Luckman (Eds.), *Introduction to pest management.* Wiley, New York.

National Academy of Sciences (NAS). 1975. *Pest control: An assessment of present and alternative technologies.* Vol. 1. The Report of the Executive Committee, NAS, Washington, D.C.

Pimentel, D. 1971. *Ecological effects of pesticides on non-target species.* Executive Office of the President, Office of Science and Technology.

Pimentel, D. 1976. World food crisis. *Bull. Entom. Soc. Amer.* **22**: 20–25.

Regev, U., A. P. Gutierrez, and G. Feder. 1976. Pest as a common property resource: a case study of alfalfa weevil control. *Amer. J. Agr. Econ.* **58**: 188–196.

Ruesink, W. G. 1976. *Modeling of pest populations in the alfalfa ecosystem with special reference to the alfalfa weevil.* In modeling for pest management. R. L. Tummala, D. L. Haynes, and B. A. Croft (Eds.), Mich. State Univ. Press, East Lansing.

Shoemaker, C. A. 1973. Optimization of agricultural pest management III: Results and extensions of a model. *Math. Biosci.* **18**: 1–22.

Shoemaker, C. A. 1977. *Optimal integrated pest management of age distributed populations.* School of Civil Engineering Technical Report. Cornell University.

Smith, G. E. and C. A. Shoemaker. 1977. *Stepwise linear estimation of insect oviposition and mortality.* School of Civil Engineering Technical Report. Cornell University.

Stern, V. M., R. F. Smith, R. van den Bosch, and K. S. Hagen. 1959. The integrated control concept. *Hilgardia* **29**(2): 81.

United States Department of Agriculture (USDA). 1965. *Losses in agriculture.* Agric. Handbook No. 291, Agr. Res. Serv. U.S. Govt. Printing Office, Washington, D.C.

Waggoner, P. E., J. G. Horsfall, and R. J. Lukens. 1972. EPIMAY, *a simulation of southern corn leaf blight.* Bulletin 729., Connecticut Agricultural Experiment Station, New Haven.

Watt, K. E. F. 1964. The use of mathematics and computers to determine optimal strategy and tactics for a given insect control problem. *Can. Entom.* **96**: 202–20.

Woodwell, G. M., C. F. Wurster, and P. A. Isaacson. 1967. DDT Residues in an east coast estuary: a case of biological concentration of a persistent insecticide. *Science* **156**: 821–824.

23
Gradient Modeling: A New Approach to Fire Modeling and Resource Management

STEPHEN R. KESSELL

O n July 25, 1975 at 2:55 PM MDT, the dispatch radio in Glacier
National Park's Communications Center came to life.

720 Control, this is 739 Swiftcurrent Lookout, with a smoke report.
739 Swiftcurrent, this is 720 Control, go ahead.
720, I have a smoke visible on West Flattop Mountain, base of smoke is below my line
of sight. Approximate location 5412 North, 280 East.
Roger, 739, smoke at 5412 North, 280 East. KOE 720 Control at 14:56.

A few minutes later the park's aerial observer was dispatched in a Cessna
182 to check out the report. At 3:41 PM MDT, he verified the presence of a
fire at UTM* (Universal Transverse Mercator) coordinates 5414.5 North,
282.6 East, just south of Redhorn Lake.

A few years ago, the park communicator would have notified the district
ranger, chief ranger, and fire control officer, and an immediate effort to
suppress the fire would have been initiated. But research in the intervening
years has shown that fire has a very natural and often beneficial role in such
a forest system. Fires can threaten life, cause erosion, and burn commercial
timber. But they also increase biotic diversity, provide browse for animals,
and recycle dead fuels that accumulate in northern forests. This decreases
the intensity of future fires. Thus the old policy of total fire control is both
economically unrealistic and ecologically unsound, while a total "let-burn"
policy that ignores other management contingencies is also unrealistic. The
fire manager needs to know what that particular fire will do—what its short-
and long-term effects on the ecosystem will be.

However, since early 1975 a different option has been available. On July
25th, Glacier's Fire Ecologist had been monitoring the Redhorn Fire. At
3:42 PM MDT, he telephoned an IBM 370/168 digital computer located in
McLean, Virginia, using a remote terminal located at park headquarters in
West Glacier, Montana. He typed in the location of the fire and the latest
weather report. Two minutes later, for a cost of less than $5, the following
information on the Redhorn Fire came off the terminal (Kessell 1975a):

WELCOME TO THE GLACIER NATIONAL PARK BASIC RESOURCE
AND FIRE ECOLOGY SYSTEMS MODEL, ACTIVATED FOR THE
SQUARE KILOMETER LOCATED AT UTM 5414 NORTH, 282
EAST.

An adaption of this article appeared in *Environmental Management*, **1**, 1, 39–48.

* The reader unfamiliar with the special language of foresters is advised to look through the
glossary at the end of this chapter.

INVENTORY ANALYSIS FOR HECTARE AT 5414.5 NORTH, 282.6
    EAST—THIS HECTARE HAS UNIFORM SITE CONDITIONS AND
    THEREFORE IS NOT BLOCKED INTO SUBUNITS.
    THIS STAND HAS AN ELEVATION OF 1768 METERS (5800 FEET)
    MSL.
    IT IS A SHELTERED SLOPE WITH SOUTHEAST ASPECT. THE
    SLOPE STEEPNESS IS 5 DEGREES.
    ITS VEGETATIVE COVER TYPE IS SHRUB, AND ITS PRIMARY
    SUCCESSION (SOIL DEVELOPMENT CONDITION) IS CODE 70
    (MIXED MEADOW & SHRUB).
    THE STAND IS MATURE.
    NO OTHER DISTURBANCES ARE RECORDED.
    THE STAND IS NOT AFFECTED BY THE ALPINE WIND-SNOW
    GRADIENT.
    THIS STAND INCLUDES LOCALIZED COVER TYPES,
    INCLUDING—TYPICAL (UPRIGHT) FOREST WHICH COVERS
    10% OF THE STAND.
    IT HAS A PRIMARY SUCCESSION CODE OF 99, WHICH IS
    TYPICAL FOREST.
    ITS CONTAGION VALUE (VARIANCE/MEAN RATIO ON A
    20 × 20 M GRID) IS 1.0.
    THIS STAND HAS NO RECORDED SPECIAL FEATURES.

GRADIENT ANALYSIS OF TREE SPECIES FOR HECTARE AT 5414.5
    NORTH, 282.6 EAST—
    (SPECIES WITH A RELATIVE DENSITY OF LESS THAN 1% ARE
    NOT PRINTED).
    LOCALIZED FOREST COVERS 10% OF THE STAND; SPECIES
    INCLUDE—

PICEA HYBRID COMPLEX (SPRUCE) WITH A RELATIVE DENSITY OF
                               75–95%
ABIES LASIOCARPA (SUBALPINE   WITH A RELATIVE DENSITY OF
    FIR)   1–5%
PSEUDOTSUGA MENZIESII   WITH A RELATIVE DENSITY OF
    (DOUGLAS-FIR)   1–5%

GRADIENT ANALYSIS OF FUEL LOADINGS FOR HECTARE AT
    5414.5 NORTH, 282.6 EAST—
    THE CALCULATED AVERAGE LOADINGS INCLUDE THE
    EFFECTS OF ALL LOCALIZED COVER.

LITTER	10.20 METRIC TONS PER HECTARE
GRASS & FORBS	11.62 METRIC TONS PER HECTARE
1 HOUR DEAD & DOWN	0.41 METRIC TONS PER HECTARE
10 HOUR DEAD & DOWN	3.02 METRIC TONS PER HECTARE
100 HOUR DEAD & DOWN	3.15 METRIC TONS PER HECTARE
GREATER THAN 100 HOUR DEAD & DOWN	6.70 METRIC TONS PER HECTARE

FIRE BEHAVIOR PREDICTIONS FOR HECTARE AT 5414.5 NORTH, 282.6 EAST—

WIND AT MIDFLAME HEIGHT (KPH = MPH = M/MIN)			DIRECTION OF SPREAD	RATE OF SPREAD (M/MIN)	FLAME LENGTH (M)	INTENSITY (KCAL/MIN/ METER)
0.0	0.0	0.0	UPSLOPE (NW)	0.01	0.01	0.35
			ACROSS (NE)	0.01	0.01	0.31
			DOWNSLOPE (SE)	0.00	0.01	0.26
			ACROSS (SW)	0.01	0.01	0.31
8.0	5.0	134.09	UPSLOPE (NW)	0.04	0.03	2.23
			ACROSS (NE)	0.04	0.03	2.19
			DOWNSLOPE (SE)	0.00	0.00	0.00
			ACROSS (SW)	0.00	0.00	0.00

This information was turned over to the park's Fire Management Specialist, who realized he had a slow-spreading, low-intensity, easy-to-control fire. Four men were dispatched by helicopter—two regular fire guards and two members of the fire ecology staff. They verified the fire's low spread rate (0 to 2.5 cm min^{-1} measured, compared to 0 to 4 cm min^{-1} predicted). Since it turned out that this particular fire was not beneficial according to the management strategy of Glacier, it was decided to initiate fire-suppression procedures. The Redhorn fire was extinguished early the following day at a total size of 0.6 hectares. However, other fires may be left burning.

Glacier National Park is currently the only large wilderness area where this kind of resource data retrieval, synthesis, and integration system is operational so that real-time fire behavior predictions can be made (Kessell,

1975a). The resource package, known as the *Glacier National Park Basic Resource and Fire Ecology Systems Model*, links (1) a hectare-by-hectare resource inventory coded from aerial photos, (2) gradient models of the vegetation and fuel, both derived from field sampling, (3) a microclimate model, and (4) a fire-behavior model, (Kessell, 1973, 1975b, 1976). The technique of linking these components to provide information on the site, vegetation, fuel, and fire behavior is called *gradient modeling*. This chapter describes the development, implementation, and operation of the model in Glacier National Park.

THE NEEDS OF ENLIGHTENED RESOURCE MANAGEMENT FOR NATURAL AREAS

Resource management of natural areas requires action to preserve, maintain, and often restore pristine ecosystems (National Park Service and USDA Forest Service policy guidelines). To accomplish this goal, managers must have a good general knowledge of the areas, an accurate and quantitative inventory of the biota, an understanding of the interaction among the various components of the dominant ecosystems, and the ability to predict or simulate changes to the system resulting from management action.

The manager daily must face questions like:

What is the distribution and relative abundance of rare or endangered species in an area proposed for recreational development?
What buildings or trails in a national park are in areas subjected to frequent flooding?
What predictions can be made for the behavior of a prescribed or natural fire? Can these predictions be generated quickly enough if fire control decisions must be reached in a few minutes? What could the fire destroy? How much fuel is present? What will be the post-fire successional communities? What is the anticipated natural fuel build-up if the fire is suppressed? Where can stands with abundant huckleberries or other prime grizzly bear habitat be found?

Questions like these are becoming increasingly important to the resource manager. All of these questions demonstrate the need for an inventory of the resources and natural communities and the ability to synthesize component data to provide specific answers to complex questions.

Inventory and modeling systems that attempt to meet these needs must be:

1   *Comprehensive.* Include all types of data and parameters necessary to understand the problem.

2   *Accurate*. Contain a detailed ecological description showing the dynamic ecological processes within an area with fine resolution.

3   *Flexible*. Applicable to a wide range of problems by easily manipulating and integrating various kinds of information.

4   *Synthetic*. Capable of drawing together diverse information to make predictions, such as integrating site, fuel, vegetation, weather, and cultural information to simulate the effects of a forest fire.

5   *Problem oriented*. Permit selective retrieval of only that information needed to solve a particular management problem.

6   *Accessible*. Available to the manager in an easily and quickly interpretable form when he needs it.

7   *Economical*. Both development and application costs must be within the manager's budget.

The traditional approach to resource inventories has been type-mapping and land-classification systems, which are usually presented in graphic form as overlay maps (reviewed in Kessell, 1976; cf. Whittaker, 1973a). Even when such maps are digitized and stored on a computer, they are unable to meet the requirements of a broad-based resource information system. There are several reasons why they cannot meet management needs.

1   *Insufficient resolution*. Overlay maps on a scale of 1:24,000 (standard $7\frac{1}{2}$-min topographic maps) cannot offer the detailed resolution needed for many management purposes. Inventories used for fuel distribution, fire behavior analysis, and animal habitats often need a resolution of 20 m or better. Overlay maps that do attempt to offer fine resolution usually suffer from:

2   *Incomprehensible graphic complexity*. Mapping systems with detailed resolution offer it at the expense of clarity. The old Glacier National Park vegetation map is a melange of 21 colors and 114 special symbols, looks like a plate of chop suey, and is virtually unusable to the manager.

3   *Inappropriate classification criteria*. The habitat-type classification of vegetation must divide natural communities into distinct, discreet units, even though most natural vegetation varies continuously in both space and time in response to various environmental gradients. In addition, even the best classification of the vegetation is of marginal or lesser value to classify fuel loadings, fuel moisture, elk or wolf range, or areas of potential human-grizzly bear conflict. Reclassification of the land by each new criteria is difficult, needlessly expensive, and due to new modeling developments, obsolete.

4   *Failure to use environment-biota correlations*. Habitat classifications do not permit use of natural correlations among the environment and the

biota. Since they cannot use basic site information to predict vegetation, fuel, and so forth, they have difficulty in efficiently using aerial photography or other remote sensing methodology to accomplish their purpose.

5   *Lack of integration of components.* No type-mapping system can efficiently integrate a large number of components to solve a particular problem. For example, no classification system can synthesize site, vegetation, fuel, and weather data to make real-time fire behavior predictions.

The next section develops a computer-based inventory system that avoids these problems and extends the inventory system to the development of real-time fire modeling.

THE GRADIENT MODELING APPROACH

Gradient modeling is a computer-based resource modeling system designed in Glacier National Park to meet the needs for a resource-information system, a resource inventory, and a fire-modeling package. It provides these capabilities by linking four major components:

1   A terrestrial resource inventory system
2   Gradient models of the vegetation and flammable fuel
3   Weather and micrometeorology models
4   Fire behavior models

Its unique linkage of these components provides modeling capabilities unavailable from any other resource management system. A very simplified flowchart of the model is shown in Figure 1.

Gradient modeling is an application of the techniques of gradient analysis and ordination (rather than habitat classification) to resource modeling. It builds on the wide theoretical and analytical base of gradient analysis developed over the past two decades by Whittaker, a group at the University of Wisconsin lead by Bray and Curtis, and many others (Whittaker, 1956, 1960, 1967, 1970a, 1970b, 1973b, 1973c; Whittaker and Niering, 1965; Bray, 1956, 1960, 1961; Bray and Curtis, 1957; for a review of techniques see Whittaker, 1973b; Whittaker and Gauch, 1973). Rather than dealing with the landscape and its vegetation as sharp, discontinuous units, gradient analysis describes and quantifies continuous variation in the landscape and its biota that corresponds to various spatial and temporal environmental gradients. We have found gradient concepts particularly applicable to the modeling and resource-management problems that we develop in this chapter.

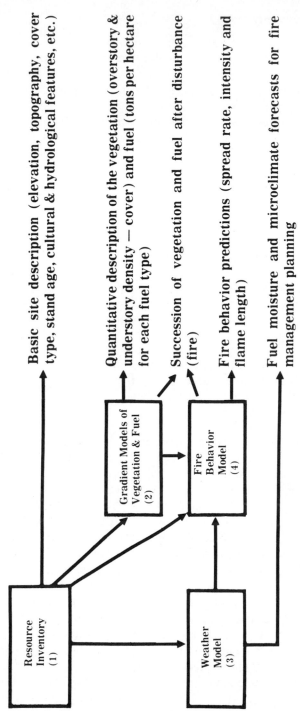

Basic site description (elevation, topography, cover type, stand age, cultural & hydrological features, etc.)

Quantitative description of the vegetation (overstory & understory density — cover) and fuel (tons per hectare for each fuel type)

Succession of vegetation and fuel after disturbance (fire)

Fire behavior predictions (spread rate, intensity and flame length)

Fuel moisture and microclimate forecasts for fire management planning

Gradient Models of Vegetation & Fuel (2)

Fire Behavior Model (4)

Resource Inventory (1)

Weather Model (3)

Figure 1 A very simplified flowchart of the resource and fire model.

A Simplified Example

Perhaps the best way to visualize the system is to consider a simplified, hypothetical ecosystem (based on Kessell, 1976). A 25-hectare area of this hypothetical ecosystem is shown in Figure 2.

The 25 individual hectares are referenced by the Universal Transverse Mercator (UTM) coordinates at the southwest corner of each square. The land is forested, and the canopy is composed of three tree species. We call these three species Alpha, Bravo, and Charlie. The numbers plotted in each hectare indicate the estimated number of individuals of each species. The problem is to design an inventory system that offers both refined resolution and application to diverse management problems.

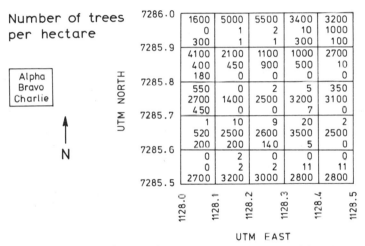

Figure 2   A 25-hectare area of a hypothetical forest ecosystem containing three overstory species (Alpha, Bravo, and Charlie). Coordinates are Universal Transverse Mercator (UTM) in kilometers; each square is one hectare (100 × 100 m). Plotted numbers are absolute densities of each species in each hectare.

A traditional approach would classify each hectare by some standard method and then plot the results on what is called a type map. For example, classification by predominant overstory species is shown in Figure 3. Unfortunately, we have lost a great deal of information by introducing this classification system; yet if we attempt to classify the hectares by several criteria, we will achieve it only at the expense of producing a vast number of little squiggles and quirks plotted on a topographic overlay map. In a real system, composed of many tree, shrub, and herb species, several fuel

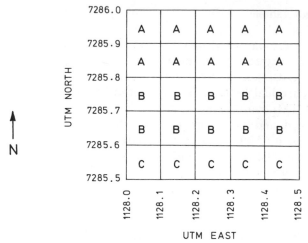

Figure 3 A classification system based on overstory dominance for the hypothetical ecosystem.

types, and numerous animal species, any type-mapping approach of sufficient resolution produces a totally unreadable and unacceptable mosaic.

As an alternative, we might consider the underlying environmental causes of the distribution of the various biota. A logical starting point would be a topographic map of the area (e.g., Figure 4). Two environmental variables

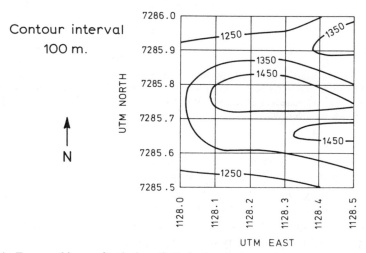

Figure 4 Topographic map for the hypothetical ecosystem.

that might well influence biotic communities may be determined from this simplified map—the elevation and aspect for each 1-hectare plot. We may use these two variables as the basis of a hectare inventory system and record each hectare's elevation and aspect as shown in Table 1. We assume that for this simplified scheme these two continuous environmental variables determine the biotic composition of the stands—that is, the presence or absence of each different tree species is due to some response of that species to aspect and elevation.

Table 1   Hectare Inventory for the Hypothetical Ecosystem.

UTM North	UTM East	Elevation (m)	Aspect
7285.5	1128.0	1250	S
7285.5	1128.1	1275	S
7285.5	1128.2	1290	SW
7285.5	1128.3	1310	SW
7285.5	1128.4	1340	SW
7285.6	1128.0	1370	SW
7285.6	1128.1	1400	S
7285.6	1128.2	1410	S
7285.6	1128.3	1470	W
7285.6	1128.4	1500	S
7285.7	1128.0	1410	W
7285.7	1128.1	1480	SW
7285.7	1128.2	1500	S
7285.7	1128.3	1470	S
7285.7	1128.4	1460	E
7285.8	1128.0	1330	NW
7285.8	1128.1	1360	N
7285.8	1128.2	1400	N
7285.8	1128.3	1390	NE
7285.8	1128.4	1320	NE
7285.9	1128.0	1240	NW
7285.9	1128.1	1250	N
7285.9	1128.2	1260	N
7285.9	1128.3	1290	NW
7285.9	1128.4	1370	NW

We now perform a gradient analysis on the area by sampling 0.1-hectare plots within the study area and recording the density of the three tree species. We now may construct the population nomogram shown as Figure 5. This nomogram plots aspect (arranged from the wettest northeast slopes to the driest southwest slopes) on the abcissa and elevation on the ordinate. After we plot the individual field samples, we may contour them and construct *isodens*, that is, lines connecting areas of equal species density. This is our gradient model for the three species. We see, for example, that species Bravo is most abundant at higher elevations, Alpha is most abundant at lower, wetter altitudes, and so on.

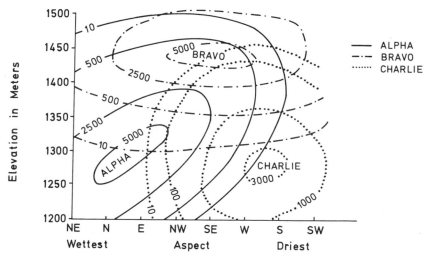

Figure 5   Gradient model for species Alpha, Bravo, and Charlie that which expresses absolute densities as functions of two environmental gradients (elevation and aspect). The contour lines (isodens) connect areas of equal species density, while the plotted numbers are absolute densities (trees per hectare).

By linking gradient models to the inventory scheme, we may predict the density of the three species in unsampled stands by using only the data presented in Table 1, that is, elevation and aspect. Using the hectare located at 7285.7 North, 1128.0 East as an example, the inventory gives an elevation of 1410 meters and west aspect. Turning to Figure 5, we find the intersection of these two points on the graph and estimate the following densities for Alpha, 500 trees hectare^{-1}; for Bravo, 2800 trees hectare^{-1}; and for Charlie, 400 trees hectare^{-1}. These results can be compared to actual measurements given in Figure 2.

Suppose we also desire quantitative information on three shrub species, Delta, Echo, and Foxtrot. We construct a gradient model for them based on field sampling as shown in Figure 6. We may now use the same inventory (Table 1) to predict their species composition; for hectare 7285.7 North, 1128.0 East we predict a density of less than 100 shrubs hectare^{-1} for Delta, 150 shrubs hectare^{-1} for Echo, and 1000 shrubs hectare^{-1} for Foxtrot. We also can use the same inventory scheme to construct gradient models of nonliving components, such as flammable fuel. For example, we might predict the fuel composition for hectare 7285.7 North, 1128.0 East to be 17.2 tons hectare^{-1} for dead large branches and 13.2 tons hectare for pine needles.

If we have properly designed the inventory system, we may continue to construct gradient models for other plant species, fuel characteristics, or animal species, and predict their abundance by linking the inventory and the gradient models. Furthermore, we may enter the site, fuel, and weather parameters into appropriate fire-behavior models to predict the behavior of a fire in any hectare under any weather conditions.

The above example with only two environmental gradients and hypothetical species needs considerable expansion and refinement for application to a real wilderness system. Yet the reader sees that the actual Glacier National Park model is simply a logical expansion of this basic approach.

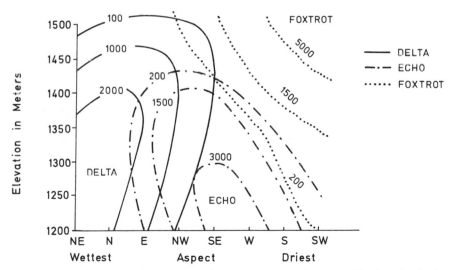

Figure 6 Gradient model for shrub species Delta, Echo, and Foxtrot (absolute density in shrubs per hectare).

THE GLACIER NATIONAL PARK MODEL

Sitting atop the continental divide in northwestern Montana, Glacier National Park is an area of tremendous relief, spectacular scenery, alpine wildflowers, luxurious forests, numerous glaciers, hanging valleys and waterfalls, and dozens of fascinating animal species. Communities range from xeric prairie intrusions and remnants to alpine tundra, from ponderosa pine savanna to mesophytic cedar and hemlock (*Thuja plicata* and *Tsuga heterophylla*) forests. This diverse habitat mosaic is further complicated by numerous natural disturbances, including natural fires, slides, fellfields, and avalanches, and by patterns resulting from hydric, glacial, and floodplain successions. Over a thousand species of vascular plants are present (Kessell, 1974). Fire has played a critical role in shaping the forests of the lower elevations and drier sites; man's suppression of fire over the past several decades has led to community changes and unnatural fuel buildups (Kessell, 1976; Habeck, 1970a, 1970b). Heavy fuel buildups following many years of fire suppression may result in much more destructive fires at later times. An intense effort currently is being made to restore natural fire to the park in such a way as to minimize the detrimental aspects of such fires, such as the destruction of historical or cultural resources, or escape from park boundaries. At the present time it is obvious to many managers that some fires should be allowed to burn while others should not. But which fires should be suppressed? We have found that computer models can be very helpful in deciding this question. But first we need to develop some background information.

Gradient Models of the Vegetation and Fuel

The gradient models of Glacier's vegetation and fuel are derived from over 2000 field samples taken during the 1972 to 1975 period (Kessell et al., 1975). Overstory species are recorded by density and basal area, understory species are recorded by total cover, and fuel loadings are recorded by dry weight per unit area for each category (live or dead) and size class [using an expansion of the planar intersect method (Kessell, 1976)]. Detailed site descriptions and UTM coordinates are recorded for each sample.

Extensive analyses, ordination, gradient analysis, and classification of the field data revealed ten major environmental influences on these communities. They include:

1 Elevation
2 Topographic-moisture (topography and aspect)
3 Primary succession (soil development following glacial retreat and rock weathering)

4   Watershed (drainage) area
5   Alpine wind exposure and snow accumulation
6   Secondary fire succession (time since the last burn)
7   Intensity of the last burn
8   Slide, fellfield, and avalanche areas
9   Hydric successions
10 Heavy ungulate winter use

Our analyses indicated that, if our computer inventory was to have at least as much accuracy as normal within-stand variation found from field sampling, the first six effects must be treated as continuous environmental gradients, while the latter four may be treated as discrete categories. Knowledge of a stand's location on each gradient and within each variation category allows prediction of its vegetational and fuel composition without visiting the stand from the ground. The predictions are within sampling error for about 93% of the several hundred stands test-predicted by the model.

It is difficult to plot a six-dimensional diagram on two-dimensional paper. However, we may hold several gradients constant and view the effects of varying any two at a time. Some real examples are shown in Figure 7 for the McDonald drainage (from Kessell, 1976). Figure 7a shows a mosaic classification of mature forest communities in terms of the first two gradients, elevation and topographic moisture. Figure 7b plots the distribution of Abies lasiocarpa (subalpine fir) in mature stands on these same two gradients; the isodens connect areas of equal relative density. Figure 7c holds elevation constant and plots time since burn against topographic-moisture for Pinus contorta (lodgepole pine). The species' seral nature is clearly shown; its peak density occurs about 25 years after a fire on the driest sites. Its short lifespan is shown by its total replacement within 150 years after the burn. Figure 7d plots the relative density of Tsuga heterophylla (western hemlock) on the same two gradients. Here we see a "climax" species that first enters a stand about 50 years after a fire and does not reach peak density until nearly 200 years after the burn.

Figure 8 plots the loadings (metric tons dry weight per hectare) of two fuel types on the elevation and topographic-moisture gradients for mature forested stands. (Fuels are classified by their time lag in responding to a humidity change. One-hour fuels are sticks less than 0.64 cm diameter; 10-hour sticks range from 0.64 to 2.54 cm diameter. Dead and down refers to dead material that has dropped to the forest floor.) We observe peak loading of the 1-hour fuels on both the higher slopes (1700 to 1900 m) and the lower elevation (below 1300 m) forests. Ten-hour fuels peak in the 1400- to 1600-m slopes with a higher total loading (about 5.5 tons/hectare).

Figure 9 shows the response of six fuel types to the time-since-burn gradient for slopes at 1000 m. Note that after 100 years of age, a stand's

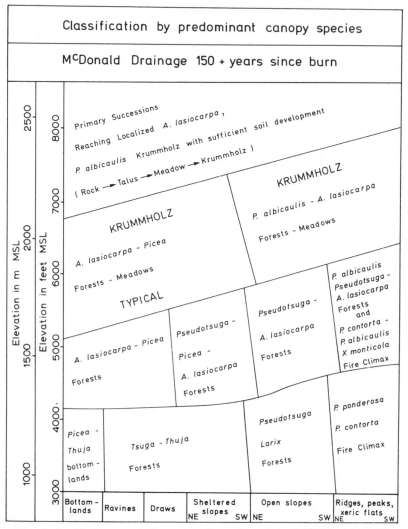

Figure 7a A classification mosaic for mature forested stands in the McDonald drainage of Glacier National Park plotted on two gradients, elevation and topographic moisture.

fuel continues to increase at the rate of about 1 ton hectare^{-1} year^{-1} indefinitely. As noted above, suppression of natural fires allows this buildup to continue unchecked, leading to fires with higher intensities and greater destructive potential in the future.

The reader will realize that several hundred illustrations would be required to show the response of all species and fuels to all gradients. The author hopes

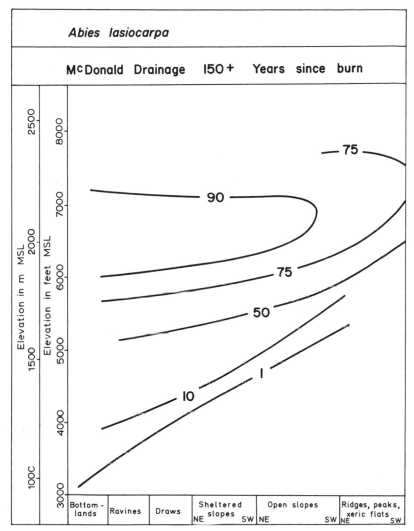

Figure 7b The gradient model for *Abies lasiocarpa* (subalpine fir) in mature stands in the McDonald drainage on the same two gradients. Isodens connect areas of equal relative density.

the above samples convey a feeling for the distribution of various plants and fuels along some of the gradients.

The Resource Inventory

As noted above, the gradient models allow retrieval and prediction of quantitative plant and fuel composition for each stand if one knows the

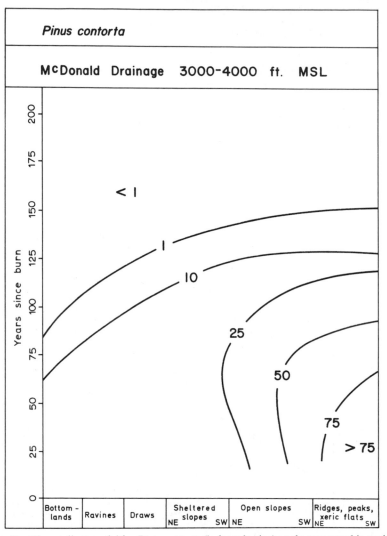

Figure 7c The gradient model for *Pinus contorta* (lodgepole pine) on the topographic-moisture and time-since-burn gradients in the McDonald drainage. Isodens connect areas of equal relative density.

stand's location on each gradient and within each category of variation. We designed an inventory system that allows determination of each stand's location on each gradient and within each category using only aerial photos (false color infrared obliques and vertical black and white), topographic maps, and fire history maps (Dwyer and Kessell, 1975). The system permits

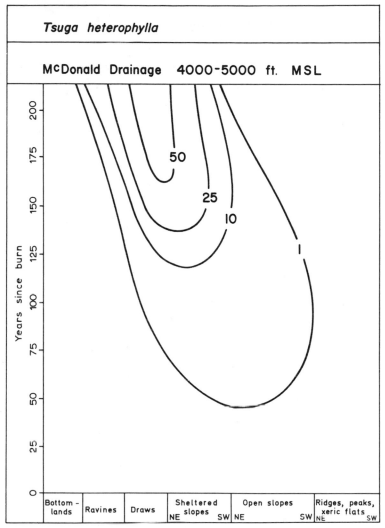

Figure 7d The gradient model for *Tsuga heterophylla* (western hemlock) on the topographic-moisture and time-since-burn gradients in the McDonald drainage. Isodens connect areas of equal relative density.

10-m resolution within each hectare and also records any hydrological or cultural features of each hectare. The within-hectare resolution is provided by two methods. If a major discontinuity along any gradient exists within any hectare (such as both ravine and slope topography, or forest and meadow cover), a separate record is used for each portion of the hectare.

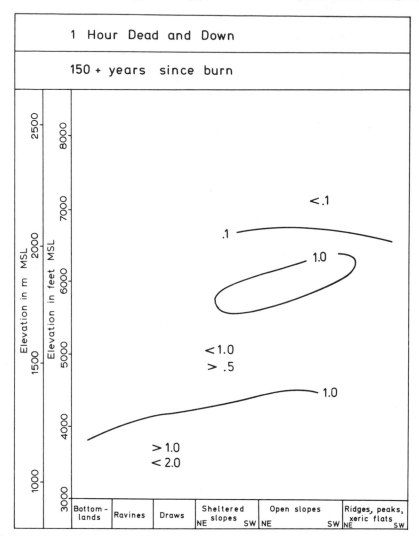

Figure 8a Distribution of 1-hr time lag (sticks less than 0.64 cm diameter) dead-and-down fuel on the elevation and topographic moisture gradients for moisture, forested stands. Contour lines connect areas of equal fuel loadings; numbers show loadings in metric tons per hectare dry weight.

If the discontinuities are small (under 20 m), such as what might be caused by a scattering of rock outcrop and shrubs on a meadow, the total percent cover by these localized cover types is recorded, along with a measure of their clumpedness (variance/mean cover ratio on a 20 × 20 m grid), as shown in the introduction for the localized trees at the Redhorn Fire site.

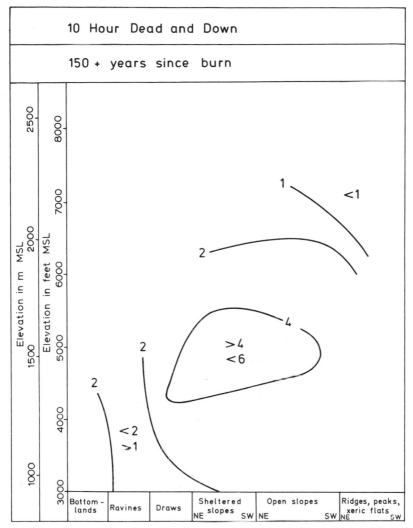

Figure 8b Distribution of 10-hr (sticks from 0.64 to 2.54 cm diameter) dead and down fuel on the same two gradients. Loadings in metric tons per hectare.

The inventory method thus provides an efficient system for locating each stand, hectare by hectare, on the various gradients. The inventory records only these gradient indices (plus the cultural and hydrological data). It need not record species density or fuel loadings—these are retrieved from the gradient models using the inventory site data.

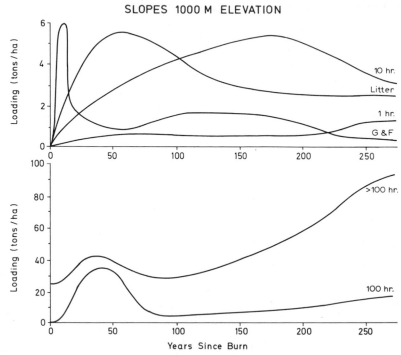

SLOPES 1000 M ELEVATION

Figure 9 Distribution of six flammable fuels along the time-since-burn gradient for forested slopes at 1000 m elevation. Fuels are 1-hr dead and down, 10-hr dead and down, 100-hr dead and down (sticks between 2.54 and 7.62 cm diameter), greater than 100-hr dead and down (sticks over 7.62 cm diameter), standing grass and forbs (both living and dead), and litter (non-branchwood dead material above the duff). Loadings are in metric tons per hectare.

Prediction of Secondary Succession

The inventory and gradient models also can be used to provide deterministic predictions of the succession of plants and fuels following a fire (Kessell et al., 1975). This is accomplished by solving the "time-since-last-fire" gradient. When a stand burns, its age in the computer inventory reverts to zero. The model may solve for different values of this gradient, and thus predict composition at any age after the fire. For example, refer to Figure 7c (*Pinus contorta*) and Figure 9. For a 1000-m elevation, south-aspect open slope, we predict that 30 years after the burn the lodgepole density will be 60%, the litter loading will be 1.4 tons hectare^{-1}, the 100-h dead-and-down loading will be 28 tons hectare^{-1}, and so forth. Stochastic elements influencing succession are being considered but have not yet been incorporated into the model.

Fire-Behavior Models

Recent advances in modeling the behavior of forest fires enable us to predict the spread rate and intensity of fires if we know certain properties of the fuel matrix and the conditions under which it will burn (Rothermel, 1972). For example, it is necessary to know the total dry weight, the degree of compaction for each combustion and size category, and the mineral and moisture content. It is also necessary to know the wind speed and direction and the steepness of slope. All of the required parameters have either been derived by empirical studies (e.g., Rothermel, 1972) or may be retrieved from the gradient models. A simple derivation of the current Rothermel (1972) fire-behavior model follows.

How does a fire spread from one burning volume to another, similar in composition but not yet burning? Consider two unit volumes of fuel adjacent to each other within a larger fuel array. The two volumes are as small as possible with the condition that they exhibit the same bulk density, packing ratio, and fuel particle array as the larger fuel matrix within which they are located. That is, they have the smallest cubic volume possible that still maintains the bed's physical characteristics. How does a propagating fire spread from one of these unit volumes to the next, adjacent volume?

Frandsen (1971) derived a formula for this fire spread between volumes, based upon a consideration of the law of conservation of energy. The burning volume will produce a certain Reaction Intensity, IR. Some of this total heat flux will become available to pass into and raise the temperature of the fuel in the second unit volume. This available heat is termed the Propagating Intensity, IP. The fire will spread into the second volume by raising a critical amount of the fuel, the Effective Bulk Density (RHO_{be}), to the required level for ignition, Q_{ig}. The effective bulk density is the density of the fuel array heated to ignition (g cm^{-3}), and Q_{ig} is the unit heat requirement for ignition (kcal g^{-1}). Together, they represent the volume heat requirement to raise the second unit volume to ignition. This heat requirement is in units of kcal cm^{-3}.

From this treatment, fire spread can be viewed as the ratio of the Propagating Intensity from the burning volume to the volume heat requirement posed by the second volume:

$$\text{Spread Rate} = \frac{IP}{RHO_{be}Q_{ig}} = \frac{\text{Heat Flux}}{\text{Heat Sink}} \frac{\text{kcal m}^{-2}\,\text{min}^{-1}}{(\text{g cm}^{-3})(\text{kcal g}^{-1})} \tag{1}$$

This relationship alone will not allow for a prediction of fire spread rate unless the Propagating Intensity can be determined prior to ignition.

The Reaction Intensity can be expressed as the product of the load loss rate in a burning fuel array times the heat content per unit mass:

$$IR = h\left(\frac{-dw}{dt}\right) \tag{2}$$

where h is kcal g^{-1} and $-dw/dt$ is g m^{-2} min^{-1}. This is approximated by

$$IR = \frac{h(-(w_0 - w_f))}{tr} = hw_0\frac{((w_f - w_0)/w_0)}{tr} \tag{3}$$

where w_0 and w_f are the initial and final fuel loadings (g m^{-2}) and tr is the reaction time (minutes). Note that the expression, $((w_f - w_0)/w_0)/tr$, is the proportion of mass consumed per unit time, or the Reaction Velocity, γ.

What controls the reaction velocity? For dry wood with no mineral content, it is solely a function of the availability of air. This in turn depends upon the fuel bed porosity and the exposed surface area of the fuels through which pyrolysis occurs. Thus for a perfect fuel particle, the maximum reaction velocity, (γ') is

$$\gamma' = f(\beta,\sigma) \tag{4}$$

where β is the packing ratio (the proportion of the fuel array that is actually occupied by fuel) and σ is the fuel particle surface-area-to-volume ratio. Both may be determined empirically.

However, we generally do not have a perfect fuel, so γ' must be decreased according to coefficients for the fuel's moisture content (k_w) and mineral dampening (k_m) coefficient

$$\gamma = \gamma'k_wk_m \tag{5}$$

The moisture-dampening coefficient is a function of how close the moisture content is to the moisture content of extinction (kx, the moisture content at which the fuel will not burn).

We have arrived at Eq. 3 for the reaction intensity (IR) based only on the characteristics of the static fuel bed. Since these parameters can be measured before a fire occurs, predictions of fire behavior are possible. The fire intensity formula is now

$$IR = hwi\gamma'k_wk_m \tag{6}$$

Rothermel (1972) actually constructed a large number of fuel beds of different, but known, wi, h, β, and σ. He ignited the beds, used thermocouples to measure the actual reaction intensity (IR), and recorded the rate of spread (R). From these empirical studies the heat flux could be calculated from Eq. 1 if R, P_{be}, and Q_{ig} are known for a specific fuel bed. The heat flux then was related to reaction intensity (IR) by the formula

$$IP = XI \cdot IR, \quad \text{or} \quad XI = \frac{IP}{IR} \tag{7}$$

where XI is an empirically derived constant. This constant, XI, changes for each fuel bed but it has been measured for many types and mixes of fuels. These studies found that

$$XI = f(\beta, \sigma) \qquad (8)$$

Now to go back to our original question, we can calculate the heat flux, R, from parameters that can be measured in a forest before a fire, as

$$R = \frac{IP}{P_{be}Q_{ig}} = \frac{XI \cdot IR}{P_{be}Q_{ig}} = \frac{XIhwi\gamma'k_w k_m}{P_{be}Q_{ig}} \qquad (9)$$

This approaches the final Rothermel equation. However, we need to make a few final adjustments. We know that the heat to ignition (Q_{ig}) is virtually the same for all fuels except for the moisture content (MC) that must be vaporized to raise a fuel particle to ignition temperature. Thus

$$Q_{ig} = f(MC) \qquad (10)$$

Now, what about P_{be}, the amount of fuel that must be raised to Q_{ig}? This was taken into account by expressing it as a ratio, ε, where

$$\varepsilon = \frac{P_{be}}{P_b} = f(\sigma) \qquad (11)$$

where P_b is the "effective" bulk density of the bed, defined as the initial wood mass divided by the vertical depth of that wood mass.

Slope and wind are introduced as multiplying factors ($\phi_w + \phi_s$) on the rate of spread. By bringing the flame closer to the fuel (as does a higher wind), the rate of spread increases. Since these two factors also alter the airflow characteristics through a fuel bed, they also must depend on the fuel's σ and β, as well as the magnitude of the wind and slope.

Thus we finally arrive at the Rothermel mathematical model of fire spread but we have done so using field-measurable parameters:

$$R = \frac{XIhwi\gamma'k_w k_m}{\varepsilon \cdot P_b \cdot Q_{ig}}(1 + \phi_w + \phi_s) \qquad (12)$$

The original Rothermel fire spread model was empirically developed using fuel beds of uniformly sized and uniformly spaced fuel particles. When the fuel array is a mixture of various particle sizes, it is necessary to determine weighted values of σ and MC. The current procedure is to weight fuels by their surface-area-to-volume ratios. Thus, if a bed is 75% 1-hr fuels (by surface area) and 25% 10-hr fuels (by surface area), the corrected σ_c is

$$\sigma_c = 0.75\sigma(1 \text{ hr}) + 0.25\sigma(10 \text{ hr}) \qquad (13)$$

Combining the Inventory Analysis and Firespread Model

We now may take the slope steepness derived from the inventory, solve for fuel loadings and packing ratio by linking the inventory with the fuel gradient models, enter the current measured wind speed and fuel moisture from the nearest weather station, and calculate fire spread rate and intensity, hectare by hectare. We may also calculate flame length (FL) by

$$FL = aIB^b \tag{14}$$

where IB is the intensity per unit length (not area) of fireline per unit time and a and b are constants. From here we may calculate other useful parameters, such as the scorch height (the height above the ground where living foliage will be killed), the probability of the fire igniting the forest crown, and so forth.

A major limitation of the Rothermel fire-behavior model is its necessary assumption of a uniform horizontal and vertical spatial distribution of fuels. However, the Glacier group, in cooperation with William Frandsen and Richard Rothermel at the Northern Forest Fire Laboratory, is circumventing this problem in the following way. A system of small grids (grids 2 m across) is mathematically imposed on the fuel array. One must assume that within each grid the fuel is uniformly distributed, but one may stochastically assign different quantities of fuel, and/or different packing ratios, to each cell. This assignment is based on the actual measured contagion of localized cover (from the inventory) and the field measurements of fuel loadings by category and size class. In this fashion, one may calculate the spread rate and heat flux for each grid cell, and simulate the behavior of a fire through a non-uniform fuel array. The limiting assumption of a uniform vertical fuel distribution is being solved at Glacier by developing a multistrata fuel model, where each strata has an individual fuel loading, packing ratio and moisture content (Bevins, personal communication).

Weather Model

A new addition to the Glacier package is a microclimate and fuel moisture model (Mason and Kessell, 1975). It is designed to extrapolate base-station fuel moisture and weather information to remote sites with different elevations, topographies, aspects, cover types, and so forth, through empirical regressions using the site parameters recorded in the resource inventory. It is currently operational for one large drainage area in the park.

Animal-Gradient Models

Work is currently underway to develop gradient models of important mammal species in Glacier, including grizzly and black bears, white-tailed and mule deer, elk, moose, mountain goats, and bighorn sheep. Results show that the vegetation-fuel gradient scheme, supplemented by two new gradients (slope steepness and winter snow depth) may be used to quantify winter ungulate ranges (Singer, Kessell, and Ackerman, in preparation). This information is being integrated with the fire model to predict animal populations in a given region following a real or projected burn.

Programming and Execution of the Models

The entire model is written as a single FORTRAN IV G1 program using numerous computational and input/output subroutines. The total length of the program (including subroutines) is about 4500 lines. It is currently being run on an IBM 370/168 VS system under IBM 370 JCL.

Execution of the program, entry of all specifications and options, job submission, and output retrieval are accomplished through a series of instructions that closely simulate interactive FORTRAN. The user gives English language commands, answers Yes or No to various options, enters UTM coordinates and weather parameters, and sets job priority. Personnel with no computer training can be taught to use the system in a few hours. A User's Manual is available (Kessell, 1975a).

Cost and Distribution

Total cost to implement the model for a large area (over 250,000 hectares) is about $1 per hectare. This cost is comparable to many type-mapping systems that offer considerably less capability, flexibility, and precision. The money saved by not suppressing a single major fire that previously would have been suppressed, but that was not actually necessary to suppress, would be enough to pay for our entire model-development program. We developed the model in Glacier National Park on a prototype test basis; the National Park Service is currently evaluating its application to other areas.

Copies of programs, data files and documentation, and a Systems Manual, are available.

CONCLUSIONS

New modeling methodology and the availability of computer services to remote natural and wilderness areas appear to be revolutionizing traditional wilderness resource management. Data retrieval, simulation, and modeling capabilities are now available that were undreamed of by ranger and management personnel even a decade ago.

Reaction to such developments has been mixed. Some managers welcome the new systems and are very excited by their applications. Others remain interested but a bit skeptical. Still others see computer modeling as anathema to wilderness.

Yet the central problem is that our nation must preserve her wilderness heritage while providing for the public enjoyment of those areas. We cannot allow degradation of this heritage, nor can we lock the public out of our national parks and wilderness areas in order to protect the wilderness. Because of this traditional conflict between preservation and use, highly competent and more aggressive management is imperative. New modeling methods should offer an invaluable tool for both minimizing this conflict and improving management quality.

ACKNOWLEDGMENTS

I am especially grateful to William Colony, Donald Dwyer, Robert Whittaker, and Charles Hall for numerous ideas, comments, and suggestions. Richard Rothermel and William Frandsen of the Northern Forest Fire Laboratory have provided considerable assistance on the fire modeling portions of the project. Most of the section on the fire-behavior model was based on material prepared by Collin Bevins. Most of all, my thanks go to the 42 underpaid, overworked, and dedicated assistants and colleagues who have made this work possible.

This research was supported by National Science Foundation grants to Amherst College and Cornell University and by a National Park Service contract to Gradient Modeling, Inc., a Montana nonprofit research corporation.

REFERENCES

Bray, J. R. 1956. A study of the mutual occurrence of plant species. *Ecology* **37**: 21–28.

Bray, J. R. 1960. The composition of the savanna vegetation of Wisconsin. *Ecology* **41**: 721–732.

Bray, J. R. 1961. A test for estimating the relative informativeness of vegetation gradients. *J. Ecol.* **49**: 631–642.

Bray, J. R., and J. T. Curtis. 1957. An ordination of the upland forest communities of southern Wisconsin. *Ecol. Monogr.* **27**: 325–349.

Dwyer, D. B., and S. R. Kessell. 1975. Gradient based resource basic inventory (RBI). *Bull. Ecol. Soc. Amer.* **56**(2): 26–27, 49–50.

Frandsen, W. H. 1971. Fire spread through porous fuels from the conservation of energy. *Combustion and Flame* **16**: 9–16.

Habeck, J. R. 1970a. The vegetation of Glacier National Park, Montana. National Park Service, West Glacier, MT. 132 pp. (Mimeo.)

Habeck, J. R. 1970b. *Fire ecology investigations in Glacier National Park*. University of Montana, Missoula. 80 pp.

Kessell, S. R. 1973. A model for wilderness fire management. *Bull. Ecol. Soc. Amer.* **54**(1): 17.

Kessell, S. R. 1974. *Checklist of vascular plants of Glacier National Park, Montana.* Glacier Natur. Hist. Assn., West Glacier, Mont. 79 pp.

Kessell, S. R. 1975a. *Glacier National Park basic resources and fire ecology systems model: User's manual.* Gradient Modeling, Inc., West Glacier, Mont. 87 pp.

Kessell, S. R. 1975b. The Glacier National Park basic resources and fire ecology model. *Bull. Ecol. Soc. Amer.* **56**(2): 49.

Kessell, S. R. 1976. Wildland inventories and fire modeling by gradient analysis in Glacier National Park. Joint 1974 Tall Timbers Fire Ecology Conf.—Intermountain Fire and Land Symp.: 115–162.

Kessell, S. R., D. B. Dwyer, and W. M. Colony. 1975. Gradient analysis and resource management. *Bull. Ecol. Soc. Amer.* **56**(2): 49.

Mason, D. L., and S. R. Kessell. 1975. A fuel moisture and microclimate model for Glacier National Park. *Bull. Ecol. Soc. Amer.* **56**(2): 50.

Rothermel, R. C. 1972. A mathematical model for predicting fire spread in wildland fuels. *USDA Forest Service Res. Paper INT-115.* 40 pp.

Whittaker, R. H. 1956. Vegetation of the Great Smoky Mountains. *Ecol. Monogr.* **26**: 1–80.

Whittaker, R. H. 1960. Vegetation of the Siskiyou Mountains, Oregon and California. *Ecol. Monogr.* **30**: 279–338.

Whittaker, R. H. 1967. Gradient analysis of vegetation. *Biol. Rev.* **42**: 207–264.

Whittaker, R. H. 1970a. *Communities and ecosystems.* MacMillan, New York. 162 pp.

Whittaker, R. H. 1970b. The population structure of vegetation. In R. Tuxen, (Ed.), *Gesellschaftsmorphologie.* Ber. Int. Sympos. Rinteln 1966: 39–62. Junk, The Hague, The Netherlands.

Whittaker, R. H. 1973a. Approaches to classifying vegetation. In R. H. Whittaker (Ed.), *Handbook of vegetation science 5: Ordination and classification of communities.* Junk, The Hague, The Netherlands.

Whittaker, R. H. 1973b. Direct gradient analysis: techniques. In R. H. Whittaker (Ed.), *Handbook of vegetation science 5: Ordination and classification of communities.* Junk, The Hague, The Netherlands.

Whittaker, R. H. 1973c. Direct gradient analysis: results. In R. H. Whittaker (Ed.), *Handbook of vegetation science 5: Ordination and classification of communities.* Junk, The Hague, The Netherlands

Whittaker, R. H. and W. A. Niering. 1965. Vegetation of the Santa Catalina Mountains, Arizona: a gradient analysis of the south slope. *Ecology* **46**: 429–452.

Whittaker, R. H. and H. G. Gauch, Jr. 1973. Evaluation of ordination techniques. In R. H. Whittaker (Ed.), *Handbook of vegetation science 5: Ordination and classification of communities.* Junk, The Hague, The Netherlands.

GLOSSARY

Classification of communities. Attempts to understand biotic communities by assigning them to defined categories or "community types" based on shared characteristics. Classification criteria might include formation type, dominant overstory or understory species, presence of certain "indicator" species, and so forth. All communities would be assigned to the various class types and viewed as possessing the characteristics of the class definition.

Climax or mature community. The theoretical endpoint of ecological succession. Usually the oldest and most stable communities, a climax is defined as a community where gross productivity equals respiration, and thus net productivity equals zero. Many ecologists argue that the concept of a fixed climax has outlived its usefulness, and one should rather speak of communities in terms of relative maturity.

Dead and down fuel. A flammable fuel that is both dead and located on the forest floor. It is also possible to have live and standing fuels (living plants) and dead and standing fuels (grasses, forbs, or shrubs that are dessicated but still standing). Dead and down fuels are divided among size classes based on the average time lag of their fuel moisture content when relative humidity changes. A 1-hr time-lag fuel would require about 1 hr to increase its fuel moisture after a humidity increase, a 10-hr fuel would require about 10-hr, and so forth.

Fellfield. A topographic feature and community type that is caused by frequent snow (avalanche), mud, or rock slides. These communities are usually very wet (hydrophytic) for at least part of the year, seldom contain a tree stratum, and often exhibit characteristic species adapted to survive such extreme conditions.

Fireline. The advancing flaming front of a fire; also a natural or man-made strip cleared of flammable fuels which will prevent a fire from crossing the strip.

Gradient analysis. An alternative to community classification that views communities as continuous or nearly continuous along spatial and temporal gradients (e.g., elevation, moisture, stand age, etc.). Rather than assign communities to fixed types, gradient analysis views and quantifies community characteristics and species composition along these gradients. *Direct gradient analysis* arranges samples along predefined direct gradients, such as elevation, stand age, and so forth. *Indirect gradient analysis* or *ordination* uses various mathematical techniques to arrange samples based on their compositional similarities; thus on an ordination, samples with similar species composition lie close together on the axis, dissimilar samples are placed further apart on the axis.

Habitat mosaic. Describes the total ecosystem's community patterns. Natural variance in the environment produces an entire mosaic of communities that reflect topographic, moisture, stand age, and other variables. It is this natural mosaic that plant ecologists attempt to describe and model using the methods of *classification, gradient analysis,* or other techniques.

Hydric.   An extremely wet, often submerged, environment. Plants characterizing such an environment are called *hydrophytic*.

Mesic.   A moist environment, but drier than a hydric environment and seldom submerged. Characteristic plants are called *mesophytic*.

Ordination.   See *Gradient analysis*.

Overstory species.   Species that grow in the top stratum of a community such as a forest's canopy or the top shrub stratum in a shrubfield.

Seral.   A successional community that has not yet reached maturity or *climax*. Successional communities usually exhibit an increase in standing crop as gross productivity exceeds respiration, and thus net productivity is greater than zero.

U.T.M.   Universal Transverse Mercator coordinate system, a metric system that covers the entire earth. The coordinates for a unit of land are expressed in kilometers north and east from a fixed reference point in one of 15 UTM zones. Thus a square of land enclosed by incremental UTM coordinates (between 5355 and 5356 North and between 243 and 244 East) is 1 km on a side and contains an area of 100 hectares.

Understory species.   Species that grow under the top stratum of a community.

Ungulates.   A hoofed mammal of the orders *Perissodactyla* or *Artiodactyla*, including deer, elk, moose, sheep, and goats.

Xeric.   A dry environment with low available moisture. Characteristic species are called *xerophytic*.

24
The Problem of Population Optima

S. FRED SINGER
with the assistance of
JAMES T. MORRIS

It may seem intuitively obvious that a country should have an optimum population level; that is, the citizens of the country would be better off at this level of population than if the population were less or if it were greater. But intuition can be misleading; therefore, a more careful analysis is needed.

How does a geophysicist get involved with population? Well, through a progression of positions with the Department of the Interior, the Environmental Protection Agency, and the Brookings Institution. Having been concerned with natural resources and with environmental problems, I asked myself the following question: Which of many factors provides a limit to population growth? Is it fuel resources and energy, mineral resources, food, water, or even simply land? What will we run out of first? What determines the carrying capacity of the world—or of the United States? After organizing a stimulating conference in 1969 under the auspices of the American Association for the Advancement of Science (Singer, 1971), my search turned out to be inconclusive. *All* of these factors determine the carrying capacity—to a greater or lesser extent. It is rare that a single factor can be pinpointed. But in the process of searching I made some important discoveries. (1) First, my question was misleading. I had confused optimum level of population with maximum level. Many people still continue to do this. I have now learned that the optimum is always less than the maximum. (2) I also came to realize that the question is too simpleminded. Optimum population depends not only on level, but on the spacial distribution and rate of growth of population, on technological progress, and on a wide variety of parameters that enter into demography and economics. (3) Finally, the optimum population level is not fixed as time changes; very likely it decreases with time.

I gradually learned that to address the problem of optimum population, I would have to develop a method that is much broader and can answer many other questions. Therefore, I rephrased my goal as follows:

What methods can we develop to assess the general societal consequences of governmental policies, of major technological advances, or of private decisions such as a reduction in fertility. In particular, how can we determine an optimum level of population for a country?

This essay gives a progress report on our efforts to develop such methods.

SCOPE OF THE STUDY

The study deals with the United States over a time span of the next 30 to 50 years. The justification for neglecting the rest of the world at this stage is as follows. We are interested in methodology, and for this purpose the United

States presents a sufficiently complicated example. Second, this is an empirical study requiring real data; U.S. data are often better developed and certainly more accessible. Third, the United States is reasonably homogeneous and has a large enough economy, so that the influence of the rest of the world is actually quite small, at least for the next 50 years.

Our justification for not going much beyond 50 years has to do with our belief that radically new technologies can and will arise that will invalidate longer-range projections. For example, the development of nuclear fusion power would lead to a situation in which energy would be practically inexhaustible, which in turn would alter many other considerations.

AXIOM

In the study we assume that people behave rationally, that they attempt to maximize their utility, that they want more and not less of any good, and that, therefore, a growth of welfare is desirable. People express their behavior by and large in the market place and in the political arena. Their purchases determine what manufacturers will produce and their voting behavior determines political decisions. Our study is not a normative study. We do not try to suggest that one kind of behavior is better than another. We accept society as it is: a society that is interested in material welfare and a high standard of living.

We can, therefore, rephrase our original question in the following terms: What are the welfare consequences of various modes of population growth and economic growth, and what are the welfare consequences of the various methods that are employed to influence economic growth?

It should be recognized that economic growth can be influenced by a variety of methods. If, for example, we wish to stop economic growth, the method employed is important because the welfare consequences are determined by it. Consider for example the following ways in which growth of GNP can be affected.

1   By influencing population growth.
2   By changing the savings rate, and thereby the rate of investment.
3   By stimulating or depressing technical innovation and technological progress.
4   By introducing a rationing or a progressive taxation of energy.
5   By introducing more leisure. For example, by early retirement or by late entry into the work force.
6   One might even postulate a situation in which work is outlawed for men, thus making up for centuries of discrimination against females.

7   Finally, government can influence growth through fiscal and monetary policies. This is often done unintentionally by causing unemployment and depressions.

OBJECTIVES OF THE STUDY

The objectives are twofold: (1) The first is to construct an objective function that measures the aggregated amount of welfare per individual in the nation. The definition must be appropriate to our society and cultural patterns. It must also be an operational definition so that welfare can be calculated from the kind of data that are available in the national statistics. (2) The second task is to develop a mathematical model that relates this index of welfare to demographic and economic parameters, thus allowing us to project a time "stream" of welfare indices as a function of various assumptions concerning population and concerning the economy.

CONSTRUCTION OF A WELFARE INDEX

One might object that the material index of welfare that we are defining does not really measure happiness. In my view it measures an important component of happiness: the "global" component. The other component is intensely local and determined by interpersonal interactions with a very few people, with family, coworkers, friends, and so forth. This latter component of quality of life should be reasonably independent of overall demographic and economic parameters in the United States. If this is so, then we can neglect it in a partial analysis.

The construction of a welfare index follows in concept the ideas of Nordhaus and Tobin (1972) and others who would "amputate and impute the GNP." But we depart considerably from previous work. Like most others, we define welfare as consumption in households. We include non-market production (but not yet illegal services and goods that are not counted by the national income statistician). We add the value of leisure—not at the opportunity cost of wages foregone, but using an empirical utility function, based on what people indicate their leisure time is worth. While we count much of educational expense as investment and, therefore, not as consumption, we depart by ascribing health costs to consumption, rather than to investment or to "regrettable necessities." Of course, such items as defense expenditures, police expenditures, and commuting to work, are all regrettable necessities and do not contribute directly to welfare. In this way we treat every item that enters into the gross national product, that is, every item of governmental and private expenditures. In addition, however, we also subtract certain disamenities that are produced by population and economic growth.

1 The three principal items are pollution control costs, which rise faster than the GNP (in spite of the fact that there is a trend away from goods production and towards services, and in spite of the fact that processes are being introduced that create less pollution). The limited assimilative capacity of the air and water environment introduces an important non-linearity that cannot be avoided.

2 Resource costs also increase as such items as fossil fuels become exhausted. In a perfect market there would of course be immediate substitutions or the immediate introduction of new technology. In an imperfect market such as ours there is a kind of "stickiness" (akin to static friction) that slowly "ratches" up the cost. We are seeing this phenomenon now applied to oil and gas prices.

3 Finally, we have the costs of agglomeration of population, related to the very uneven distribution of people. Large cities become increasingly less efficient as traffic jams increase internal distribution costs; as land prices and rents rise, thereby adding to the costs of all products; and as interpersonal disamenities such as crime and all kinds of urban problems increase. Of course, much can be done by technology, for example, by a better transportation system. But we have to take account of the increasing inefficiences that raise the cost of living and, therefore, diminish per capita welfare as population and GNP grow.

Of course, there is much arbitrariness in the definition of an index of welfare. I have called it a Q index rather than "quality of life," simply because many people have already formed their definitions or opinions. While definitions are always a matter of taste, I hope that reasonable people may agree that it is a better measure of welfare than GNP, and that in spite of arbitrariness it can be useful. I hope, however, that I have not been guilty of what St. Augustine confessed: "For so it is, oh Lord my God, I measure it; but what it is that I measure I do not know."

As an aside I should mention that it would be quite difficult to carry out the analysis I have just described in a socialist country, that is, one having a controlled market economy, or in an underdeveloped country where a market is not well developed and where barter and household production are relatively important.

CONSTRUCTION OF THE MATHEMATICAL MODEL

Our mathematical model consists of three parts. The demographic part
(Figure 1) is rather complicated, yet conceptually quite straightforward. It
considers native and immigrant population; their age distributions; allows
for different fertility assumptions; projects characteristics of households,
geographic distribution of population, and income distribution; projects
labor productivities in different sectors, such as agriculture, manufacturing,
services, and government, and ends up with effective labor after having
considered labor participation rates and trends.

The economic model (Figure 2) is simple, essentially a neoclassical model
with a one-sector output—GNP. Part of the GNP is reinvested in capital
cohorts that have a rising productivity because of technological progress.
The model allows for different modes of investment, including one that
preserves a constant capital-labor ratio.

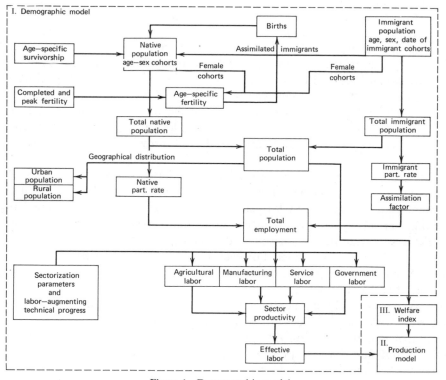

Figure 1   Demographic model.

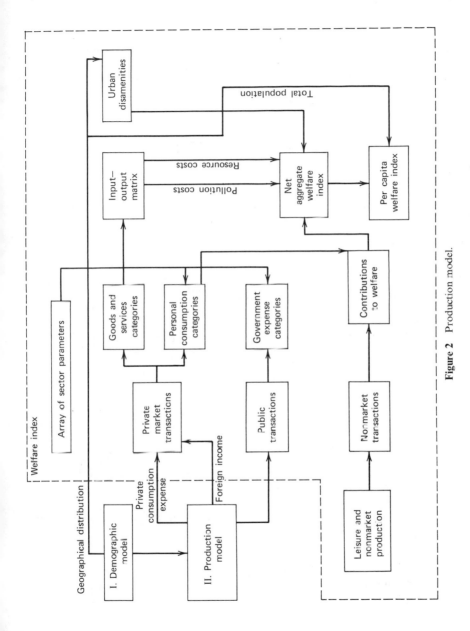

Figure 2 Production model.

613

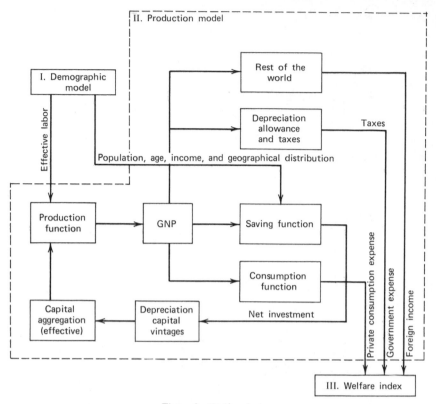

Figure 3   Welfare index.

The third part of the model (Figure 3) is a diagnostic that calculates the welfare index. It sectors the output of the economy according to demand. It subjects each sector to an analysis that follows the definition of our welfare index. It also sectors the output according to production so as to be able to calculate the resource and environmental implications, and, therefore, the resource and environmental costs.

I do not draw any comparisons here between our model and that of the Club of Rome group at MIT. I have explained my objections elsewhere.*

Suffice it to say that our model does not attempt a world-wide aggregation, nor do we aggregate all resources, all pollution, and so forth. Instead, we have done our aggregation at the economic level using dollars as the common unit. This essentially is the philosophy of the GNP, which aggregates

---

* In EOS, Transactions of the American Geophysical Union, 53, 697–700, 1972. The results of the Club of Rome Study have been published (Meadows, 1972).

goods and services of various kinds. Our model attempts to simulate the operation of our market economy as closely as possible.

RESULTS

There are many kinds of results, and these can best be visualized by glancing at a typical printout from a computer run (Singer, 1972).

The assumptions are shown in the legend of Figure 4, but we comment here on some of the prominent results. The main one certainly is that per

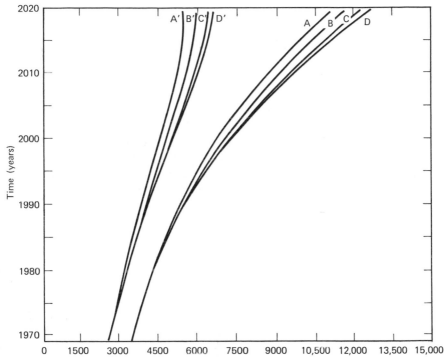

Figure 4 Per capita GNP (A, B, C, D) and per capita welfare (A′, B′, C′, D′) with four different fertility projections: (A) 1968 values of fertility, f_0, held constant. (B) Reduced to an asymptotic value of $f_{00} = 0.85\,f_0$ with a half-time of 5 years. (C) $f_{00} - 0.66\,f_0$ with a half-time of 10 years. (D) $f_{00} = 0.50\,f_0$ with a half-time of 20 years. The decrease in fertility is exponential; reaching halfway to the asymptotic f_0 value in one half-time T, that is, $f(t) = f_0 - (f_0 - f_{00})[1 - \exp - (t/T)]$. For all four runs we assumed an initial age distribution as of 1969; we held constant the following items: (a) age-specific survivorship rates (mortality), (b) immigration as a proportion of total population, (c) savings (gross investment) as a fraction of GNP, (d) rate of capital depreciation. We trended leisure time, labor-participation rates, labor productivity, and applied appropriate forecasts for all consumption sectors, for pollution control costs, resource cost increases, and urban disamenities.

capita welfare is increasing today and will continue to increase for some time to come before it reaches a maximum, presumably diminishing beyond this point. In the meantime, however, much can happen in the way of technological change that makes projections running 30 to 50 years quite uncertain. Of particular interest, of course, are differential results, obtained by running the model with one of the main assumptions slightly changed. Since we are primarily interested in the effects of population growth, we present the results form four runs having different fertility assumptions. Run A assumes the 1968 fertility remaining unchanged. Run C assumes that this fertility diminishes to 66% of this value asymptotically, with a half-time of 10 years, that is, teaching halfway, to 83%, in 10 years.

I have plotted the results in Figure 4 with comments reserved to the figure legends. The main result, comparing the Q index, shows that the slower rate of population growth leads to higher benefits at all times in the future.

GENERAL COMMENTS ON RESULTS

What the model does for us is to provide a diagnostic that allows us to total up the effects of any particular assumption, or of a governmental policy, a private decision, and so forth. It sums the pluses and the minuses and compares them eventually in the Q index, which is the important objective function. The result is a time run of Q indices (Figure 5). The larger values are clearly more desirable; the smaller ones are less desirable. The interesting situations are those in which the Q index as a result of one policy is both larger and smaller at different times with respect to the Q index that one obtains without this policy (Figure 5 b). There are many practical examples for such situations. A large capital investment over the next few years would

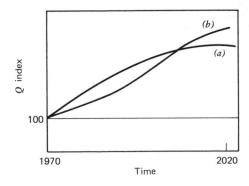

Figure 5 Two possible time streams of the Q index. Which is to be preferred?

reduce consumption and, therefore, lower the Q index, but would presumably raise it some years hence and, therefore, benefit a future generation.

One value of our model is that it allows an explicit method of looking at the future and evaluating the consequences of a policy decision. In our example, one would then compare two streams of Q indices that have been calculated by the computer model. To make such a comparison we need, of course, to assume a certain discount rate. Our model cannot determine the discount rate. It is set by other considerations that are largely political and determined by how much we feel we owe to a future generation. (Some have argued that since future generations will be wealthier and better off than we are and certainly have a higher welfare index, then taxing ourselves now to benefit a future generation amounts to regressive taxation. As some have put it "What has posterity ever done for us?")

There are, of course, many examples where such matters need to be considered. In the pollution control field, for example, the separation of sanitary and storm sewers in U.S. cities may cost between 30 and 50 billion dollars. The restoration of the Great Lakes may be a project of equal magnitude. Should we undertake such a project and over what time scale?

There are still many weaknesses in our model. We are aware of some of these. Missing at this stage is distribution of income and distribution of welfare; the model deals only with averages. We have no theory of how growth affects distribution of income, nor do we have a theory of how distribution of income affects growth (perhaps by way of productivity or through the absence of strikes and social upheavals). Knowledge of the distribution of income is important because it affects consumption patterns. Higher income means more luxury goods and more services, relative to basic necessities.

We have not quite discovered how to compare one distribution of welfare with another. Research needs to be done on the marginal utility of welfare; we have no way as yet of taking into account the "Duesenberry effect."* Until then we can use the concept of cardinal welfare value. We plan to do some experimentation on discretionary welfare, defined as welfare beyond the basic necessities of life.

At this stage the model has no regional or local detail. It is aggregated over the United States with some disaggregation according to urban versus rural population.

At this stage the input/output table has not yet been introduced. We are, therefore, not yet in a position to have consistent interindustry forecasts of the kind carried out by Almon (1975). It is our hope that we may be able to introduce this feature very soon.

---

* Which argues that individual consumption (and welfare) is affected by that of others, through a "demonstration effect."

SUMMARY

We have taken some steps to construct a diagnostic tool to measure the consequences of decision making, technology assessment, population policies—in short, every kind of human decision and in all of its ramifications. By focusing on a single index that measures welfare as well as possible we hope to be able to answer the question whether we as a society will be better off or worse off under one set of assumptions as against another set of assumptions, under one set of policies as against another set of policies, with one technology as against another technology, and with one set of private decisions, for example, as regarding fertility, as with another set. In particular, we should be able to investigate which combinations of demographic parameters lead to a maximum value of the welfare index.

ACKNOWLEDGMENTS

I appreciate the assistance of Harry Burt, and Giorgio Canarella. The research was supported by the Center of Population Research, National Institutes of Health under contract NIH-NICHD-72–2052.

REFERENCES

Almon, C. 1966. *The American economy to 1975*, Harper & Row, New York.

Ehrlich, P. R., and A. H. Ehrlich. 1972. *Population, Resources, Environment*. Freeman, San Francisco.

Meadows, D. H., D. L. Meadows, J. Randers, and W. W. Behrens III. 1972. The Limits To Growth, Universe, New York.

Colloquium of the National Bureau of Economic Research. Columbia, New York.

Population and the American Future, New American Libary, New York, 1972.

Singer, S. F. Ed. 1971. *Is there an optimum level of population?* McGraw-Hill, New York.

Singer, S. F. 1972. A study of optimum population levels—a progress report. *Proc. Nat. Acad. Sci. USA* **69**: 3839–3848.

APPENDIX   APPLICATION OF THE U.S. MODEL TO OTHER COUNTRIES

It should be clear that the U.S. model just described cannot be applied to other countries directly and for several reasons. First, the concept of an optimum population may not be reasonable in all cases. Consider, for example, the extreme case of a country that has great mineral wealth but nothing else, little arable land, little water, no capital, no factories, and so forth.

It is intuitively obvious that the optimum population is synonymous with minimum population, since then the per capita return from the sale of the minerals would be a maximum. While admittedly this is an extreme case, it does show that the matter of imports and exports has to be considered very carefully, especially for small underdeveloped countries.

Next, we should examine the concept of the "target." Is the objective in all cases a high level of material welfare? While this may be appropriate to many societies, one can imagine a cultural pattern, say a monastic one, that does not lay great stress on material welfare. In that case, the definition of the Q index would be different from the one I have adopted for the United States. But the general methodology could still be applied; that is, after an appropriate Q index has been defined, one could still construct a mathematical model that calculates it and projects it into the future, under a given set of initial assumptions.

The matter of assumptions has to be examined very carefully also. What are important assumptions for the United States may not be important for other countries and vice versa. Also, the type of information required to run the mathematical model may be different for other countries. In general, however, one would want good statistical data that are appropriate to the type of mathematical model that is constructed.*

As mentioned earlier, the Q index may be difficult to calculate for a country that does not have a well-developed market economy. If barter constitutes a large part of the economy, then prices may have to be "imputed" in order to be able to measure welfare. Even in the U.S. model imputation is done for nonmarket production, for example, for goods and services produced in households.

For similar reasons it may be difficult to apply our model to a socialist country. In our model the welfare value of goods and services is determined by the price which people pay in a free market. In a controlled market the welfare value of goods and services would have to be obtained by other methods.

Aside from the above exceptions, it seems to me that our methods should be applicable to other countries whether they are sparsely or densely populated, whether they are large or small, whether they are heavily industrialized or agricultural, urbanized or rural, whether they have a high fertility or a low fertility, or whether they have a high mortality or a low mortality. Consider, for example, an agricultural country having a high fertility and high mortality. The Q index might be defined in a very similar fashion to that of the United States. A mathematical model would be constructed in a

* In fact, an important application of our mathematical model would be to establish what accuracies are required for demographic surveys and data, by empirically testing the effects of small errors.

similar fashion. But the parameters entering into it would be quite different. There probably would be no migration to the country. The amount of schooling received would be less and the age of entry into the work force would be lower. Educational expenses therefore would be less on a per capita basis. However, the number of children would be relatively higher because of an age structure with a high dependency ratio. On the other hand, medical costs might be less per capita since people do not survive to the age where degenerative diseases are important. For such a country, pollution control costs and urban disamenities would be less important, but not the increased cost of natural resources. In particular, one would want to know the productivity of marginal agricultural land and the cost of working it in order to produce acceptable crops.

It is clear that a great deal of research would have to be done in order to develop satisfactory methods of building demographic models for different countries. Yet, I believe that the task is essential and important. In order to influence policy decisions and policy makers it is important to present them with quantitative arguments. It is important to indicate to them the effects of different policies and the consequences of different assumptions. Hopefully, this will not only lead to an optimum population, but also to an optimum path for reaching an optimum population.

25
Narragansett Bay—The Development of a Composite Simulation Model for a New England Estuary

SCOTT W. NIXON
JAMES N. KREMER

E stuaries are difficult to study and even harder to understand. They are strongly influenced by the geography of the surrounding land and by tidal flows from the sea. They are the interface systems between rivers and the ocean, often characterized by complex water masses and changing salinities, by large seasonal variations and migrating populations, by high silt loads, complex chemistry, and, increasingly, by unnatural inputs from man's activities. Ever since early settlements were concentrated along the coasts, estuaries have served as vital resources supplying food, transportation, harbors, and waste-disposal needs. They have also been systems of great natural beauty and a source of recreation and escape. They are far too important not to attempt our best efforts to resolve their workings. Their complexity requires that we attempt to apply what may be our best hope for ecological systems management, the simulation model, to help plan the difficult future of man and nature in the coastal zone.

Economic models such as that by Isard and Romanoff (1967) and Laurent and Hite (1971) have been generated for man's activities in the coastal zone, and a variety of hydrodynamic and hydraulic models of estuarine circulation have been developed. Many marine laboratories along the coasts of the United States, Europe, and Japan have attempted or now carry out coordinated programs to study local estuarine systems. Among the more successful of these were the early studies of Gordon Riley's (1955) Long Island Sound group in the late 1940s and 1950s, H. T. Odum's (1967) analysis of the Texas Bays some 10 years later, and Okuda's (1960) studies of Matsushima Bay. Some of the most promising recent efforts are described in other chapters. While the state of estuarine modeling has recently been reviewed (Environmental Protection Agency, 1971), it is noteworthy that an often-cited volume on estuarine ecology (Lauff, 1967) contained almost no work on the systems analysis of estuaries, simulation modeling, or other synthetic treatments. As with most environments, the traditional approach has been a diverse array of detailed analyses. In many cases, the water-quality engineers have been far ahead of the coastal ecologists in developing approaches to the study of estuarine ecosystem dynamics. An excellent review of their approach and some water-quality modeling techniques has recently been given by Thomann (1972).

However impressive and encouraging many of these studies are, they have usually been confined to consideration of water chemistry, circulation, or biology. Seldom, if ever, have all of these aspects of the estuarine ecosystem been brought together into a dynamic coupled model of economics, water circulation, water quality, and ecology. Yet it is difficult to understand the present state of the system, let alone make predictions, without such a tool. At the University of Rhode Island, the results of our efforts in this direction have been encouraging. As part of the Sea Grant College program in resource economics, ocean engineering, and oceanography, we have been working together to develop multidisciplinary models of the Narragansett Bay ecosystem. The project is now nearing completion, though much remains to be done. And, in an important sense, ecosystem models should never really be "finished," but should serve as a source of feedback for guiding future research on the system, so that they can themselves be continuously refined and modified as new knowledge is gained. In developing our model, we are attempting not only to produce a management tool that can be applied to pressing decisions of power plant siting, oil spills, eutrophication, and so forth, but also to use the model to increase our basic understanding of the bay system. It has been and continues to be a very difficult, frustrating, yet exciting synthesis of effort. But before becoming too enmeshed in the matrices and differential equations that are the structure and function of our model system, we should give some feeling for the real Narragansett Bay (frontispiece 5), for its history, and for the diversity, abundance, and changing patterns of the life that it supports.

NARRAGANSETT BAY

Historical Note

The coupling of man and the sea has been a strong but evolving tradition in New England, where the rich coastal waters first helped to compensate for the often insufficient yields of the land. A brief history of the changing relationship between man and nature in Narragansett Bay has been given by Alexander (1966). The abundant shellfish of the Bay sustained the Indians and early settlers alike, and its protected waters made Newport and Providence leading centers of early American trade for rum, shipbuilding, candles, fish, ironwork, and textiles. Later, near the turn of this century, colorful fleets of excursion boats and freight steamers linked Narragansett Bay with major ports along the Atlantic seaboard, and a "golden age" of bay resorts and casinos grew up for the summer season out of Boston and New York. At its peak around 1900, over 1,250,000 passengers a year rode the bay steamers. But the region's popularity gradually declined, and the government took advantage of the bay's narrow openings and strategic location to develop major defense installations and a home port for much of the Atlantic Fleet in both world wars.

The bay has always served as a recepticle for the domestic, agricultural, and industrial wastes generated along its shores. Increasing population and industrial growth has brought the total now to over 100 million gallons of waste each day. Shellfish are still taken commercially from the bay, although pollution levels have closed many of the productive beds, and the most important species has shifted from oysters to clams. However, even the clam catch has declined sharply from its peak in 1955, when Rhode Island provided over half of the total New England catch. Commercial finfish from the bay, largely flounder, scup, and menhaden, now make only a small and variable contribution to the total Rhode Island landings, though a sport fishery is growing rapidly. Newport never recovered from the Revolutionary War as a commercial center, but the port of Providence is the third largest in New England and serves mainly as a point of entry for oil and lumber and of export for scrap metal. Only one excursion boat survives, and the upper-bay resorts have almost vanished. However, a large and rapidly growing fleet of some 14,000 pleasure boats covers the bay from May through October.

Changing technology and shifting national priorities have brought the area's 50-year association with the Navy largely to a close and are freeing parts of the coastline for new uses. While Narragansett and Newport are no longer the centers of summer fashion, expanding tourist and recreation trade is once again bringing numbers of summer visitors to the bay and increasing the concern of residents, as well as local, state, and federal agencies for regulating future growth to protect and improve the bay ecosystem.

Economy of the Bay Area

Population, Employment, and Income

Growth of the population in Rhode Island has been roughly exponential since 1790 and reached a total of 892,700 by the 1965 census. A detailed analysis of recent changes in population distribution by Rorholm et al. (1969) has shown a strong movement from urban centers on the Providence River to more rural communities along the lower bay. Even though there is a marked increase in growth along the lower bay, the greatest population density still remains in the upper bay and river.

Location	Area (km^2)	Population Density (No. km^{-2})
Upper Bay	196.6	2073
Lower Bay	587.2	360

Alexander (1966) has estimated that only about 3% of the state labor force is directly involved in the exploitation of Narragansett Bay resources through fishing, clamming, and so forth. However, a great many more workers are also directly or indirectly involved with the bay. A survey conducted in 1965 to 1966 found 76 marine-oriented firms employing well over 4000 people located around the bay with total annual sales of over 60 million dollars (Rorholm et al., 1967).

Economic Activity and the Impact of the Bay

The 1800-mi^2 Narragansett Bay drainage basin (Figure 1) includes over 90% of Rhode Island's economic activity and slightly less than 10% of that of Massachusetts. It also contains 91% of Rhode Island's population and 93% of the total business and manufacturing firms (Feld and Rorholm, 1973; Griffiths, 1970). It has been estimated by the Governor's Technical Committee on the Coastal Zone (1970) that the waste effluents from 90% of the households and almost all of the industries in Rhode Island drain into

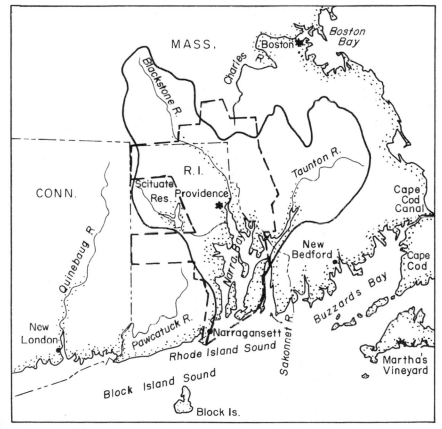

Figure 1 Narragansett Bay drainage basin (solid line) and the area included in the economic model (heavy broken line).

Narragansett Bay. The use of the bay as a mechanism for sewage disposal has important economic as well as ecological consequences for the region. In an extensive study of the socioeconomic characteristics of the area, Rorholm et al. (1969) identified waste disposal as the fourth largest quantifiable aspect of gross revenues or savings associated with the bay. By far, the greatest economic impact of the bay has been associated with the U.S. Navy, until recently the state's largest single employer, and with the use of the bay to float ships, cool equipment, and carry wastes. While esthetic considerations have not been included and recreational values were underestimated, the Rorholm study suggests that those activities responding most strongly to water-quality constraints are of secondary economic importance.

Physical Setting

Narragansett Bay runs roughly north to south in the Rhode Island coast between Long Island, N.Y. and Cape Cod, Mass. (Figure 1). Immediately offshore are the waters of Rhode Island Sound and Block Island Sound. A drainage basin of some 1800 mi^2 in Rhode Island and Massachusetts provides an average fresh water input of 1600 cfs to the system. The general oceanography of the bay has been described by Hicks (1959) and in a series of reports edited by Fish (1953). The geological development and sediment characteristics have been discussed and studied by Shaler et al. (1899) and by McMaster (1960). In general, silt-clay sediments dominate the upper bay, with fine sands near the mouth. The bay model encompasses an area of 265 km^2 with a length of 45 km and a maximum width of 18 km. Mean depth in the bay is about 9 m, with averages of 7.5 m in the West Passage and 15.2 in the East Passage. The hypsographic curve for the model area shows that 75% of the bay is shallower than 12 m. Tides are semidiurnal with a mean range of 1.1 m at the mouth and 1.4 m at the head. The mean tidal prism is about 13% of the mean volume and over 250 times the mean total river flow during a tidal cycle. In addition to large energy inputs associated with the tides (Levine, 1972), the waters of the bay are well mixed by winds that may completely dominate the short-term circulation pattern in some sections of the bay (Weisberg and Sturges, 1973). The wind pattern shifts markedly from northwest in the winter to southwest during summer with greatest speed during December and January (Figure 2). Water temperatures range from about -0.5 to $24°C$ with a well-developed thermocline present only in the upper bay and river during summer. During extreme conditions, the temperature range of the surface water within the bay may be on the order of $10°C$. The annual temperature cycle lags solar radiation by about 40 days (Figure 2). Rainfall in this area is about 1 m yr^{-1} and is relatively evenly distributed throughout the year, though river discharge

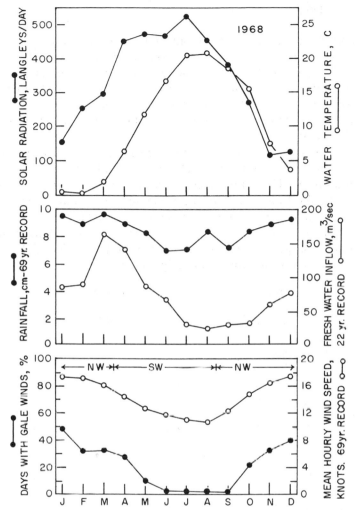

Figure 2 General climatological data for Narragansett Bay (from Fish 1953; light data from Eppley Labs.; water temperature from Jeffries, unpublished).

reflects lags and shows a peak in March and April (Figure 2). Detailed measurements of light extinction coefficients throughout the bay have been made by Schenck and Davis (1972) and show clear gradients that reflect distance from the river and the tidal circulation pattern. Values range from lows of about 0.20 m^{-1} in the lower East Passage to highs of almost 1.0 m^{-1} in the river.

Water Chemistry

The relatively small fresh-water input and large tidal volume of the bay result in a well-mixed water column and small salinity gradients down the bay. There is about a 12‰ range from salinities of about 20‰ in the Providence river to a maximum of 32 or 33‰ at the mouth of the bay. In general, there is no well-defined halocline except in the upper bay and river. Seasonal variations in salinity are slight except in the uppermost portion of the system.

There are strong seasonal cycles and sharp gradients in the distribution of biologically important nutrients, including ammonia, nitrite, nitrate, phosphate, and silicate. The annual cycle for each of these nutrients at a midbay station in the West Passage is shown in Figure 3. Maximum concentrations are found in the upper bay and river and reflect the large inputs of sewage in this section of the system. A seasonal analysis of sewage effluents and flow rates from the largest of these plants, located near the city of Providence, gave the following values for mean daily inputs:

Input	kmoles day^{-1}
Ammonia	69.6
Nitrite	0.5
Nitrate	13.1
Phosphate	9.0
Organic phosphorus	7.2
Silicate	9.0

The sewage effluents are also a major source of input for petroleum hydrocarbons and heavy metals. Analysis of the effluent from the same treatment plant by Farrington and Quinn (1973) indicated that from 0.4 to 2.0 metric tons of petroleum hydrocarbons per day enter the bay from that one plant alone. Measurements of heavy metals in the effluent from a smaller treatment plant on the bay have been given by Ryther et al. (1972).

Since the Port of Providence handles substantial volumes of petroleum, amounting to over 7.5 million metric tons in 1968 and the U.S. Navy makes the bay home port for some 70 ocean-going ships, there is a continuing occurrence of oil spills in the bay. In the past few years there have been over 50 spills in the bay and its tributaries, at least one amounting to about 10,000 barrels.

Biology

Narragansett Bay is a phytoplankton-based ecosystem in which water depths and turbidity, as well as a lack of firm substrate, have minimized the

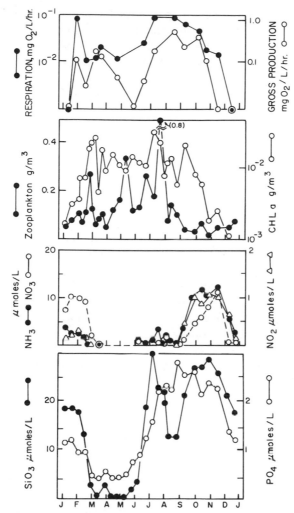

Figure 3 Seasonal patterns of metabolism and the abundance of organisms and nutrients at one station in the West Passage of the bay.

importance of attached algae and vascular plants. The phytoplankton populations have been described by Smayda (1957) and Pratt (1959, 1965) and are characterized by much greater standing crops than are found in adjacent waters and by an annual winter bloom (Figure 3). The winter population is usually dominated by the diatoms; *Skeletonema, Thalassiosira, Nitzchia,* and *Detonula,* while the summer flora is composed predominately of flagellates such as *Olisthodiscus* and μ-flagellate species. The winter bloom appears to begin in the upper West Passage and then spreads or is carried throughout the bay. While there is a sharp increase in the metabolism of the plankton community associated with this bloom, algal production levels are also high in summer during a second bay-wide bloom (Figure 3). Occasional localized blooms with large standing crops and high production are also found associated with periods of heavy runoff or other disturbances in spring and summer.

The major consumers of the bay are the zooplankton, which are dominated alternately in winter and summer by the ubiquitous copepods, *Acartia clausi* and *A. tonsa*, which may make up 95% of the total population (Martin, 1965; Jeffries and Johnson, 1973). Zooplankton biomass is greatest in early summer, and Martin (1968) has suggested that the increasing grazing pressure during the spring terminates the winter phytoplankton bloom in the bay (Figure 3). His preliminary work, now being investigated in detail by Vargo (personal communication), also suggests that the excretion of nitrogen by the zooplankton may be important in providing this element for the summer phytoplankton (Martin, 1968).

The zooplankton are subjected to predation by fish larvae and meroplankton (Herman, 1958) and late in the summer by a large pulse of *Mnemiopsis leidyi*, a carnivorous ctenophore whose population levels, grazing rates, growth and nutrient excretion have been studied as part of the modeling program (P. Kremer, 1975).

Other pelagic consumers that become abundant for short periods include dense schools of Atlantic menhaden, *Brevoortia tyrannus*, that feed on plankton and detritus in the bay (Jeffries, 1973) and important carnivorous sport fish such as striped bass, *Morone saxatilis*, and bluefish, *Pomatomus saltatrix*. Some aspects of the ecological impact of the menhaden have been described (Oviatt et al., 1972), and their role in grazing and nutrient excretion is being studied (Durbin and Durbin, 1975; Durbin, 1976). The demersal fish of the bay are also abundant, with a relatively constant biomass of some 0.05 g m^{-2} dry weight that is largely sustained by feeding on the infauna and epifauna of the sediments (Oviatt and Nixon, 1973). The feeding and excretion rates of the dominant species, winter flounder and sand dab flounder, are being measured for inclusion in the model. The community of larger benthic animals, however, is not dominated by the finfish, but by dense populations of the hard clam, *Mercenaria mercenaria*. These animals appear to have a standing crop of some 0.6 g m^{-2} dry meat weight (Russell, 1972) and can exert a strong grazing pressure on the phytoplankton when the water temperatures are over $8°C$ (Loosanoff, 1939). We are now beginning to measure the grazing rates and nutrient regeneration of the clam community *in situ*. The bay bottom consists largely of empty clam, mussel, scallop, and oyster shells, with an epifauna of starfish, lobsters, conchs,

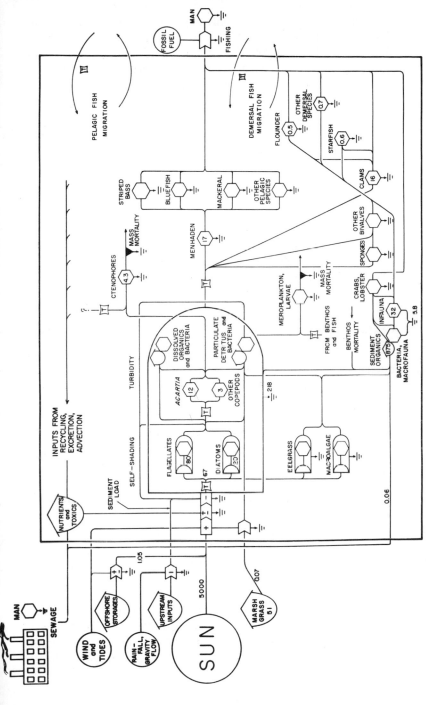

Figure 4 Preliminary energy-flow diagram for Narragansett Bay on a late summer day. Because of strong seasonal and geographical variation, many of the values change markedly at other times and in specific areas of the bay. Storages are in kcal m^{-2} and flows in kcal m^{-2} day^{-1}. Values have been calculated from the many sources cited in text.

631

scattered sponge beds, crabs, and swarms of grass shrimp and sand shrimp. There is an abundant infauna that has been described by Stickney and Stringer (1957) and Phelps (1958). The major communities are dominated by the bivalves, *Nucula proxima* and *Yoldia limatula*, and the polychaete, *Nephthys incisa*, with densities of about 8.0 g m^{-2} dry weight.

On the basis of the standing crop maintained, it appears that much of the high phytoplankton production of the bay is directed to benthic food chains and the support of an abundant infauna and large populations of bivalve molluscs and flounder. A preliminary energy-flow diagram for the Narragansett Bay ecosystem indicates the importance of these flows and storages (Figure 4). Inputs of energy from marsh detritus and sewage may also be important in the upper bay. Studies of marsh production along the bay (Nixon and Oviatt, 1973a,b) indicate a potential total input of detritus of about 1.9×10^6 kg yr^{-1}. The secondary production of pelagic bacterial biomass in the bay also appears substantial in preliminary calculations by Sieburth (personal communication) and may serve as an important food source for some of the smaller species.

THE NARRAGANSETT BAY MODEL

The computer-modeling program for the bay is a hybrid consisting of three large efforts to deal with economic activity and effluent loads, the dynamics of water movement, and the chemistry and biology of the bay ecosystem (Figure 5). Ideally, each of these three submodels would interact dynamically, but at present only a more simplified approach is possible. Each of the three can be further divided and requires its own extensive set of equations, assumptions, data inputs, and forcing functions, which in combination far exceed the hardware limitations of the university's IBM 370/155 computer. Thus the combined submodels are not intended to be run simultaneously as a interacting "megamodel."

As is discussed in the following section, the economic model is designed to relate composition and magnitude of effluents for the total watershed to economic activity by large industrial sectors. The hydrodynamic and ecological models, in comparison, deal with day-to-day patterns and processes occurring in specific regions of the bay. These differences in spatial and temporal detail make direct interfacing with the economic model difficult. Future work will attempt to extend the economic model, allowing evaluation of specific industries and sites around the bay. While substantial modifications were necessary to adapt the hydrodynamic model for use in the ecological model, this coupling has been successful.

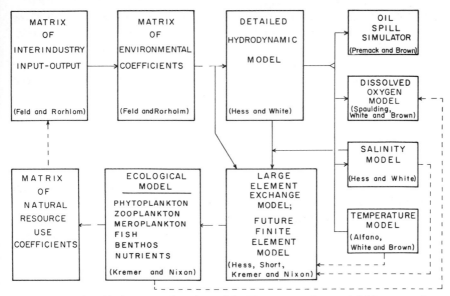

Figure 5 Relationship between subsections of the Narragansett Bay model. Names are those most responsible for their development. Broken lines indicate logical connections between three submodels. Dashed lines indicate possible future interactions.

Submodel I: An Input-Output Model of Economics and Waste Effluents

The submodel dealing with interactions among Narragansett Bay area industries and the resulting generation of waste effluents and demand on natural resources has been developed and described (Feld, 1973; Feld and Rorholm, 1973). Since the economic data required for such a model is compiled according to political or socioeconomic divisions, rather than according to natural ecosystem boundaries, the economic model region does not coincide exactly with the Narragansett Bay watershed. This problem will continue to hinder large ecosystem studies until government information-retrieval systems can be made more flexible. However, in this case the Providence Standard Metropolitan Statistical Area plus Aquidneck Island included almost all of the Rhode Island portion of the watershed and followed the hydrodynamic-model boundaries closely (Figure 1).

The economy of the study area was apportioned into 59 endogenous sectors and 4 exogenous final demand* sectors based on data for the reference

* Final demand refers to that portion of an industry's market demand not otherwise included in the model sectors, that is, nonindustrial consumers.

year 1967. As shown in Figure 5, the economic model consists of three basic parts: interindustry-dependency coefficients, environmental or waste-discharge coefficients, and natural-resource-use coefficients. These component matrices are ultimately combined mathematically to produce estimates of the total emission load imposed by each of the segments on the bay system in response to changes in the economic community.

Interdependency Coefficients

A matrix of interdependency coefficients, the "Leontief inverse," for the 59 endogenous sectors is derived from an input-output analysis. This matrix is then multiplied by the final demand matrix to calculate the impact on output of a unit change in final demand.

An example of the process for a three-sector economy from Feld and Rorholm (1973) begins with the basic input-output analysis expressed as a system of linear equations:

$$
\begin{aligned}
X_1 &= x_{11} + x_{12} + x_{13} + Y_1 \\
X_2 &= x_{21} + x_{22} + x_{23} + Y_2 \\
X_3 &= x_{31} + x_{32} + x_{33} + Y_3
\end{aligned}
\tag{1}
$$

where X_i = total gross output of the ith sector

x_{ij} = intermediate demands for gross output, that is, purchase made by the jth sector from the ith sector

Y_i = final demand (consumption by nonindustrial sectors)

A series of technical coefficients are then calculated as the ratio of purchases made by sector j of sector i output over the gross output of sector j:

$$
A_{ij} = \frac{x_{ij}}{X_j}
\tag{2}
$$

where A_{ij} = the technical coefficient

Substituting $A_{ij}X_j$ for x_{ij} and solving for Y gives

$$
\begin{aligned}
Y_1 &= X_1 - A_{11}X_1 - A_{12}X_2 - A_{13}X_3 \\
Y_2 &= X_2 - A_{21}X_1 - A_{22}X_2 - A_{23}X_3 \\
Y_3 &= X_3 = A_{31}X_1 - A_{32}X_2 - A_{33}X_3
\end{aligned}
\tag{3}
$$

Or in the matrix form,

$$
\begin{bmatrix} Y_1 \\ Y_2 \\ Y_3 \end{bmatrix} =
\begin{bmatrix}
1 - A_{11} & -A_{12} & -A_{13} \\
-A_{21} & 1 - A_{22} & -A_{23} \\
-A_{31} & -A_{32} & 1 - A_{33}
\end{bmatrix}
\begin{bmatrix} X_1 \\ X_2 \\ X_3 \end{bmatrix}
\tag{4}
$$

so that

$$Y = (I - A)X \tag{5}$$

where Y = the final demand matrix
     $(I - A)$ = an identity matrix less a matrix of technical coefficients
     X = the matrix of gross outputs

Solving for X gives the total impact on output of a unit change in final demand:

$$X = (I - A)^{-1} \tag{6}$$

where $(I - A)^{-1}$ is the matrix of interdependency coefficients.

Waste-Discharge Coefficients

As part of the metabolism of industry, waste effluents containing a great variety of compounds are produced. Just as one may relate the production of CO_2 in animal respiration to temperature or other variables, coefficients may be developed that relate the production of waste by industry to total output or to other aspects of economic activity. A few sets of such environmental or waste-discharge coefficients have been determined that reflect the generation of specific wastes per dollar of output for a variety of economic sectors (Isard and Romanoff, 1967; Cumberland and Stram, 1972). Unfortunately, coefficients derived in other study areas may often be of limited value. The great numbers of individual industries, each operating with differing technology even when producing the same product, make it necessary to develop coefficients for types of industries according to the Standard Industrial Classifications (SIC). Since this system is based on similarities of product or raw materials rather than on analysis of effluents produced, important differences in effluent characteristics may remain within sectors even when a local industry is placed in one of the detailed SIC groups. This same problem appears in slightly different form when dealing with data on revenues. Since cost and profit information for individual industries is almost always confidential, statistical data is only made available by the government according to SIC categories. After examinations of the available coefficients, it was felt that only six of the 59 bay model manufacturing sectors agreed well enough to be accommodated by literature values.

     For the 20 largest and most important sectors a relatively new source of data was explored—the reported analysis of effluents dumped into navigable waters as required by the U.S. Army Corps of Engineers under the Federal Refuse Act. Firms submitting permit applications were placed in the appropriate SIC category, and the analysis of their waste was extrapolated to other industries in that group in proportion to their output and employment. Coefficients for 15 to 20 waste constituents could be assigned for each sector in this manner. The method was limited, of course, by the number of firms in each SIC reporting actual effluent analysis and by the degree of aggregation within the SIC. Fortunately, data for the largest and most important firms in the Narragansett Bay area had been provided.

Of the remaining 33 sectors, only 2 were manufacturing. For the 31 non-manufacturing sectors, human waste of the employees constituted most, if not all, of the direct effluent generated. Accordingly, analysis of per-capita-per-day human wastes and employment statistics were used to calculate their environmental coefficients. No values were entered for the two remaining manufacturing sectors, but they are not significant in the bay-area economy.

Natural-Resource-Use Coefficients

The development of the third area of this submodel is conceptually similar to the waste-discharge matrix. In fact, the two could be related by simply assigning negative values to the effluent loadings. At this time, only the coefficients for human water use have been derived, but one could expand the matrix to include fish, lobsters, clams, thermal differences, or any other component of the bay system.

Operation of the Model—Results and Implications

When the matrix of input-output interdependency coefficients is multiplied by the table of environmental coefficients and resource use coefficients, a matrix of direct and indirect waste effluent generation and natural resource demand is produced. For a model as large and detailed as this one (59 sectors and 35 environmental coefficients), the extensive calculations require the digital computer. The sum of the direct effluent loadings by each of the sectors represents the total industrial pollution load for the economy. A more interesting result, however, is that the subtle effect of economic interactions in determining the ultimate effluent loading that results from *changing* activity *in any one* sector can be very large (Table 1). The environmental impact of any industry may be as much a consequence of the composition and interdependency of the area's economy as it is of the industry's own technology. However, even those sectors with high direct plus indirect loadings per unit of output may play a small role in the total input of effluents to the bay because of a correspondingly small role in the area's economy. The model takes this into account and provides an economic sensitivity analysis that makes it possible to identify the largest sources of input for particular residuals in the effluents entering the bay. The five economic sectors responsible for the largest percentage of the total direct and indirect emissions of ten common residuals are shown in Table 2. For all 35 parameters examined, the five most important sectors always accounted for 50 to 100% of the total loading associated with an expansion in final demand. In 13 cases, one sector alone accounted for 50% or more. The implication for management is that highly discriminatory measures aimed at particular economic sectors can be very effective in reducing particular effluents in the bay.

It is also possible to use the model to relate waste loadings to the local income generated by economic activity. This trade-off may have more meaning to local communities than simply relating effluents to output. For example, the relationship between total waste and local income generation in the Narragansett Bay system is shown for 25 economic sectors in Table 1. The ultimate set of decisions for management of the bay must be based on this sort of data, and that in Table 2, as well as on the results of hydrodynamic and ecological studies that can anticipate the fate, distribution, and environmental impact of each residual after it enters the bay.

Submodel II: The Hydrodynamic Model

In an estuary like Narragansett Bay, the interaction of the local tides, river flows, winds, and geography make it difficult to predict the effects of storms, the fate of pollutants, or the dynamics of ecological systems without a hydrodynamic model that simulates currents, flow rates, and flushing times throughout the system. A numerical tidal model has been developed for Narragansett Bay by Hess and White (1974) through an extension and adaptation of the basic two-dimensional long-wave propagation models of Leendertse (1967). Since most of the bay is not strongly stratified, the present model is averaged in the vertical dimension using equations whose mathematical development has been summarized by Pritchard (1971). With the recent aquisition of an IBM 370/155 computer, the development of a three-dimensional model is now underway.

Development of the Model

A grid system was selected to fit the complicated geometry of the bay and to give as detailed a picture of the hydrodynamics as practical. The result was a 19 × 48 field with 324 operational elements each 0.5 nautical miles

Table 1 Comparison of Model Outputs of Water-Borne Emission Levels with Economic Activity of Selected Manufacturing and Nonmanufacturing Sectors.[a]

Sector	Total Residuals Pounds Generated Per $1000 of Output[b]			Total Residuals Pounds Generated Per $1000 of Local Income[b]
	Direct Emissions	Direct and Indirect Emissions	Increase (Indirect Emissions)	Direct and Indirect Emissions
Primary nonferrous metal manufacturing	6	120	114	392
Jewelry manufacturing	1	132	131	267
Rubber and miscellaneous plastics	11	163	152	363
Printing and publishing	3	189	186	285
Electronic components	11	196	185	348
Metal-working machinery and equipment	3	215	212	297
General industrial equipment	3	231	228	408
Construction and mining	8	239	231	317
Equipment and machines	3	278	275	448
Motor vehicles and equipment	4	304	300	647

Residential households	101	335	234	–
United States Navy	12	346	334	270
Leather	4	377	373	870
Glass, stone, and clay products	124	406	282	641
Plastics and synthetics	485	587	102	1517
Food and kindred products	538	673	135	1626
Petroleum refining and related industry	881	1088	207	3150
Miscellaneous fabricated textile products	957	1161	204	2751
Miscellaneous textile goods	1188	1361	173	2706
Apparel	1767	1936	169	4288
Hotels and personal services	2048	2344	296	4030
Lumber and wood products	2345	2934	589	4858
Primary iron and steel	4012	4214	202	7525
Chemicals, drugs, and paints	8544	8740	196	22107
Paper and allied products	9618	9748	130	22347

[a] Modified from Feld and Rorholm (1973). 1 lb = 0.45 kg
[b] 1967 base year data.

639

Table 2 Direct and Indirect Emissions Attributed by the Input-Output Model to the Five Most Prominent Sectors by Type of Emission.[a]

B.O.D.$_5$

Lumber and wood products	39.26%
Paper and allied products	5.81%
Food and kindred products	5.45%
Chemicals, drugs, etc.	4.26%
Miscellaneous textile goods	2.90%

Total solids

Paper and allied products	19.02%
Primary iron and steel	12.87%
Chemicals, drugs, etc.	7.62%
Hotels and personal services	7.14%
Electricity, gas, water, and sanitation	5.94%

Organic nitrogen

Apparel	40.66%
Miscellaneous fabricated textile goods	40.31%
Chemicals, drugs, etc.	14.33%
Other miscellaneous manufactured	1.48%
Broad and narrow fabrics	0.53%

C.O.D.$_5$

Lumber and wood products	53.01%
Primary iron and steel	11.88%
Miscellaneous textile goods	5.77%
Chemicals, drugs, etc.	5.29%
Apparel	3.69%

Organic carbon

Chemicals, drugs, etc.	36.43%
Apparel	26.50%
Miscellaneous fabricated textile goods	26.33%
Broad and narrow fabrics	1.35%
Electrical lighting and wiring equipment	1.23%

Ammonia

Glass, stone, and clay	40.15%
Chemicals, drugs, etc.	32.57%
Primary iron and steel	5.67%
Miscellaneous fabricated textile goods	4.39%
Apparel	4.35%

Nitrate

Primary iron and steel	70.48%
Stamp, screw, and bolt production	5.84%
Chemicals, drugs, etc.	2.68%
Electrical lighting and wiring equipment	2.52%
Other miscellaneous manufacturing	1.83%

Copper

Chemicals, drugs, etc.	60.45%
Other miscellaneous manufacturing	16.97%
Primary iron and steel	8.40%
Broad and narrow fabrics	2.22%
Office supplies	1.59%

Phosphorus

Apparel	31.14%
Miscellaneous fabricated textile goods	30.90%
Chemicals, drugs, etc.	13.34%
Hotels and personal services	6.09%
Primary iron and steel	2.42%

Lead

Glass, stone, and clay	91.44%
Chemicals, drugs, etc.	2.31%
Maintenance and repair	1.48%
Construction and mining	0.71%
State and local government enterprises	0.34%

[a] From Feld and Rorholm (1973).

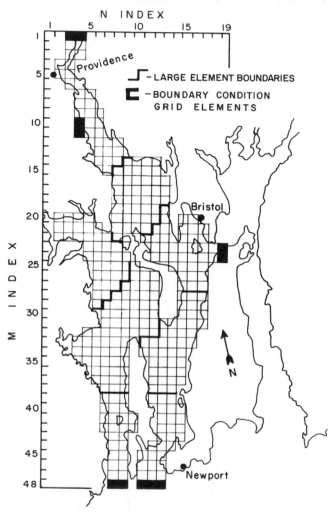

Figure 6 Grid pattern used in the hydrodynamic model and the boundries of the eight large elements used as ecological subsystems.

($=0.926$ km) on a side (Figure 6). Basic Navier-Stokes momentum equations plus a conservation of mass equation were applied to each of the vertical cells defined by the grid system, using a right-handed coordinate system with the Z-axis directed upward. After simplification and the addition of the Chezy relationship for bottom friction and a quadratic approximation for wind stress (Hess and White, 1974), the final differential equations for the vertically averaged velocities, U, and V, are

$$\frac{\partial U}{\partial t} + U\frac{\partial U}{\partial x} + V\frac{\partial U}{\partial y}$$

$$= -g\frac{\partial \eta}{\partial x} + fV + k\frac{\rho_a}{\rho}\frac{W_x|W_x|}{H} - g\frac{U(U^2 + V^2)^{1/2}}{C^2H} \quad (7)$$

$$\frac{\partial V}{\partial t} + U\frac{\partial V}{\partial x} + V\frac{\partial V}{\partial y}$$

$$= -g\frac{\partial \eta}{\partial y} - fU + k\frac{\rho_a}{\rho}\frac{W_y|W_y|}{H} - g\frac{V(U^2 + V^2)^{1/2}}{C^2H} \quad (8)$$

$$\frac{\partial n}{\partial t} + \frac{\partial}{\partial x}(HU) + \frac{\partial}{\partial y}(HV) = 0 \quad (9)$$

where h = depth at mean sea level
  η = water level above mean sea level
  f = coriolis parameter, $2\Omega \sin \phi$
  k = dimensionless wind drag coefficient
  g = gravitational acceleration
  W = wind speed components
  ρ = water density
  ρ_a = density of air
  C = Chezy coefficient for bottom friction, derived from sediment types in the bay

The equations are solved for each cell square using the "semiimplicit" approach of Leendertse (1967) in which variables U and V are staggered in both space and time. Using this scheme, the momentum and mass equations are transformed to finite difference equations and then solved by a "multioperation" or "leap-frog" method. A very helpful consequence of the implicit solution is that the model is numerically stable regardless of the time step used. The time step is, however, an important factor in the accuracy of the solution. Using the high-accuracy criteria developed by Leendertse (1967), the grid size and depths of the Narragansett Bay model require a time step of 4 min or less. Since the model is vertically averaged, only the gross effect of wind drag on total transport is calculated. The effect of wind driven circulation in the surface layers is not included.

    Driving forces at the bay boundaries consist of the astronomical tide at the mouth, discharge from the two largest rivers, and the tidal velocity at the mouth of Mt. Hope Bay. The tidal forcing function is represented by

$$\eta(t) = \sum_n f_n(t)H_n \cos\left[\omega_n t + (V_0 + u)_n - K_n\right] \tag{10}$$

where t = time from reference time (hours)

        $\eta(t)$ = water level above mean sea level

and for each tidal constituent, n,

    $f(t)$ = amplitude factor depending on the position of the moon's line of nodes

     H = amplitude of the tidal constituent

     ω = angular speed

$V_0 + u$ = value of the equilibrium argument at $t = 0$

     K = angular phase difference from Greenwich

The values of H_0, H_n, and k_n are available from the Coast and Geodetic Survey for Newport, Bristol, and Providence, R.I. The astronomical motions are used to calculate W, f_n, and $(V_0 + u)$. The river discharge is taken simply from gauge data or may be related statistically to rainfall. Tidal flows at the Mt. Hope bridge are approximated by an empirical equation fit by least-squares regression to observed flow rates:

$$q = \sum_{k=1}^{3} q_k \cos\left[\frac{2\pi k}{12.42}(t - T_k)\right] \tag{11}$$

where q = flow rate

     T = time to first flood after high water

The entire model has been programmed in G level FORTRAN IV and was originally run on the IBM 360/50 digital computer. A new IBM 370/155 as well as some software modification has reduced the computer-time to real-time ratio in simulations from about 1:30 to a more practical value of 1:200.

Verification and Application

The hydrodynamic model has been verified with field observations of tide heights at Newport where data is available (Figure 7). Current velocity observations from near-surface and near-bottom meters and from drifting poles were also available for comparison with model outputs in the lower East and West Passages (Sturges and Weisberg, 1973; Levine, 1972). The computed velocity compares favorably with that observed, especially in the West Passage (Figure 8). The model has been applied to the historical simulation of hurricanes in the bay, where damage was so great in the storms of 1938, 1944, and 1954 that the Army Corps of Engineers proposed

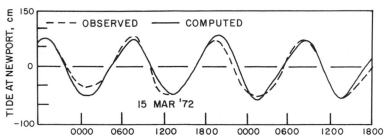

Figure 7 Comparison of observed and simulated tide heights at Newport (from Hess and White, 1974).

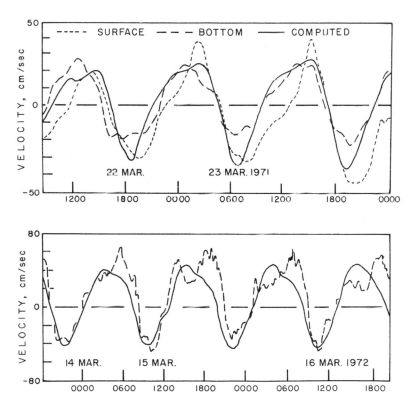

Figure 8 Comparison of observed and predicted current velocity in the West (top) and East (bottom) Passages (from Hess and White, 1974).

construction of an extensive series of barriers across the bay (Hicks et al., 1956). Using the approach described by Bodine (1971) for analyzing storm surge on the coast, the winds and tidal inputs for the great hurricane of September 1944 were modeled to predict the resulting water levels in Narragansett Bay. The storm had a radius of some 200 nautical miles, traveled at an average speed of 30 knots, and had maximum winds of 80 knots. It moved northeast up the Atlantic Coast and passed directly over the bay. Comparison of the results of the simulated and real storms in affecting surge at Providence indicates the value of the model in predicting future storm impact (Figure 9).

The detailed hydrodynamic model enabled the development of temperature (Alfano, 1973) and salinity models (Hess, personal communication) for the bay (Figure 5). It has also been modified to simulate the movement of oil slicks (Premack and Brown, 1973) and serves as the basis for a multi-layered water-quality model of dissolved oxygen, BOD, and coliform bacteria distribution (Spaulding, 1972). The temperature model has been used to predict near- and far-field effects of thermal effluents from a proposed nuclear power plant on the West Passage (Alfano, 1973).

Unfortunately, the short time step and long running time required in the hydrodynamic model make it difficult to use directly in the ecological model, where long real-time simulations are needed and all of the rate constants are determined for time intervals on the order of days instead of minutes. A working solution to this problem has been found by dividing the detailed bay grid into eight large elements or ecological subsystems (Figure 6). The advective transport of particles and dissolved materials among the large

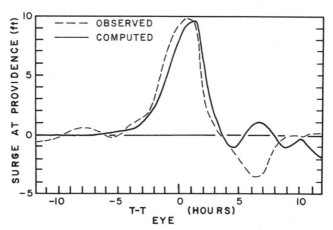

Figure 9   Comparison of observed and calculated water surge at Providence during an historical simulation of the Great Atlantic Hurricane of 1944 (from Hess and White, 1974).

elements has been estimated using the detailed hydrodynamic-salinity model. All of the fine grids in one of the large elements at a time were assigned an initial tracer concentration of 1.0 arbitrary units. All other grids were initialized at a concentration of zero with no further input to the system. At the end of a two-tidal-cycle simulation, the net daily transport to each element was calculated as a percentage of the test element. The calculations were made for each of the elements for five different tidal amplitudes, and regression analyses indicated that the relationships between exchange coefficients and tide at Newport were linear. The final model is based on a set of linear equations that calculate net daily transport between any two elements as a function of tabular tidal amplitude for Newport. This scheme rapidly predicts tidal disperson estimates for any conservative substance throughout the eight large elements of the bay model. One-year simulation requires only 15 sec of computer time. The model adequately reproduces observed salinity patterns given boundary conditions, and simulated dye studies approach the results of an earlier Corps of Engineers' hydraulic model of the bay. Such a model is proving useful not only for mixing of biological and chemical species in the context of the ecological model, but also as an independent unit for indications of heavy metal, hydrocarbon, and particulate dispersion connected with other studies at the university.

Submodel III: The Ecological Model*

The goal of the ecological model is to depict both the spatial variations throughout the bay and the temporal patterns of change during the year for the biological compartments. To accomplish this, the total ecological system diagrammed in Figure 10 functions simultaneously within each of eight large spatial elements into which the bay has been divided (Figure 6). Characteristic parameters such as depth and extinction coefficient as well as

---

* The equations in this section use a symbolic notation designed to have intuitive mnemonic meaning. In most cases, transition to the conventions suggested by H. T. Odum (1975) for energetic equations is possible. Biomass or energy storages are denoted by Qs, transfers of energy or fluxes by Js, and constants by ks, each with appropriate identifying subscripts. Thus the equations for juvenile zooplankton (Eq. 28) could be rewritten:

$$Q_{zjuv} - J_{gzjuv} - J_{jmat} - J_{gzadult} - J_{gcarn} + J_{qzjuv}$$

where $J_{gzjuv} = k_{FOODLIM} \, k_{gmax} \, Q_{zjuv}$, the growth of juveniles

$J_{gcarn} = k_{gcarn} \, Q_{zjuv} \, Q_{carn}$, the grazing loss to carnivores, and so forth

In our model the ks are constant only within the time step of computation and may vary, for example, as a function of temperature. These conversions have not been made here in order to retain a notation consistant with earlier authors or publications and with the coding of the computer model.

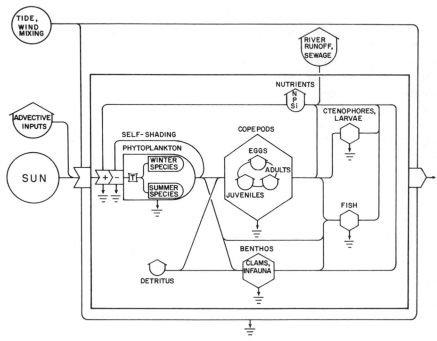

Figure 10 Flow diagram representing the ecological model within each of the eight large elements shown in Figure 6.

initial conditions of biomass and nutrients reflect the spatial separation of the elements. All external forces operating on the system vary on a daily basis. However, the distribution of daily incident radiation over the bay is uniform, while temperature, tidal mixing, and sewage input differ appropriately among the eight elements. The biotic compartments reflect the planktonic nature of the dominant parts of the bay system. The phytoplankton compartment incorporates a seasonal shift in population dominance from cold-water, primarily diatom, to warm-water, flagellate species groups. Choices of growth parameters are made so as to reflect the respective advantage of these characteristic populations. Herbivores consist of smaller zooplankton, primarily copepods, along with the clam component of the benthos. Additional food sources are also available in a density dependent manner to the copepods. Carnivorous organisms include seasonal pulses of meroplankton, primarily larval fish, and ctenophores. Aside from the menhaden, pelagic and demersal fish are not presently included, although future modifications may add the predation of flounder on the infauna. All compartments and the flows between them are expressed in terms of carbon, and conservation of mass is observed in all processes. Since nutrients are essential influences on phytoplankton growth, balance and conservation of these compounds is also maintained throughout the model.

Treatment of the processes within the biological compartments may be divided into two types: those with deterministic formulations concerned with mechanistic detail within the compartment and those treated as forces operating on the system but not dealt with in substantial detail. Deterministic treatments represent statistical formulations based on theoretical considerations. Forcing processes represent important influences not well enough understood or considered outside the scope of this model for detailed treatment. For example, water temperature, while modeled dynamically in the hydrodynamic-thermal model, is treated here as a force programmed as input. A simple sinusoidal function has been determined to describe the seasonal trend in water temperature of one element in Narragansett Bay (after Busser, 1967):

$$TEMP = 10.75 - 10.25 \cos\left(\frac{2\pi(DAY - 30)}{365}\right) \tag{12}$$

where DAY = Julian calander date (Jan. 1 = 001)

Based on field observations, appropriate correction factors are applied to these values to obtain a realistic temperature gradient in the other elements.

Pertinent biological processes are formulated as rate functions on a daily basis. Frequently, the approach is to specify a maximum rate under optimum conditions and then to determine the extent to which existing conditions are suboptimal. The optimum rates may then be multiplied by the unitless limitation factors, resulting in a series of differentials that are integrated each day as finite differences. A flow diagram summarizing the equations and functional relationships of the submodel is given in Figure 11.

The final model is a combination of statistically supported formulations based on experimental data and current theory. An appealing consequence is that the physiological and experimental data used to develop the model is of a completely different form than the time-series field data used to test or verify it.

The program for the ecological submodel was developed in modular form, with separate subroutines for the major biological compartments. Determination of forcing functions, physical mixing of the conservative properties, and proper sequencing of the program modules is controlled by a main program. This main module also handles input of program control parameters and initial conditions as well as final output of time-series data in plotted format.

PHYTOPLANKTON COMPARTMENT

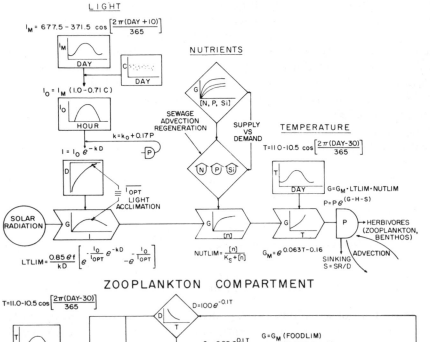

ZOOPLANKTON COMPARTMENT

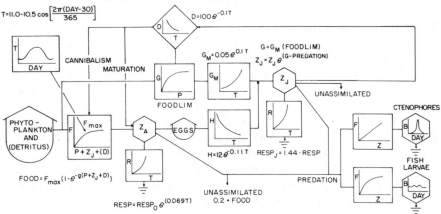

Figure 11 Mathematical and symbolic representation of the flows and controls operating on the phytoplankton and zooplankton compartments in each large spatial element of the ecological submodel. Details in text (with minor modifications from Kremer and Nixon, 1975).

The Phytoplankton Compartment

The keystone of the phytoplankton is the productive rate. A maximum growth rate as a function of temperature was chosen as theoretically reasonable and appears to be supported by experimental data (Eppley, 1972). Nonoptimum illumination is a major factor resulting in lowered productivity of the phytoplankton. Steele (1965) presented a formulation relating photosynthesis of phytoplankton at a given depth to light intensity (Figure 12). This response has been substantiated and modified by numerous workers (Ryther, 1956; Jitts et al., 1964; Vollenweider, 1965; Fee, 1969) and demonstrates the tendency for productivity to increase approximately in proportion to light at low intensities, approach a maximum rate at an optimum intensity, and decrease again at high light levels. Of course, for the total water column the integrated plankton production does not show light inhibition. Our formulation includes various factors that effect the light intensity to which the phytoplankton are exposed, and the photosynthetic response of phytoplankton is mediated through this relationship. A weakness of earlier uses of this type of formulation is the arbitrary choice of I_{opt}, which has been experimentally demonstrated to be highly variable within and between species (McAllister et al., 1964; Smayda, 1969; Ignatiades and Smayda, 1970). Further, acclimation of populations to varying light intensities is well documented and is potentially an important consideration (Steemann Nielsen et al., 1962; Yentsch and Lee, 1966).

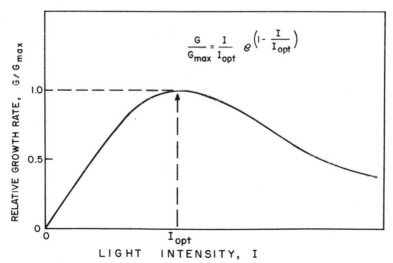

$$\frac{G}{G_{max}} = \frac{I}{I_{opt}} \, e^{\left(1 - \frac{I}{I_{opt}}\right)}$$

Figure 12 Hypothetical growth response of phytoplankton to light intensity. Equation according to Steele (1965).

In view of light acclimation, it was desirable to introduce some variation in the daily light levels. Based on sun-altitude data for this latitude obtained from Navy hydrographic tables, theoretical calculations were made for the maximum clear-sky, total daily incident radiation. A cosine function was then developed that agreed with the solstices both in period and amplitude:

$$\mathrm{RADN_{max}} = 677.5 - 371.5 \cos\left(\frac{2\pi(\mathrm{DAY} + 10)}{365}\right) \tag{13}$$

where $\mathrm{RADN_{max}}$ = clear-sky radiation (ly day^{-1})

To simulate variable cloud cover, monthly means and standard deviations were determined for 12 years of U.S. Weather Bureau data for the Narragansett Bay area. A stochastic cloudiness factor produced daily by a random number generator conforming to the monthly distributions is then combined with the clear-sky maximum according to the equation (Sverdrup et al., 1942):

$$\mathrm{RADN} = \mathrm{RADN_{max}}(1.0 - 0.71C) \tag{14}$$

where C = cloud cover in tenths
$\quad$ RADN = simulated light incident at sea level (ly day^{-1})

The result is a pattern of incident radiation, variable on a day-to-day basis within the model, that agrees closely with average seasonal trends on the bay. This formulation had the added facility of allowing easy simulation of atypical years or seasons by simply altering the cloud cover statistics for one or more months.

Returning to the biological aspects, Steemann Nielsen et al. (1962) reported that acclimation to light intensity by *Chlorella* cultures under constant illumination occurs within 2 or 3 days, with the most rapid change during the first day. Our experimental data suggests that fully acclimated natural bay populations frequently demonstrate maximum production at a depth of one meter. Based on this, we modeled the I_{opt} parameter as a moving, weighted average of the previous 3-day mean light intensity at the 1-m depth. In other words, this model includes the effects of the adaptation of organisms. Let

$$I' = I_0 e^{-kz} \tag{15}$$

where I_0 = 0.9 RADN, assuming 10% average albedo
$\quad$ k = extinction coefficient (m^{-1})
$\quad$ I' = light at depth Z (ly day^{-1})

then

$$I_{\mathrm{opt}} = 0.7\bar{I}'_1 + 0.2\bar{I}'_2 + 0.1\bar{I}'_3 \tag{16}$$

where $\bar{I}'_j$ = average light level at 1 m, j days earlier

The extinction coefficient is affected by the phytoplankton present in the water column. The self-shading effect of an increase in phytoplankton on k has been expressed by Chen (1970):

$$k = k_0 + 0.17P \qquad (17)$$

where k_0 = the coefficient with no phytoplankton (m^{-1})
$\quad\;\; P$ = phytoplankton concentration (mg dry wt liter^{-1})

This equation agrees with the more complex one of Riley (1956)—at least for the levels of phytoplankton in Narragansett Bay.

The remainder of the light-productivity formulation concerns the integration of the photosynthesis-light equation over time of day and throughout the water column. Di Toro et al. (1971) have developed an equation for the exact double integral of Steel's equation, taking into account the extinction coefficient of the water, k, the photoperiod length as a fraction of the day, f, and the depth of the water column (assuming complete mixing), z. Their formulation, however, assumed a square-wave input during the photoperiod equal to the daily average. Our analysis of this assumption suggest that, over a wide range of photoperiods, radiation levels, and I_{opt} values, their formulation overestimates production by about 15% when compared to a sinusoidal daily light pattern with the same total energy input. With this correction, the expression for the limitation of the production of the phytoplankton due to nonoptimum light, integrated over the water column throughout the day, is

$$\text{LTLIM} = \frac{0.85ef}{kz}\left[\exp\left(-\frac{I_0}{I_{opt}}(e^{-kz})\right) - \exp\left(-\frac{I_0}{I_{opt}}\right)\right] \qquad (18)$$

The second potential factor operating to reduce productivity is nutrient limitation. In our model we consider phosphate, silicate, and two pools of nitrogen (ammonia, and nitrite plus nitrate). Enzyme kinetics have long been described by Michaelis-Menton or Monod formulations (Figure 13). While subject to some criticism, this theory may be extended to represent nutrient limitation on the growth of phytoplankton (MacIsaac and Dugdale, 1969; Eppley et al., 1971; Fuhs et al., 1972; Paasche, 1973; and others). Based on the choice of the half-saturation constant, K_s, for growth for each nutrient and each species, the growth rate possible with a given nutrient concentration is defined relative to the maximum rate with abundant nutrients:

$$\text{NUTLIM} = \frac{G}{G_{max}} = \frac{[n]}{K_{s_j} + [n]} \qquad (19)$$

where NUTLIM = the fraction of the maximum growth possible with the nutrient concentration, $[n]$

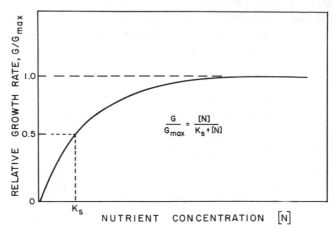

Figure 13 Hypothetical growth response of phytoplankton to nutrient concentration.

It has been suggested that the limitation factors for additional nutrients be multiplied together to yield the combined effect (Di Toro et al., 1971; Kelly, in press; see Chapter 4). This approach is subject to a number of problems. For example, since the relationship is assymtotic for each nutrient, the respective limitation terms will always be less than 1.0, even with relatively high levels of the nutrient. Thus simply including many nutrients in the model necessarily lowers the productive rate. Further, Williams (1973) has pointed out potential errors from the use of Monod equations to represent uptake or growth kinetics of mixed species, though this is less of a problem in the low-diversity systems that frequently dominate the bay. Complex nutrient interaction schemes, such as the multiplicative one mentioned above or the ones suggested recently by Droop (1973) or Grenney et al. (1973) may be introduced at a later date, allowing more complete analysis of the problem. Presently, however, the model determines which nutrient is in shortest supply and uses the appropriate K_s to calculate NUTLIM. Biomass-related uptake limitation is also considered, so that growth cannot exceed available nutrients even if the concentration related NUTLIM would allow growth. Even this simplified scheme allows complexities of interspecies competition to develop, and the programming details of following four nutrient pools are formidable.

The final expression for phytoplankton growth is a simple combination of light and nutrient limitations (Eqs. 18 and 19) acting on the potential maximum rate as a function of temperature (Figure 11):

$$GTH = (G_{max})(LTLIM)(NUTLIM) \qquad (20)$$

Other models (Riley, 1946; DiToro et al., 1971) have attempted to include respiration and sometimes the accompanying nutrient excretion by the phytoplankton. However, since experimental determinations of both growth rate and nutrient uptake represent *net* rate processes, these are already included implicitly in the formulation presented above. Moreover, the variation of reported respiration rates is so great that the choice of any one is problematic. If nutrient excretion is in a form other than that utilized by the plants (i.e., organically bound in some way), excretion in proportion to respiration or growth itself can be introduced simply, with a time lag before these nutrients return to the pool available for growth. The only factor that remains to complete the phytoplankton equations is a negative term to express losses due to sinking and to grazing by copepods, clams, and menhaden. While many laboratory measurements of sinking rate have been made (Smayda, 1970), turbulence makes its assessment in the estuary difficult. Initially a conservative value of 0.25 m day^{-1} has been used. Grazing losses are discussed in the following sections.

$$\frac{dP}{dt} = P(\text{GTH}) - H_{\text{cope}} - H_{\text{clam}} - H_{\text{menh}} - S + Q_p \qquad (21)$$

or integrated,

$$P = Pe^{(\text{GTH}-H-S)} + Q_p$$

where P = phytoplankton biomass
$\quad\ H$ = grazing due to herbivorous copepods, clams and menhaden
$\quad\ S$ = sinking rate
$\quad Q_p$ = advective change

The Copepods

The smaller zooplankton are the most conspicuous herbivores in the system. Increasing evidence suggests that these copepods are general omnivores consuming smaller zooplankters, including juveniles and eggs of their own species, particulate organic material, and phytoplankton (Petipa, 1966; Zillioux, 1970; Heinle, personal communication).

Under conditions of abundant food, the maximum daily ration obtainable by the zooplankton is a function of temperature. Petipa (1966) has presented data for observed ration with temperature of *Acartia clausi* from which the temperature coefficient for feeding can be determined (Figure 14). The actual level of the zero degree maximum ration, R_0, as a fraction of body weight is somewhat uncertain, perhaps as much as 10 to 20% at 0°C. The equation for maximum or preferred ration is then

$$R_{\text{max}} = R_0 e^{0.12 \text{ TEMP}} \qquad (22)$$

The actual ration of copepods follows a saturation pattern with respect to available food supply (Figure 15; Parsons et al., 1967; Conover, 1968; Sushchenya, 1970), as suggested by the rectangular hyperbola of Ivlev (1945) for fish. For the adults, the available food includes carbon sources other than

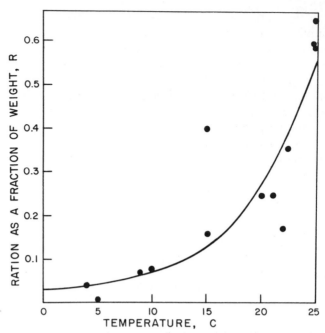

Figure 14 Daily ration of adult copepods as a function of temperature. Data from Petipa (1966) with least-squares fit $R = 0.024\,e^{0.12\text{temp.}}$.

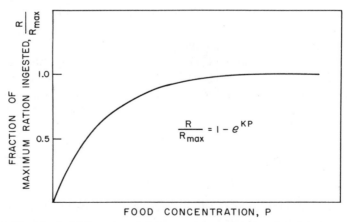

$$\frac{R}{R_{max}} = 1 - e^{KP}$$

Figure 15 Hypothetical daily ration of adult copepods as a function of available food. Equation according to Ivlev (1945).

656

phytoplankton, as mentioned above. Some workers report a feeding threshold, which may be incorporated into the equation as follows (McAllister, 1970):

$$\text{FOODLIM} = \frac{R}{R_{\max}} = 1.0 - e^{-K(P - P_0)} \tag{23}$$

where $R_{\max}$ = the preferred ration
$\quad\quad R$ = the actual ration achieved
$\quad\quad P$ = the available food concentration
$\quad\quad P_0$ = the feeding threshold

This equation represents the limitation imposed by the available food supply upon the ability of the zooplankton to achieve the maximum ration of Eq. 22. In the expression, K is a parameter determined experimentally, controlling the degree of curvature of the hyperbola. The model is sensitive to the value of K, which unfortunately shows wide variation in the literature for a variety of reasons (Parsons et al., 1967; McAllister, 1970; Suschenya, 1970; Zillioux, 1970; Frost, 1973; Petipa, personal communication). Nevertheless, the combination represents the ration of food ingested by the adults depending upon temperature and the food available.

$$\text{RTN} = R_{\max}(\text{FOODLIM}) \tag{24}$$

Assuming an assimilation efficiency of 80% (Marshall and Orr, 1955; Petipa, 1966; Conover, 1968; Mullin, 1969), this fraction of the ingestion is available for metabolic processes, the remainder returning to the system for regeneration of nutrients. Respiration decreases the assimilated ration, and an exponential equation with a Q_{10} of 2 was chosen that agrees satisfactorily with the diverse data reported in the literature (Riley, 1947; Gauld and Raymount, 1953; Anraku, 1964).

$$\text{RESP} = \text{RESP}_0 e^{0.069 \text{ TEMP}} \tag{25}$$

Any excess unrespired assimilation enters the pool of incubating eggs. The time required for eggs to hatch has been determined for a number of estuarine species (Figure 16), and an exponential equation fits this data well.

$$H = 12.0 e^{-0.110 \text{ TEMP}} \tag{26}$$

where H = hatching time (days).

After this time lag, the appropriate amount of carbon enters the juvenile section of the compartment, where growth begins. Maturation to adulthood involves a daily growth rate operating for a specified development time, both of which vary with temperature. Data for some estuarine species is available for these factors (Greeze and Baldina, 1967; Heinle, 1969) (Figure 16), and equations for the relationships have been determined:

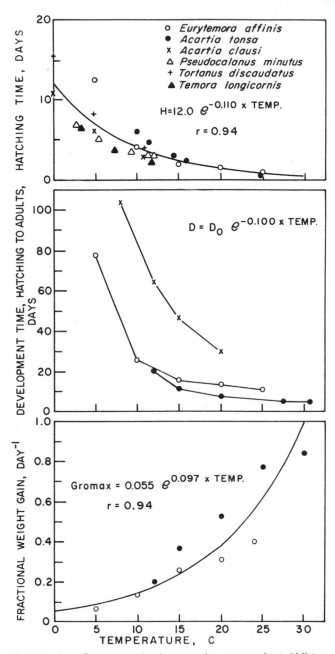

Figure 16 Hatching time of eggs (top), development time to maturity (middle), and maximum daily growth rate of juveniles (bottom) for representative estuarine copepods. Equation fit by least-squares regressions (data from McLaren et al., 1969; Heinle, 1966).

$$D = 100e^{-0.10 \text{ TEMP}}$$
$$G_{max} = 0.05e^{0.10 \text{ TEMP}}$$

(27)

where D = development time from hatching to maturity (days).
G_{max} = maximum daily growth rate (day^{-1})

While the exponential pattern is strongly supported, the actual choice of the equation parameters is arbitrary within limits imposed by the data. Although the compartment grows as a homogeneous carbon pool in the model, choice of the parameters was facilitated by an extensive analysis of the pattern of growth of an individual from egg to adult. This analysis also revealed a simple estimate of the effect of constantly changing temperature on the predicted total development time. Little error was introduced when the projected development time was taken as the average of the initial estimate and the estimate based on the temperature at that later time (Eq. 27). Thus if on a given day D = 50 days, and 50 days hence D = 20 based on the regular temperature function (Eq. 12), the estimated actual development time would be 35 days. Continuing this example, the pool of juveniles would be increased by the daily growth increment throughout this 35-day period at the end of which the appropriate mass would be transferred to the adult section of the compartment.

As with the adults, a food-limitation effect, based in this case only on the abundance of phytoplankton, may reduce the daily growth from the temperature dependent maximum potential rate. The net effect of this decreased daily growth factor is simulated by slowing the development to adulthood in proportion to the degree of food limitation. Since this correction is not predictable in the context of the model, as was the effect of changing temperature on the development time, a small error is unavoidably introduced in the accuracy of the actual development time estimate.

The model also includes terms for the feedback between the zooplankton, the phytoplankton, and the nutrients. Herbivorous grazing by juveniles and adults decreases the standing stock of phytoplankton. Similarly, adults consume an appropriate proportion of the juvenile pools as well. Except for the time-lag considerations, then, the changes in the zooplankton compartment are described:

$$\frac{dZ_{adult}}{dt} = J_{mat} - g_{carn}[-\text{RESP}] + Q_Z$$

(28)

$$\frac{d\,\text{Egg}}{dt} = [+\text{URA}] - \text{Egg}_{hatch} - g_{adult} - g_{carn} + Q_{Egg}$$

$$\frac{dZ_{juv}}{dt} = Z_{juv}(G_{mat})(\text{FOODLIM}_{juv}) - J_{mat} - g_{adult} - g_{carn} + Q_{Zjuv}$$

where Z, Egg = biomass of adult, egg, or juvenile compartments

$\quad\quad\quad\quad J_{\text{mat}}$ = portion of juveniles maturing

$\quad\quad\quad\quad\quad\quad g$ = grazing attributed to adults or other carnivores

$\quad\quad\quad$ RESP = respiration of adults (juvenile respiration is implicit in G_{max})

$\quad\quad\quad$ URA = unrespired assimilation, if any

$\quad\quad\quad\quad\quad\quad Q$ = advective influence on each compartment

$\quad\quad\quad$ [ ] = represent a contribution depending on the balance of assimilation and respiration

Carnivorous Zooplankton

A variety of species exert predation pressure on the smaller zooplankton in the bay. The most important members of the compartment and those considered in the model are larval fish, the ctenophore, *Mnemiopsis leyidi*, which are present in a few large peaks of abundance during the summer and menhaden. The three populations are input to the model as arbitrary functions that approximate field observations (Matthiessen, 1973; P. Kremer and Nixon, 1976; Durbin, 1976). Empirical values of excretion rates for each nutrient are applied to the biomass estimates to include the carnivores in nutrient recycling. Laboratory data of feeding of the ctenophores (P. Kremer, 1975) suggest that they are passive filterers, so that a constant volume-swept-clear for each unit of biomass is reasonable. Larval fish are more complex.

The literature on larval fish feeding is quite limited, although extensive aquaculture research with juvenile fish is useful. An instantaneous feeding rate based on a maximum daily ration of 25% of the body weight per day is assumed for an average size larvae (Laurence, National Marine Fisheries Service, personal communication; Sorokin and Panov, 1965). This zero-degree rate is adjusted for temperature using a $Q_{10} = 2$, which is supported by the fish-culture guidelines of Ghittino (1972). A hyperbolic relationship between food abundance and ration consumed such as used in the zooplankton feeding formulation (Eq. 23; Figure 15) is also employed here, according to the original suggestion of Ivlev (1945). Evidence supporting this formulation and aiding in the choice of the crucial exponent for the equation for larval fish has been presented by Laurence for fresh and salt water species (Laurence, personal communication). Finally, as the larvae are primarily if not exclusively visual feeders the temperature-dependent maximum rate corrected for abundance limitation is further reduced by the photoperiod. Thus

$$FR_{\text{max}} = FR_0 e^{0.069 \text{ Temp}}$$
$$FSHRTN = FR_{\text{max}} f (1 - e^{-K_F Z_{\text{TOT}}})$$

$$(29)$$

where FR_0 = maximum daily ration (mg C/mg C day^{-1})
  FR_{max} = temperature corrected maximum ration
  f = photoperiod as the daylight fraction
  K_F = hyperbolic limitation parameter, see Eq. 23 (liter/mg C)
  Z_{TOT} = total available food, zooplankton adults, juveniles, and eggs (mg C/liter)
  FSHRTN = resulting feeding rate of the larval fish (mg C larvae^{-1} day^{-1})

Fish

The necessary rate constants are not yet available to include this compartment in the working model with the exception of preliminary values for menhaden. As mentioned earlier, experiments are now underway to provide the values needed for the feeding, respiration, and excretion rates of menhaden (Durbin, 1967) and winter flounder. Again, these terms will be combined with an empirical function approximating fish biomass throughout the year in each of the eight elements, (Oviatt and Nixon, 1973).

Benthos

Narragansett Bay is marked by its large populations of the hard clam, *Mercenaria mercenaria*. Unfortunately, there is little data available on the dynamics of this or other estuarine benthic communities, and an active program to measure feeding, respiration, and nutrient regeneration rates *in situ* is now underway as part of the modeling effort. As an intermediate solution to consumption by the benthos, we have modified an earlier conceptual model of clams and the clam fishery in the bay developed by Lampe and Nixon (1970). The biomass of clams in each element is input according to population surveys by the Marine Fisheries Section of the State Department of Natural Resources (personal communication; Russell, 1972) and attributed to clams of the mean size in the bay. The grazing pressure of clams on the phytoplankton is estimated by calculating the amount of time the clams spend pumping as a response to water temperature (Loosanoff, 1939). The pumping rate of the animals as a function of size has been studied by Coughlan and Ansell (1964), Loveland and Chu (1969), and Walne (1972). The pumping rate appropriate for the mean size estimate is used with the time function from Loosanoff to calculate the daily grazing rate on the phytoplankton.

Nutrient Regeneration

Excretion of nutrients by the copepods is of two forms. Unassimilated ingestion returns to the available pool all the nutrients in 20% of the ingested material in the ratios of the food source. Also included in this are the unused nutrients resulting from different carbon to nutrient ratios of the compartments, that is, all of the silicate and a portion of N and P. The second form of excretion is that due to metabolic activity. For this, nutrients are released in proportion to the respiratory rate and in the elemental ratio of the zooplankton themselves. Phosphorus is returned directly as inorganic PO_4 (Peters and Lean, 1973). Nitrogen returns as ammonia, which is oxidized to nitrite and nitrate in a temperature-dependent reaction (Jaworski et al., 1972):

$$[NO_2 + NO_3] = [NH_3]e^{K_t(\text{time})} \qquad (30)$$

where $K_t = K_{20}(\theta^{\text{TEMP}-20})$
$\quad K_{20} = 0.068$
$\quad\quad \theta = 1.188$
$\quad$ time $= 1$ day

A similar process is followed for the excretion of carnivorous zooplankton. Nutrient regeneration from the benthic community is based on *in situ* measurements at three stations in the bay over an annual cycle (Hale, 1974; Nixon et al., 1976). At all times there is a net flux of phosphate, silicate, and ammonia from the sediments to the water that varies as an exponential function of water temperature. The fluxes of nitrate and nitrite were much less important and too erratic to be included in the model at this stage. The summary equation for nutrient conservation is then

$$\frac{d\text{Nutr}}{dt} = (\text{GTH}_{\text{Phyto}})(R_{\text{Nutr:Phyto}}) + E_Z + E_{\text{carn}} + E_{\text{bent}} + \text{Sewage} + Q_{\text{Nutr}}$$

where for each of the nutrients, Nutr,

R = ratio of carbon to Nutr in the phytoplankton
E = excretion of the zooplankton, carnivores, and benthos

Simulation and Analysis

The ecological model is stable and reproduces most seasonal and spatial patterns, and the observed magnitudes, of all the compartments around the bay, although not all fluctuations are exactly in phase (Figures 17 and 18). While this kind of agreement is rewarding and exciting, especially in light of the large amount of biological and physical detail in the model, we are more impressed with the potential of the model as an ecological research tool than as a direct management tool. We emphasize *direct* because anything that helps us to understand the dynamics of a system and to define new areas where research is needed is certainly contributing to management goals. However, since many management decisions are tied to questions of large and/or long-term perturbations of the system, it is not appropriate to

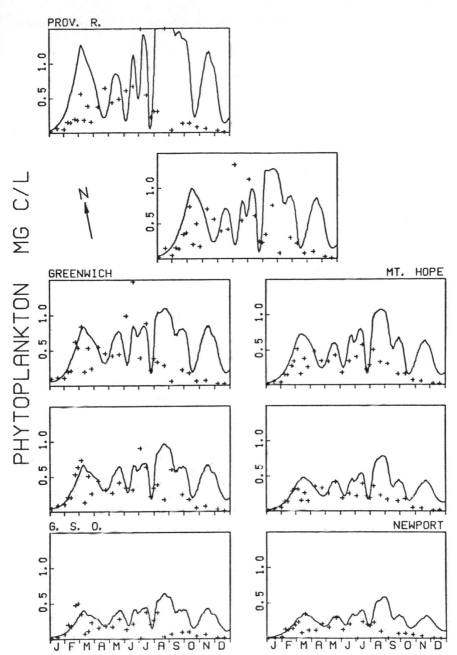

Figure 17 Simulated (−) and observed (+) patterns in Narragansett Bay phytoplankton from August 1972 to August 1973. In general the agreement between model and data are within the measured variability of the data. The arrangement of the plots reflects the geography of the bay and the eight large spatial elements the model described in Figure 6. Disagreement during the fall suggests that factors not included in the model are important at that time (see text and Figure 18). All parameters used are well within literature values, including a detrital food supplement for the zooplankton when phytoplankton are below 0.2 mg C/liter.

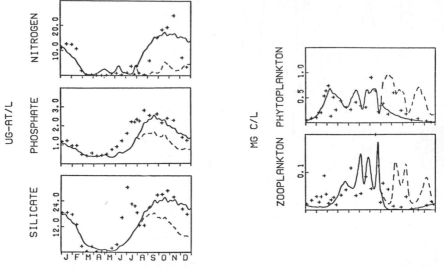

Figure 18   Effect of a simulated forced decline in phytoplankton beginning in August on all model compartments for spatial element 4 in the middle of the West Passage (—). The simulated pattern without the forced phytoplankton decline is also included (———), along with observed data (+). Agreement of all the compartments is improved markedly by changing the phytoplankton compartment. Similar results are found in all of the elements.

use a model that is constrained by being "closed" and "nonevolving" to extrapolate very far from the present into the future. The real system is coupled to a vast and infinitely varied assortment of potential compartments and is made up of parts that adapt and evolve. New forms, new physiologies, and new interrelationships give it stability in its present state as well as the potential to evolve gradually or in response to strong perturbation. The whole question of the limits and dynamics of ecosystem stability and resiliency is a largely undeveloped area of ecology that is just beginning to be studied experimentally (Holling, 1973).

These constraints become less important if the model is used to explore the sensitivity of the *present* system to parameters and processes that are specifically dealt with in the model. For example, it is a frustrating but inescapable fact that careful laboratory measurements by numerous workers have produced widely varying results for many of the biological coefficients used in the model. The values of the K and $R_{\max}$ terms in the Ivlev formulation describing zooplankton feeding each vary over an order of magnitude (Kremer and Nixon, 1975). When placed in the ecological model, extreme choices of these parameters can result in predictions of zooplankton that range from rapid extinction to values ten times greater than ever observed in the bay. The situation with respect to the K_s values for phytoplankton nutrient uptake is almost as variable (Eppley et al., 1969). However, we have found that the addition of spatial heterogeneity through the advective coupling of the eight subsystems (Figure 6) can substantially reduce the sensitivity of the model to the variation in biological coefficients. It is very encouraging that the simulations shown in Figures 17 and 18 were obtained without using extreme values for any coefficients. In fact, the choices used for all of the coefficients are well supported by the literature.

As a research tool, the model may also be most interesting and helpful when its predictions do not agree with observed data. Such disagreement may tell us that we are missing something, or that the value of a coefficient is unreasonable, or that we do not yet understand some of the processes that are operating in the system. With this in mind, some interesting exceptions to the generally good agreement between observed and predicted concentrations emerge from a closer examination of Figure 18. The model consistently predicts higher levels of phytoplankton in the late summer and fall than were observed during the year in which our field samples were taken. Going back to the field data, it became clear that during the time in which phytoplankton were showing a marked decline, the observed zooplankton biomass was low, all of the nutrients measured (NH_4, NO_2, NO_3, PO_4, Si) were high, and light and temperature, though declining, were both higher than observed during the spring bloom. In short, these data and classical theory combine to suggest that a fall bloom should occur in Narragansett Bay. And, in fact, analysis of phytoplankton data from one station in the bay collected regularly over 15 years shows that such a bloom often does occur (Smayda, personal communication). It appears that in some years, including the one in which our field program was active, some critical factor may be limiting in the bay or an inhibitor of some kind may be present. At this point, we do not have any idea what is regulating the fall bloom, and the phytoplankton ecology group here is beginning to study the problem experimentally. In the model system, it is possible to force a decline in the model phytoplankton populations that matches that observed. If this is done, the late-summer and fall simulations of zooplankton and nutrients correspond much more closely with the levels for these compartments that were observed during our sampling program (Figure 18).

Another revealing shortcoming of the model concerns zooplankton feeding. During the early stages of the spring phytoplankton bloom, the model zooplankton are unable to sustain levels that agree with the low, but consistent, observed biomass. This winter die-off propagates a delay that distorts the seasonal pattern of zooplankton throughout the simulation. No combination of reasonable feeding and respiration parameters avoids this problem. If, however, an additional source of ingestable carbon is simulated during these few winter months, the model zooplankton pattern approaches the field data. This suggests that zooplankton must be using an alternative food source for maintenance if not for growth during these low phytoplankton periods. Such a proposal is not unusual in view of recent reports of detritus consumption by zooplankton (Heinle et al., 1974);

in fact, Narragansett Bay copepods have recently been reported to have large amounts of detrital material in their guts during the winter (Gerber and Marshall, 1974). In addition, the photosynthesis/respiration data of Figure 3 indicate the importance of allochthonous inputs. While these results are only a very small sample of the output from the model, they do serve to illustrate some of the ways in which such a model can provide exciting feedback into field, laboratory, and theoretical work in ecological systems analysis.

CONCLUDING REMARKS

The three basic models described in this chapter represent but a few of the many possible approaches that may be taken in the analysis and simulation of systems. Each has its own set of assumptions and limitations. For example, the economic model is not dynamic, is not designed for optimizing, and does not have spatial detail. Many of the limitations of the hydrodynamic model stem from wind effects and have already been discussed. The biological model has problems with boundary conditions, variability in biological coefficients, and the importance of stochastic elements. In addition, there is a fundamental assumption that should be pointed out. In effect, this type of mechanistic model is a synthetic test of the reductionist philosophy that has dominated much of biological and ecological research. Can the emergent properties of a complex system be simulated in a mechanistic fashion by summing the best data that is available on its parts in a dynamic model? There are ecologists with a great deal of experience in this area (e.g., Patten, 1972; Odum, 1973; Mann, 1975) who feel that the answer is "No," and that those of us taking this approach are doomed, like "all the king's horses and all the king's men" never to put the system back together again. While we share many of their reservations, we also feel that the ecological model described here, as well as the results of a number of other modeling projects, provide evidence that this approach may be quite successful in providing a powerful tool for ecosystem research (Dugdale, 1970; DiToro et al., 1971; Jaworski et al., 1972; O'Brien and Wroblewski, 1972; Nixon and Oviatt, 1973; Walsh, 1975). In any case, there remains the challenge of developing methods and models to study biological, chemical, and physical rates and processes in a manner that H. T. Odum has termed "*in ecos,*" as part of the intact and functioning system.

ACKNOWLEDGMENT

Development of the Narragansett Bay Model has been a group effort involving close cooperation among economists, engineers, and ecologists at the University of Rhode Island. At various times during the 4-year project, this group has included Sidney Feld, Niels Rorholm, Harlan Lampe, and Thomas Grigalunas in Resource Economics; Frank White, Kurt Hess, George Brown, John Alfano, Malcom Spaulding, Tad Kowalski, and Robert Begg in Ocean Engineering; and Candace Oviatt, Fred Short, Perry Jeffries, and Nelson Marshall in Oceanography. Figures were prepared by M. Leonard. The work has been supported by the Office of Sea Grant Programs, U.S. Department of Commerce, NOAA.

REFERENCES

Alexander, L. M. 1966. *Narragansett Bay: a marine use profile.* Final report, Geography Branch, ONR, Kingston, R.I., University of Rhode Island.

Alfano, J. J. 1973. A two-dimensional depth-averaged numerical temperature model of Narragansett Bay. Ph.D. thesis, University of Rhode Island, Kingston.

Anraku, M. 1964. Influences of the Cape Cod Canal on the hydrography and on the copepods in Buzzards Bay and Cape Cod Bay, Massachusetts. II: Respiration and feeding. *Limnol. Oceanogr.* **9**: 195–206.

Bodine, B. R. 1971. *Storm surge on the open coast: fundamentals and simplified prediction.* Coastal Eng. Res. Center, Tech. Memo No. 35.

Busser, J. 1967. Number language in natural history using examples of the phytoplankton and Narragansett Bay. Ph.D. thesis, University of Rhode Island, Kingston.

Chen, C. W. 1970. Concepts and utilities of ecologic modeling. *J. Sanit. Eng. Div., ASCE,* **96** (SA5).

Conover, R. J. 1968. Zooplankton—life in a nutritionally dilute environment. *Am. Zool.* **8**: 107–118.

Coughlan, J., and A. D. Ansell. 1964. A direct method for determining the pumping rate of siphonate bivalves. *J. Conseil Int. Explor. Mer,* **29**: 205–213.

Cumberland, J. and R. Stram. 1972. Economic flows and environmental coefficients. Paper presented at the Southern Regional Science Association, Williamsburg, Virginia.

Droop. M. R. 1973. Nutrient limitation in osmotrophic protista. *Am. Zool.* **13**: 209–214.

DiToro, D. M., D. J. O'Connor and R. V. Thomann. 1971. A dynamic model of phytoplankton populations in the Sacramento-San Joaquin delta. *Adv. Chem. Ser.* **106**: 131–180.

Durbin, A. G. 1976. Some ecological implications of the migratory behavior of the Atlantic Menhaden, *Brevoortia tyrannus*, and the Alewife, *Alosa pseudoharengus*. Ph.D. thesis, University of Rhode Island, Kingston.

Durbin, A. G. and E. Durbin. 1975. Grazing rates of Atlantic Menhaden, *Brevoortia tyrannus*, as a function of particle size and concentration. *Mar. Biol.* **33**: 265–277.

Environmental Protection Agency. 1971. *Estuarine modeling: an assessment, capabilities and limitation for resource management and pollution control.* Tracor, Inc., U.S. Govt. Printing Office, Washington, D.C.

Eppley, R. W., 1972. Temperature and phytoplankton in the sea. *Fish. Bull.* **70**: 1063–1085.

Eppley, R. W., A. F. Carlucci, O. Holm-Hansen, D. Kiefer, J. J. McCarthy, E. Venrick, and P. M. Williams. 1971. Phytoplankton growth and composition in shipboard cultures supplied with nitrate, ammonium, or urea as the nitrogen source. *Limnol. Oceanogr.* **16**: 741–751.

Eppley, R. W., J. N. Rogers, and J. J. McCarthy. 1969. Half-saturation constants for uptake of nitrate and ammonium by marine phytoplankton. *Limnol. Oceanogr.* **14**: 912–920.

Farrington, J. W., and J. G. Quinn. 1973. Petroleum hydrocarbons and fatty acids in wastewater effluents. *J. Water Poll. Cont. Fed.* **45**: 704–712.

Fee, E. J. 1969. A numerical model for the estimation of photosynthetic production, integrated over time and depth, in natural waters. *Limnol. Oceanogr.* **14**(6): 906–911.

Feld, S. E. 1973. An economic-waste generation linkage model for Narragansett Bay. Ph.D. thesis, University of Rhode Island, Kingston.

Feld, S., and N. Rorholm. 1973. *Economic growth and generation of waterborne wastes.* URI Marine Tech. Report No. 12, University of Rhode Island, Kingston, Rhode Island, 17 p.

Fish, C. J. (Ed.), 1953. *Physical oceanography of Narragansett Bay—R. I. Sound.* I.S.P. Final Report, Narragansett Marine Laboratory, University of Rhode Island, Kingston.

Frost, B. W. 1972. Effects of size and concentration of food particles on the feeding behavior of the marine planktonic copepod *Calanus pacificus. Limnol. Oceanogr.* **17**: 805–815.

Fuhs, G., S. Demmerle, E. Canelli, and M. Chen. 1972. Characterization of P—limited plankton algae. Nutrients and Eutrophication. *Limnol. Oceanogr. Special Symposia,* **1**: 113–132 p.

Gauld, D. T., and J. E. G. Raymont. 1953. The respiration of some planktonic copepods II. The effect of temperature. *J. Mar. Biol. Assoc. U.K.* **31**: 447–460.

Gerber, R. P., and N. M. Marshall. 1974. Ingestion of detritus by the lagoon pelagic community at Eniwetok Atoll. *Limnol. Oceanogr.* **19**: 815–824.

Ghittino, P. 1972. The diet and general fish husbandry. In J. E. Halver (Ed.), *Fish nutrition.* Academic, New York.

Governor's Technical Committee on the Coastal Zone. 1970. *State of Rhode Island Report of the Governor's Committee on the Coastal Zone.* State House, Providence, R.I.

Grenney, W. J., D. A. Bella, H. C. Curl, Jr. 1973. A theoretical approach to interspecific competition in phytoplankton communities. *Amer. Nat.* **107**: 405–425.

Greze, V. N., and E. P. Baldina. 1964. Population dynamics and annual production of *Acartia clausi* Gieshr. and *Centrophages* Kroyeri Giesbr. in the neritic zone of the Black Sea. *Trudy Sevastopol skoi Biologicheskio stantsii, Akad. Nauk Ukrain. SSR* **17**: 299–261. (*Fish. Res. Bd. Canada Trans.* 1967).

Griffiths, L. 1970. *Indication of economic changes for Rhode Island cities and towns 1957–1968.* Occasional Paper No. 70–59. University of Rhode Island, Kingston.

Guillard, R. R., and J. H. Ryther. 1962. Studies of marine planktonic diatoms. I. *Cyclotella nana* Hustedt, and *Detonula confervacea* (Cleve.) Gran. *Can. J. Microbiol.* **8**: 229–239.

Heinle, D. R. 1969. The effects of temperature on the population dynamics of estuarine copepods. Ph.D. thesis, University of Maryland.

Heinle, D. R., D. A. Flemer, J. F. Ustach, R. A. Murtagh, and R. P. Harris. 1974. *The role of organic debris and associated micro-organisms in pelagic estuarine food chains.* Tech. Rep. No. 22, Water Resources Research Center, University of Maryland, College Park, 54 pp. plus app.

Herman, S. S. 1958. The planktonic fish eggs and larvae of Narragansett Bay. M.S. thesis, University of Rhode Island, Kingston.

Hess, K., and F. White. 1974. *A numerical tidal model of Narragansett Bay.* University Rhode Island Marine Tech. Rep. No. 20. Kingston. 141 pp.

Hicks, S. D. 1959. The physical oceanography of Narragansett Bay. *Limnol. Oceanogr.* **4**: 316–327.

Hicks, S. D., D. E. Frazier, and L. E. Garrison. 1956. *Physical oceanographic effects of proposed hurricane protection structures on Narragansett Bay under normal conditions.* Ref. No. 56, 12. Hurricane Protection Project. Pell Library, U.R.I., Kingston.

Holling, C. S. 1973. Resilience and stability of ecological systems. In R. F. Johnston, P. W. Frank, and C. D. Michener (Eds.), *Annual Review of Ecology and Systematics.* Annual Reviews, Palo Alto, Calif. Pp. 1–23.

Ignatiades, L., and T. J. Smayda. 1970. Autecological studies on the marine diatom *Rhizosolenia fragilissima* Bergon. I. The influence of light, temperature, and salinity. *J. Phycology* 6: 332–339.

Isard, W., and E. Romanoff. 1967. *Water use and water pollution coefficients: preliminary report.* Tech. Paper No. 6, Regional Sci. Res. Inst., Cambridge, Mass.

Ivlev, V. S. 1945. The biological productivity of waters. *Uspekhi Sovrem. Biol.* 19: 98–120.

Jaworski, N. A., D. W. Lear, Jr.   O. Villa, Jr. 1972. Nutrient management in the Potomac estuary. In G. E. Likens (Ed.). Nutrients and Eutrophication: The Limiting-nutrient controvercy. Amer. Soc. Limnol. Ocean. Special Symposia Vol. I. Pp. 246–272.

Jeffries, H. P. 1973. Utilization of detritus by estuarine filter feeders. *Abstracts N. E. Branch Meeting, Amer. Soc. Agron.*, University of Rhode Island, Kingston.

Jeffries, H. P., and W. C. Johnson. 1973. Distribution and abundance of zooplankton. In *Coastal and offshore environmental inventory, Cape Hatteras to Nantucket Shoals.* Marine Pub. Series No. 2, University of Rhode Island, Kingston.

Jitts, H. R., E. D. McAllister, K. Stephens, and J. D. H. Strickland. 1964. The cell division rates of some marine phytoplankters as a function of light and temperature. *J. Fish. Res. Bd. Can.* 21: 139–157.

Kain, J. M., and G. E. Fogg. 1958. Studies on the growth of marine phytoplankton. I. *Astenonella japonica* Graw. *J. Mar. Biol. Assoc. U.K.* 37: 397–413.

Kelly, R. A. 1976. Conceptual ecological model of Delaware estuary. In B. C. Patten (Ed.), *Systems analysis and simulation in ecology,* Vol. 4. (In press.)

Kremer, J. N., and S. W. Nixon. 1975. An ecological simulation model of Narragansett Bay—the plankton community. In J. Cronin (Ed.), *Recent Advances in Estuarine Research. Proc. 2nd Intl. Est. Res. Conf., Myrtle Beach, S. C. 1973.* Academic, New York.

Kremer, P. 1975. The ecology of the ctenophore *Mnemiopsis leidyi* in Narragansett Bay. Ph.D. thesis, University of Rhode Island, Kingston.

Kremer, P., and S. Nixon. 1976. The distribution and abundance of the ctenophore, *Mnemiopsis lediyi* in Narragansett Bay. *Est. Coast. Mar. Sci.* 4: 627–639.

Lampe, H. L., and S. W. Nixon. 1970. A bioeconomic model of Narragansett Bay. Paper presented at the 33rd Annual Meeting, ASLO, Kingston, R.I.

Laurent, E. A., and J. C. Hite. 1971. *Economic-ecologic analysis in the Charleston Metropolitan region: an input-output study.* Water Research Institute, Clenson University, Clemson, South Carolina.

Leendertse, J. J. 1967. *Aspects of a computational model for long-period wave propagation.* Mem. RM5294–PR, Rand Corporation, Santa Monica, California.

Levine, E. 1972. The tidal energetics of Narragansett Bay. M.S. thesis, University of Rhode Island, Kingston.

Loosanoff, V. L. 1939. Effect of temperature upon shell movement of clams, *Venus mercenaria* (L). *Biol. Bull.* 76: 171–182.

Loveland, R., and D. Chu. 1969. Oxygen consumption and water movement in *Mercenaria mercenaria. Comp. Biochem. Physiol.* 29: 173–184.

Mann, K. H. 1975. Relationship between morphometry and biological functioning in three coastal inlets of Nova Scotia. In J. Cronin (Ed.), *Recent advances in estuarine research. Proc. 2nd Int. Est. Res. Conf., Myrtle Beach, S.C. 1973.* Academic, New York.

Marshall, S. M., and A. P. Orr. 1965. On the biology of *Calanus finmarchicus* VIII. Food uptake, assimilation and excretion in adult and stage V *Calanus. J. Mar. Biol. Assoc. U.K.* **34:** 495–529.

Martin, J. H. 1965. Phytoplankton-zooplankton relationships in Narragansett Bay. *Limnol. Oceanogr.* **10:** 185–191.

Martin, J. H. 1968. Phytoplankton-zooplankton relationships in Narragansett Bay. III. Seasonal changes in zooplankton excretion rates in relation to phytoplankton abundance. *Limnol. Oceanogr.* **13:** 63–71.

Matthiessen, G. C. 1973–74. *Rome Point investigations.* Quarterly Report to the Narragansett Electric Co. from Marine Research, Inc., East Warham, Mass.

Miller, R. J. 1970. Distribution and energetics of an estuarine population of the ctenophore, *Mnemiopsis lediyi.* Ph.D. thesis, North Carolina State University, Raleigh.

Mullin, M. M. 1969. Production of zooplankton in the ocean: the present status and problems. *Oceanogr. Mar. Biol. Ann. Rev.* **7:** 293–314.

MacIsaac, J. J., and R. C. Dugdale. 1969. The kinetics of nitrate and ammonia uptake by natural populations of marine phytoplankton. *Deep Sea Res.* **16:** 45–57.

McAllister, C. D. 1970. Zooplankton rations, phytoplankton mortality and the estimation of marine production. p. 419–457. In J. H. Steele (Ed.), *Marine food chains.* University of California Press, Berkeley.

McAllister, C. D., N. Shah, and J. D. H. Strickland. 1964. Marine phytoplankton photosynthesis as a function of light intensity: A comparison of methods. *J. Fish. Res. Bd. Can.* **21:** 159–181.

McLaren, I. A., C. J. Corkett, and E. J. Zillioux. 1969. Temperature adaptation of copepod eggs from the Arctic to the Tropics. Unpub. manuscript, Nat. Mar. Water Qual. Lab., EPA, Narragansett, Rhode Island.

McMaster, R. L. 1960. Sediments of Narragansett Bay System and Rhode Island Sound, Rhode Island. *J. Sed. Petrol.* **30:** 249–274.

Nixon, S. W. and C. A. Oviatt. 1973. Ecology of a New England saltmarsh. *Ecolog. Monogr.* **43:** 463–498.

Nixon, S. W. and C. A. Oviatt. 1973. Analysis of local variation in the standing crop of *Spartina alterniflora. Bot. Marina.* **16:** 103–109.

Nixon, S. W., C. A. Oviatt, and S. S. Hale. 1976. Nitrogen regeneration and the metabolism of coastal marine bottom communities. In J. Anderson and A. Macfaydew (Eds.), *The role of terrestrial and aquatic* organisms in decomposition processes. Blackwell, London. Pp. 269–283.

O'Brien, J. J. and J. S. Wroblewski. 1972. *An ecological model of the lower marine trophic levels on the continental shelf off West Florida.* Tech. Rep., Geophysical Fluid Dynamics Institute, Florida State University. 170 pp.

Odum, H. T. 1967. Biological circuits and the marine systems of Texas. p. 99–158. In T. A. Olson and F. J. Burgess (Eds.), *Pollution and marine ecology.* Wiley Interscience, New York.

Odum, H. T. 1975. Marine ecosystems with energy circuit diagrams. In J. C. J. Nihoul (Ed.), *Modeling of marine systems.* Elsevier, Amsterdam. Pp. 127–151.

Okuda, T. 1960. Metabolic circulation of phosphorus and nitrogen in Matsushima Bay (Japan) with special reference to exchange of these elements between sediment water and sediments. *Trabalhos Inst. Biol. Marit. Oceanog.* **2:** 7–153.

References

671

Oviatt, C. A., A. L. Gall, and S. W. Nixon. 1972. Environmental effects of Atlantic Menhaden on surrounding water. *Ches. Sci.* **13**: 321–323.

Oviatt, C. A. and S. W. Nixon. 1973. The demersal fish of Narragansett Bay: an analysis of community structure, distribution and abundance. *Estuarine Coastal Mar. Sci.* **I**: 361–378.

Paasche, E. 1973. Silicon and the ecology of marine plankton diatoms. I. *Thalassiosira pseudonana* grown in a chemostat with silicate as limiting nutrient. *Mar. Biol.* **19**: 117–126.

Parsons, T. R., R. J. Le Brasseur and J. D. Fulton. 1967. Some observations on the dependence of zooplankton grazing on the cell size and concentration of phytoplankton blooms. *J. Ocean. Soc. Jap.* **23**: 10–17.

Peters, R. and D. Lean. 1973. The characterization of soluble phosphorus released by limnetic zooplankton. *Limnol. Oceanogr.* **18**: 270–279.

Petipa, T. S. 1966. Relationship between growth, energy metabolism, and ration in *A. clausi*. In *Physiology of marine animals, Akademiya Nauk SSSR, Oceanographical Commission*, (Trans. M. A. Paranjape). Pp. 82–91.

Phelps, D. K. 1958. A quantitative study of the infauna of Narragansett Bay in relation to certain physical and chemical aspects of their environment. M.S. thesis, University of Rhode Island, Kingston.

Pratt, D. M. 1959. The phytoplankton of Narragansett Bay. *Limnol. Oceanogr.* **4**: 425–440.

Pratt, D. M. 1965. The winter-spring diatom flowering in Narragansett Bay. *Limnol. Oceanogr.* **10**: 173–184.

Premack, J., and G. A. Brown. 1973. Predictions of oil slick motions in Narragansett Bay. In *Proceedings of Joint Conference on Prevention and Control of Oil Spills.* Washington, D.C. Pp. 5531–5540.

Pritchard, D. W. 1971. Two-dimensional models. In *Estuarine modeling: An assessment.* Tracor, Inc., Austin, Texas.

Riley, G. A. 1946. Factors controlling phytoplankton populations on Georges Bank. *J. Mar. Res.* **6**: 54–73.

Riley, G. A. 1947. A theoretical analysis of the zooplankton population of Georges Bank. *J. Mar. Res.* **6**: 104–113.

Riley, G. A. 1955. Review of the oceanography of Long Island Sound. *Deep-Sea Res. Suppl.* **3**: 224–238.

Riley, G. A., H. Stommel and D. F. Bumpus. 1949. Quantitative ecology of the plankton of the western North Atlantic. *Bull. Bingham Oceanogr. Coll.* **12**: 1–169.

Rorholm, N., H. C. Lampe, and J. F. Farrell. 1969. A socio-economic study of Narragansett Bay, Rhode Island. Final report to the Federal Water Pollution Control Administration. University of Rhode Island, Kingston. 196 pp.

Russel, H. J., Jr. 1972. Use of a commercial dredge to estimate a hardshell clam population by stratified random sampling. *J. Fish. Res. Bd. Can.* **29**: 1731–1735.

Ryther, J. H. 1956. Photosynthesis in the ocean as a function of light intensity. *Limnol. Oceanogr.* **1**: 61–70.

Ryther, J. H., W. M. Dunstan, K. R. Tenore, and J. E. Huguenin. 1972. Controlled eutrophication—increasing food production from the sea by recycling human wastes. *BioScience* **22**: 144–152.

Schenck, H. and A. Davis. 1972. *A turbidity survey of Narragansett Bay.* Mimeo report, Dept. Ocean Engineering, University of Rhode Island, Kingston.

Shaler, N. S., J. B. Woodworth, and A. F. Foerste. 1899. *Geology of the Narragansett Basin.* U.S. Government Printing Office, Washington, D.C.

Smayda, T. J. 1957. Phytoplankton studies in lower Narragansett Bay. *Limnol. Oceanogr.* **2:** 342–359.

Smayda, T. J. 1969. Experimental observations on the influence of temperature, light and salinity on cell division of the marine diatom, *Detonula confervacea* (Cleve) Gran. *J. Phycology* **5:** 150–157.

Smayda, T. J. 1970. The suspension and sinking of phytoplankton in the sea. *Oceanogr. Mar. Biol. Ann. Rev.*, **8:** 353–414.

Sorokin, Y. I. and D. A. Panov. 1965. Balance of consumption and expenditure of food by larvae of bream at different stages of development. *Doklady Akad. Nauk. SSSR* **165:** 797.

Spaulding, M. L. 1972. Two dimensional, laterally-integrated estuarine numerical water quality model. Ph.D. thesis, University of Rhode Island, Kingston.

Spencer, C. P. 1954. Studies on the culture of a marine diatom. *J. Mar. Biol. Assoc. U.K.* **33:** 265–290.

Steele, J. H. 1965. Notes on some theoretical problems in production ecology, p. 383–398. In C. R. Goldman (Ed.), *Primary productivity in aquatic environments.* Mem. Ist. Ital. Idrobiol., 18 Suppl., University of California Press, Berkeley.

Steemann Nielsen, E., V. K. Hansen and E. G. Jorgensen. 1962. The adaptation to different light intensities in *Chlorella vulgaris* and the time dependence on transfer to a new intensity. *Physiologia Plantarum* **15:** 505–1517.

Stickney, A. P. and L. D. Stringer. 1957. A study of the invertebrate fauna of Greenwich Bay, Rhode Island. *Ecology* **38:** 111–122.

Sushchenya, L. M. 1970. Food rations, metabolism and growth of crustaceans. In J. H. Steele (Ed.) Marine Food Chains, University of California Press, Berkeley.

Sverdrup, H. V., M. W. Johnson and R. H. Fleming. 1942. *The oceans*, Prentice-Hall, Englewood Cliffs, N.J.

Thomann, R. V. 1972. *Systems analysis and water quality management.* Environmental Science Services Division, Environmental Research and Applications, Inc. New York.

U.S. Dept. of Interior. 1968. *The cost of clean water*, Vol. I. Summary report. U.S. Government Printing Office, Washington, D.C.

Verduin, J. 1969. Hard clam pumping rates: energy requirement. *Science* **166:** 1309–1310.

Vollenweider, R. A. 1965. Calculation models of photosynthesis-depth curves and some implications regarding day rate estimates in primary production measurements, p. 427–457. In C. R. Goldman (Ed.), *Primary productivity in aquatic environments.* Mem. Ist. Ital. Idrobiol. 18 Suppl. University of California Press, Berkeley.

Walne, P. R. 1972. The influence of current speed, body size, and water temperature on the filtration rate of five species of bivalves. *J. Mar. Biol. Assoc. U.K.* **52:** 345–374.

Walsh, J. J. 1975. A non-linear spatial approach to simulation of aquatic ecosystems. In J. Cronin (Ed.), *Recent advances in estuarine research. Proc. 2nd Intl. Est. Res. Conf., Myrtle Beach, S.C., 1973,* Academic, New York.

Walsh, J. J., and R. C. Dugdale. 1971. A simulation model of the nitrogen flow in the Peruvian upwelling sustem. *Analysis of upwelling systems, Investigacion Pesquera* **35:** 309–330.

Weisberg, R. H. and W. Sturges III. 1973. *The net circulation in the West Passage of Narragansett Bay.* Tech. Rep. No. 3–73, Graduate School of Oceanography, University of Rhode Island, Kingston. 96 p.

Williams, P. J. Le B. 1973. The validity of the application of simple kinetic analysis to heterogeneous microbial populations. *Limnol. Oceanogr.* **18:** 159–164.

Yentsch, C. S., and R. W. Lee. 1966. A study of photosynthetic light reactions, and a new interpretation of sun and shade phytoplankton. *J. Mar. Res.* **24**: 319–337.

Zillioux, E. 1970. Ingestion and assimilation in laboratory cultures of *Acartia*. Unpublished technical report, National Marine Water Quality Laboratory, EPA, Narragansett, Rhode Island.

Author Index

Botkin, D., 213
Boynton, W., 477
Brown, M., 393

Day, J., 5, 37, 235, 381
Dohan, M., 133

Friedland, E. I., 115

Gray, C., 477

Hall, C., 5, 37, 345, 365
Hawkins, D., 477
Homer, M., 507
Hopkinson, C., 235, 381

Jansson, B-O, 323

Kelly, R., 419
Kemp, M., 507
Kessel, S., 575
Kremer, J., 621

Lehman, M., 507

Littlejohn, C., 451
Loesch, H., 381

MacBeth, A., 197
McKellar, H., 507
Morris, J., 607

Nixon, S., 621

Odum, H. T., 37, 173, 507
Overton, S., 49

Richey, J. E., 267

Shoemaker, C., 75, 545
Singer, F., 607
Smith, W., 507
Spofford, W., 419

Wartenberg, D., 365
Wiegert, R., 289
Wulff, F., 323

Young, D., 507

Subject Index

Adaptation, 177, 410, 457
Aggregation, 19–21, 76
  Baltic model, 337
  Crystal River power plant model, 520
  and evaluation, 127
  population model, 617
  salt marsh model, 243
  thermal spring model, 290, 309–310
  by trophic level, 20
Aggregation problem, 67
Aggregative indices, 125
Alfalfa weevil management model, 556
Algorithm, 32, 72
Analog computer, 32, 403, 475, 481
  patching, 32
Analytical solution techniques, 76, 91–92
Analytic models, 9
Appalachicola Bay, Fla., 478
Appalachicola Bay model, conceptual, 484
  hierarchial arrangement, 482–483
  management guidelines, 492
  mathematical, 481–483
  simulation, 481
  validation, 492–493
Aquatic ecosystem model, 85, 242, 245, 267, 332, 421, 483, 509, 647
Arrays, 28
Asko laboratory, 324–336
Asko project, 331
Assessment, 13

Atchafalaya River, 382

Baltic Sea, 325–331
  physical processes, 329
Baltic Sea model, aggregation, 337
  conceptual, 334
  pelagic, 336
  sensitivity analysis, 340
  state variables, 333–339
  validation, 340
Barataria Bay, 235
Bellman's principle, of optimality, 109
Benefits of ecosystems, evaluation, 152
Biology of pest control, 546–551
Biome, 7
Biosphere, 7
Biospheric productivity model, conceptual, 370
  results, versus assumptions, 375
  sensitivity analysis, 374
  state variables, 373
  utility, 376

Carbon dioxide, atmospheric, 367
Carbon model, La. salt marsh, 243
Carrying capacity, 77
Castle Lake, Ca., 267, 271
Center for Wetland Resources, LSU, 238
Channelization, 452
Circuit language, 44

cycling receptor, 44
forcing function, 40
heat sink, 40
sources, 39
state variables, 41
storage, 41
work gate, 41
Citrus County, Fla., 518
Climax community, 604
Closed form solution, 91, 92
Coastal food model, 386
Coefficient, 22
  half-saturation, 79, 88, 653
Coefficient determination, by correlation,
    24
  by isolation, 24
  by stocks and flows, 23, 61
Coefficient of determination, 26
Collective goods, 148
Common goods, 148
Common property, 148
Community, 7
Community classification, 603
Compartmental model, 18
Competition, 218
Competitive exclusion, 180
Computer model, 30
Computer programming, 31
Conceptual model, 17
  Appalachicola Bay, 484
  biospheric productivity, 370
  coastal food, 387
  Crystal River power plant, 520
  Mississippi flood control, 383
  phosphorus cycle, 268
  salt marsh, 241
  thermal spring, 291
  war, 395
Constants, 22
Constraint equations, 102
Consumer, 138
Convergence, differential equation,
    84
Correlation analysis, 26
Correlation coefficient, 27
Cost benefit analysis, 152, 159
  coastal food production, 388
  limitations, 166
  noneconomic, 166
  option demand, 166
  power plants, 518

wetland use, 164–165
Costs, transaction, 149
Court trial, 200
Criterion function, 102
Crops, model of, 546
Crystal River, Florida, 508–518
Crystal River power plant model,
    aggregation, 520
  conceputal, 520
  hierarchical development, 520
  mathematical, 521–522
  simulation, 524
  validation, 525–526
Cultural inertia, 183
Cycling receptor, 44
Cypress swamp, 452

Decision making, 360
Decision theory, 116
Decision variable, 79
Decomposition, model, 22, 56, 70, 111,
    483, 511, 582, 612, 633
Delaware estuary, 420
Delaware estuary model, constraints, 421
  environmental response matrix, 435
  linear management, 432–433
  mathematical, 423–427
  nonlinear management, 433–435
  state variables, 423
  validation, 429
Delaware River Basin Commission, 421
Demand curve, 141
Dependent variables, 21
Derivative, 84
Deterministic equations, 97
Deterministic model, 34
Diagrammatic model, 18
Difference equation, 76, 83
Differential equation, 76, 84
  solution of, 92–93
Discount rate, 126
  social, 617
DiToro, D., 85
Do loop, 33
Donor control, 23
Donor-recipient control, 23
Dredge and fill, 452
Driving variable, 18, 40
Dynamic model, 96
Dynamic programming, 109–112
Dynamic programming model, 556

Economic concepts, goods, 137
  individual welfare, 139
  individual well being, 136
  marginal utility, 138
  market value, 144
  money values, 140
  net social benefit, 143
  pareto, 139
  social welfare, 139
  utility, 136
Economic theory of social value, 136
Economic value, 134
  model of, 432, 482, 608, 612, 633
  and natural ecosystems, 133, 452
  theory, 136
Ecosystem, 7
  and failure of market, 147
  stability, 664
Edge effect, 239
Emergent properties, 13
Endogenous variable, 18
Energy, 173
  disordering, 95, 394
  high quality versus low quality, 184, 398, 509
  ordering, 95, 394
  planetary control of, 194
Energy certificate, 182
Energy cost-benefit analysis, 190, 508–509
Energy flow, and money flow, 182
Energy laws, 174–177
Energy matching, 194
Energy quality, 185, 189, 510
  cultural value, 189
Energy quality ratio, 513
Energy quality scale, 188
Energy system generator model, 121
Energy values, 186
Entrainment, 357, 539
Entropy, 177
Environmental problems, government regulation, 198
  public versus private, 198
Environmental response matrix, 435
Equation, descriptive, 12
  difference, 76, 83
  differential, 76, 84
  exponential growth, 77, 85, 92
  linear, 10, 72
  logistic growth, 85, 92

mechanistic, 12
nonlinear, 10, 425, 433
predation, 78, 92
predictive, 83, 92
Equilibrium system, 92, 96
Estuarine organisms, life histories, 346
Estuary, 236, 346, 420, 478, 518, 622
  and market failure, 149
Euler method, 94
Eutrophication, Baltic Sea, 324
Evaluation, criterion, 121, 124
  general, 123
  metric, 124
  standard, 124
Exogenous variable, 79
Exponential growth equation, 85, 92
Externalities, 152
  and resource use, 152

Facts, 116
Failure of the market, 147
Feedback, 15, 179
Fire modeling, 576
FLEX program, 64
Flood control, Florida, 452
  Louisiana, 382
Florida, 452, 478, 508
Flow rate equation, 77
Forcing function, 18, 37
Forest ecosystem, 213
  role of fire, 576
Forest fire, 576
Forest fire model, components, 579, 581
  cost, 601
  fire behavior submodel, 597
  resource inventory, 591
  submodel coupling, 600
  validation, 589
  weather submodel, 600
Forest model, algorithm, 217
  interactive graphics, 218
  mathematical, 224–227
  nonintuitive results, 230–231
  validation, 223
Fossil fuel equivalent, 189, 398, 527
Franklin County, Fla., 478

General feeding and growth equation, 294–297
General systems theory, 55

Glacier National Park, 290, 576
GNP, 609, 610
  and energy flow, 510
Goals, of models, 52
Goods, collective, 148
  common, 148
  economic, 137
  private, 137
  public, 148
Gordon River, Fla., 452
Gradient analysis, 576, 604
Gradient modeling, forest fires, 576, 581
Greenhouse effect, 367
Growth rate equation, 77, 92

Habitat mosaic, 604
Half-saturation coefficient, 79, 88, 653
Happiness, 610
Heat sink, 37, 177
Hierarchial arrangement of systems, 4, 6,
    54, 82, 511, 517, 520
Holism, 6, 241
Hubbard Brook, 217
Hudson River, 346
  power plants, 206
Hudson River power plant model, and
    decision making, 360
  electric utility consultant, 346, 358
  implications of, 360
  mathematical, 352
  sensitivity analysis, 358
  validation, 356
Human population, 366, 608
Hurricane simulation, 646
Hybrid, 632

Income distribution, 617
Independent variable, 21
Inflation, 191
Input-output, 153, 633
Interactive graphics, 218
Internal variable, 64
Investment ratio, 192, 194, 540
Iterative processes, 29

Jabowa, 217–229

Land reclamation, 452
Land use, 198
Law of diminishing marginal utility, 142

Law of sea, 205
Legal use of models, assumptions, effects
    of, 203
  best opportunity, 205
  candor, value of, 203
  clarity, value of, 203
  limitations, 205
  sensitivity analysis, 203
Leontief inverse, 634
Linear equation, 10, 72
Linear programming, 106
Linear regression, 72
Litigation versus science, 200
Logistic growth equation, 85, 92
Lotka principle, 174, 177, 480, 509
Lotka-Volterra equations, 85, 92, 96
Louisiana Coastal Zone, 12, 236, 240, 381
  man's impact, 240
Loyalties, of modelers, 362

Malthusian, 180
Management model, nonlinear, 433–435
  solution methods, 103
Mangroves, 400
Marginal cost, 142
Marginal utility, 138
Marginal value, 146
Market failure, 147
  and wetlands, 149
Market value, 144
Marx, K., 174
Mathematical model, 76
  Appalachicola Bay, 481–483
  Crystal River power plant, 521–522
  Delaware estuary, 423–427
  forest, 224–227
  Hudson River power plant, 352
  human population, 608
  Mississippi flood control, 384
  pest management, 552–554
  phosphorus cycle, 275–280
  salt marsh, 249
  thermal spring, 294–303
  war, 403
  water budget, 467
Matrices, 29, 439
Mature community, 604
Maximum power principle, 173, 174, 177,
    480, 509–510
Megamodel, 632

Metric problem, 134
Michaelis-Menten, 92, 429
Migration, 349
Mississippi River, 236
  flood control, 382
Mississippi flood control model, 382
  conceptual, 383
  mathematical, 384
Model, analog, 30
  analytic, 9
  assumptions, 13
  Baltic Sea, 324
  biospheric productivity, 366
  coastal food production, 387
  compartmental, 18
  computer, 30
  conceptual, 17
  and decision making, 360
  definition, 7
  Delaware water management, 420
  deterministic, 34
  diagrammatic, 18–21
  digital, 30
  dynamic, 96
  flood control, 382
  forest, 214
  forest fire, 576
  gradient, 576
  human population, 608
  legal use, 202
  linear, 68
  management, 102
  mathematical, 75
  multidimensional, 76
  Narragansett Bay, 622
  oyster fishery, 478
  pest management, 546
  phosphorus cycle, 265
  power plant, 346, 508
  predictive, 102
  regional water management, 452
  salt marsh, 236
  simulation, 9
  steady state, 96
  stochastic, 35
  thermal spring, 290
  uses, 13–15
  war, 394
Model behavior, 54
Model construction, 21–28, 49, 54,

  112, 482, 612
  algorithm, 32
  chronology, 15–17
  state variable relations, 21–28
  strategy, 54, 119
  subsystem structure, 54
  validation, 56
Model decomposition, 22, 56, 70, 111,
    483, 511, 582, 612, 633
Model failure, 280, 366
Model formulation, 112
Modeling, and administrative agencies, 204
  approaches, 9–12
  and choices, 116
  legal considerations, 198, 200
  and the political processes, 128–130
Model mechanism, 53
Model objectives, 51, 52
Model prediction, 14
Model purpose, 52
Model structure, 63
Model success, 118
Model values, 116
Money, 173
Money exchange, 191
Money values, 140
Monte Carlo simulations, 100
Multidimensional model, 76
Multiple benefit resource, 151

Naples, Florida, 452
  salt water intrusion, 456
Narragansett Bay, 623
  biology, 628–632
  economy, 624–626
  water chemistry, 628
Narragansett Bay ecological submodels,
    benthos, 661
  copepods, 655
  fish, 661
  phytoplankton, 651
Narragansett Bay hydrodynamic submodel,
    637
  grid size, 637, 642
  mathematical, 643
  validation, 644
Narragansett Bay input-output submodel,
    633
  management implications, 636
  natural resource use coefficients, 636

waste discharge coefficients, 635
Narragansett Bay model, ecosystem
    stability, 664
  emergent properties, 666
  hybrid, 632
  mechanistic simulation, 666
  megamodel, 632
  sensitivity analysis, 664
  validation, 644, 664
National Environmental Policy Act, 136,
    199–204, 318, 508
Natural resource use coefficients, 636
Natural selection, 174, 457, 510
Net energy analysis, 361, 508
New England estuary, 622
New York State, 546
Nitrogen model, La. salt marsh, 244
Nonintuitive results, biospheric
    productivity model, 376
  forest model, 230–231
Numerical solution method, 76, 91–96
Nursery grounds, estuarine, 239, 329,
    347, 386

Objective criterion, 509
Objective function, 126, 432
Objective question, 51
Objectivity, 119
Opportunity costs, 148
Optimal management policy, 110
Optimization, 13–14, 104, 432, 446
  Bellman's principle, 109
  pest management model, 555
  and values, 120–121
Option demand, 166
Order, degradation, 174
  generation, 174
Order-disorder ratio, 397
Oyster fishery, 480

Paradigm, 72
Parallel modularity, 70
Parameter, 72
  distributed or lumped, 18
Parameter estimation, 59
Pareto criterion, 139
Peace, 394
Penalty function, 432, 435
Pest control, economic threshold, 549
Pest management, 546

Pest management model, decision, 155, 559
  optimization, 555
    versus simulation, 517
  population models, 559
Phosphorus model, conceptual, 268
  enrichment, effect, of, 282
  mathematical, 275, 278
  results, 280–281
  validation, 280–281
Population, 7
  human, 366, 608
Population growth, limiting factors, 608
Population model, aggregation, 617
  general applicability, 618
  income distribution, 617
  mathematical, 612
  Q index, 611, 616
  scope, 608
  sensitivity analysis, 616
  striped bass, 351, 355
  welfare index, 610
Population optima, human, 608
Power explosion, 195
Power plant, cooling tower versus once
    through water cooling, 518
  Crystal River, Florida, 518
  Hudson River, 346
Powershed, 518
Predation rate equation, 78
Prediction, 14
Predictive equation, 83, 92
Predictive model, 102
Present discounted value, 156
  of fishery productivity, 160
  or wetland benefits, 148
Prey-predator equation, 78, 85, 92
Price, 191
Processes, condition dependent, 30
  time dependent, 30
Programming, 32
Proportionality constant, 22
Public goods, 148
Public opportunity costs, 161

Q index, 611–616
Quality, energy, 185, 189, 398, 508

Random number, pseudorandom, 101
Random number generator, 101
Random variable, 98, 229, 305, 373, 501

Rate of change, 81
  prey-predator, 82
Rate equation, growth, 77, 92
  predation, 78, 92
Rate function, measurement of, 79
  multiplicative relationship, 80–81
Regional planning, 478
Regional waste-management models, 432
Regional water management, 452
Regression analysis, least squares, 26
  linear, 26
  multiple, 26
Resource management, deficiencies of
      traditional approach, 580
  forest fire, 576
  needs, 579
Reward loop, 179
  amplification factor, 180
  compensation with, 180
Runge-Kutta, 95, 256, 305

Salt marsh, 236
Salt marsh model, aggregation, 243, 261
  conceptual, 241
  mathematical, 249
  sensitivity analysis, 260
  simulation, 247
  validation, 256
San Joaquin River model, 85–91
Self design, 182
Sensitivity analysis, 34, 95, 60, 68, 155
  Baltic model, 340
  biospheric productivity model, 374
  Hudson River model, 358
  human population model, 616
  legal, 203
  Narragansett Bay model, 664
  salt marsh model, 260
  thermal spring model, 305, 309–310
Shadow prices, 155
Simulation, and values, 120
Simulation model, 9, 247, 303, 403, 481
Social discount rate, 157, 159, 617
Social value, 144
Solution of management models,
      optimization, 103–112
  trial and error, 103–112
Solution techniques, analytical, 76, 91
  closed form, 91–92
  numerical, 76, 93

Stable equilibrium point, 97
State variables, 19, 41, 64, 76, 293, 333,
      339, 373, 423
Steady state, world, 195
Steady state model, 96
Stochastic model, 35, 229, 305, 373, 501
Stochastic predictive equation, 97
Stress adapted ecosystem, 397, 576
Striped bass, life history, 348
  population model, 351, 355
Subsystem structure, 54
Succession, 215, 222
Systems, basic principles, 5–6
Systems analysis, 6
Systems inertia, 183
Systems theory, 55

Taylor series, expansion, 94
Thermal spring community, 290–291
  trophic relationships, 289–291
Thermal spring model, aggregation,
      309–310
  flowchart, 304–305
  initial conditions, 305–307
  mathematical, 297–303
  model condensation, 309–310
  sensitivity analysis, 305, 309–310
  simulation, 303–313
  state variables, 293
  substrate variables, 300
  validation, 307–309
Thermodynamics, 175
Transaction costs, 149
Transfer coefficient, 22
  linear, 22
  nonlinear, 22
Trial and error, 103
Trophic level, 20
Truth, legal versus scientific, 200, 362

United States energy system, 122
Utility function, 126, 137

Validation, 13, 33, 53, 56, 70, 290,
      307–309, 492–493
  Baltic model, 340
  comparison, 32
  Crystal River power plant model,
      525–526
  Delaware estuary model, 429

empirical, 57
extended, 58
forest fire model, 589
forest model, 223
Hudson River model, 356
Narragansett Bay model, 644, 664
phosphorus model, 280–281
and prediction, 34
salt marsh model, 256
war model, 394, 405–407
Value, 116, 173
economic, 134
energy measure of, 174
energy theory of, 184
marginal, 146
market, 144
metric, 140
of modelers, 130
and modeling choice, 119
money, 140
and power, 119
social, 136
social factor, 127
unit of measure, 139
useful work, 174
Value and ethics, general theory, 184
Value functions, 126

Variable, exogenous, 79
random, 98
Vector, 76
Vietnam, 394
order/disorder ratio, 397
Vietnam war, 394
consequences, 402

War adaptations, 410
War model, conceptual, 395
mathematical, 403
simulation, 403
validation, 394, 405–407
Waste discharge coefficients, 635
Water budget, Naples, Florida, 458
Water budget model, analog simulation, 459
mathematical model, 467
Water management, natural, 456
regional, 452
Water quality management, 420
Water quality management model, 420
Welfare, individual, 139
Welfare index, 610
Wetlands, 134
benefits, present discounted value, 160
channelization, 452

Yellowstone National Park, 290